W0245783

Electrified Interfaces in Physics, Chemistry and Biology

NATO ASI Series

Advanced Science Institutes Series

A Series presenting the results of activities sponsored by the NATO Science Committee, which aims at the dissemination of advanced scientific and technological knowledge, with a view to strengthening links between scientific communities.

The Series is published by an international board of publishers in conjunction with the NATO Scientific Affairs Division

A Life Sciences	Plenum Publishing Corporation
B Physics	London and New York
C Mathematical	Kluwer Academic Publishers
and Physical Sciences	Dordrecht, Boston and London
D Behavioural and Social Sciences	
E Applied Sciences	
F Computer and Systems Sciences	Springer-Verlag
G Ecological Sciences	Berlin, Heidelberg, New York, London,
H Cell Biology	Paris and Tokyo
I Global Environmental Change	

NATO-PCO-DATA BASE

The electronic index to the NATO ASI Series provides full bibliographical references (with keywords and/or abstracts) to more than 30000 contributions from international scientists published in all sections of the NATO ASI Series.
Access to the NATO-PCO-DATA BASE is possible in two ways:

– via online FILE 128 (NATO-PCO-DATA BASE) hosted by ESRIN,
Via Galileo Galilei, I-00044 Frascati, Italy.

– via CD-ROM "NATO-PCO-DATA BASE" with user-friendly retrieval software in English, French and German (© WTV GmbH and DATAWARE Technologies Inc. 1989).

The CD-ROM can be ordered through any member of the Board of Publishers or through NATO-PCO, Overijse, Belgium.

Electrified Interfaces in Physics, Chemistry and Biology

edited by

R. Guidelli

Department of Chemistry,
Florence University, Florence, Italy

Springer-Science+Business Media, B.V.

Proceedings of the NATO Advanced Study Institute on
Electrified Interfaces in Physics, Chemistry and Biology
Varenna, Italy
July 23–August 3, 1990

ISBN 978-94-010-5132-3   ISBN 978-94-011-2566-6 (eBook)
DOI 10.1007/978-94-011-2566-6

All Rights Reserved
© 1992 Springer Science+Business Media Dordrecht
Originally published by Kluwer Academic Publishers in 1992

No part of the material protected by this copyright notice may be reproduced or
utilized in any form or by any means, electronic or mechanical, including photo-
copying, recording or by any information storage and retrieval system, without written
permission from the copyright owner.

CONTENTS

Preface   vii

An overview of electrified interfaces
R. Guidelli   1

Structure and electronic properties of metal surfaces
K. Wandelt   65

Physics of Surfaces
R. Del Sole   117

Ab-Initio Molecular Dynamics:
Selected Applications to Disordered Systems and Surfaces
G. Galli, F. Ancilotto and A. Selloni   133

The Problem of Schottky Barrier
P. Perfetti   153

The Semiconductor/Electrolyte Interface
L.M. Peter   179

Stark Effect on Adsorbates at Electrified Interfaces
A. Stella   201

Thermodynamics of Adsorption
R. Parsons   213

Electrode Potentials and Energy Scales
S. Trasatti   229

Phenomenological Approach to Metal/Electrolyte Interfaces
S. Trasatti   245

The Application of Scanning Tunneling Microscopy to Electrochemistry
D.M. Kolb, R.J. Nichols and R.J. Behm   275

Single Crystal Electrodes
R. Parsons   293

Modeling of Metal-Water Electrified Interfaces
R. Guidelli and G. Aloisi   309

Molecular Models of Organic Adsorption from Water at Charged Interfaces
R. Guidelli and G. Aloisi   337

The Interface Between a Metal and a Solution in the Absence of Specific Adsorption
W. Schmickler   369

vi

Adsorption at the Metal/Solution Interface
W. Schmickler   399

Electron-Transfer Reactions at Metal-Solution Interfaces:
an Introduction to Some Contemporary Issues
M.J. Weaver   427

The Solid-Electrolyte Interface as Exemplified by Hydrous Oxides;
Surface Chemistry and Surface Reactivity
W. Stumm   443

Discrete Charges on Biological Membranes
R.T. Mathias, G.J. Baldo, K. Manivannan and S. McLaughlin   473

Structural Rearrangements in Lipid Bilayer Membranes
Y. Chizmadzhev   491

Evaluation of the Surface Potential at the Membrane-Solution Interface
of Photosynthetic Bacterial Systems
A. Corazza, B.A. Melandri, G. Venturoli and R.Casadio   509

Electrical Currents of the Light Driven Pump Bacteriorhodopsin.
The Role of Asp 85 and Asp 96 on Proton Translocation
E. Bamberg, H-J. Butt, J. Tittor and D. Oesterhelt   533

The Electrochemical Relaxation at Thylakoid Membranes
W. Junge, A. Polle, P. Jahns, G. Althoff and G. Schönknecht   551

Evaluation of the Electric Field in a Protein by Dynamic Measurements
of Proton Transfer
R. Yam, S. Kiryati, E. Nachliel and M. Gutman   565

List of Participants   577

List of Contributors   582

PREFACE

Electrified interfaces span from metal/semiconductor and metal/electrolyte interfaces to disperse systems and biological membranes, and are notably important in so many physical, chemical and biological systems that their study has been tackled by researchers with different scientific backgrounds using different methodological approaches.

The various electrified interfaces have several common features. The equilibrium distribution of positive and negative ions in an electrolytic solution is governed by the same Poisson-Boltzmann equation independent of whether the solution comes into contact with a metal, a colloidal particle or a biomembrane, and the same is true for the equilibrium distribution of free electrons and holes of a semiconductor in contact with a different conducting phase. Evaluation of electric potential differences across biomembranes is based on the same identity of electrochemical potentials which holds for a glass electrode and which yields the Nernst equation when applied to a metal/solution interface. The theory of thermally activated electron tunneling, which was developed by Marcus, Levich, Dogonadze and others to account for electron transfer across metal/electrolyte interfaces, is also applied to light induced charge separation and proton translocation reactions across intercellular membranes. From an experimental viewpoint, the same electrochemical and *in situ* spectroscopic techniques can equally well be employed for the study of apparently quite different electrified interfaces.

This volume contains a series of contributions related to the lectures given at the NATO Advanced Study Institute on "Electrified Interfaces in Physics, Chemistry and Biology", held in Varenna from July 23 to August 3, 1990, which brought together scientists with different backgrounds, such as solid state and surface physicists, electrochemists and biophysicists, in an attempt to overcome the barriers created by the traditional organization of science and to stimulate creative collaborations and untraditional approaches.

In view of the highly interdisciplinary nature of the field covered by this volume, I have deemed it convenient to start with a short survey of electrified interfaces at an introductory level, followed by the various contributions of a more or less specialized character. These contributions cover, in the order, electrified interfaces of physical, chemical and biological interest, although a sharp separation is neither possible nor desirable.

The structure and electronic properties of metal surfaces and the most important experimental and theoretical methods employed in surface physics are described first, including some applications of *ab initio* molecular dynamics to the study of electronic properties of surfaces. Even though metal and semiconductor surfaces *in vacuo* are not electrified interfaces in the strict sense of the word, their work function can be modulated by varying the amount of heteroatoms and adsorbed molecules.

Some fundamental aspects of metal/semiconductor and semiconductor/electrolyte interfaces are then surveyed, and the Stark effect on adsorbates at $Si-SiO_2$ and Si-electrolyte interfaces is examined. The use of thermodynamics in obtaining information about liquid/liquid and solid/liquid interfaces is subsequently outlined.

Metal/electrolyte interfaces are then dealt with. The physical meaning of single electrode potentials is disclosed, and a phenomenological approach to metal/electrolyte interfaces is provided, based on correlations between the potential of zero charge and the work function of metals. The application of scanning tunneling microscopy to *in situ* investigations of electrode surfaces is outlined, and the fundamental role played by studies at single crystal electrodes in elucidating the behaviour of solid metal/electrolyte interfaces is stressed. Electronic and molecular models of metal/electrolyte interfaces are reviewed, both in the absence and in the presence of ionic or organic specific adsorption, and some pivotal aspects of electron-transfer reactions at metal/electrolyte interfaces are outlined.

Those features of disperse systems which permit us to regard them as electrified interfaces are emphasized, and exemplified by hydrous oxides. Finally, several aspects of biological membranes and their models are considered: discreteness of charge on biological membranes, structural rearrangements in bilayer lipid membranes such as electroporation and electrofusion, evaluation of the ionic surface potential at membrane/solution interfaces, binding of protons to proteins, and some photosynthetic aspects such as the functional properties of the light-driven proton pump of bacteriorhodopsin and the electrochemical relaxation at thylakoid membranes.

As is usual in multiauthored books, a certain lack of homogeneity in the length and depth of the various contributions is also present here, but the abundance of cited papers and reviews at the end of each contribution should be helpful to readers wishing to deepen their understanding of the topics treated.

I wish to express my gratitude to all lecturers for the quality of their oral presentations and their active interactions with students and other colleagues during the Institute, and to the authors of the contributions to this volume for the care in preparing their typescripts. The contributions cover all the lectures delivered in Varenna, with the only exclusion of the lectures on liquid/liquid interfaces given by Prof. David J. Schiffin and those on *in situ* vibrational spectroscopies given by Prof. Alan Bewick. I apologize to the participants in the Institute for having failed to obtain the typescripts of these lectures, which would have contributed to provide a more complete picture of the field.

My sincere thanks are also due to all members of the Scientific Committee of the NATO ASI, including Profs. H. Gerischer and J. Lyklema who were unable to participate in the Institute, for invaluable advice in its scientific organization.

I wish to express my deep gratefulness to the NATO Scientific Affairs Division for the particularly generous support to the Institute.

Special thanks go to Dr. Maria Rosa Moncelli for her invaluable help in the local organization of the Institute and to Dr. Giovanni Aloisi for his assistance in the editorial work.

Rolando Guidelli
Department of Chemistry
Florence University
Florence, Italy
September, 1991

AN OVERVIEW OF ELECTRIFIED INTERFACES

ROLANDO GUIDELLI
Dipartimento di Chimica - Universita' di Firenze
Via G. Capponi, 9 - 50121 FIRENZE- Italy

ABSTRACT. Electrified interfaces are briefly surveyed at an introductory level. A comparative description of the electrochemical potential of electrons in metals, in semiconductors and in redox systems is provided. Experimentally accessible quantities such as work function, contact potential and absolute electrode potential are examined. Interfacial charge distributions due both to coulombic forces and to specific noncoulombic forces are described. The methods employed to control the electrical state of disperse systems and of biomembranes and their models are briefly reviewed. Finally, the current flowing across electrified interfaces is examined.

1. The Electrochemical Potential

Consider an electronic or ionic conductor in contact with vacuum. The work required to bring a charged species k from a charge-free vacuum at an infinite distance from the conductor into its interior, passing across the vacuum/conductor interface, is called the *electrochemical potential* $\tilde{\mu}_k$ of the charged species. The $\tilde{\mu}_k$ value in the remote charge-free vacuum is taken conventionally equal to zero, whereas its value inside the conductor is negative, due to the greater stability of the charged species k there.

The electrochemical potential $\tilde{\mu}_k$ consists of three contributions:

(*i*) A first contribution is the work required to bring the charged species k from the remote charge-free vacuum to a position *in vacuo* just outside the conductor surface; this position must be close enough to the surface for the species to experience the maximum coulombic field produced by any net charge on the conductor surface, but far enough for it to escape any short range interactions with the conductor surface. In practice this requirement is usually fulfilled at distances from the surface ranging from 10^{-3} to 10^{-5} cm. This work is given by $z_k e \psi$, where e is the absolute value of the electron charge, $z_k e$ is the charge of the species k and ψ is the electric potential as measured relative to the remote charge-free vacuum. The quantity ψ, called the *outer* or *Volta potential*, is the potential difference between two points in the same medium (the vacuum); as such, it can be defined thermodynamically and measured experimentally without having recourse to modelistic assumptions. Clearly, when the net charge on the conductor surface equals zero, the same is true for the outer potential ψ.

(*ii*) A second contribution is the electrical work required to bring the charged

R. Guidelli (ed.), Electrified Interfaces in Physics, Chemistry and Biology, 1–64.
© 1992 *Kluwer Academic Publishers.*

species k across the thin shell at the conductor surface, where the anisotropy of the forces determines a distorted charge distribution. Note that such a distortion does not necessarily result in a net macroscopic charge, σ, per unit area of the conductor surface. The electric potential difference across this thin shell is called the *surface potential* χ, and the resulting electrical work equals $z_k e\chi$. At the surface of a metal crystal, the mobile conduction electrons tend to spill over the lattice of metal ion cores due to their very small mass, leaving behind an excess of positive charge. This creates a potential difference (the *electronic surface potential*) with the positive side towards the metal. This electronic surface charge distribution is smoother than that of the electron clouds formed by the conduction electrons around the ion cores inside the metal. Such a smoothing leaves the surface ion cores out of electrostatic equilibrium. The net force on these ion cores points primarily into the metal crystal, so that the surface lattice plane tends to approach the immediately underlying lattice plane (*relaxation*; see Fig.8 in Wandelt's contribution to this volume) until equilibrium is reestablished. In a semiconductor, where truly directional chemical bonds exist between the constituent atoms, a highly unstable state originates when these bonds are broken to form the surface. The surface and subsurface atoms will therefore tend to reach a structure facilitating new bond formation. This *surface reconstruction* yields geometrical structures which are often quite different from that in the bulk material. Surface reconstruction is also frequently observed with metal crystals. This distorted atomic distribution makes a further contribution to the surface potential χ and, in metals, is closely related to the surface electronic distribution. In semiconductors the alteration in the distribution of free electrons and of their vacancies (the so called *holes*, see further) extends over a large distance from the surface; the resulting *space charge* makes a further contribution to the surface potential χ. In ionic conductors such as electrolytic solutions, the solvent dipoles at the surface tend to assume a net preferential orientation under the influence of the anisotropic forces acting there; this generates a contribution to the surface potential χ called the *dipole surface potential*. A further contribution to χ, called the *ionic surface potential*, may stem from an alteration in the distribution of the anions and cations of the electrolyte relative to the bulk; this alteration extends over a relatively wide region adjacent to the surface, called the *diffuse layer*.

(*iii*) Finally, to bring the charged species k deep into the electronic or ionic conductor, a further work is required to establish the short-range interactions between this species and the surrounding atoms, electrons or molecules. This work is usually denoted by the symbol μ_k and is called the *chemical potential*.

Summarizing, the electrochemical potential $\tilde{\mu}_k$ can be written as the sum of the three above contributions:

$$\tilde{\mu}_k = \mu_k + z_k e\psi + z_k e\chi \equiv \mu_k + z_k e\phi \quad \text{with } \phi \equiv (\psi + \chi) \tag{1}$$

The sum ϕ of the outer and surface potentials is called the *inner* or *Galvani potential*.

Note that no net volume charge density ρ can exist in the bulk of a conductor at equilibrium. The *Poisson equation* of electrostatics states that the volume charge density $\rho(x)$ at a given position x=x is proportional to the gradient of the electric field -dϕ/dx at that position:

$$\frac{d^2\phi(x)}{dx^2} = -\frac{4\pi\rho(x)}{\varepsilon}$$

(2)

where ε is the dielectric constant of the medium. In other words, a nonzero volume charge density $\rho(x)$ is the source of an electric field, which will move mobile charges until equilibrium is attained. This equilibrium situation will be reached when the net charge is distributed on the conductor surface in such a way as to set the electric field (and hence the volume charge density) to zero throughout the whole bulk phase. If the conductor is in contact with vacuum, any surface charge density σ so formed will generate in the surrounding vacuum an electric field whose lines of force will terminate into some remote sink. However, if we bring this conductor into contact with a further conductor, the net charges at the surface of both conductors will reorganize in such a way as to set the electric field to zero in both bulk phases. If the two conductors have a mobile charged species in common, the attainment of equilibrium will also involve the flow of this species across the interface being formed. At equilibrium a charge separation will generally exist at the interface. However, the overall charge enclosed in an ideal closed surface extending from the bulk of one phase to the other must be equal to zero. In view of *Gauss theorem* of electrostatics, the flux of the electric field across this closed surface equals zero only provided the overall charge enclosed within the surface is also equal to zero. In other words, the conductor/ conductor interface, in spite of the presence of a charge separation, is electroneutral as a whole.

Strictly speaking, the *interface* between two different conductors α and β is to be regarded as *electrified* if the inner potential difference ϕ^α-ϕ^β≡$\phi^{\alpha-\beta}$ between the two conductors can be varied at will and if its changes (not its absolute value!) are experimentally accessible.

If the same charged species k is present to a detectable extent in the bulk of two phases α and β which are being brought into contact, then it will momentarily flow across the resulting interface until equilibrium is attained. The equilibrium situation is realized when no work has to be spent to move the common species k from one phase to the other across the interface. Imagine performing this trasport by following an alternative pathway, namely by bringing the species k from the bulk phase α to a remote charge-free vacuum "across the α/vacuum interface" and then from this vacuum into the bulk phase β "across the vacuum/β interface". In view of the definition of electrochemical potential, the work spent in the former step equals the opposite, $-\tilde{\mu}_k{}^{\alpha}$, of the electrochemical potential of the species k in phase α, whereas the work spent in the latter step equals $\tilde{\mu}_k{}^{\beta}$. Since the sum of these two works equals zero at equilibrium, the equilibrium condition requires the electrochemical potential of the common charged species k to take the same value in the two contacting phases:

$$\tilde{\mu}_k^{\alpha} = \tilde{\mu}_k^{\beta} \tag{3}$$

In view of eqns.(1) and (3), the inner potential difference $\phi^{\alpha-\beta}$ between phase α and phase β at equilibrium takes the well defined value:

$$\phi^{\alpha} - \phi^{\beta} \equiv \phi^{\alpha-\beta} = (\mu_k^{\beta} - \mu_k^{\alpha})/(z_k e) \tag{4}$$

Note that if the short-range interactions of the charged species k within the bulk phase α are more attractive than those within the bulk phase β, then $\mu_k{}^{\alpha}$ is more negative than $\mu_k{}^{\beta}$, and hence the inner potential difference $\phi^{\alpha-\beta}$ has the same sign as the charge $z_k e$ of the species k, in view of eqn.(4). This is due to the fact that, when forming the interface, the species k flows temporarily from phase β to phase α until the inner potential ϕ^{α} becomes sufficiently positive (if the species k is positively charged) or negative (if k is negatively charged) relative to ϕ^{β} as to block such a flow by electrostatic repulsion; in practice, the flow of charge required to attain this equilibrium situation is extremely small.

The above situation is encountered at the interface between two different metals, between a metal and a semiconductor, or between two different semiconductors. This is because both metals and semiconductors are characterized by the presence of a common charged species, namely the free electron. An analogous situation is also encountered at the interface between a metal (or a semiconductor) and an electrolytic solution which contains a redox couple ox/red in a condition of exchanging electrons with the metal (or the semiconductor). The solution of a salt of a metal ion in contact with the corresponding metal (or one of its alloys) represents a further notable example, in which the metal ion can be formally regarded as the charged species common to the two contacting phases. In all other cases, the interface between an electrolytic solution and a metal or a semiconductor is not subject to the equilibrium condition of eqn.(3). Such an interface, which is referred to as an *ideal polarized electrode*, attains an *electrostatic equilibrium* for any of the infinite $\phi^{\alpha-\beta}$ values corresponding to a relatively wide range of *applied potentials* (namely potentials applied with respect to a given reference electrode).

2. The Electrochemical Potential of Electrons

In view of the relevance of the electrochemical potential $\tilde{\mu}_e$ of electrons in determining the equilibrium conditions at several electrified interfaces, we will consider it in some detail.

According to the *band theory* [1], a solid (metal, semiconductor or insulator) can be thought to be formed starting from a group of identical atoms located at the sites of a lattice with such a large lattice constant as to exclude any interactions between them. In this initial configuration the electrons are completely localized at the different lattice sites and occupy identical energy levels ε_1, ε_2, etc., as shown in Fig.1. If we now imagine shrinking this artificially large lattice constant gradually, so as to assemble the atoms into a regular lattice, the progressive increase in their mutual interactions splits each of the energy levels in the isolated atoms into a number of levels equal to the number of the atoms being assembled.

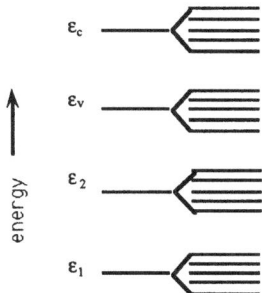

Figure 1. Electronic energy levels in a single atom and in a crystal.

For a very high number of atoms, the electronic levels generated from a given energy level in the atom are very close to each other so as to form an almost continuous *energy band* (see Fig.1).

In metals, the energy bands generated from the highest energy levels still occupied by electrons in the ground state of the isolated atoms, overlap. Now, for an applied electric field to produce an orderly motion of electrons in the solid, namely an electric current, the electrons must increase their energy by moving from a lower to a higher level. When the above overlapping of energy bands takes place, an electric current may flow across the material even at very low temperatures, as is actually the case for metals. However, if the band generated from the highest occupied energy level in the ground state of the isolated atoms does not overlap with that generated from the lowest unoccupied energy level, at T=0 the former band (the *valence band*) in the solid is also fully occupied, whereas the latter (the *conduction band*) is empty. In this case, for an electric field to move electrons in the solid and to generate a current, the energy of thermal agitation must be high enough to excite a sufficient number of electrons from the top of the valence band to the bottom of the conduction band, across the so called *energy gap* E_g. Materials with an energy gap less than 1÷2 eV exhibit a detectable electrical conductivity at room temperature or, at any rate, below their melting point, and are called *intrinsic semiconductors*. Instead, materials with E_g values appreciably higher than 1 eV do not exhibit an observable electrical conductivity and are called *insulators*.

In semiconductors, the thermal excitation of electrons and their transition into the bottom of the conduction band generates empty electron levels (called *holes*) at the top of the valence band. The electrical conductivity of semiconductors is therefore produced by the motion of electrons not only in the conduction band, but also in the valence band. Here, an electron filling a vacancy leaves behind another vacancy, which in turn may be filled. The drift of the negatively charged electrons in one direction within the valence band, as produced by an electric field, is equivalent to the drift of the positively charged holes in the opposite direction. The motion of holes in the valence band is more conveniently dealt with than that of electrons. Hence, in a formal way we may state that a semiconductor has two types of charge carriers, namely free electrons in the conduction band and holes in the valence band. In an intrinsic semiconductor the number of holes equals that of the free electrons.

The properties of semiconductors are extremely sensitive to impurities, since the latter contribute to the density of conduction electrons and holes [1]. Impurities are called *donors* if they supply additional electrons to the conduction band, and *acceptors* if they supply additional holes to the valence band. In general, donors are atoms with a higher chemical valence than the atoms making up the pure (intrinsic) semiconductor, while acceptors have a lower chemical valence. As an example, consider the case of a crystal of pure germanium, a group IV semiconductor, and imagine replacing one germanium atom in the crystal by an arsenic atom, of group V. The germanium ion core has a charge $+4e$ and contributes four valence electrons, while the arsenic ion core has a charge $+5e$ and contributes five valence electrons. The presence of the impurity atoms of arsenic generates an additional energy level ε_d within the energy gap in the proximity of the bottom ε_c of the conduction band, as shown in Fig. 2a. Hence, at room temperature thermal ionization of the arsenic atom is highly probable. In this respect, if we ignore the difference in structure between the arsenic and germanium ion cores, we may formally regard an arsenic atom in the lattice of the germanium crystal as a germanium atom with a positive charge, plus an electron in the conduction band. By analogous arguments, *acceptor impurities* of gallium, a group III element, in a germanium crystal introduce an additional electronic level ε_a which is higher than the top ε_v of the valence band by an amount that is small compared with the energy gap E_g (see Fig. 2b); these impurities may therefore readily accept an electron from the valence band, so as to be regarded as equivalent to germanium atoms with a fixed negative charge $-e$, plus a hole in the valence band. Indeed, due to the close vicinity of the donor levels ε_d to the bottom ε_c of the conduction band and of the acceptor levels ε_a to the

(a)   **(b)**

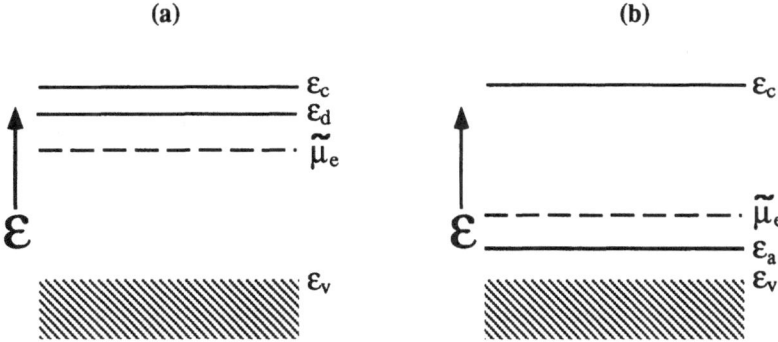

Figure 2. Energy levels in an *n*-type (a) and a *p*-type semiconductor (b).

top ε_v of the valence band, for all practical purposes both acceptor and donor impurities are completely "ionized" at room temperature. Unless the concentration of these impurities is very small, they therefore represent a much more important source of charge carriers than the intrinsic mechanism of thermal excitation of electrons across the whole energy gap.

2.1. THE ELECTROCHEMICAL POTENTIAL OF ELECTRONS IN METALS

The Pauli exclusion principle states that we may place at most one electron in each single electronic level. The probability $f(\varepsilon_i)$ of an electron being in the level of energy ε_i is expressed by the *Fermi-Dirac statistics*:

$$f(\varepsilon_i) = \frac{1}{e^{(\varepsilon_i - \tilde{\mu}_e)/kT} + 1} \tag{5}$$

where k is the Boltzmann constant, T is the absolute temperature, and $\tilde{\mu}_e$ is the electrochemical potential of electrons. It is evident from eqn.(5) that this probability equals unity (namely the energy level ε_i is certainly occupied by a single electron) when ($\varepsilon_i - \tilde{\mu}_e$) is negative and its absolute value is >>kT, whereas it equals zero (namely the energy level is certainly unoccupied) when ($\varepsilon_i - \tilde{\mu}_e$) is positive and >>kT. In practice, at room temperature kT is <<$|\tilde{\mu}_e|$, and hence only a relatively narrow range of energies about $\tilde{\mu}_e$ is characterized by probabilities different from either 1 (for $\varepsilon_i < \tilde{\mu}_e$) or zero (for $\varepsilon_i > \tilde{\mu}_e$). In the ground state, as attained at T=0 , $f(\varepsilon_i)$ is constantly equal to unity for $\varepsilon_i < \tilde{\mu}_e$, and constantly equal to zero for $\varepsilon_i > \tilde{\mu}_e$, so that $\tilde{\mu}_e$ marks the boundary between occupied and unoccupied electronic levels.

In the *independent electron approximation*, electrons in a metal are assumed to move without interacting with each other. According to this approximation, the ground state of a system of N electrons enclosed in a volume V can be found by first finding the energy levels of a single electron in the volume V, and by then filling these levels up, starting from the lowest one, in a manner consistent with the exclusion principle. By so doing, the energy level below which the one-electron levels are occupied and above which they are unoccupied in the ground state of the metal is called the *Fermi level*, and the corresponding energy is denoted by ε_F. It is therefore evident that ε_F is the limiting value of the electrochemical potential $\tilde{\mu}_e$ of electrons for T->O. In metals, $\tilde{\mu}_e$ remains practically equal to the Fermi level ε_F all the way up to room temperature, to a high degree of accuracy. For this reason, no distinction is usually made between $\tilde{\mu}_e$ and ε_F when dealing with metals, although a slight difference between the two quantities emerges in precise calculations.

It should be noted that the Fermi level is sometimes measured relative to the lowest one-electron level, whereas the electrochemical potential $\tilde{\mu}_e$ is constantly measured relative to the vacuum level. It is evident that in this case the two quantities cannot be identified. As an example, consider the *Sommerfeld theory of metals*, according to which electrons are assumed to be confined in the interior of the metal and to move in the absence of both electron-ion and electron-electron interactions (*free electron approximation*). This confinement is brought about by trapping the electrons into a square-potential well whose depth U is so great as to prevent any spill over. According to this model the Fermi energy ε_F as measured from the bottom of the potential-energy well equals $\hbar^2(3\pi^2 n)^{2/3}/(2m)$, where *m* is the electron mass and *n* is the electron number density. Broadly speaking, this ε_F value expresses the maximum kinetic energy of electrons at T=0 and is clearly positive. Its relation to the electrochemical and chemical potentials previously defined (cf. eqn.1) is therefore:

$$\tilde{\mu}_e \cong U + \varepsilon_F - e\phi, \quad \text{with } U + \varepsilon_F \equiv \mu_e \qquad (6)$$

This is because, in the framework of this model, the sum of the negative potential energy U and of the positive *kinetic Fermi energy* ε_F expresses the negative work μ_e spent in establishing the short range interactions of the electron within the metal lattice. Note that in the literature concerned with solid state physics, the electrochemical potential is often referred to as the chemical potential, and viceversa.

If the metal contains N free electrons in a volume V, the $\tilde{\mu}_e$ value can be estimated by noting that the probability $f(\varepsilon_i)$ of eqn.(5) also measures the mean number of electrons in the one-electron level of energy ε_i. Since the total number of electrons N is just the sum over all possible levels of their mean number in each level, we get:

$$N = \sum_i f(\varepsilon_i) = \sum_i \frac{1}{e^{(\varepsilon_i - \tilde{\mu}_e)/kT} + 1} \qquad (7)$$

which expresses $\tilde{\mu}_e$ implicitly as a function of N and T.

2.2. THE ELECTROCHEMICAL POTENTIAL OF ELECTRONS IN SEMICONDUCTORS

Let us now turn to semiconductors. Their electrochemical potential lies almost invariably in the energ gap between the top ε_v of the valence band and the bottom ε_c of the conduction band, as shown in Fig.2. Even when dealing with semiconductors, it is common practice to refer to $\tilde{\mu}_e$ as the "Fermi level". This terminology is somewhat misleading since, strictly speaking, the Fermi level is the well defined energy below which the one-electron levels are occupied and above which they are unoccupied in the ground state of a metal. Now, in a semiconductor any energy in the gap separates occupied from unoccupied levels at T=0. Hence, whenever used in connection with semiconductors, the term "Fermi level" should be regarded as nothing but a synonym for the electrochemical potential $\tilde{\mu}_e$.

Just as in metals, $\tilde{\mu}_e$ determines the probability $f(\varepsilon_i)$ of occupation of the electronic levels both in the valence and in the conduction band *via* the Fermi-Dirac distribution law of eqn.(5). However, the position of $\tilde{\mu}_e$ within the energy gap often satisfies the condition:

$$\varepsilon_i - \tilde{\mu}_e \gg kT \quad \text{for } \varepsilon_i > \varepsilon_c ; \qquad \tilde{\mu}_e - \varepsilon_i \gg kT \quad \text{for } \varepsilon_i < \varepsilon_v \tag{8}$$

at room temperature, even for energy gaps $E_g = \varepsilon_c - \varepsilon_v$ of a few tenths of an electron volt. In this case the distribution law of eqn.(5) simplifies as follows:

$$f(\varepsilon_i) = \frac{1}{e^{(\varepsilon_i - \tilde{\mu}_e)/kT} + 1} \approx e^{-(\varepsilon_i - \tilde{\mu}_e)/kT} \quad \text{for } \varepsilon_i > \varepsilon_c \tag{9}$$

for the electrons in the conduction band. As concerns the holes in the valence band, the probability of a hole occupying a level $\varepsilon_i < \varepsilon_v$ equals $1 - f(\varepsilon_i)$, with $f(\varepsilon_i)$ given by eqn.(5), since a level can only contain either 0 or 1 electron. Hence, from eqns.(5) and (8) it follows that this probability is given by:

$$1 - \frac{1}{e^{(\varepsilon_i - \tilde{\mu}_e)/kT} + 1} = \frac{1}{e^{-(\varepsilon_i - \tilde{\mu}_e)/kT} + 1} \approx e^{-(\tilde{\mu}_e - \varepsilon_i)/kT} \quad \text{for } \varepsilon_i < \varepsilon_v \tag{10}$$

The distribution laws of eqn. (9) and (10) clearly satisfy the well-known *Boltzmann statistics*. In other words, in spite of the fact that electrons (and holes) are *fermions*, namely particles satisfying the Fermi-Dirac statistics, they practically behave as though they were *boltzons*; in this respect they are sometimes referred to as *corrected boltzons*. By far the majority of semiconductors satisfy the conditions of eqn.(8), and hence the distribution laws of eqns.(9) and (10); they are referred to as *nondegenerate semiconductors*. Instead, with *degenerate semiconductors* the conditions of eqn.(8) are not satisfied to a sufficient degree of accuracy, and the Fermi distribution law must be retained.

Semiconductors doped with donor impurities, which supply electrons to the conduction band, are called *n-type semiconductors*, where *n* stands for "negative"; conversely, semiconductors doped with acceptor impurities, which generate holes in the valence band, are called *p-type semiconductors*, where *p* stands for "positive". The presence of impurities affects neither the density $\rho_v(\varepsilon)$ of the electron levels in the valence band nor that, $\rho_c(\varepsilon)$, in the conduction band to an appreciable extent; here $\rho(\varepsilon)$ denotes the number of one-electron levels within an infinitesimal energy interval $d\varepsilon$ about ε, over $d\varepsilon$. However, impurities shift the position of the electrochemical potential $\widetilde{\mu}_e$ upwards in the case of donors, and downwards in the case of acceptors. Let us see how this happens.

The number *n* of electrons in the conduction band is obtained by multiplying the density $\rho_c(\varepsilon)$ of electron levels there by the probability $f(\varepsilon)$ of their being occupied, and by integrating this product over ε between the bottom ε_c of the conduction band and infinity. Upon using the Boltzmann expression of eqn.(9) for $f(\varepsilon)$ we get:

$$ n = \int_{\varepsilon_c}^{\infty} \rho_c(\varepsilon)\, e^{-\beta(\varepsilon-\widetilde{\mu}_e)}\, d\varepsilon = e^{-\beta(\varepsilon_c-\widetilde{\mu}_e)} \int_{\varepsilon_c}^{\infty} \rho_c(\varepsilon)\, e^{-\beta(\varepsilon-\varepsilon_c)}\, d\varepsilon \equiv N\, e^{-\beta(\varepsilon_c-\widetilde{\mu}_e)} \tag{11} $$

where $\beta \equiv 1/kT$. Analogously, the number *p* of holes in the valence band is obtained by multiplying the density $\rho_v(\varepsilon)$ of electron levels there by the probability of their being unoccupied as expressed by eqn.(10), and by integrating over ε between $-\infty$ and the top ε_v of the valence band:

$$ p = \int_{-\infty}^{\varepsilon_v} \rho_v(\varepsilon)\, e^{-\beta(\widetilde{\mu}_e-\varepsilon)}\, d\varepsilon = e^{-\beta(\widetilde{\mu}_e-\varepsilon_v)} \int_{-\infty}^{\varepsilon_v} \rho_v(\varepsilon)\, e^{-\beta(\varepsilon_v-\varepsilon)}\, d\varepsilon \equiv P\, e^{-\beta(\widetilde{\mu}_e-\varepsilon_v)} \tag{12} $$

In practice, the values N and P of the integrals in eqns.(11) and (12) turn out to be very close. Upon multiplying the expressions of *n* and *p* in these two equations, we obtain:

$$ n\,p = N\,P\, e^{-\beta E_g} \qquad \text{with } E_g \equiv \varepsilon_c - \varepsilon_v \tag{13} $$

This expression, which is sometimes called the "law of mass action", states that the product of the numbers of conduction electrons and holes is a constant which depends exponentially upon the energy gap E_g, but is almost independent of the presence of impurities.

In the case of an intrinsic semiconductor, n equals p:

$$n = N e^{-\beta(\varepsilon_c - \widetilde{\mu}_e^{in})} = p = P e^{-\beta(\widetilde{\mu}_e^{in} - \varepsilon_v)} \equiv n_{in} \tag{14}$$

where n_{in} denotes the common value of n and p for the intrinsic case, and $\widetilde{\mu}_e^{in}$ is the corresponding electrochemical potential. Upon rearranging terms, the following expression of $\widetilde{\mu}_e^{in}$ is obtained:

$$\widetilde{\mu}_e^{in} = \frac{\varepsilon_v + \varepsilon_c}{2} + \frac{kT}{2} \ln \frac{P}{N} = \varepsilon_v + \frac{E_g}{2} + \frac{kT}{2} \ln \frac{P}{N} \tag{15}$$

Since P and N take close values, this equation shows that $\widetilde{\mu}_e^{in}$ for the intrinsic case lies practically in the middle of the energy gap.

Let us now consider a semiconductor doped with N_d donor impurities and N_a acceptor impurities per unit volume, and let us confine ourselves to the usual situation in which these impurities are completely ionized. The electroneutrality condition in the bulk semiconductor is expressed by the equation:

$$p + N_d - n - N_a = 0 \tag{16}$$

since the ionized donors are positively charged, whereas the ionized acceptors are negatively charged. To determine the expression for the electrochemical potential $\widetilde{\mu}_e$ in the extrinsic case, we derive N and P, which are practically unaffected by the presence of impurities, from eqn.(14) as a function of n_{in}, and we replace the corresponding expressions into the general equations (11) and (12), thus obtaining:

$$n = n_{in} e^{-\beta(\widetilde{\mu}_e^{in} - \widetilde{\mu}_e)} \quad ; p = n_{in} e^{-\beta(\widetilde{\mu}_e - \widetilde{\mu}_e^{in})} \tag{17}$$

Upon substituting n and p from this equation into eqn.(16) and rearranging terms, we obtain:

$$\widetilde{\mu}_e = \widetilde{\mu}_e^{in} + kT \text{ arcsinh} \frac{N_d - N_a}{2 n_{in}} \tag{18}$$

According to this equation, if the semiconductor is heavily doped with donor impurities $(N_d \gg n_{in}; N_a \approx 0)$, the electrochemical potential $\widetilde{\mu}_e$ is shifted upwards, closely approaching the bottom of the conduction band; conversely, if the semiconductor is heavily doped with acceptor impurities $(N_a \gg n_{in}; N_d \approx 0)$, $\widetilde{\mu}_e$ is shifted downwards, closely approaching the top of the valence band. This is shown schematically in Fig.2.

2.3. THE ELECTROCHEMICAL POTENTIAL OF ELECTRONS IN REDOX SYSTEMS

We have seen that the Boltzmann distribution law of eqns.(9) and (10) applies to fermions whenever the number of these particles is much less than the number of the one-particle energy levels which they are in a condition of occupying with a sufficiently high probability. It is intuitive that the same distribution law holds, without any restriction, when a single energy level may be occupied by any number of particles, no matter how high; these particles are referred to as *boltzons*, as distinct from corrected boltzons. Examples of boltzons are the solute molecules of an electrolytic solution. In the case of validity of the Boltzmann statistics, the electrochemical potential of a species k can be readily expressed as a function of the number N of k particles. Just as in the case of eqn.(7), N is obtained by summing the probability $f(\varepsilon_i)$ of occupation of an energy level ε_i over all possible energy levels. Using for $f(\varepsilon_i)$ the Boltzmann expression $f(\varepsilon_i)=\exp[(\tilde{\mu}_k-\varepsilon_i)/kT]$ of eqn.(9) we obtain:

$$N = \Sigma_i \, e^{(\tilde{\mu}_k-\varepsilon_i)/kT} = q \, e^{\tilde{\mu}_k/kT} \quad \text{with } q \equiv \Sigma_i \, e^{-\varepsilon_i/kT} \tag{19}$$

where the summation q of the Boltzmann exponential factors over all energy levels is called the *partition function*. Equation (19) applies to all boltzons; when applied to a solute of concentration $c=N/V$, it takes the familiar logarithmic form:

$$\tilde{\mu}_k = -kT \ln q + kT \ln N = \text{constant} + kT \ln c \tag{20}$$

Let us now consider the electrochemical potential $\tilde{\mu}_e$ of electrons in a solution containing a redox couple ox/red exchanging electrons with an inert electrode:

$$\text{ox} + e \Leftrightarrow \text{red} \tag{21}$$

Clearly, in a homogeneous redox system electrons are not free as they are in a metal; rather, they are bound to the reduced species red. Crudely speaking, we may regard the red molecules as analogous to "occupied electron levels" in a metal, and the ox molecules as analogous to "unoccupied electron levels". Since all red molecules in the solution are a priori equivalent, and each of them may accept only one electron, the electron levels provided by the red molecules are one-electron levels of equal energy. In metals and semiconductors, however, a given energy level has the same energy independent of whether it is occupied or not. This is also true for a redox couple *in vacuo*. On the contrary, in the presence of a polar solvent, the solvent molecules will polarize to a different extent around a red and an ox molecule, due to the different charge carried (and hence the different electric field generated) by the two molecules. In turn, this difference in the most probable, *equilibrium polarization* of the solvent molecules will affect the electron levels, causing the electron level in ox, ε_{ox}, to be different from that in red, ε_{red}. To proceed further in the analogy with the free electrons in a metal, we must therefore refer to a non-equilibrium solvent polarization, in between the two equilibrium polarizations around the ox and red molecule respectively, for which the electron energy is the same in ox and in red. We will denote this energy, sometimes referred to as the *standard redox Fermi level*, by $\varepsilon^\circ_{redox}$. Thanks to temporal fluctuations of the solvent molecules, this non-equilibrium polarization is actually attained, although with a very low probability. After the uptake of one electron by ox, the solvent molecules around the resulting red molecule will rapidly relax to the equilibrium polarization corresponding to the electron energy ε_{red}. Analogously, after the release of an electron by red, the solvent molecules around the resulting ox molecule will rapidly relax to the equilibrium polarization corresponding to the electron energy ε_{ox}.

If we may disregard the contribution to polarization from the solvent or ligand molecules in the inner coordination sphere of the red and ox molecules, the standard redox Fermi level $\varepsilon^\circ_{redox}$ lies exactly midway between ε_{ox} and ε_{red} (see Fig.3):

$$\varepsilon^\circ_{redox} - \varepsilon_{red} = \varepsilon_{ox} - \varepsilon^\circ_{redox} = \lambda \tag{22}$$

This common value of the energy of solvent relaxation is called the *reorganization energy*, and plays an important role in the kinetics of simple electrode reactions. According to the classical theory of electron transfer developed by Marcus [2], in which the ox and red molecules are represented as hard spheres in a dilectric continuum, λ takes the simple form:

$$\lambda = \frac{e^2}{2a}\left(\frac{1}{\varepsilon_{opt}} - \frac{1}{\varepsilon_{stat}}\right) \tag{23}$$

where ε_{opt} and ε_{stat} are the values of the optical and static dielectric constants of the medium and a is the radius of the redox species. More accurate expressions have been derived by Levich [3], Dogonadze [4] and others [5], in which the molecules of red, ox, solvent, ligands and other ions (the slow subsystem) are treated by the semi-classical approximation, whereas electrons are treated quantum-mechanically. The probability of the electron level in ox having a given energy ε is expressed by a Gaussian distribution about the equilibrium value ε_{ox}, and that for the electron level in red by an analogous Gaussian distribution about the equilibrium value ε_{red}. More precisely, the corresponding Gaussian distribution functions are given by:

$$D_{ox} = \exp\left[-\frac{(\varepsilon-\varepsilon_{ox})^2}{4kT\lambda}\right] = \exp\left[-\frac{(\varepsilon-\varepsilon^\circ_{redox}-\lambda)^2}{4kT\lambda}\right] \tag{24}$$

and:

$$D_{red} = \exp\left[-\frac{(\varepsilon-\varepsilon_{red})^2}{4kT\lambda}\right] = \exp\left[-\frac{(\varepsilon-\varepsilon^0_{redox}+\lambda)^2}{4kT\lambda}\right] \qquad (25)$$

Some typical plots of $D_{ox}(\varepsilon)$ and $D_{red}(\varepsilon)$ vs. ε are shown in Fig.3. The two distribution functions D_{ox} and D_{red} overlap just at the standard redox Fermi level ε^0_{redox}, where the occupied and the unoccupied electron level have an equal, albeit quite low, probability.

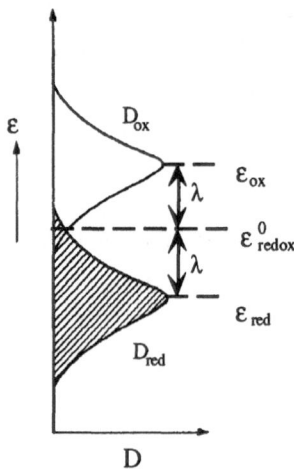

Figure 3. Distribution functions D_{ox} and D_{red} for a redox system in solution.

In general, the above distribution functions are reported along the abscissas in plots of ε against $D_{ox}(\varepsilon)$ and against $D_{red}(\varepsilon)$, side by side with plots of ε against the density $\rho(\varepsilon)$ of electron levels in the metal (or semiconductor) which exchanges electrons with the redox system (see further, Figs.18 and 20). The usefulness of these combined plots will be more apparent in the following, when we shall examine the faradaic current flowing through a metal/electrolyte or a semiconductor/electrolyte interface. However, it is worth noting right now that the distribution functions D_{red} and D_{ox} should by no means be confused with the densities of electron levels: they just express the probabilities that the occupied electron level in red and the unoccupied electron level in ox may take energy values ε different from their most probable values, ε_{red} and ε_{ox} respectively.

Upon referring to a situation analogous to that encountered with metals and semiconductors, namely that in which the energy of the electronic levels in the electrolytic solution takes the same value independent of whether they are occupied or not (i.e., the non-equilibrium value ε^0_{redox}), the Fermi distribution of eqn.(7) can still be applied. Naturally, in the present case the number N of electrons equals that,

N_{red}, of the red molecules in solution, whereas the summation Σ_i over all levels, no matter if occupied or not, equals the sum, $N_{ox}+N_{red}$, of the numbers of red and ox molecules. Hence, eqn.(7) becomes:

$$N_{red} = \sum_{i=1}^{N_{ox}+N_{red}} \frac{1}{e^{(\varepsilon_{redox}^o-\widetilde{\mu}_e)/kT} + 1} = \frac{N_{ox}+N_{red}}{e^{(\varepsilon_{redox}^o-\widetilde{\mu}_e)/kT} + 1} \tag{26}$$

Upon rerranging terms we get:

$$\widetilde{\mu}_e = \varepsilon_{redox}^o - kT \ln \frac{N_{ox}}{N_{red}} = \varepsilon_{redox}^o - kT \ln \frac{c_{ox}}{c_{red}} \tag{27}$$

where c_{ox} and c_{red} are the molar concentrations of ox and red. This equation shows that ε_{redox}^o equals the electrochemical potential of electrons in the solution when $c_{ox}=c_{red}$, which justifies its name of standard redox Fermi level.

We may attach a more formal significance to ε_{redox}^o upon considering the reaction of eqn.(21) as a self-exchange redox reaction taking place in the solution. The equilibrium condition for this reaction is obtained by setting the sum of the electrochemical potentials of reactants equal to that of products, each electrochemical potential being multiplied by the corresponding stoichiometric coefficient in the equation of the reaction. Hence:

$$\widetilde{\mu}_{ox}+\widetilde{\mu}_e = \widetilde{\mu}_{red} \tag{28}$$

If we separate $\widetilde{\mu}_{ox}$ and $\widetilde{\mu}_{red}$ into a chemical and an electrical contribution on the basis of eqn.(1), and we express the concentration dependence of the chemical potentials μ_{ox} and μ_{red} explicitly through eqn.(20), we obtain:

$$\widetilde{\mu}_{ox} = \mu_{ox}+z_{ox}e\phi = \mu_{ox}^o+kT \ln c_{ox}+z_{ox}e\phi$$
$$\widetilde{\mu}_{red} = \mu_{red}+z_{red}e\phi = \mu_{red}^o+kT \ln c_{red}+z_{red}e\phi \tag{29}$$

Here, μ_{ox}^o and μ_{red}^o are the values of the chemical potentials μ_{ox} and μ_{red} for $c_{ox}=1$ and $c_{red}=1$, respectively, and are therefore referred to as the *standard chemical potentials*. Upon substituting $\widetilde{\mu}_{ox}$ and $\widetilde{\mu}_{red}$ from eqn.(29) into eqn.(28) and rearranging, we obtain once again the expression for $\widetilde{\mu}_e$ of eqn.(27), provided we set:

$$\varepsilon_{redox}^o = (\mu_{red}^o-\mu_{ox}^o)-e\phi \tag{30}$$

This expression relates the standard redox Fermi level to the standard chemical potentials of the two redox species and to the inner potential of the bulk solution.

3. Experimentally accessible quantities: work function, contact potential and absolute electrode potential

Let us now examine which of the quantities so far considered is amenable to experimental measurement without recourse to modelistic assumptions. A quantity which plays a crucial role in the description of surface phenomena and satisfies the latter requirement is the *work function* Φ, defined as the minimum energy required to remove an electron from the interior of a solid across the solid/vacuum interface to a "position just outside" [1]. By a position just outside we mean a position which is reached by the electron upon overcoming the surface potential χ but not the outer potential ψ. In view of eqn.(1) as applied to electrons, Φ is therefore given by:

$$\Phi = -(\mu_e - e\chi) = -(\tilde{\mu}_e + e\psi) \tag{31}$$

where the minus sign indicates that the work made *on* the system is positive, whereas $\tilde{\mu}_e$ is negative. Several methods are used to measure Φ. One of them exploits the photoelectric effect, measuring the minimum photon energy required to eject an electron through a crystal face, and setting it equal to Φ. Another method is based on the thermoionic emission, and measures the temperature dependence of the electron current flowing from the face of a hot metal.

In general, Φ measurements are carried out on an uncharged solid; hence, one might conclude that $\psi=0$ and Φ is just the opposite of the electrochemical potential $\tilde{\mu}_e$. This conclusion is not entirely correct. The distribution of the ion cores and of the electron charge near the surface differs from that in the bulk and depends on the orientation of the crystal face under examination with respect to the crystallographic axes. Hence, the surface potential χ is also expected to be different at inequivalent crystal faces. On the other hand, the chemical potential μ_e depends exclusively upon the interactions of the electron within the bulk solid (cf., for instance, eqn.6), and hence is not affected by the choice of the crystal face. The same conclusion holds for the electrochemical potential $\tilde{\mu}_e$, which measures the work required to bring the electron from a remote charge-free vacuum into the interior of the crystal. In view of eqn.(31), if we denote two inequivalent faces of the same crystal by a single and a double prime, the difference in work function at the two faces is given by:

$$\Phi'' - \Phi' = e(\chi'' - \chi') = -e(\psi'' - \psi') \tag{32}$$

Hence, a change in χ is reflected by an equal change in Φ/e and is exactly compensated for by an opposite change in ψ, so as to maintain the electrochemical potential of electrons in the solid constant, independent of the chosen face. The existence of a difference in ψ between inequivalent faces of a crystal which is uncharged as a whole requires a modest redistribution of charge among the different crystal faces.

Note that the equality between the first and third member of eqn. (32):

$$\Phi''-\Phi' = -e(\psi''-\psi') \tag{33}$$

holds not only for two faces of the same metal, but also for those of two different metals in contact. When equilibrium is attained, the electrochemical potential must be the same in the two metals (cf. eqns.3 and 31). The difference $(\psi''-\psi')$ in the outer potentials of the two metals, called the *contact potential*, can be measured experimentally, thus permitting an estimate of the difference in the corresponding work functions through eqn.(33). The contact potential can be measured by arranging the two metal samples so that the two crystal faces form the plates of a parallel capacitor with vacuum in between. An adjustable external potential bias is applied between the plates, and one of them is caused to vibrate so as to vary the distance d between the plates continuously. The potential difference between the plates equals $V=4\pi|Q|d$, where $|Q|$ is the absolute value of the charge on any of the two oppositely charged plates. For a constant potential bias, the current which flows in the wire joining the plates equals $i=dQ/dt=(V/4\pi)\,d(1/d)/dt$, and hence vanishes only when V, and hence Q, equals zero. In the absence of a potential bias, V equals the contact potential $(\psi''-\psi')$; hence, if we adjust the external potential bias until no current flows when d is varied, the final potential bias will be just equal and opposite to the contact potential. The same method, due to Kelvin, can also be used to measure the outer potential difference between a metal and an electrolytic solution.

The work function $\Phi=-(\tilde{\mu}_e+e\psi)$ of semiconductors is experimetally measurable just as that of metals, e.g. by the thermoionic emission method. In metals, Φ is the minimum energy required to remove an electron from the interior of the metal to a position *in vacuo* just outside the metal; for an uncharged surface, Φ coincides with $-\tilde{\mu}_e$. This definition seems to be hardly applicable to semiconductors, since the $\tilde{\mu}_e$ "level" lies in the energy gap, and hence no electron can be removed from that level. Conversely, the electrons which can be removed from the semiconductor by spending a minimum of energy are those at the bottom ε_c of the conduction band; this is the reason why $-\varepsilon_c$ is also called the *electron affinity*, often denoted by χ. However, what we actually measure in a thermoemission experiment is a thermoemission current; this is indeed dependent upon the minimum energy $\chi=-\varepsilon_c$ through the familiar Boltzmann factor $\exp(-\chi/kT)=\exp(\varepsilon_c/kT)$, but it is also proportional to the number of free electrons about the ε_c level, which is given by $N\exp[-(\varepsilon_c-\tilde{\mu}_e)/kT]$ in view of eqn.(11). Hence, the net result is that the thermoemission current due to the free electrons is proportional to $\exp(\tilde{\mu}_e/kT)\approx\exp(-\Phi/kT)$ and provides a method to measure $\tilde{\mu}_e$, rather than the electron affinity χ.

The way in which the electrochemical potential $\tilde{\mu}_e$ of electrons for a redox system in solution has been introduced is by no means related to the presence of an electrode in contact with the solution. Thermodynamic quantities are not affected by the specific mechanism by which the relevant equilibrium state is attained; hence, the electron exchange between ox and red may well be be thought to proceed without involving any electrode. In practice, however, the *electrode potential* of a redox couple is necessarily measured by immersing an inert metal in the solution and by combining the resulting half-cell with a reference electrode, chosen at will. It is gratifying that the electromotive force of the cell so formed turns out to be independent of the nature of the inert metal, for a given reference electrode. It is, however, desirable to express such an electrode potential in a form which is also independent of the choice of the reference electrode, and depends exclusively upon the redox couple and the electrolytic solution in which it is dissolved. To achieve this goal imagine immersing an inert metal M in the solution of the redox couple. Attainment of equilibrium will be accompanied by a redistribution of the surface charge on the two phases not only at the newly formed metal/solution interface, but also at the "free surfaces" of the metal and of the solution (i.e., at the corresponding "interfaces with vacuum"). In view of eqn.(3) equilibrium is achieved when the electrochemical potential of electrons in the metal, $\tilde{\mu}_e{}^M$, is equal to that in the solution, $\tilde{\mu}_e{}^S$:

$$\tilde{\mu}_e^M = \tilde{\mu}_e^S \tag{34}$$

The electrochemical potential in the metal is given by the work spent in carrying an electron from a remote charge-free vacuum into the bulk metal *across the metal/vacuum interface*, and in view of eqns.(1) and (31) is given by:

$$\tilde{\mu}_e^M = \mu_e^M - e\chi^M - e\psi^M = -\Phi^M - e\psi^M \tag{35}$$

Note that ψ^M is now the outer potential of the metal after equilibration with the electrolytic solution. Analogously, the electrochemical potential of electrons in the solution is given by the work spent in carrying an electron from a remote charge-free vacuum into the solution *across the solution/vacuum interface*. Hence, it is likewise given by:

$$\tilde{\mu}_e^S = \mu_e^S - e\chi^S - e\psi^S \equiv -\Phi^S - e\psi^S \tag{36}$$

where $\Phi^S \equiv -(\mu_e^S - e\chi^S)$ may be regarded as a "work function of the solution" and ψ^S is the outer potential of the solution after equilibration with the inert metal. Upon comparing eqns. (34) to (36) we obtain:

$$\Phi^S = \Phi^M + e(\psi^M - \psi^S) \tag{37}$$

Both Φ^M and $(\psi^M - \psi^S)$ are amenable to experimental measurement; Φ^M can be measured by exploiting the photoelectric or thermoionic emission, whereas the outer potential difference $(\psi^M - \psi^S)$ can be measured by the method of the vibrating condenser. Hence, in spite of the experimental difficulties encountered in its measurement, Φ^S is an experimentally measurable quantity which depends on the redox system but not on the inert metal electrode; it is often referred to as the *absolute electrode potential* (see Trasatti's contribution to this volume).

The surface potential χ^M at the interface between a metal M and vacuum cannot be directly measured. However, it is possible to estimate variations in the difference between χ^M and the surface potential, χ^{M-S}, at the interface between the metal M and a given solution S, as the nature of the metal is varied. (Henceforth, the symbols χ^α and ϕ^α will be used to denote the surface and inner potential differences at the interface between phase α and vacuum, whereas the symbols $\chi^{\alpha-\beta}$ and $\phi^{\alpha-\beta}$ will denote the corresponding quantities at the interface between phases α and β.) To this end, consider the cell:

$$\alpha \mid M \mid S \mid Ref \mid \alpha' \tag{38}$$

where the metal/solution interface M/S does not allow the flow of electrons (i.e., it is an ideal polarized electrode), and Ref is a charge-transfer reference electrode. Let the two leads α and α' of the cell consist of the same metal. The potential applied at the two leads for which the surface charge density σ at the M/S interface equals zero is an experimentally measurable quantity, called the *point* (or *potential*) *of zero charge* relative to the chosen reference electrode. It can be estimated from interfacial tension measurements or, more directly, from capacitative charge measurements. This quantity, denoted by E_z, is clearly equal to the sum of the inner potential differences across all interfaces encountered in passing from lead α to lead α':

$$E_z = \phi^{\alpha-M} + \phi^{M-S} + \phi^{S-Ref} + \phi^{Ref-\alpha'} \tag{39}$$

At equilibrium the electrochemical potential of electrons is the same in the two metals α and M in contact. Hence, in view of eqn.(4), $\phi^{\alpha-M}$ is given by:

$$\phi^{\alpha-M} = (\mu_e^\alpha - \mu_e^M)/e \tag{40}$$

where μ_e^α and μ_e^M are the chemical potentials of electrons in α and M. Let us now replace the metal M by a different metal M' in the cell of eqn.(38). This replacement will leave all inner potential differences of eqn.(39) unchanged, except for those at the interfaces α/M and M/S which are involved in the replacement. Hence the new value, E'_z, of the point of zero charge will be given by:

$$E'_z = \phi^{\alpha\text{-}M'} + \phi^{M'\text{-}S} + \phi^{S\text{-}Ref} + \phi^{Ref\text{-}\alpha'} \qquad (41)$$

Upon subtracting eqn.(41) from eqn.(39) and expressing in the resulting equation $\phi^{\alpha\text{-}M}$ through eqn.(40) and $\phi^{\alpha\text{-}M'}$ through an analogous relationship, we obtain:

$$E_z\text{-}E'_z = \left(-\frac{\mu_e^M}{e} + \phi^{M\text{-}S}\right) - \left(-\frac{\mu_e^{M'}}{e} + \phi^{M'\text{-}S}\right) \qquad (42)$$

Since at the potential of zero charge no net charges contribute to the inner potential differences $\phi^{M\text{-}S}$ and $\phi^{M'\text{-}S}$, these are commonly identified with the corresponding surface potential differences, $\chi^{M\text{-}S}$ and $\chi^{M'\text{-}S}$ respectively. This identification is justified if we imagine moving the two sides of the interface apart, with a resulting interposition of vacuum, while leaving the distribution of the interfacial particles unaltered. In this case the outer potentials of the two disjointed parts of the interface (i.e., ψ^M and ψ^S or else $\psi^{M'}$ and ψ^S) are actually equal to zero. This by no means implies that the outer potentials at the "free surfaces" of the two contacting phases M and S, or M' and S, are equal to zero. Indeed, the opposite is usually true.

Upon taking the metal M' as the reference metal, eqn.(42) can be written in a form which stresses the dependence of E_z upon the nature of the metal:

$$E_z = \left(-\frac{\mu_e^M}{e} + \chi^{M\text{-}S}\right) + constant \qquad (43)$$

where the constant depends on the choice of the reference metal M'. If we compare this expression of E_z with the expression of eqn.(35) for the work function of the metal, $\Phi^M = (-\mu_e{}^M + e\chi^M)$, we obtain:

$$E_z = \frac{\Phi^M}{e} + (\chi^{M\text{-}S} - \chi^M) + constant \qquad (44)$$

This equation relates E_z, which is a typical electrochemical quantity, to the work function Φ^M, which is a typical physical quantity. If the surface potential $\chi^{M\text{-}S}$ at the M/S interface were just the difference of the surface potentials χ^M and χ^S at the M/vacuum and S/vacuum interfaces, then, upon plotting the points of zero charge E_z for different metals in contact with the same electrolytic solution S against the corresponding Φ^M/e values, one would obtain a straight line of unit slope in view of eqn.(44). As a matter of fact, slopes different from unity are generally obtained. This is not surprising because, when we bring a metal/vacuum interface and a vacuum/solution interface into contact to form a metal/solution interface, the anisotropic forces which are responsible for the χ^M and χ^S surface potentials at the two separate free surfaces interact with each other. The above E_z vs. Φ^M/e plots may therefore provide valuable information about the nature of these forces and how they change with the nature of the metal (see Trasatti's contribution to this volume).

4. Interfacial Charge Distributions Due to Coulombic Forces

Let us now briefly consider how the anisotropic forces acting at interfaces affect the charge distribution therein. A general, albeit approximate treatment, is available when these anisotropic forces are purely coulombic in nature. In this case, due to the long-range character of coulombic forces, the resulting charge distribution often extends over distances of tens or even hundreds of Angstroms from the boundary of the two contacting phases. Convesely, short-range anisotropic forces affect the charge distribution over a few Angstroms from the boundary, but are by far more specific in nature, and hence depend dramatically upon the nature of the interface. For the sake of definiteness, we will constantly use the term *boundary* to denote the region of transition from a net predominance of the main component of one of the two contacting phases (e.g., metal ion cores in a metal, solvent molecules in an electrolytic solution, etc.) to a net predominance of the main component of the other phase. In general the boundary, so defined, is only a few Angstroms thick, with the possible exception of the boundary between two immiscible liquids. In this respect the boundary is generally much thinner than the interface, by which we mean the whole inhomogeneous region enclosed between two perfectly homogeneous bulk phases.

We shall first examine the interfacial charge distribution due to coulombic forces, starting from extrinsic semiconductors. The charge distribution in intrinsic semiconductors and in electrolytic solutions of 1,1-valent electrolytes may then be formally derived as a particular case of that in extrinsic semiconductors.

4.1. THE SPACE CHARGE IN EXTRINSIC SEMICONDUCTORS

In the interfacial region of a semiconductor the "local" electroneutrality condition, which holds for any bulk phase, does not apply; in other words, the negative charges of the mobile conduction electrons and of the fixed acceptor impurities do not compensate exactly the positive charges of the mobile holes and of the fixed donor impurities. A nonzero local volume charge density:

$$\rho(x) = e[p(x) + N_d - n(x) - N_a] \tag{45}$$

will therefore exist in the proximity of the phase boundary [6-8]. Here x is the distance from the phase boundary, $n(x)$ and $p(x)$ are the local number densities of conduction electrons and holes, whereas the number densities N_d and N_a of donors and acceptors are regarded as uniform, for simplicity. This charge distribution, called *space charge*, is related to the electric potential gradient through the Poisson equation (2).

To estimate the profile of the inner electric potential $\phi(x)$ in the *space charge region* we must consider that the one-electron energy levels are altered in this region, due to the space dependence of $\phi(x)$. If we refer the electric potential $\phi(x)$ at a given distance x=x to its constant value $\phi(x=\infty)$ in the bulk semiconductor, taken as zero, then each one-electron energy level will vary by an amount $-e\phi(x)$ with respect to its bulk value. The same is also true for the bottom of the conduction band and the top of the valence band, so that we may write:

$$\varepsilon_c(x) = \varepsilon_c - e\phi(x); \quad \varepsilon_v(x) = \varepsilon_v - e\phi(x) \tag{46}$$

This implies that the energy bands exhibit a bending within the space charge region (*band bending*), as shown in Fig. 4.

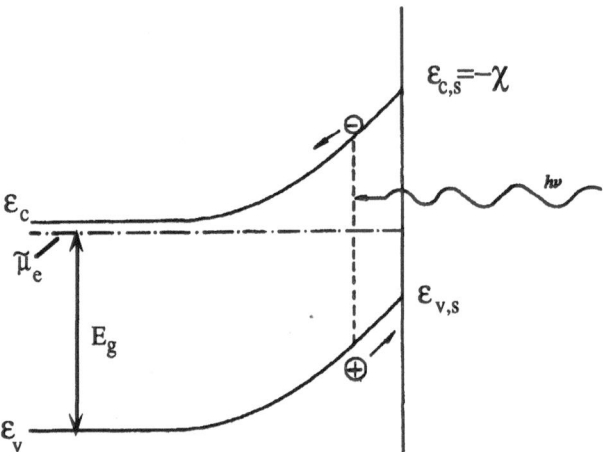

Figure 4. Band bending in an *n*-type semiconductor. The figure also shows that an incident photon of sufficient energy will promote electrons from the valence band to the conduction band, generating an electron-hole pair. The electric field in the space charge region moves the electron into the bulk semiconductor and the hole towards the surface.

Note that while the band edges $\varepsilon_c(x)$ and $\varepsilon_v(x)$ show a bending, the electrochemical potential $\tilde{\mu}_e$ remains rigorously uniform in space, as required by the equilibrium conditions. In other words, at equilibrium a local increase in the electron energy ε is accompanied by a corresponding local decrease in the electron number density such as to maintain $\tilde{\mu}_e$ uniform.

The expressions for $n(x)$ and $p(x)$ are obtained by replacing the bulk values of the energies ε_c and ε_v with the corresponding x-dependent values $\varepsilon_c(x)$ and $\varepsilon_v(x)$ in the expressions of eqns.(11) and (12) for the bulk values of n and p. Hence:

$$n(x) = N\, e^{-\beta[\varepsilon_c - e\phi(x) - \tilde{\mu}_e]} = n\, e^{\beta e\phi(x)}$$
$$p(x) = P\, e^{-\beta[\tilde{\mu}_e - \varepsilon_v + e\phi(x)]} = p\, e^{-\beta e\phi(x)} \tag{47}$$

where n and p retain their significance of bulk number densities. Upon substituting $n(x)$ and $p(x)$ into the $\rho(x)$ expression of eqn.(45), and then the resulting expression into the Poisson equation (2), the following *Poisson-Boltzmann equation* is obtained:

$$\frac{d^2\phi(x)}{dx^2} = -\frac{4\pi e}{\varepsilon_{sc}}(pe^{-\beta e\phi(x)} - ne^{\beta e\phi(x)} + N_d - N_a) \tag{48}$$

where ε_{sc} is the dielectric constant of the semiconductor. A first integration of this differential equation is carried out upon multiplying both members by $2(d\phi/dx)$. By so doing, integration of the first member over x yields immediately $(d\phi/dx)^2$, whereas integration of the second member is readily performed upon changing the integration variable from x into ϕ. If we integrate over x from x=x to x=∞ (and hence over ϕ from $\phi=\phi$ to $\phi=0$) and we note that $(d\phi/dx)$ equals zero for x-> ∞, we get:

$$\left(\frac{d\phi}{dx}\right)^2_{x=x} = \frac{8\pi}{\beta\varepsilon_{sc}}[p(e^{-y}-1) + n(e^y-1) + (N_a-N_d)y] \quad \text{with } y \equiv \beta e\phi(x) \tag{49}$$

The Gauss theorem of electrostatics states that the component, $E_x(x) = -(d\phi/dx)_x$, of the electric field along the positive direction of the x-axis at x=x equals $-(4\pi/\varepsilon_{sc})$ times the whole charge enclosed between the x=x plane and the bulk phase (where the volume charge density and the electric field are both equal to zero) as referred to the unit area of the interface. Hence the electric field $-(d\phi/dx)_{x=0}$ at the boundary x=0 of the semiconductor is just equal to $-(4\pi\sigma_{sc}/\varepsilon_{sc})$, where σ_{sc} is the whole space charge as referred to the unit surface, namely the *surface charge density* due to the space charge. In view of eqn. (49) we may therefore write:

$$\left(\frac{d\phi}{dx}\right)_{x=0} = \frac{4\pi\sigma_{sc}}{\varepsilon_{sc}} = \pm\sqrt{\frac{8\pi}{\beta\varepsilon_{sc}}[p(e^{-y_o}-1)+n(e^{y_o}-1)+(N_a-N_d)y_o]} \quad \text{with } y_o\equiv\beta e\phi_{sc} \quad (50)$$

where the plus sign holds when the electric potential $\phi_{sc}\equiv\phi(x=0)$ at the semiconductor surface is negative with respect to its bulk value and hence increases with increasing x, whereas the minus sign holds when ϕ_{sc} is positive. This equation relates the surface charge density to the inner potential difference ϕ_{sc} across the space charge region and to the bulk concentrations of both mobile and fixed charges.

4.1.1. The Space-Charge Differential Capacity. An interfacial property which is quite sensitive to the fine structure of electrified interfaces and is readily amenable to experimental measurement is represented by the differential capacity $C=|d\sigma/d\phi^{\alpha-\beta}|$, where $\phi^{\alpha-\beta}$ is the inner potential difference across the interface between two phases α and β, and σ it the surface charge density on any of the two sides of the given interface. In practice, by combining the interface of interest with a suitable reference electrode, the applied potential E across the resulting cell differs from $\phi^{\alpha-\beta}$ by an unknown constant depending on the choice of the reference electrode. Hence we may write:

$$C = \left|\frac{d\sigma}{d\phi^{\alpha-\beta}}\right| = \left|\frac{d\sigma}{dE}\right| \quad (51)$$

If the whole potential difference across the interface between a semiconductor and a different phase (e.g., a metal or an electrolytic solution) is almost completely localized across the space charge region of the semiconductor, then the experimental differential capacity is directly obtained from eqn.(50) and can be regarded as a *space-charge differential capacity* C_{sc}:

$$C_{sc} = -\frac{d\sigma_{sc}}{d\phi_{sc}} = \left(\frac{\beta e^2\varepsilon_{sc}}{8\pi}\right)^{1/2} \frac{|-pe^{-y_o}+ne^{y_o}+N_a-N_d|}{\sqrt{p(e^{-y_o}-1)+n(e^{y_o}-1)+(N_a-N_d)y_o}} \quad (52)$$

Semiconductor interfaces often meet the above requirement. The negative sign in front of $d\sigma_{sc}/d\phi_{sc}$ accounts for the fact that a positive (negative) ϕ_{sc} value attracts free electrons (holes) from the bulk semiconductor, making σ_{sc} negative (positive).

The relatively complicated expression of eqn.(52) simplifies in three interesting limiting cases. If y_o is so large and positive that only the terms in $\exp(y_o)$ are retained, eqn.(52) takes the form:

$$C_{sc} = \left(\frac{\beta e^2 \varepsilon_{sc}}{8\pi}\right)^{1/2} n^{1/2} e^{y_0/2} \qquad (53)$$

In this limiting case the differential capacity increases exponentially with increasing $y_0 \equiv \beta e \phi_{sc}$, namely as the potential applied to the semiconductor relative to the adjacent phase is made progressively more negative. (If ϕ_{sc} is very positive, the electric potential in the bulk semiconductor is very negative with respect to the semiconductor boundary.) In this case the conduction electrons move from the bulk of the semiconductor to the surface, accumulating in the space charge region. Two subcases can be distinguished, depending on whether the semiconductor is of the n-type ($N_d \gg N_a$) or of the p-type ($N_a \gg N_d$). In the former case ionization of the donor impurities generates a number of conduction electrons which is by several orders of magnitude higher than that of the holes, also in view of the law of mass action of eqn.(13). We then state that electrons are the *majority carriers*, whereas holes are the *minority carriers*. The opposite is true in the case of p-type semiconductors. When the mobile charges which accumulate at the interface are the majority carriers, the space charge region is called the *accumulation layer*, whereas the name *inversion layer* is reserved to the case in which the minority carriers accumulate at the inteface. The situation expressed by eqn.(53) gives rise to an accumulation layer with n-type semiconductors, and to an inversion layer with p-type semiconductors.

The opposite limiting situation is encountered when y_0 is so large and negative that only the terms in $\exp(-y_0)$ are retained, in which case eqn.(52) takes the form:

$$C_{sc} = \left(\frac{\beta e^2 \varepsilon_{sc}}{8\pi}\right)^{1/2} p^{1/2} e^{-y_0/2} \qquad (54)$$

At the very positive applied potentials at which this expression holds, holes move from the bulk of the semiconductor to the interface, accumulating there. This gives rise to an accumulation layer with p-type semiconductors, and to an inversion layer with n-type semiconductors. A comparison of eqns.(53) and (54) shows that the curves of C_{sc} against ϕ_{sc} are not symmetrical for extrinsic semiconductors (see Fig.5b), since n and p take notably different values. It should be noted that eqns. (53) and (54) are not verified exactly on experimental grounds, because the accumulation of electrons or holes may require the Boltzmann statistics to be abandoned in favour of the Fermi-Dirac statistics.

By far the most interesting case is encountered in an intermediate situation, which we shall consider in connection with an n-type semiconductor. In this case N_d is $\gg N_a$; moreover, the bulk number density n of conduction electrons originates almost exclusively from the donor impurities, so that we may set $n \approx N_d \gg p$. If ϕ_{sc} is negative enough for the condition $n \exp(y_0) \ll N_d$ to hold, but not extremely negative, so as to permit the further condition $p \exp(-y_0) \ll N_d$ to be satisfied, then eqn.(52) takes the simplified form:

$$C_{sc} = \left(\frac{\beta e^2 \varepsilon_{sc}}{8\pi}\right)^{1/2} \frac{N_d^{1/2}}{\sqrt{-y_0-1}} \quad \text{or else} \quad \frac{1}{C_{sc}^2} = \frac{8\pi}{\beta e^2 \varepsilon_{sc}} \frac{-y_0-1}{N_d} \tag{55}$$

This situation is clearly encountered when the local concentrations of both the holes and the conduction electrons in the space charge region are much less that the concentration of the fixed donor impurities; the space charge region is then called the *depletion layer*.

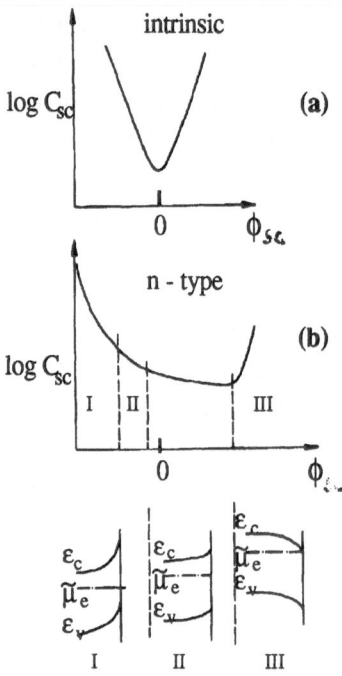

Figure 5. Space-charge differential capacity vs. ϕ_{sc} for an intrinsic semiconductor (**a**) and for an n-type extrinsic semiconductor (**b**). The regions I, II and III in (**b**) roughly correspond to the inversion region, the depletion region and the accumulation region. The band bending in these regions is schematically depicted in (**c**).

According to eqn.(55) a plot of $1/C_{sc}^2$ against $\beta eE = -\beta e\phi_{sc} + constant$, where E is the applied potential, yields a straight line whose intercept on the $1/C_{sc}^2 = 0$ axis equals the applied potential at which ϕ_{sc} equals $-kT/e$. This plot, called the *Mott-Schottky plot*, provides a convenient method for determining the applied potential E at which $\phi_{sc} = 0$, namely at which the space charge is vanishingly small. Since at this potential the band edges do not show any curvature throughout the interfacial region, this potential is called the *flat-band potential*. The slope of the Mott-Schottky plot provides the density N_d of the impurities. It should be mentioned, however, that this ideal behaviour is not exactly satisfied by real semiconductors; thus, the slope of Mott-Schottky plots often depends upon the frequency of the a.c. voltage used to measure the differential capacity.

The qualitative behaviour of the log C_{sc} vs. ϕ_{sc} curve for an extrinsic n-type semiconductor is shown in Fig.5b. At extreme potential values, C_{sc} depends exponentially upon potential; in the intermediate range of potentials, corresponding to the depletion layer, C_{sc} changes slowly with potential and exhibits a quite asymmetric behaviour.

4.2. THE SPACE CHARGE IN INTRINSIC SEMICONDUCTORS

The asymmetry in the C_{sc} vs. ϕ_{sc} curves for an extrinsic semiconductor is due to the notable difference in the bulk densities n and p of its conduction electrons and holes. Intrinsic semiconductors are characterized by the same bulk density n_{in} for electrons and holes, and hence are expected to exhibit symmetric C_{sc} vs. ϕ_{sc} curves. This is immediately verified by setting $N_a=N_d=0$ and $n_{in}=n=p$ in the general expression for C_{sc} of eqn.(52):

$$C_{sc} = \left(\frac{\beta e^2 \varepsilon_{sc}}{8\pi}\right)^{1/2} \frac{n_{in}^{1/2}(e^{y_o}-e^{-y_o})}{\sqrt{(e^{y_o}+e^{-y_o}-2)}} = \left(\frac{\beta e^2 \varepsilon_{sc}}{2\pi}\right)^{1/2} n_{in}^{1/2} \cosh\left(\frac{y_o}{2}\right) \tag{56}$$

The last member of this equation is obtained by noting that $e^x-e^{-x}=(e^{x/2}-e^{-x/2})(e^{x/2}+e^{-x/2})$, that $(e^x+e^{-x}-2)=(e^{x/2}-e^{-x/2})^2$, and that $\cosh(x)\equiv(e^x+e^{-x})/2$. The $\cosh(y_o/2)$ function is indeed symmetric with respect to the $y_o=0$ axis and has the shape of an inverted parabola with the minimum at $\phi_{sc}=0$. Hence this is also true for the C_{sp} vs. ϕ_{sc} curve (see Fig.5a), whose minimum locates the flat band potential.

The charge density σ_{sc} for the intrinsic case is obtained by setting once again $N_a=N_d=0$ and $n_{in}=n=p$ in the more general equation (50):

$$\sigma_{sc} = -\left(\frac{\varepsilon_{sc}}{2\pi\beta}\right)^{1/2} n_{in}^{1/2}\sqrt{e^{y_o}+e^{-y_o}-2} = -\left(\frac{2\varepsilon_{sc}}{\pi\beta}\right)^{1/2} n_{in}^{1/2}\sinh\left(\frac{y_o}{2}\right) \tag{57}$$

The minus sign has been chosen because the potential ϕ_{sc} at the semiconductor boundary causes the accumulation of a positive charge σ_{sc} at the interface if it is negative with respect to the bulk, and viceversa.

4.3. THE DIFFUSE LAYER IN THE SOLUTION OF A 1,1-VALENT ELECTROLYTE

At the interface of an electrolytic solution the local concentrations of the constituent ions of the electrolyte are generally altered with respect to their bulk values, giving rise to a nonzero volume charge density $\rho(x)$ which vanishes at a sufficient distance from the phase boundary. The region where $\rho(x)$ is different from zero is called the *diffuse layer*. $\rho(x)$ is once again related to the inner potential $\phi(x)$ within the solution through the Poisson equation (2), whereas the ions obey the Boltzmann statistics. The case of a 1,1-valent electrolyte, in which the bulk concentration c_+ of the cation and that, c_-, of the anion are equal, is therefore perfectly equivalent to that of an intrinsic semiconductor. The expression of eqn.(57) for the surface charge density and that of eqn.(56) for the differential capacity can therefore be directly applied to the case of a 1,1-valent electrolyte, provided we replace σ_{sc} with the surface charge density σ_d due to the diffuse layer, ε_{sc} with the dielectric constant ε_s of the solution, C_{sc} with the *diffuse-layer capacity* C_d, and the common value $n_{in}=n=p$ of the bulk densities of conduction electrons and holes with the common value, $c\equiv c_+=c_-$, of the bulk concentrations of cation and anion:

$$\sigma_d = -\left(\frac{2\varepsilon_s}{\pi\beta}\right)^{1/2} c^{1/2} \sinh\left(\frac{\beta e\phi_d}{2}\right) \tag{58}$$

$$C_d = \left(\frac{\beta e^2 \varepsilon_s}{2\pi}\right)^{1/2} c^{1/2} \cosh\left(\frac{\beta e\phi_d}{2}\right) \tag{59}$$

To conform to the common use, the potential difference across the diffuse layer in these equations has been denoted by ϕ_d. However, the symbol ϕ_2 is also used, whereas biophysicists often adopt the symbol ψ_0. Equations (58) and (59) were derived by Gouy and, independently, by Chapman; the theory which underlies their derivation is therefore referred to as the *Gouy-Chapman theory*. Figure 6 shows the dependence of ϕ_d upon $-\sigma_d$ for different c values. It can be seen that $|\phi_d|$ increases with $|\sigma_d|$ the less rapidly the higher is $|\sigma_d|$; moreover, it is depressed by a gradual increase in the electrolyte concentration c at constant σ_d.

5. The Inner Portion of the Interfacial Region: the Helmholtz Layer

So far, in considering the charge distribution due to coulombic forces we have focused our attention on only one side of an interface (the semiconductor side in the case of the space charge, the side of the electrolytic solution in the case of the diffuse layer). However, within the interfacial region between two bulk phases α and β, the narrow boundary region as previously defined is the site of short-range anisotropic forces which generate an additional potential difference. For interfaces between an electrolytic solution β and a different phase α, this region is called the *inner, compact* or *Helmholtz layer*, and is inaccessible to the centres of charge of the ions of the solution phase β. The plane of closest approach of these centres of charge to the other phase α is called the *outer*

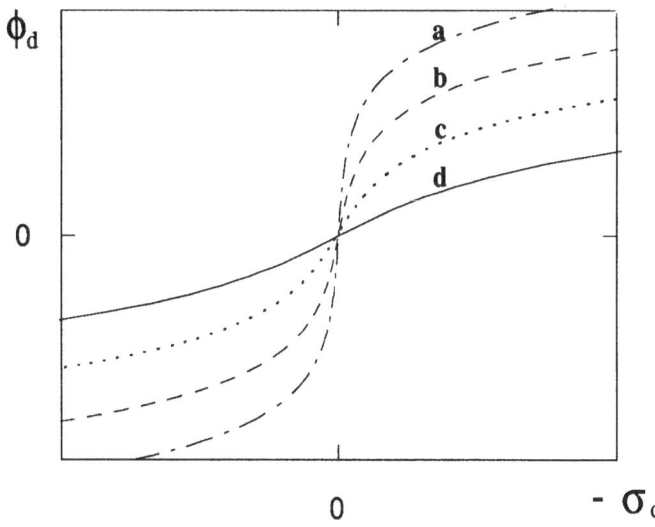

Figure 6. Plots of ϕ_d vs. $-\sigma_d$; curves **a** to **d** refer to electrolyte concentrations c which increase by one order of magnitude in passing from one curve to the successive one.

Helmholtz plane or the *inner Helmholtz plane* depending on whether the ions preserve their inner solvation sheath in the direction of phase α or not; in the latter case, the ions at the inner Helmholtz plane are considered to be held there by specific, noncoulombic forces, and are therefore referred to as *specifically adsorbed ions*.

It is interesting to consider the contribution of the Helmholtz layer to the potential difference $\phi^{\alpha-\beta}$ across the whole interface. To this end we shall first refer to the case of a semiconductor/ solution interface, which is particularly illuminating in this respect.

5.1. THE SEMICONDUCTOR / SOLUTION INTERFACE

Figure 7 depicts schematically the structure of this interface for a negative space charge and a positive diffuse layer charge, as well as the profiles of the electric potential $\phi(x)$ and of the volume charge density $\rho(x)$ across the interface. For simplicity, ionic specific adsorption is excluded, and hence the charge density σ_{sc} due to the space charge region equals the opposite, $-\sigma_d$, of the charge density due to the diffuse layer, in view of the electroneutality of the interface as a whole. The potential difference $(\phi^\alpha-\phi^\beta)\equiv\phi^{\alpha-\beta}$ of the semiconductor phase α relative to the solution phase β is given by:

$$\phi^{\alpha-\beta} = -\phi_{sc} + \phi_H + \phi_d \tag{60}$$

where ϕ_{sc} denotes the potential difference across the space charge region, ϕ_H that across the the Helmholtz layer, and ϕ_d that across the diffuse layer. The minus sign in front of ϕ_{sc} is due to the fact that ϕ_{sc} denotes the potential at the semiconductor boundary relative to the corresponding bulk phase, whereas we are now measuring potential differences in the opposite direction (see Fig.7).

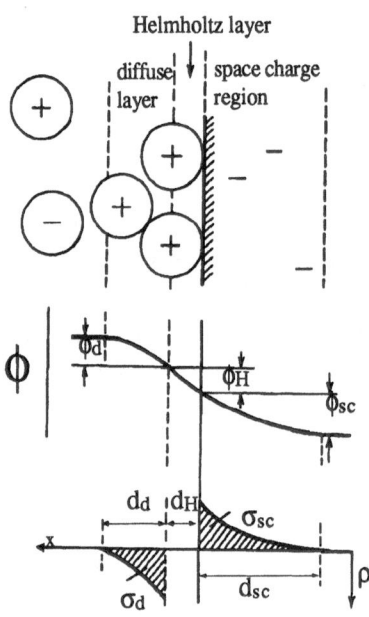

Figure 7. Schematic picture of a semiconductor/solution interface, and profile of the inner potential $\phi(x)$ and of the volume charge density $\rho(x)$ across this interface.

To estimate the relative contributions to $\phi^{\alpha-\beta}$, we shall roughly represent the Helmholtz layer as a parallel plate capacitor with a charge density σ_{sc} on one plate and a charge density $\sigma_d = -\sigma_{sc}$ on the other. The potential difference ϕ_H will then be equal to:

$$\phi_H = 4\pi d_H \sigma_{sc}/\varepsilon_H \tag{61}$$

where d_H is the thickness of the Helmholtz layer and ε_H is its dielectric constant. The latter quantity is somewhat difficult to estimate, or even to define. If we regard ε_H as a measure of the capability of the material inside the Helmholtz layer to oppose the electric field $4\pi\sigma_{sc}$ which the charge densities on the two ideal plates would create *in vacuo*, then ε_H is expected to decrease with an increase in the absolute value of this *external electric field*. Thus, if the Helmholtz layer is populated by water dipoles, the latter will align in the direction of the external field $4\pi\sigma_{sc}$ the more the higher is its absolute value, and they will therefore gradually decrease their capability to oppose this field. Ultimately, when they are fully aligned (a situation which is referred to as *dielectric saturation*), they will be able to oppose the external field only through their distortional polarization. The dielectric constant ε_H in eqn.(61) must account for both distortional and orientational polarization; in the case of a Helmholtz layer of water molecules, ε_H is usually ascribed values ranging from 10 to 20.

As concerns the potential differences ϕ_{sc} and ϕ_d, we shall refer for simplicity to the case in which their absolute values are both much less than $kT \equiv 1/\beta$, which implies that σ_{sc} is also small. In this case we may set $\sinh(y_o/2) \approx \beta e \phi_{sc}/2$ in eqn.(57) and $\sinh(\beta e \phi_d/2) \approx \beta e \phi_d/2$ in eqn.(58). Upon substituting the resulting expressions for ϕ_{sc} and ϕ_d in eqn.(60), we get:

$$\phi^{\alpha-\beta} = -\phi_{sc} + \phi_H + \phi_d = \left(\frac{2\pi kT}{\varepsilon_{sc}e^2 n_{in}}\right)^{1/2}\sigma_{sc} + \frac{4\pi d_H}{\varepsilon_H}\sigma_{sc} + \left(\frac{2\pi kT}{\varepsilon_s e^2 c}\right)^{1/2}\sigma_{sc} \qquad (62)$$

where we have set $\sigma_d = -\sigma_{sc}$. The first term in the last member of eqn.(62) holds for an implicit semiconductor. If we consider an explicit semiconductor of the *n*-type, we may set $N_a=0$, $p=0$ and $n=N_d$ in eqn.(50); if we also assume once again that $|y_o| \ll 1$, we may set $\exp(y_o) \approx 1 + y_o + y^2/2 + \cdots$ in eqn.(50), thus obtaining:

$$-\phi_{sc} = \left(\frac{4\pi kT}{\varepsilon_{sc}e^2 N_d}\right)^{1/2}\sigma_{sc} \qquad (63)$$

which differs from the corresponding expression for an intrinsic semiconductor by the presence of a factor $2^{1/2}$ and by the replacement of n_{in} with the density N_d of donor impurities.

For ease of comparison of ϕ_{sc} and ϕ_d with ϕ_H in eqn.(62), we may set $-\phi_{sc} \equiv 4\pi d_{sc}\sigma_{sc}/\varepsilon_{sc}$ and $\phi_d \equiv 4\pi d_d\sigma_{sc}/\varepsilon_s$, where d_{sc} and d_d are equivalent lengths for the charge distributions satisfying the Poisson-Boltzmann equation, often called *Debye lengths*. We then have:

$$d_{sc} = \left(\frac{\varepsilon_{sc}kT}{8\pi e^2 n_{in}}\right)^{1/2} \text{ for intrinsic semiconductors}$$

$$d_{sc} = \left(\frac{\varepsilon_{sc}kT}{4\pi e^2 N_d}\right)^{1/2} \quad \text{for extrinsic } n\text{-type semiconductors} \tag{64}$$

and

$$d_d = \left(\frac{\varepsilon_s kT}{8\pi e^2 c}\right)^{1/2} \tag{65}$$

If we put numbers in these equations we see, for instance, that a germanium sample with $N_d \approx 10^{14}$ cm^{-3} and a dielectric constant $\varepsilon_{sc} = 16$, has a Debye length $d_{sc} \approx 10^{-4}$ cm. On the other hand, a 10^{-2}M aqueous solution of a 1,1-valent electrolyte ($c \approx 10^{19}$ ions/cm^3 and $\varepsilon_s = 81$) has a Debye length $d_d \approx 10^{-6}$ cm. If we consider that the thickness d_H of the Helmholtz layer is of the order of 10^{-8} cm, we may conclude that by far the largest part of the potential difference $\phi^{\alpha-\beta}$ across a semiconductor/solution interface is located in the space charge region of the semiconductor. Hence, the space-charge differential capacity $C_{sc} \equiv -d\sigma_{sc}/d\phi_{sc}$ practically coincides with the capacity $C \equiv d\sigma_{sc}/d\phi^{\alpha-\beta}$ of the whole interface. In drawing these conclusions we must not forget, however, that eqn.(62) holds only for very small σ_{sc} values.

If we differentiate eqn.(60) with respect to σ_{sc} without making any assumptions about its magnitude, we obtain:

$$\frac{d\phi^{\alpha-\beta}}{d\sigma_{sc}} \equiv \frac{1}{C} = -\frac{d\phi_{sc}}{d\sigma_{sc}} + \frac{d\phi_H}{d\sigma_{sc}} + \frac{d\phi_d}{d\sigma_{sc}} \equiv \frac{1}{C_{sc}} + \frac{1}{C_H} + \frac{1}{C_d} \tag{66}$$

where C_H is the *Helmholtz layer capacity*. According to this equation the interface behaves as though it consisted of three capacitors in series, so that its overall capacity C is controlled by that of the three capacitors which has the lowest capacity. Now, C_{sc} and C_d are not independent of the corresponding potential differences ϕ_{sc} and ϕ_d, as predicted by the approximate eqn.(62), but rather depend parabolically upon potential, increasing on both sides of the flat-band potential (see eqns.56 and 59 and Fig.5a). On the contrary, C_H depends much less critically upon potential. The contributions from C_{sc} and C_d to C are therefore expected to decrease in both directions as we depart from $\sigma_{sc} = 0$. With semiconductors the condition $C \approx C_{sc}$ is often satisfied at all accessible potentials, due to the high d_{sc} value with respect to the d_H and d_d values; however, with heavily doped semiconductors, d_{sc} may not be sufficiently high as to satisfy the above conditions (with $N_d = 10^{18}$ cm^{-3}, d_{sc} is of the order of 10^{-6} cm).

Let us consider an ideal polarized semiconductor/solution interface, and let us assume, for simplicity, that the applied potential $E = (\phi^{\alpha-\beta} + \text{constant})$ is initially set to the flat-band potential, as shown in Fig. 8 [7]. When E is shifted by an amount $\Delta E = \Delta\phi^{\alpha-\beta}$ in the positive direction, all one-electron levels in the bulk semiconductor will decrease by an

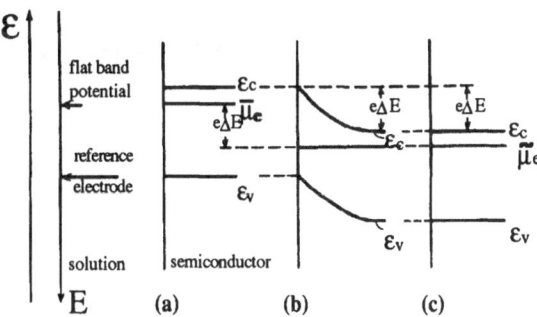

Figure 8. Energy diagram of an ideal polarized solution/semiconductor interface. The effect of a shift ΔE in the applied potential E relative to the flat-band potential is shown both in the case of band-edge pinning (transition from **a** to **b**) and of Fermi-level pinning (transition from **a** to **c**).

amount $e\Delta E$ with respect to the reference electrode. If by far the major part of $\Delta\phi^{\alpha-\beta}$ is located within the space charge region, then Fig. 8 shows that the band edges at the surface(namely, the bottom $\varepsilon_c(x=0)\equiv\varepsilon_{c,s}$ of the conduction band and the top $\varepsilon_v(x=0)\equiv\varepsilon_{v,s}$ of the valence band at the boundary of the semiconductor) will maintain the same position relative to the reference electrode; we then state that the band edges are *pinned* at the surface and we refer to this frequently encountered situation as the *band-edge pinning*. For a positive E shift, the bands are bent upwards at the boundary; this implies that the electric potential ϕ_{sc} at the boundary is negative relative to the bulk giving rise to a predominance of holes over conduction electrons (cf. eqn.47), namely to a positive space charge. The opposite is true if the applied potential is shifted in the negative direction with respect to the flat band potential.

5.2. THE METAL / SOLUTION INTERFACE

A typical value for the density of free electrons in a metal is $n\approx10^{22}$ cm^{-3}. Hence, if we were allowed to apply eqn.(64), which holds for boltzons, to metal electrons, we would predict a thickness d_{sc} of the space charge region of the order of 10^{-8} cm. This prediction turns out to be substantially correct, in spite of the fact that eqn.(64) does not apply to electrons in a metal. Indeed, these obey the Fermi-Dirac statistics and must be treated quantum-mechanically in view of their extremely low mass.

A method widely employed to treat the *inhomogeneous electron gas* at a metal surface is the *density functional method*, which is based on the rigorous demonstration that the ground state energy E of a system of interacting electrons is a *unique functional* of the ground state electron density $n(\mathbf{r})$ [9,10]:

$$E[n(\mathbf{r})] = - \sum_{R} Ze \int \frac{n(\mathbf{r})}{|R\text{-}\mathbf{r}|} d\mathbf{r} + \frac{1}{2} \int \int \frac{n(\mathbf{r})n(\mathbf{r}')}{|\mathbf{r}\text{-}\mathbf{r}'|} d\mathbf{r}d\mathbf{r}' + T[n(\mathbf{r})] + E_{xc}[n(\mathbf{r})] \qquad (67)$$

Incidentally, $E[n(\mathbf{r})]$ is defined as a functional of $n(\mathbf{r})$ because, as distinct from a function, it depends on all values assumed by $n(\mathbf{r})$ from the bulk metal to the bulk of the contacting phase. The first term in eqn.(67) is the ion-electron coulombic interaction energy, where Ze is the charge of the ion cores in the metal. The second term is the average electrostatic potential energy of electrons. $E_{xc}[n(\mathbf{r})]$ is the *exchange-correlation energy*, which embodies all the many-body quantum mechanics of the problem and accounts for short range Coulomb interactions; crudely speaking, the potential energy of each electron is lowered because neighbouring electrons tend to stay away. $T[n(\mathbf{r})]$ is the kinetic energy of a reference system of noninteracting electrons in a potential v_{eff} such that its electron density is the same as the electron density $n(\mathbf{r})$ of the system of interacting electrons under study. This allows us to write for the reference system the following set of one-electron Schrödinger equations:

$$-\frac{1}{2}\nabla^2\psi_i(\mathbf{r}) + v_{eff}(\mathbf{r})\psi_i(\mathbf{r}) = \varepsilon_i\psi_i(\mathbf{r}) \qquad (68)$$

with

$$v_{eff}(\mathbf{r}) = - Ze^2 \sum_{R} \frac{1}{|R\text{-}\mathbf{r}|} + \int \frac{n(\mathbf{r}')}{|\mathbf{r}\text{-}\mathbf{r}'|} d\mathbf{r}' + v_{xc}(\mathbf{r}) \qquad (69)$$

where $n(\mathbf{r})=\Sigma|\psi_i|^2$ and $v_{xc}[n(\mathbf{r})]=\delta E_{xc}[n(\mathbf{r})]/\delta n(\mathbf{r})$. Naturally, the electron-electron interactions are hidden in the *exchange-correlation potential* $v_{xc}[n(\mathbf{r})]$. The density $n(\mathbf{r})$ which minimizes the ground state energy $E[n(\mathbf{r})]$ in eqn.(67) is found by solving the set of ordinary differential equations (68). The electron density $n(\mathbf{r})$ is needed to construct v_{eff} whereas, in turn, v_{eff} is required to solve the set of differential equations (68) which yields the one-electron functions $\psi_i(\mathbf{r})$ whose sum of squares provides $n(\mathbf{r})$. A self-consistent solution to the equations must be found. So far the procedure is exact. However, E_{xc} is written as a functional of $n(\mathbf{r})$, and practical application of this method requires a good approximation for this quantity. To this end the simple, yet satisfactory *local density approximation* (LDA) is commonly adopted. According to this approximation, the exchange-correlation energy density in each infinitesimal region of the inhomogeneous electron distribution $n(\mathbf{r})$ is regarded as identical to the exchange-correlation energy density of a homogeneous electron gas with the same density n as the corresponding infinitesimal region.

To avoid the difficulties inherent in the solution of the above semi-infinite lattice problem, the situation is simplified by having recourse to the *jellium model*. Here the discrete ion cores of the metal lattice are replaced by a uniform, positive background charge (the *jellium*) with the same density as the spatial average of the charge of the ion cores. This leads to a much simpler problem in one dimension. Denoting the axis normal to the metal surface by z, the electrostatic potential V(z) created by the jellium replaces the ion-electron potential, as expressed by the first term in eqn.(69). Likewise, the ion-electron coulombic interaction energy in eqn.(67) is replaced by $\int V(z)n(z)dz$. A further acceptable approximation consists in using directly the energy functional $E[n(\mathbf{r})]$ of eqn.(67), as simplified in the framework of the jellium model, taking into account that the electron density distribution $n(z)$ must minimize it; $n(z)$ is approximately represented by a trial function containing one or more parameters, which are adjusted to minimize $E[n(z)]$. Some examples of trial functions and their applications are reported in Schmickler's contributions to the present volume.

The jellium model completely neglects the ionic lattice. However, some cristallinity can be embodied in the jellium model by using *pseudopotentials*. To recover the cristalline case exactly starting from the jellium model, one should add to the jellium potential V(z) a potential $\delta V(\mathbf{r})$ equal to the difference between the exact potential due to the ion cores and V(z). If we average $\delta V(\mathbf{r})$ over planes normal to z (i.e., parallel to the crystal surface), we obtain a quantity $\delta V(z)$ which becomes strongly attractive in the proximity of the lattice planes parallel to the surface. By removing the strong attractive part of $\delta V(z)$, which would be untractable, a weaker pseudopotential is obtained, which can be treated by standard perturbative methods. The use of pseudopotentials brings the jellium theory into much better accord with experiment. To deal with metal/solution interfaces, the jellium model is sometimes combined with a system of both dipolar and charged hard spheres, simulating solvent molecules and ions, which are allowed to approach the jellium edge.

The differential capacity C of a metal/solution interface can be written in a form analogous to that of eqn.(66):

$$\frac{d\phi^{\alpha-\beta}}{d\sigma_M} \equiv \frac{1}{C} = \frac{1}{C_M} + \frac{1}{C_H} + \frac{1}{C_d} \tag{70}$$

where $\sigma_M = -\sigma_d$ is the charge density on the metal side of the interface and C_M is the contribution to the differential capacity C from the metal electrons. We have seen that the Debye length d_d for the diffuse layer is not enormously greater than that of the Helmholtz layer ($d_d \approx 10^{-6}$ cm for $c = 10^{-2}$M and $\approx 10^{-7}$ for $c = 1$M; cf. eqn.65). If we also consider that the dielectric constant $\varepsilon_s = 81$ of an aqueous solution is greater than that, $\varepsilon_H = 10 \div 20$, of the Helmholtz layer, we can safely state that in no case can the differential capacity C_d of the diffuse layer be identified with that, C, of the whole interaface.

In view of eqn.(70) the contribution of C_d to C is more appreciable in the proximity of the potential at which $\sigma_M = -\sigma_d$ equals zero, where C_d is predicted to exhibit a minimum (see Fig.5a), and at low electrolyte concentrations (say, for $c < 10^{-2}M$; see eqn.59). Experimental C vs. E curves with a minimum are indeed observed at the interface of a number of *sp* and *sd* metals with aqueous solutions of salts whose constituent ions are not specifically adsorbed. The minimum is commonly used to locate the applied potential at which σ_M equals zero. This potential, called the *point of zero charge* and denoted by E_z, plays for metals the same role which the flat band potential plays for semiconductors.

If the charge density σ_M is kept constant and ionic specific adsorption can be excluded, both C_M and C_H turn out to be practically independent of the electrolyte concentration c; conversely, C_d increases progressively with an increase in c, as predicted by the Gouy-Chapman theory (cf. eqn.59). The validity of this theory and of the assumption about the concentration independence of C_M and C_H in eqn.(70) are often checked simultaneously by plotting $1/C$ vs. $1/C_d$ for different electrolyte concentrations at the potential E_z of the capacity minimum, where $\sigma_M = 0$; C_d is estimated from eqn.(59). When the above requirements are met, the $1/C$ vs. $1/C_d$ plot, called the *Parsons-Zobel plot*, shows a unit slope and intersects the vertical axis at $1/C = (1/C_M + 1/C_H)$. When the applied potential E is sufficiently removed from the point of zero charge E_z, C_d increases to such an extent as to contribute practically nothing to C even at low electrolyte concentrations. Nonetheless, the Gouy-Chapman theory can be verified even under these conditions, taking advantage of the fact that an increase in c at constant σ_M "compresses" the diffuse layer by decreasing ϕ_d (see Fig.6), while leaving the other contributions to the whole potential difference $\phi^{\alpha-\beta}$ across the interface unaltered. Hence, if the Gouy-Chapman theory is satisfied, the changes in the applied potential E which are required to maintain the charge density σ_M contant while varying the electrolyte concentration are expected to be equal to the corresponding changes in ϕ_d, as expressed by eqn.(58); obviously, this procedure cannot be applied in the proximity of E_z, where ϕ_d is vanishingly small at all electrolyte concentrations, as shown in Fig.6. In spite of the crudeness of some of the underlying assumptions, the Gouy-Chapman theory has been verified several times not only at metal/solution interfaces, but also at the interfaces of electrolytic solutions with colloidal particles and biological membranes. With the latter systems the Gouy-Chapman theory is also conveniently exploited to control the electrical state of the interface (see further).

5.3. THE SEMICONDUCTOR / SEMICONDUCTOR INTERFACE

When a *p*-type semiconductor and an *n*-type one are brought into contact to form what is called a *p-n junction*, electrons diffuse from the *n*-side (where their concentration is high) to the *p*-side (where their concentration is low), whereas holes diffuse in the opposite direction. The resulting transfer of charge builds up an electric field opposing further diffusion, until an equilibrium configuration is reached in which the effect of the field cancels the effect of diffusion. At this point the *n*-side of the junction is depleted of its mobile majority carriers (i.e., the conduction electrons), and only contains the immobile positive donor impurities. Analously, the *p*-side is depleted of holes and only contains the immobile negative acceptor impurities. It is this charge separation which generates the electric field in the junction and causes an upward bending of the bands on its *n*-side and a downward bending on its *p*-side (see Fig. 9). After equilibrium is attained, the electrochemical potential $\tilde{\mu}_e$ is the same in the two bulk phases and the band edges vary monotonically through the boundary of the *p-n* junction (this is exactly true only in the absence of surface states; see further).

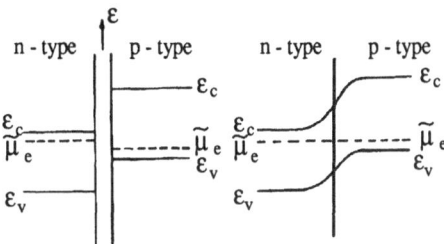

Figure 9. Band edges before and after formation of a *p-n* junction.

6. Non-Coulombic Adsorption of Charged Species at the Boundary Between two Phases

So far we have only considered charge distributions which satisfy the Poisson-Boltzmann equation and result from a balance between ordering coulombic forces and the randomizing forces of thermal agitation. When the short-range anisotropic forces acting in the proximity of the Helmholtz layer attract charged species there, the picture becomes much more complicated.

Let us first consider a metal/solution interface whose boundary contains ions which are held there by specific, non-coulombic forces. We will denote the surface charge density due to these specifically adsorbed ions by σ_i. The ordering coulombic forces acting on the diffuse layer ions will now be generated not only by the charge density σ_M on the metal, but also by that, σ_i, from the specifically adsorbed ions. Moreover, in view of the electroneutrality of the interface as a whole, we will have:

$$\sigma_M + \sigma_i = -\sigma_d \tag{71}$$

Note that only the charge σ_M is amenable to experimental determination, e.g. from capacitative charge measurements. This is because only the net amount of charge required to charge the interface can be measured; this charge results from a flow of electrons within the metal, and by a flow of ions (carrying an equal and opposite charge) within the solution. Now, it is evident that it is impossible to determine by direct means which of these flowing ions are held in the Helmholtz layer by specific forces and which in the diffuse layer by coulombic forces. This also implies that it is now impossible to verify the validity of the Gouy-Chapman theory directly, e.g. by varying the electrolyte concentration c at constant σ_M and by checking whether the resulting shift in the applied potential E matches the shift in ϕ_d as expressed by eqn.(58). In fact, σ_d is no longer equal to $-\sigma_M$ and, moreover, an increase in c will produce an unknown increase in σ_i. In these cases one usually assumes the validity of the Gouy-Chapman theory a priori, using eqn.(58) to estimate σ_d by a self-consistent procedure [11]; the specifically adsorbed charge σ_i is then obtained from σ_d and the directly measured charge density σ_M through eqn.(71).

Practically all anions are more or less specifically adsorbed on metals, at least when it is possible to apply to the metal sufficiently positive potentials without causing its oxidation. In general, for a constant electrolyte concentration, anionic specific adsorption increases as σ_M becomes progressively more positive, whereas it becomes vanishingly small at sufficiently negative σ_M values. Quite often, at positive σ_M values, the absolute value of the charge density σ_i due to specifically adsorbed anions exceeds σ_M; in this case the diffuse layer charge density σ_d has the same sign as the charge σ_M on the metal side of the interface, as shown in Fig.10b. This gives an idea of how much ionic specific adsorption may affect the charge distribution in the diffuse layer.

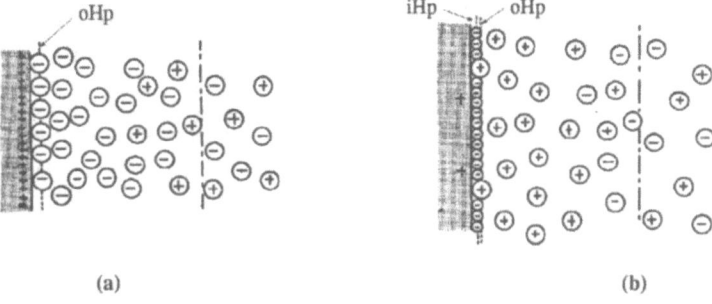

(a)   (b)

Figure 10. Schematic picture of the diffuse-layer ions near a positively charged metal surface both in the absence (a) and in the presence of anionic specific adsorption (b). Note the plane of closest approach to the metal surface for the specifically adsorbed anions (inner Helmholtz plane, iHp) and that for nonspecifically adsorbed ions (outer Helmholtz plane, oHp). In (b) the charge density σ_d due to the diffuse layer ions has the same sign as that, σ_M, on the metal surface.

Ionic specific adsorption is also frequently observed at the interface between an aqueous solution on one side, and a colloidal particle or a biomembrane on the other. We will briefly consider this point when describing these interfaces and the way in which their electrical state can be varied.

Semiconductor/ vacuum and semiconductor/ solution interfaces are also characterized by the presence of charges localized at the boundary and held there by specific non-coulombic forces. These charges, called *surface states*, stem from the breakdown of the periodicity of the lattice at the boundary; this results in the formation of electronic states whose wave functions decrease rapidly in size towards the bulk semiconductor [6-9]. From a chemical point of view, surface states can be considered to stem from the bond breaking which is required to form the surface by cleaving the solid. An unpaired electron on a surface atom may then be lost (in which case the atom has provided a donor surface state), or an electron may be added to form a pair (in which case the atom has provided an acceptor surface state). The surface of a doped semiconductor will contain both acceptor and donor atoms, capable of providing surface states. Besides this type of surface states, called *Tamm levels*, other surface states, called *Shockley levels*, are generated by the adsorption of foreign atoms or ions on the semiconductor surface.

Although there may be surface states with energies within the valence or conduction bands, those with energies in the energy gap are of most interest. Surface states can be either donors or acceptors and obey the Fermi-Dirac statistics. If we denote by $N_{a,s}$ the number of one-electron acceptor levels per unit surface, the number of surface negative charges which result from the acceptance of electrons from the bulk semiconductor is given by $N_{a,s}$ times the Fermi probability of each level being occupied (cf. eqn.5):

$$N_{a,s}^- = N_{a,s} \frac{1}{e^{(\varepsilon_{a,s}-e\phi_{sc}-\tilde{\mu}_e)/kT}+1} \tag{72}$$

In this equation $(\varepsilon_{a,s}-e\phi_{sc})$ is one of the energies of the (usually narrow) energy band of acceptor surface states, which is clearly affected by the electric potential ϕ_{sc} at the boundary relative to the bulk semiconductor; moreover, $\tilde{\mu}_e$ is the electrochemical potential of electrons in the bulk, under the assumption that electrons in the surface states are in equilibrium with those in the bulk. Likewise, upon denoting by $N_{d,s}$ the number of one-electron donor levels per unit surface, the number of surface positive charges which result from the release of electrons to the bulk semiconductor is given by $N_{d,s}$ times the Fermi probability of each level being empty (cf. eqn.10):

$$N_{d,s}^+ = N_{d,s}\left(1-\frac{1}{e^{(\varepsilon_{d,s}-e\phi_{sc}-\tilde{\mu}_e)/kT}+1}\right) = N_{d,s}\frac{1}{e^{(\tilde{\mu}_e-\varepsilon_{d,s}+e\phi_{sc})/kT}+1} \tag{73}$$

Here $\varepsilon_{d,s}-e\phi_{sc}$ is one of the energies of the narrow energy band of donor surface states.

According to this model, the donor and acceptor surface states are practically tanks of electrons and holes, which may be more or less emptied or filled by varying the applied potential, and hence the $\tilde{\mu}_e$ value.

In the presence of surface states, the whole charge density on the semiconductor side of the interface is given by the sum of the space charge density σ_{sc} and of the surface charge density σ_{ss} due to the surface states. At a semiconductor/solution interface containing both surface states and specifically adsorbed ions, the electroneutrality condition for the interface as a whole takes the general form:

$$\sigma_{sc} + \sigma_{ss} = - (\sigma_i + \sigma_d) \tag{74}$$

It is generally assumed that surface states are located on the semiconductor side of the Helmholtz layer, whereas specifically adsorbed ions are located on its solution side. The potential difference ϕ_H across the Helmholtz layer is therefore written (cf. eqn. 61):

$$\phi_H = 4\pi d_H(\sigma_{sc}+\sigma_{ss})/\varepsilon_H \tag{75}$$

It should be noted that, while the location of surface states and specifically adsorbed ions at the boundary between the two phases is a matter of modelistic speculation, their ascription to the two different phases is not. Thus, if we imagine charging or discharging the semiconductor/solution interface, the charges on the surface states come unequivocally from the bulk semiconductor, whereas specifically adsorbed ions come from the bulk solution. This is true even if the surface states are of the Shockley type, namely are generated by the adsorption of ions from the solution.

We have seen that the probability of the surface states being occupied is directly related to the position of $\tilde{\mu}_e$. It is therefore evident that, when their number per unit surface is high (say, $>10^{14}$ cm^{-2}), the major part of the change $\Delta(\sigma_{sc}+\sigma_{ss})$ in the overall charge on the semiconductor side of the interface following a change ΔE in the applied potential will localize in the surface states. In view of eqn.(75), this also implies that ΔE will be almost completely located within the Helmholtz layer, whereas the approximate constancy of σ_{sc} will determine the constancy of the potential difference ϕ_{sc} across the space charge region. Under these conditions the shift ΔE in the applied potential will produce a shift $\Delta\varepsilon_c(x=0)\equiv\Delta\varepsilon_{c,s}$ and $\Delta\varepsilon_v(x=0)\equiv\Delta\varepsilon_{v,s}$ in the band edges at the semiconductor boundary with respect to the reference electrode which is practically equal to $-e\Delta E = -e\Delta\phi^{\alpha-\beta}$, as shown in Fig.8(a) and (c). We then state that the band edges are *unpinned* from the surface. Since the shift in the electrochemical potential $\tilde{\mu}_e$ (usually called "Fermi level") is also equal to $-e\Delta E$, the band profiles and the horizontal Fermi level move as a whole along the energy axis with varying E; we then speak of a *Fermi level pinning* with respect to the band edges [7].

A different type of Fermi level pinning is encountered at a semiconductor/ vacuum interface with a high density of surface states (say, $\approx 10^{15}$ cm^{-2}). We then observe that, upon n-type doping of the bulk semiconductor, its work function Φ, which equals the opposite of $\tilde{\mu}_e$ provided the semiconductor/vacuum interface is uncharged, remains substantially constant. Figure 11 illustratates the situation. In view of eqn.(18) a gradual increase in n-type doping shifts $\tilde{\mu}_e$ upwards, towards the bottom ε_c of the conduction band, as shown in Fig.11b. Hence, electrons tend to vacate the high-lying, bulk, donor levels and populate the lower-lying surface states.

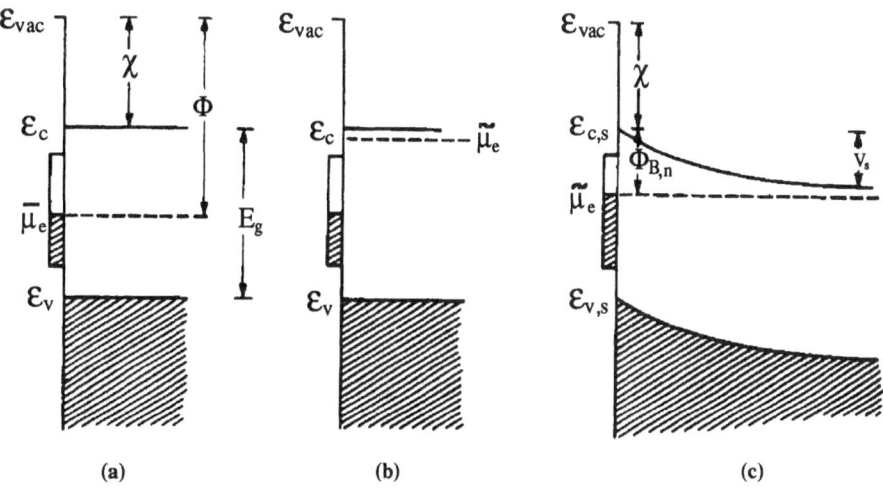

(a)   (b)   (c)

Figure 11. Band edges and surface states at a semiconductor/vacuum interface in equilibrium before doping (a), in nonequilibrium just after n-type doping (b), and after attainment of the new equilibrium condition (c). For clarity, a relatively large band of surface states has been depicted. ε_{vac} is the reference vacuum level.

When thermodynamic equilibrium is achieved (see Fig.11c), the accumulation of electrons in the surface states creates a negative potential ϕ_{sc} at the boundary relative to the bulk; this, in turn, attracts positive charges into the adjacent space charge region and determines an upward band bending. Hence, while the "Fermi level" $\tilde{\mu}_e$ has ultimately approached the bottom of the conduction band in the bulk semiconductor, its position relative both to the band edge $\varepsilon_{c,s}$ at the boundary and to the vacuum level remains practically unchanged. Figure 11 illustrates the significance of a number of quantities employed in the physics of semiconductors: the *band bending* $V_s = -e\phi_{sc}$, the surface *electron affinity* $\chi \equiv -\varepsilon_{c,s}$, and the *Schottky barrier* $\Phi_{B,n}$, defined as:

$$\Phi_{B,n} = \Phi - \chi = \varepsilon_{c,s} - \tilde{\mu}_e \tag{76}$$

The letter term is usually employed in connection with metal/semiconductor interfaces. Imagine bringing a metal into contact with an *n*-type semiconductor to form the interface. Before equilibrium is attained, the electrochemical potential of electrons is less negative in the semiconductor than in the metal. Hence electrons will diffuse from the semiconductor into the metal, until an electric field is formed which cancels the effect of diffusion, and the electrochemical potential $\tilde{\mu}_e$ becomes identical in the two contacting phases, as shown in Fig. 12. In practice, the depletion of electrons (the majority carriers) on the semiconductor side of the interface creates a positive space charge there, and an equal and opposite charge on the metal side. This is accompanied by a decrease in the inner electric potential in passing from the semiconductor to metal side of the interface, and hence by an upward band bending (see Fig. 12).

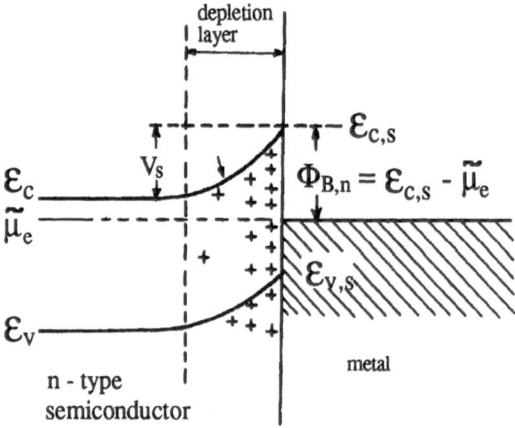

Figure 12. Band edges at an *n*-type semiconductor/metal interface.

The Schottky barrier is just the energy $(\varepsilon_{c,s} - \tilde{\mu}_e) = (\varepsilon_c - e\phi_{sc} - \tilde{\mu}_e)$ in the Boltzmann factor which appears in the expression of eqn.(47) for the number density $n(x=0)$ of conduction electrons at the semiconductor boundary, and hence is a measure of the barrier opposing the electron flow across the boundary. Likewise, when the interface between a *p*-type semiconductor and a metal is formed, holes (the majority carriers) flow from the semiconductor to the metal; in this case the Schottky barrier is just the barrier which opposes the hole flow (cf. the expression for $p(x=0)$ in eqn.47):

$$\Phi_{B,p} = E_g - (\Phi - \chi) = \tilde{\mu}_e - \varepsilon_{v,s} = \tilde{\mu}_e - \varepsilon_v + e\phi_{sc} \tag{77}$$

Referring to an n-type semiconductor/metal interface, the definition of the Schottky barrier in eqn.(76) may suggest the possibility of a control of its height by choosing metals with different work functions Φ or else by doping the semiconductor to different extents, so as to vary its electrochemical potential. However, experiments show instead that the Schottky barrier $\Phi_{B,n}$ is independent of the bulk doping of the semiconductor and is often remarkably insensitive to the type of metal used to form the Schottky barrier. This phenomenon is commonly explained by the presence of surface states induced by the contacting metal, called *metal-induced gap states* (MIGS). The density of these states is high enough to determine a Fermi-level pinning with respect to the band edges, i.e. to maintain the difference $(\varepsilon_{c,s} - \tilde{\mu}_e) = \Phi_{B,n}$ roughly constant. These *interface states* act as charge tanks which allow μ_e (i.e., the Fermi level) not to change by very much to accomodate any charge flow needed to equilibrate the interface to the bulk (see Perfetti's contribution to this volume).

7. How Do We Control the Electrical State of an Electrified Interface?

The control of an electrified interface between two conducting materials can be carried out by including the interface in a suitable chain of conductors and by applying a potential difference at the terminals of the chain by some power source. For the chain to be "suitable", any changes in the applied potential must be exclusively localized at the interface under study.

The interfaces between an electrolytic solution and the dispersed particles of a colloidal system or between an electrolytic solution and a biomembrane seem more difficult to control, because dispersed particles and membranes are both composed of substances of low conductivity and permittivity. In spite of this difficulty, the electrical state of these two types of interfaces can be controlled and varied, such as to number them with the electrified interfaces.

7.1. COLLOIDS

The classical model colloid is represented by silver iodide, because it yields stable sols and reproducible results [12,13]. Another class of model compounds of growing interest is represented by insoluble oxides. Application of the Gibbs adsorption equation to a disperse system yields a directly applicable expression which, in several respects, is analogous to that for a metal/solution interface. For the sake of definiteness, let us refer to the case of a silver iodide sol in a solution of an indifferent electrolyte (say, KNO_3). The well known potential dependent term $-\sigma_M dE$ in the expression for the differential $d\gamma$ of the interfacial tension at a metal/solution interface is now replaced by the term $-kT(\Gamma_+ - \Gamma_-)d\ln a_+$, where Γ_+

and Γ_- are the surface excesses (in ions per unit surface) of the Ag^+ and I^- ions, and a_+ is the Ag^+ bulk activity. The analogy between these two apparently different terms is evident when we consider that Ag^+ and I^- are the *constituent ions* of the AgI dispersed particles; hence, $e(\Gamma_+-\Gamma_-)$ is just the surface charge density on the particles, provided we make the reasonable, mildly modelistic, assumption that the constituent ions are adsorbed with much preference over all other ionic species; the latter species are considered to be "on the solution side" of the ionic solid/solution interface, and hence to compose the diffuse layer.

The differential $d\,ln\,a_+$ in the $-kT(\Gamma_+-\Gamma_-)d\,ln\,a_+$ term is directly related to the potential difference $\phi^{\alpha-\beta}$ across the interface, where α denotes the solid phase and β the solution phase. The presence of Ag^+ ions both in the solid phase and in the contacting solution phase requires that, once equilibrium is achieved, the electrochemical potential of Ag^+ in the solid, $\tilde{\mu}_+^\alpha$, be equal to that, $\tilde{\mu}_+^\beta$, in the solution (cf. eqn.3). Upon separating both electrochemical potentials into an electrical and a chemical part and accounting for the concentration dependence of the chemical potential μ_+^β of the Ag^+ ions in solution (cf. eqn.29), we get:

$$\tilde{\mu}_+^\alpha = \mu_+^\alpha + e\phi^\alpha = \tilde{\mu}_+^\beta = \mu_+^{o,\beta} + kT \ln a_+ + e\phi^\beta \qquad (78)$$

In writing this equation we have replaced the Ag^+ bulk concentration in phase β by its activity a_+, in order to account for ion-ion interactions, not included in the Boltzmann statistics. Upon differentiating eqn.(78) at constant temperature, and taking into account that the standard chemical potentials μ_+^α and $\mu_+^{o,\beta}$ are constant at constant temperature, we obtain the Nernst-like equation:

$$d(\phi^\alpha-\phi^\beta) \equiv d\phi^{\alpha-\beta} = \frac{kT}{e} d\,ln\,a_+ \qquad (79)$$

The term $-kT(\Gamma_+-\Gamma_-)d\,ln\,a_+$ in the Gibbs adsorption equation for the AgI sol can therefore be written in the form:

$$-kT(\Gamma_+-\Gamma_-)\,d\,ln\,a_+ = -\frac{kT}{e}\,\sigma\,d\,ln\,a_+ = -\sigma\,d\phi^{\alpha-\beta} \qquad \text{with } \sigma \equiv e(\Gamma_+-\Gamma_-) \qquad (80)$$

which is perfectly equivalent to the $-\sigma_M dE$ term in the Gibbs equation for metal/solution interfaces. In view of the solubility product of AgI ($a_+a_- = $ constant at constant T), $-d\,ln\,a_+$ in eqn.(80) can also be replaced by $+d\,ln\,a_-$, where a_- is the bulk activity of the I^- ion.

Both the surface charge density σ and the potential difference across the interface, $\phi^{\alpha-\beta} = (kT/e)ln\,a_+ + $ constant, can be varied at will by adding progressive amounts of either $AgNO_3$ or KI to the KNO_3 solution containing the AgI particles. The changes in $ln\,a_+$ can be monitored potentiometrically with an Ag/AgI indicator electrode immersed in the solution. Let us assume that the initial solution has a lna_+ value low enough to determine an eccess of I⁻ ions over Ag⁺ ions (and hence a negative charge density σ) on the dispersed particles. Upon adding a known amount of AgNO$_3$ to the solution, the resulting increase in lna_+, as monitored potentiometrically, is less than that estimated in the absence of the dispersed particles by an amount corresponding to the Ag⁺ ions being adsorbed on these particles. If we divide the amount of adsorbed Ag⁺ ions so determined by the surface area of the disperse system, as estimated by independent means, we immediately obtain the change Δσ in the surface charge density following the above AgNO$_3$ addition. Proceeding in this way by progressive additions, we obtain a curve of Δσ against lna_+ (curve 1 in Fig.13a).

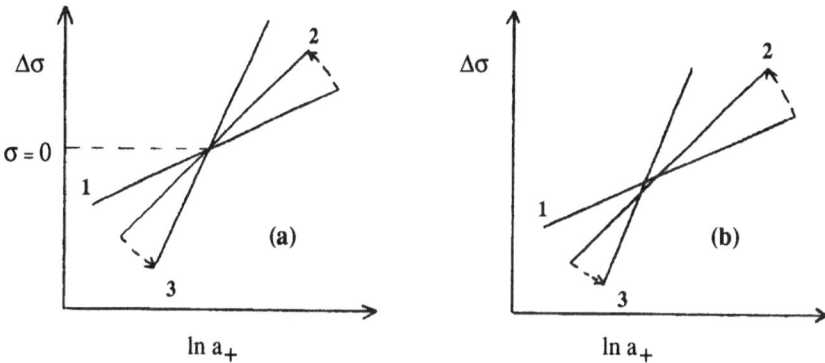

Figure 13. Curves of Δσ vs. lna_+ in the absence of ionic specific adsorption (a) and in the presence of anionic specific adsorption (b). The passage from one curve to the following results from an increase in the concentration of an inert electrolyte, e.g. KNO$_3$.

To determine absolute charge values, the above curve must be extended up to lna_+ values high enough to ensure a surface excess of Ag⁺ ions higher than that of I⁻ ions, and hence a positive σ. At this point an amount of the inert electrolyte KNO$_3$ is added to the solution. The resulting increase in ionic strength compresses the diffuse layer and hence decreases the potential difference ϕ_d across it, in view of the Gouy-Chapman theory (see Fig.6). If the potential difference $\phi^{\alpha-\beta}$ across the whole interface were to remain constant, the decrease in ϕ_d should be compensated for by an equal increase in the potential difference $\phi_H=4\pi d_H\sigma/\epsilon_H$ across the Helmholtz layer (cf. eqn. 61), which requires an increase in the positive charge density σ on the dispersed particles. In the present case, however, the increase in σ may only take place through a withdrawal of Ag⁺ ions from the bulk solution, and hence through a decrease Δlna_+ in lna_+, which implies a decrease $\Delta\phi^{\alpha-\beta}=(kT/e)\Delta lna_+$ in $\phi^{\alpha-\beta}$. The situation is schematically depicted in Fig. 14. The experimental measurement of the decrease in lna_+ following the addition of KNO$_3$ permits us to estimate the increase in σ and hence to locate a new point on the Δσ vs. lna_+ plot of Fig.13a. Upon keeping the ionic strength constant at the new value, lna_+ it then gradually decreased by progressive additions of KI, and a new $\Delta\sigma$ vs. lna_+ curve is obtained (curve 2 in Fig.13a). This curve intersects the curve corresponding to the lower ionic strength at a particular point. If necessary, after the second titration, the Ag^+ ion concentration can be increased again to obtain a new $\Delta\sigma$ vs. lna_+ curve (curve 3 in Fig.13a). In the absence of ionic specific adsorption, the intersection point between any two consecutive curves is constant and locates the zero charge. In fact, it is only the lna_+ value at which $\sigma=0$, called the *point of zero charge*, which is invariant with respect to an increase in ionic strength; here such an increase does not produce any compression of the diffuse layer, and hence any change in lna_+, for the simple reason that there is no diffuse layer. This procedure allows the $\Delta\sigma$ scale on the vertical axis to be replaced by an absolute charge scale.

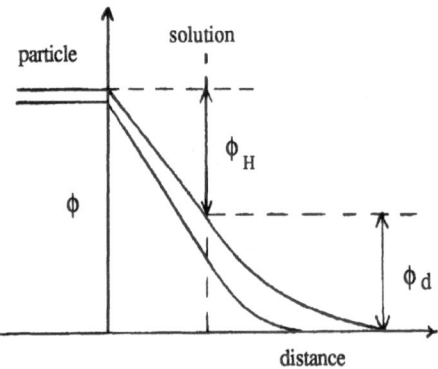

Figure 14. Inner potential ϕ vs. distance profile across the interface between a dispersed particle and the adjacent solution. For clarity, the Helmholtz layer has been enlarged with respect to the diffuse layer. The lower profile results from an increase in the concentration of an inert electrolyte.

In the absence of adsorption, the point of zero charge coincides with the *isoelectric point*, namely the lna_+ value at which the *electrokinetic* (or ζ) *potential* equals zero. Incidentally, in electrokinetic measurements it is assumed that a dispersed charged particle moving under the action of an external electric field, drags along a layer of liquid, at most a few molecules thick, which adheres to its surface. The ζ potential is actually the electric potential at this *slipping plane* relative to the bulk liquid, and is commonly regarded as almost coincident with ϕ_d. The point of zero charge plays a crucial role in colloid stability. The diffuse layer region governs the electrostatic repulsion between the dispersed particles, thus preventing their aggragation. When ϕ_d tends to zero, electrical repulsion also drops down to zero, thus producing flocculation.

In several cases ions other than the constituent ions of the dispersed particles are bound very strongly to the surface by specific forces, as with phosphate or oxalate ions on many oxides (see Stumm's contribution to this volume). We then state that these ions are specifically adsorbed. This phenomenon is often explained by surface complex formation. In the case of anionic specific adsorption, a progressive increase in the concentration of the anion produces a negative shift of the point of zero charge. Hence, the intersection point of two consecutive $\Delta\sigma$ vs. $ln a_+$ curves is not constant, but rather shifts towards lower $ln a_+$ values as the concentration of the anion is gradually increased, as shown in Fig 13b. An opposite effect is observed with specifically adsorbed cations. This phenomenon is readily explained by considering that at zero charge the potential difference ϕ_H across the Helmholtz layer equals zero in view of eqn.(61), with σ_{sc} replaced by σ. (More precisely, ϕ_H is roughly constant but different from zero, because it includes a contribution from the preferential orientation of solvent molecules which is not accounted for in eqn. 61.) A gradual increase in the bulk concentration of, say, a specifically adsorbed anion will determine a progressive accumulation of this anion on the "solution-side plate" of the capacitor by which we represent the Helmholtz layer; this accumulation is compensated for by the accumulation of an equal and opposite charge within the diffuse layer, in order to maintain the electroneutrality of the whole interface. The potential difference produced by this progressive accumulation is clearly negative when measured from the metal towards the solution, and hence determines a negative shift both in $\phi^{\alpha-\beta}$ and in $ln a_+$.

Ionic specific adsorption exerts upon the isoelectric point an effect which is opposite to that exerted upon the point of zero charge. Thus, the specific adsorption of an anion makes ϕ_d more negative, so that a higher $ln a_+$ value is required to bring ϕ_d back to zero. Hence, the isoelectric point is shifted towards positive values by specifically adsorbed anions, and towards negative values by specifically adsorbed cations. An opposite shift in the point of zero charge and in the isoelectric point with an increase in ionic strength provides strong evidence in favour of ionic specific adsorption.

To estimate the charge density σ, the specific surface area of the disperse system must first be determined [12]. One method consists in adsorbing a dye on the dispersed particles up to saturating their surface, and in measuring the resulting decrease in the bulk concentration of the dye photometrically. This provides the amount of adsorbed dye, but some reasonable guess must be made as to the surface area occupied by a single dye molecule. Another method consists in decreasing $ln a_+$ to such an extent as to produce a very high negative charge density σ. In this case anions will be almost completely expelled from the diffuse layer; this will result in an increase in their bulk concentration which is proportional to the expelling surface area and can be measured potentiometrically or by some other technique.

Within certain limits, the above considerations can be extended to insoluble oxides, provided we replace $-log a_+$ by pH. However, with oxides some difficulties arise. Thus, OH^- and H^+ ions are not usually constituent ions of the oxides, and the Nernst relation between the potential difference $\phi^{\alpha-\beta}$ and pH is not always satisfied. Some of the drastic differences in behaviour between oxides and AgI have been explained by assuming that

48

oxide surfaces are somewhat penerable to H^+ and OH^- ions.

7.2. MEMBRANES

Biological membranes envelop not only the living cells, but also various corpuscles inside the cells called *organelles*. Their fundamental role in all basic processes of life was recognized as early as at the end of the nineteenth century. The main components of biomembranes are lipids and proteins. The lipidic components of biomembranes are represented primarily by *phospholipids*. The most common phospholipids, called *phosphoglycerides*, consist of a glycerol molecule which is esterified with two higher fatty acids (stearic, palmitic, oleic, etc.) and with orthophosphoric acid; one of the acidic groups of phosphoric acid is in turn esterified with other molecules such as substituted ethanolamines, serine, inositol, etc.. Proteins are more or less embedded in the membrane structure. Some of them (*structural proteins*) simply support the texture of the membrane, whereas some others (*functional proteins*) participate directly in the membrane processes. The basic structural matrix of a number of membranes consists of a bimolecular lipid layer. The lipids are oriented with their hydrophobic alkyl *tails* directed towards the interior of the membrane, whereas their polar *heads* are directed towards the adjacent solution.

To simplify the study of the important functions served by biomembranes (passive and active ion transport, conduction of nervous impulse, conversion of light into chemical or electrical energy, etc.), well defined experimental models of biomembranes have been devised. So far the most successful model is the bilayer lipid membrane (BLM) [14]. It is obtained by using a cell containing an electrolytic solution and divided into two compartments by a Teflon septum with a small hole, as shown in Fig.15a. Upon placing a drop of a lipid solution in a suitable solvent (e.g., decane) on the hole, the layer of the lipid solution becomes progressively thinner, until a bimolecular film about 100 Å thick is formed, in which the phospholipid molecules retain the same tail-to-tail orientation as in

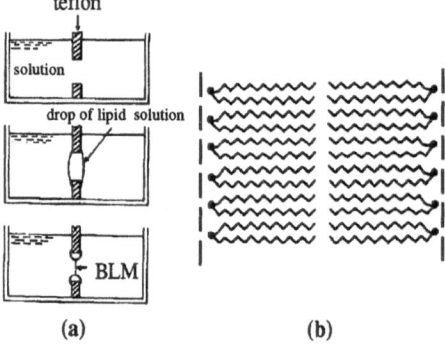

(a)   (b)

Figure 15. Cell for the preparation of a BLM (**a**) and a schematic picture of its structure (**b**): the black circles represent the polar heads and the zig-zag lines the two hydrocarbon chains bound to each polar head.

biomembranes (see Fig.15b). BLMs can be modified (*reconstituted*) by introducing some basic structural units of natural membranes; these modifiers may alter the passive electrical properties of the BLM, change its mechanical properties, impart ion selectivity, induce electric excitability or generate photoelectric effects. Other frequently used experimental models of biomembranes are the *liposomes* or lipid *vesicles*. These are small spheres with a diameter of several hundreds of Ångstroms, which consist of a bimolecular film enclosing an aqueous solution (see Fig.16). They are usually prepared by treating a lipid suspension in water with ultrasounds. This procedure is called *sonication*, and is also applied to suspensions of biomembranes, in which case the vesicles contain the proteins of the original membrane.

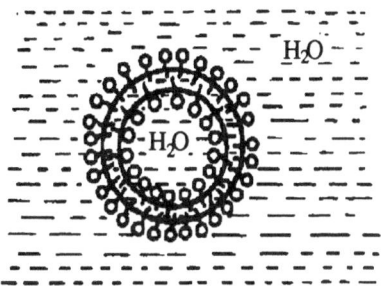

Figure 16. A lipid vesicle.

Membranes bear discrete electrical charges on their surface. These charges are carried by phospholipids with a negatively charged polar head and by proteins, most of which are also negatively charged. In all biomembranes the distribution of these charges over the two sides of the membrane is asymmetric. Hence, the charge density σ_m on one surface is generally different from that on the opposite surface. The charge densities on the two surfaces generate two diffuse layers of ions on the two adjacent (*bathing*) solutions. However, the charge density on any of the two membrane surfaces is not necessarily counterbalanced by an equal and opposite charge in the corresponding diffuse layer; in other words, we are not in the presence of two independent interfaces, each of which electroneutral as a whole. If this were the case, then the electric field along the alkyl tails of the membrane would be constantly equal to zero, because of the zero value of the overall charge on each side of these tails. The electroneutrality condition can only be legitimately applied to the whole system consisting of the membrane and of its two diffuse layers. Several charged surfactants and (mostly divalent) metal ions are specifically adsorbed on the membrane surface, sometimes through surface complex formation.

Both the potential difference across the whole membrane (*transmembrane potential*) and the ϕ_d potential (which biophysicists often call *surface potential*) play fundamental roles in metabolism, control and signaling processes in cells and organelles. The ability to measure and vary these potentials is therefore crucial to the understanding of many cellular processes. This also permits us to regard biomembranes and their models as real electrified interfaces.

7.2.1. Control of the Transmembrane Potential. With biomembranes and BLMs the transmembrane potential can be readily adjusted and varied by applying a given potential difference across two identical reference microelectrodes (say, two Ag/AgCl electrodes) immersed in the bulk solutions bathing the two sides of the membrane. This procedure is not applicable to vesicles and small cells, due to their very small size. Hence, indirect methods must be employed. In general these methods consist in adding to the solution of the vesicles (the *suspending solution*) an ion i which is permeable to them [15]. After the equilibrium distribution of the permeant ions is attained, the electrochemical potential of the ion must be equal both in the interior of the vesicle and in the suspending medium (cf. eqn.3). Upon separating the electrochemical potentials inside and outside the vesicle into an electrical and a chemical part according to eqn.(29), we obtain:

$$\tilde{\mu}_i^{int} = \mu_i^{o,int} + kT \ln a_i^{int} + z_i e \phi^{int} = \tilde{\mu}_i^{ext} = \mu_i^{o,ext} + kT \ln a_i^{ext} + z_i e \phi^{ext} \tag{81}$$

where $z_i e$ is the ion charge, whereas $\mu_i^{o,int}$, a_i^{int}, ϕ^{int} and $\mu_i^{o,ext}$, a_i^{ext}, ϕ^{ext} are the standard chemical potential, the activity and the inner potential inside and outside the vesicle, respectively. The standard chemical potential is only affected by the short range interactions between the solute species and the surrounding solvent molecules. Hence, in view of the predominance of water molecules both in the interior of the vesicle and in the suspending solution, we may set $\mu_i^{o,int}=\mu_i^{o,ext}$. Equation (81) can then be written in the form:

$$\phi^{int-ext} \equiv \phi^{int} - \phi^{ext} = \frac{kT}{z_i e} \ln \frac{a_i^{ext}}{a_i^{int}} \tag{82}$$

Note that the same Nernst equation applies to glass electrodes used in pH measurements and to ion-exchanging membranes used in industry. The $\phi^{int-ext}$ value is estimated by determining the activities of the permeant ion both inside and outside the vesicle. The ion activity a_i^{int} inside the vesicles is difficult to measure when that, a_i^{ext}, in the suspending medium is higher. Therefore, when $\phi^{int-ext}$ is negative, it is necessary to choose as the test ion a cation, which will accumulate in the vesicle, and viceversa. The accumulation of the properly chosen test ion compensates for the much smaller internal volume of the vesicle. Several synthetic organic ions readily soluble in lipids (*lipophilic ions*), such as tetraphenylborate (TPB⁻) or tetraphenylarsonium (TPAs⁺), are used to this purpose; their permeability is due to the hydrophobic shell which shields the central charge. Another strategy consists in loading the membrane with a neutral *ionophore* (namely a lipophilic ligand of a specific ion) which acts as a shuttle across the membrane for the otherwise impermeble ion. The classical example is represented by valinomycin, which is used to promote the transport of K^+ or Rb^+ ions.

To determine the distribution of the test ion between the interior of the vesicles and the suspending solution, one can monitor the decrease in the external activity a_i^{ext} of the ion following its penetration into the vesicles by a selective ion electrode inserted into the suspension. Since the volume of the medium is much larger than that of the vesicles, this method can only be applied with relatively large $\phi^{int-ext}$ values. Otherwise, the vesicles are separated from the suspending medium by fast centrifugation or filtration, and the ion content of both the vesicles and the medium is determined. To this end, radioactively labelled test ions are usually employed, although optical absorption or fluorescence detection may also be used with appropriate test ions.

A variant of the above method is based on the partition of a permeable redox couple, rather than a single ion, between the interior and the exterior of the vesicle [16]. When both the oxidized species ox and the redox species red of a given redox couple are permeable to the membrane, they equilibrate independently, until the equilibrium conditions $\tilde{\mu}_{ox}^{int} = \tilde{\mu}_{ox}^{ext}$ and $\tilde{\mu}_{red}^{int} = \tilde{\mu}_{red}^{ext}$ are both satisfied. In view of eqn.(28), we may therefore state that an equilibrium for the electron across the membrane is also established *with respect to this couple*. Upon using eqn.(27), the resulting equilibrium condition is written:

$$\tilde{\mu}_e^{int} = \varepsilon_{redox}^{o,int} - kT \ln \frac{c_{ox}^{int}}{c_{red}^{int}} = \tilde{\mu}_e^{ext} = \varepsilon_{redox}^{o,ext} - kT \ln \frac{c_{ox}^{ext}}{c_{red}^{ext}} \tag{83}$$

Since the ox/solvent and red/solvent short-range interactions are substantially the same inside and outside the vesicle, the same is true for the corresponding standard chemical potentials. Hence, in view of the expression for ε^o_{redox} in eqn.(30), eqn.(83) takes the form:

$$\phi^{int-ext} = \frac{kT}{e} \ln \left(\frac{c_{ox}^{ext} \, c_{red}^{int}}{c_{red}^{ext} \, c_{ox}^{int}} \right) \tag{84}$$

which allows the inner potential difference $\phi^{int-ext}$ to be determined provided a method is devised to measure the c_{ox}/c_{red} ratio both inside and outside the vesicle.

Equation (84) can also be applied if an equilibrium is established between a redox couple embedded in the membrane and the same redox couple in the suspending solution. In this case, however, the surrounding medium is definitely different in the two different phases (i.e., the hydrophobic interior of the membrane and the aqueous solution). Hence, the standard chemical potentials of ox and red inside the membrane are different from those in the solution. Combining eqns.(30) and (83) we now get:

$$\phi^{int-ext} = \frac{\mu_{red}^{o,int} - \mu_{red}^{o,ext}}{e} - \frac{\mu_{ox}^{o,int} - \mu_{ox}^{o,ext}}{e} + \frac{kT}{e} \ln \left(\frac{c_{ox}^{ext} \, c_{red}^{int}}{c_{red}^{ext} \, c_{ox}^{int}} \right) \tag{85}$$

The first two terms on the right-hand side of this equation are now unknown, and hence $\phi^{int-ext}$ in not experimentally accessible, as is always the case for the inner potential difference between two phases of different composition. Equation (85) can be usefully exploited for a different purpose. If we gradually change the electrolyte concentration in the solution, the resulting changes in $\phi^{int-ext}$ are due primarily to the changes $\Delta\phi_d$ in the potential ϕ_d across the diffuse layer outside the vesicle. Hence, if the changes in the last term of eqn.(85) can be monitored, they provide directly a measure of $\Delta\phi_d$. Incidentally, the c_{ox}/c_{red} ratio in the solution is usually measured potentiometrically, whereas that inside the membrane requires some spectroscopic technique. Note that $\Delta\phi_d$ measurements based on eqn.(85) can be advantageously carried out even if the redox couple in the solution is different from that embedded in the membrane; to this end the membrane must be made permeable to one redox couple and a rapid equilibrium between the inner and the outer redox couple must be established.

7.2.2. Control of the Ionic Surface Potential.

The ionic surface potential ϕ_d in vesicles and cells can be determined by measuring their velocity of migration in an electric field. The electrokinetic potential (ζ *potential*) is calculated from the *Helmholtz-Smoluchowski theory*, and is then identified with the ϕ_d potential, just as in the case of colloidal particles. More frequently, ϕ_d is determined from the ϕ_d-induced accumulation of probe ions near the charged surface [15]. Naturally, the charge $z_i e$ of a given probe ion i must be of opposite sign with respect to the charge density σ_m on the membrane surface in order to produce accumulation. At equilibrium the electrochemical potential $\tilde{\mu}_i^{oHp}$ of the ion at the outer boundary of the Helmholtz layer (the so called *outer Helmholtz plane*, oHp) equals that in the bulk, $\tilde{\mu}_i^b$:

$$\tilde{\mu}_i^{oHp} = \mu_i^o + kT \ln c_i^{oHp} + z_i e \phi_d = \tilde{\mu}_i^b = \mu_i^o + kT \ln c_i^b \tag{86}$$

where c_i^{oHp} and c_i^b are the ion concentrations at the oHp and in the bulk. As usual, the inner potential ϕ_d at the oHp is measured relative to the bulk, so that the inner potential in the bulk equals to zero. Upon rearrangement we immediately obtain a Boltzmann distribution of ions across the diffuse layer:

$$\frac{c_i^{oHp}}{c_i^b} = \exp\left(-\frac{z_i e \phi_d}{kT}\right) \tag{87}$$

Note that this equation is obtained upon assuming that the standard chemical potential μ_i° of the ion is the same at the oHp and in the bulk, namely that ion-solvent interactions remain practically unchanged in passing from the bulk to the oHp. In other words, eqn.(87) applies only in the absence of specific chemical interactions of the probe ion with the membrane surface. To estimate ϕ_d from eqn.(87), the concentration ratio on the left-hand side of this equation must be measured. Use can be made of organic ions whose accumulation on the membrane surface produces an aggregation which is accompanied by fluorescence quenching (with fluorescent ions) or by a change in the resonance raman spectrum of the ion. Unfortunately, the very aggregation of the ions on the electrode surface makes the validity of the assumptions underlying eqn.(87) somewhat questionable. In general, it is preferred to admit the existence of a chemical bonding of the probe ion to the membrane surface, and to measure the enhancement of the probe bonding that results from the increased ϕ_d-induced accumulation. However, the binding of the charged probe to the membrane surface may alter the surface charge density σ_m and, hence, the very ϕ_d value which one is supposed to measure. It is therefore advisable, whenever possible, to choose a probe that can be detected with high sensitivity at a surface concentration which is negligible with respect to σ_m.

When working with BLMs, the above difficulties in ϕ_d measurements can be avoided by using the following procedure [17]. Imagine for simplicity a negatively charged, symmetrical membrane separating two solutions of identical composition, and consider the potential-distance profile across the whole interface, when the potential difference between the two solutions equals zero (see Fig. 17). This situation is simply achieved by setting to zero the potential difference E applied between two identical reference electrodes immersed in the two solutions. Figure 17a shows the exponential decay of the inner potential $\phi(x)$ across the two identical diffuse layers on the two sides of the membrane and the presence of a zero electric field inside the membrane. A nonzero potential difference ϕ_H across any of the two Helmholtz layers adjacent to the two membrane surfaces is also to be expected, due to preferential orientation of both the polar heads of the lipids and the adsorbed water molecules. For simplicity, we shall assume that these additional potential differences remain constant during the experiment, and hence we shall not represent them in the figure. If we now increase the electrolyte concentration in the bathing solution on the right, the resulting compression of the diffuse layer will decrease the absolute value of ϕ_d there by an amount $\Delta\phi_d$. The imposed equality of the inner potentials in the two bulk solutions will therefore create a potential gradient inside the membrane, as shown in Fig17b. If the membrane is sufficiently compressible, the electric

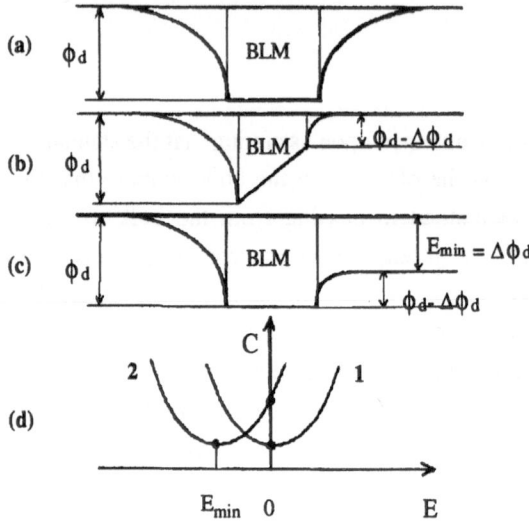

Figure 17. (a) Inner potential $\phi(x)$ vs. distance curve across a symmetrical BLM bathed by two identical solutions, at E=0. The curve of the differential capacity C vs.E provided by this BLM shows a minimum at E=0 (curve 1 in **d**). (**b**) Same BLM after increasing the ionic strength of the solution on the right, at E=0. (**c**) Same BLM as in **b** after setting E to the value E_{min} corresponding to the new differential capacity minimum (curve 2 in **d**).

field inside the membrane will produce a decrease in its thickness d_m (*electrostriction*) and hence an increase in its differential capacity $C_m = \varepsilon_m/(4\pi d_m)$, where ε_m is the dielectric constant of the membrane. Typically, we have $\varepsilon \approx 2$ and $d_m \approx 100$ Å, and hence $C_m \approx 1$ µF cm^{-2}. It is therefore evident that C_m makes by far the major contribution to the overall differential capacity C of the interface, since the differential capacities of the two diffuse layers, in series with C_m, are notably higher than C_m (cf. eqn.59). Upon varying the applied potential E between the two reference electrodes, the differential capacity $C \approx C_m$ attains a minimum for a particular value $E = E_{min}$, as shown in Fig.17d. Clearly this minimum corresponds to the maximum thickness of the BLM, and hence to a zero electric field inside the membrane. From Fig.17c it is apparent that E_{min} measures directly the decrease $\Delta\phi_d$ in the potential difference ϕ_d across the diffuse layer following the increase in the electrolyte concentration.

When ionic specific adsorption is negligible, the diffuse-layer charge density σ_d is usually found to satisfy the Gouy-Chapman theory, and the ϕ_d potential depends upon $\sigma_d = -\sigma_m$ and upon the electrolyte concentration c as predicted by eqn.(58). The theory is usually verified with BLMs prepared from mixtures of neutral and negatively charged lipids of known composition, and hence of known negative surface charge density σ_m [18]. Some deviations from the theory observed at distances less than ≈ 20 Å from the membrane surface have been ascribed to discreteness-of-charge effects, not accounted for by the simple Gouy-Chapman theory. With membranes these effects are actually expected to be more significant than with other interfaces. In some biomembranes, protein clustering, aggregation and lateral segregation of phospholipids may indeed result in nonuniform distribution of charge. Moreover, membrane proteins may project their charges into the aqueous phase up to distances of several tens of Å. It is therefore surprising that the Gouy-Chapman theory may work so well with membranes.

When the bathing solution contains relatively high concentrations of alkali metal and alkaline earth cations, the experimental ϕ_d value turns out to be less negative than predicted by eqn.(58) upon using the known σ_m value. These discrepancies, which are not observed when using monovalent tetramethylammonium or divalent dimethonium cations, are ascribed to specific adsorption of the metal ions, which decreases the "effective" negative charge density "seen" by the diffuse-layer ions. Such an adsorption is usually ascribed to surface complex formation of the metal ions with the negatively charged heads of the lipids. Experiment can once again be reconciled with theoretical predictions if such a surface complex formation is accounted for by a Langmuir isotherm. By so doing, the intrinsic association constant of Na^+ ions with negative lipids is found to be of the order of $1 \ M^{-1}$, whereas it is notably higher for Ca^{2+} ions. Naturally, at these relatively high metal ion concentrations the validity of the Gouy-Chapman theory is assumed, rather than verified. This assumption is commonly made to estimate the unknown charge density σ_m of a biomembrane from ϕ_d measurements by some of the procedures previously described. In this case ϕ_d measurements at different electrolyte concentrations, c, are carried out; σ_m is then ascribed the value which provides the best fit between the experimental dependence of ϕ_d upon c and theoretical predictions.

8. Electrified Interfaces in Nonequilibrium

Let us now briefly examine the effect of nonequilibrium conditions upon electrified interfaces in which the two contacting phases have electrons in common. This includes the interfaces resulting from the contact of metals and semiconductors between themselves and with electrolytic solutions containing a redox couple. Even though in redox systems electrons are bound to the reduced species, we have seen in Sec. 2.3 that the electrochemical potential of electrons in solution can be consistently defined. With all these interfaces, equilibrium is achieved at that particular applied potential $E=E_{eq}$ at which the electrochemical potential of electrons is the same in the two contacting phases α and β:

$$\tilde{\mu}_e^\alpha = \tilde{\mu}_e^\beta \tag{88}$$

As the applied potential is shifted with respect to E_{eq} by a given amount $\eta \equiv E-E_{eq}$, called *overpotential*, a net current flows across the interface.

Let us consider the dependence of the current density i upon η for the interface between a phase α characterized by the presence of free electrons (i.e., a metal or a semiconductor) and a phase β consisting of a solution of a redox couple. Considering the relative inertia of atoms with respect to electrons in a molecule or in an electrode material, we may safely state that during an electronic transition all the atoms in the skeleton of the molecule or of the electrode remain fixed. In other words, since the electronic motions are so much faster than the atomic motions, in a transition one electron changes state before the atoms have moved to an appreciable extent; this is the essence of the *Franck-Condon principle*. For this reason no energy exchange between the electron and the surrounding atomic system is possible during the electron transfer. Consequently, for an electron to pass from a red molecule to the electrode, it must find in the electrode an unoccupied energy level with the same energy that it has in the red molecule just before the transfer. The contribution to the resulting anodic current density i_a from an electron energy ε is therefore proportional to the "concentration" of electrons about the given energy level ε in the red molecule, times the "concentration" of unoccupied electron levels with the same energy in the electrode. Bearing in mind the significance of the distribution functions $D_{red}(\varepsilon)$ and $D_{ox}(\varepsilon)$ for a redox system (see Sec.2.3), the former "concentration" equals $c_{red} D_{red}(\varepsilon)$, where c_{red} is the concentration of the red species in the solution; conversely, the latter "concentration" is given by $\rho(\varepsilon)[1-f(\varepsilon)]$, where $\rho(\varepsilon)$ is the density of the electron levels (no matter if occupied or not) about the energy ε, whereas $[1-f(\varepsilon)]$ is the probability of their being unoccupied (cf. eqn.5 for the expression of $f(\varepsilon)$). The anodic current density i_a will therefore be proportional to the integral of the product of these two "concentrations" over all possible energy levels:

$$i_a = \text{const. } c_{red} \int_{-\infty}^{+\infty} D_{red}(\varepsilon) \, \rho(\varepsilon) \, [1-f(\varepsilon)] \, d\varepsilon \tag{89}$$

By analogous arguments, for an electron to pass from an electron level of energy ε in the electrode to an ox molecule in the solution, it must find an ox molecule with an electron level of the same energy. Hence, the resulting cathodic current density i_c will be proportional to the integral, over ε, of the product of the concentration $\rho(\varepsilon)f(\varepsilon)$ of occupied electron levels of energy ε in the electrode, by the concentration $c_{ox}D_{ox}(\varepsilon)$ of ox molecules with the same electron energy:

$$i_c = \text{const. } c_{ox} \int_{-\infty}^{+\infty} D_{ox}(\varepsilon)\, \rho(\varepsilon)\, f(\varepsilon)\, d\varepsilon \tag{90}$$

8.1. THE CURRENT DENSITY ACROSS METAL / SOLUTION INTERFACES

The contribution of the various factors in the integrands of eqns.(89) and (90) to the corresponding currents is best appreciated by examining the plots of ε vs. $D_{ox}(\varepsilon)$ and $D_{red}(\varepsilon)$ and of ε vs. $\rho(\varepsilon)$, side by side [7,8,19]. Figure 18 shows these plots for a metal/solution interface at equilibrium, when the electrochemical potential $\tilde{\mu}_e{}^M$ of electrons in the metal equals that in the solution, as expressed by eqn.(27):

$$\tilde{\mu}_e^M = \varepsilon_{redox}^o - kT \ln \frac{c_{ox}}{c_{red}} \tag{91}$$

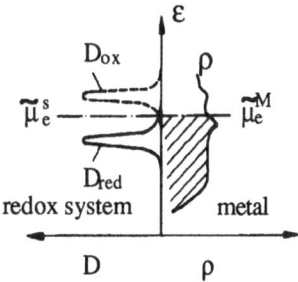

Figure 18. Distribution of energy levels in a redox system and in a metal at equilibrium.

The energy levels which make by far the major contribution to the integrals of eqns.(89) and (90) are those in a narrow energy range of $\approx kT$ width about the electrochemical potential $\tilde{\mu}_e{}^M$ (or Fermi level, as is called more commonly). Thus, if we move upwards with respect to $\tilde{\mu}_e{}^M$ along the vertical ε axis, the probability of finding occupied levels decreases rapidly both in the metal and in the redox system, whereas it increases rapidly if we move downwards. On the other hand, the cathodic (anodic) current is generated from those occupied (unoccupied) electron levels in the metal which match unoccupied (occupied) levels in the redox system. As a first approximation we may therefore replace the integrals in eqns.(89) and (90) by the values of the corresponding integrands at $\varepsilon = \tilde{\mu}_e{}^M$, times $\Delta\varepsilon \approx 1\ kT$:

$$i_a = \text{const. } kT\, c_{red}\, \rho(\tilde{\mu}_e^M)\, [1 - f(\tilde{\mu}_e^M)] \exp\left[-\frac{\left(\tilde{\mu}_e^M - \varepsilon_{redox}^o + \lambda\right)^2}{4kT\lambda} \right] \tag{92}$$

$$i_c = \text{const. kT } c_{ox} \, \rho(\widetilde{\mu}_e^M) \, f(\widetilde{\mu}_e^M) \, \exp\left[-\frac{\left(\widetilde{\mu}_e^M - \varepsilon_{redox}^o - \lambda\right)^2}{4kT\lambda}\right] \qquad (93)$$

In writing these equations, use has been made of the D_{ox} and D_{red} expressions of eqns.(24) and (25).

If we shift the applied potential from the equilibrium value E_{eq} by an amount η, the resulting potential difference is localized almost completely in the Helmholtz layer and in the diffuse layer. Hence, the electron levels in the metal shift relative to those in the solution by an amount $-e\eta$, and the same is true for the electrochemical potential $\widetilde{\mu}_e^M$ relative to the equilibrium value of eqn.(91). The situation is schematically depicted in Fig.19. Upon setting $\widetilde{\mu}_e^M = \varepsilon^o_{redox} - kT \ln(c_{ox}/c_{red}) - e\eta$ in eqns.(92) and (93), we obtain:

$$i_a = \text{const. kT } c_{red} \, \rho(\widetilde{\mu}_e^M) \, [1 - f(\widetilde{\mu}_e^M)] \, \exp\left[-\frac{\left(kT \ln\frac{c_{ox}}{c_{red}} + e\eta - \lambda\right)^2}{4kT\lambda}\right] \qquad (94)$$

and

$$i_c = \text{const. kT } c_{ox} \, \rho(\widetilde{\mu}_e^M) \, f(\widetilde{\mu}_e^M) \, \exp\left[-\frac{\left(kT \ln\frac{c_{ox}}{c_{red}} + e\eta + \lambda\right)^2}{4kT\lambda}\right] \qquad (95)$$

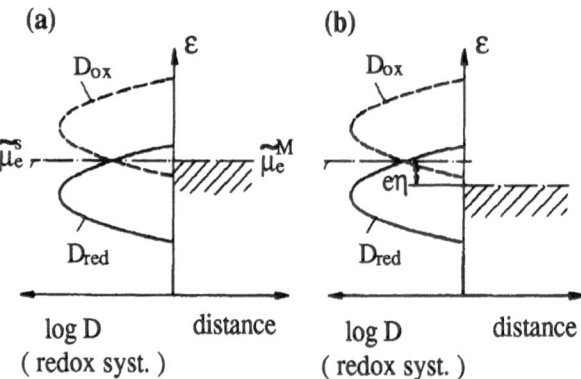

Figure 19. Energy levels in a redox system and in a metal at equilibrium (a) and at a positive overpotential (b).

In practice, the reorganization energy λ is much greater than both $e\eta$ and $kT \ln(c_{ox}/c_{red})$; hence, of the η-dependent terms in the two squares of eqns.(94) and (95), by far the leading one is $-2\lambda e\eta$ in the former equation and $+2\lambda e\eta$ in the latter. Therefore, these equations simplify as follows:

$$i_a = i_0 \, e^{e\eta/2kT} \; ; \qquad i_c = i_0 \, e^{-e\eta/2kT} \tag{96}$$

where i_0 is the common value of the anodic and cathodic components of the net current density $i = i_a - i_c$ for $\eta = 0$, namely at equilibrium; i_0 is referred to as the *exchange current density*. Experimental plots of $ln\,i_a$ vs. $e\eta/kT$ or of $ln\,i_c$ vs. $-e\eta/kT$ (called *Tafel plots*) with slopes close to 0.5 have been reported countless times in the literature. Incidentally, the experimental slopes of Tafel plots are called *charge transfer coefficients*.

8.2. THE CURRENT DENSITY ACROSS SEMICONDUCTOR / SOLUTION INTERFACES

Let us now apply the expressions for the anodic and the cathodic current densities in eqns.(89) and (90) to a semiconductor/ solution interface. Here too, combined ε vs. $D_{ox}(\varepsilon)$ and $D_{red}(\varepsilon)$ plots and ε vs. $\rho(\varepsilon)$ plots can be usefully employed to illustrate the situation. Figure 20 shows two of these plots for the equilibrium case, when the electrochemical potential $\tilde{\mu}_e^S$ in the solution equals that, $\tilde{\mu}_e^{sc}$, in the semiconductor. The first plot refers to the case of an *n*-type semiconductor, with $\tilde{\mu}_e^{sc}$ close to the bottom ε_c of the conduction band, whereas the second refers to the case of a *p*-type semiconductor, with $\tilde{\mu}_e^{sc}$ close to the top ε_v of the valence band (cf. eqn.18). By considerations analogous to those made in connection with metal/solution interfaces, the energy levels which make the major contribution to the current are those in the proximity of the band edges *at the semiconductor boundary*, $\varepsilon_{v,s}$ and $\varepsilon_{c,s}$, which are generally different from those, ε_v and ε_c, in the bulk semiconductor due to band bending.

The current density i^n due to the flow of conduction electrons towards and from the conduction band (the *electron current density*) consists of an anodic and a cathodic contribution, i^n_a and i^n_c, which compensate each other at equilibrium. Analogously, the current density i^p due to the flow of holes towards and from the valence band (the *hole current density*) consists of an anodic contribution i^p_a and of a cathodic one, i^p_c. As a rule, the electron current tends to prevail over the hole current at *n*-type semiconductor/solution interfaces, where the electrochemical potential of electrons in solution, $\tilde{\mu}_e^S$, is closer to the bottom of the conduction band at equilibrium (see Fig.20a). Conversely, the hole current tends to prevail over the electron current at *p*-type semiconductor/ solution interfaces (see Fig.20b). As a matter of fact, the above predictions only descibe general tendencies. Whether electron transfer actually occurs *via* the conduction or the valence band also depends on the ionization energy and on the band gap of the semiconductor, and must be checked in each case.

The expression for the current takes a particularly simple form when the shift η of the applied potential E relative to the equilibrium value is totally localized in the space charge region, namely when the *band edges are pinned at the surface* (cf. Sec.5.1). In this

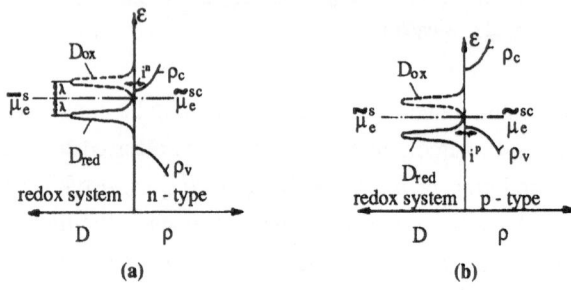

Figure 20. Distribution of energy levels in a redox system and in an *n*-type (**a**) or a *p*-type semiconductor (**b**), at equilibrium.

case the relative positions of the band edges at the semiconductor boundary and of the energy levels in the redox system remain unchanged. This situation is schematically depicted in Fig.21, where the band edges as a function of the distance are reported side by side with ε vs. $log D_{ox}$ and ε vs. $log D_{red}$ plots for the redox system. The comparison of the equilibrium situation ($\tilde{\mu}_e{}^S = \tilde{\mu}_e{}^{sc}$; Fig.21a) with the nonequilibrium situation produced by a positive overpotential η ($\tilde{\mu}_e{}^{sc} = \tilde{\mu}_e{}^S - e\eta$; Fig.21b) shows that the relative positions of the band edges $\varepsilon_{c,s}$ and $\varepsilon_{v,s}$ at the boundary are unaffected by polarization.

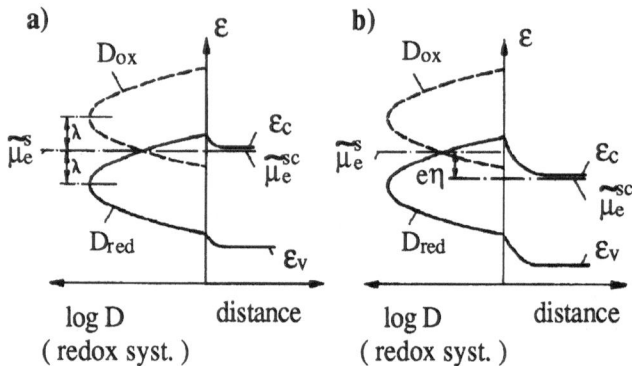

Figure 21. Energy levels in a redox system and in an *n*-type semiconductor at equilibrium (**a**) and at a positive overpotential (**b**).

As concerns the electron current density i^n, its anodic component i^n_a can be approximately derived from eqn.(89) by replacing, as usual, the integral with the product of $\Delta\varepsilon \approx kT$ by the value of the integrand at $\varepsilon=\varepsilon_{c,s}$. Now, the number of unoccupied levels at the bottom of the conduction band, $\rho(\varepsilon_{c,s})[1-f(\varepsilon_{c,s})]$, can be approximately identified with $\rho(\varepsilon_{c,s})$ because most of the electron levels are unoccupied even in an n-type semiconductor, especially when the overpotential η is positive. This implies that i^n_a is practically independent of η and is therefore equal to the *electron exchange current density* i^n_0, namely the common value of i^n_a and i^n_c at equilibrium. In practice, as soon as an electron enters the semiconductor, it is immediately swept over to its interior by the strong electric field that prevails in the space charge region. The cathodic component i^n_c of the electron current density is obtained from eqn.(90) by an analogous substitution of ε by $\varepsilon_{c,s}$ in the the integrand. In this case, however, $\rho(\varepsilon_{c,s})f(\varepsilon_{c,s})$ measures the number density of conduction electrons at the semiconductor boundary, $n(x=0)=n\ exp(e\phi_{sc}/kT)$, where n is the density of conduction electrons in the bulk and ϕ_{sc} is the potential at the boundary relative to the bulk (cf. eqn. 47). Since the application of an overpotential η decreases ϕ_{sc} by an amount η relative to its equilibrium value, we may set $n(x=0)=n_{eq}(x=0)exp(-e\eta/kT)$, where $n_{eq}(x=0)$ is the equilibrium value of $n(x=0)$. Hence, $n(x=0)$ is very sensitive to η, and the same is true for i^n_c. We may therefore write:

$$i^n = i^n_a - i^n_c = i^n_0(1 - e^{-e\eta/kT}) \qquad (97)$$

Figure 22 shows a typical i vs. η plot for an n-type semiconductor/solution interface with a net predominance of the electron current over the hole current. The strong asymmetry of the curve is responsible for the rectifying properties of this interface. The dashed curve represents schematically the increase in the anodic current obtained upon illuminating the electrode surface. If the incident photons have a sufficient energy, they will

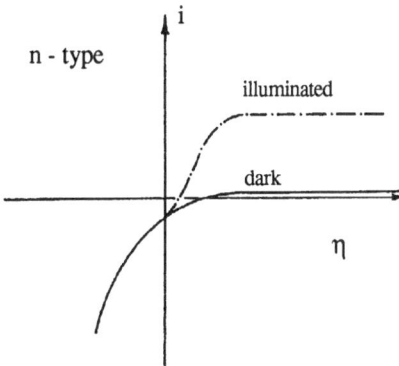

Figure 22. Current vs. overpotential curves at an n-type semiconductor/solution interface in the dark and under illumination.

promote electrons from the valence band to the conduction band, generating what is called an *electron-hole pair* (see Fig.4). The strong electric field existing in the space charge region under anodic polarization prevents *electron-hole recombination* by moving the electron into the bulk semiconductor and the hole towards the surface, where it is captured by a red molecule. (A hole moving towards the surface within the valence band is a formal way of representing an electron moving in the opposite direction.)

The behaviour of the hole current density i^P is opposite to that of the electron current density i^n. Let us replace ε with $\varepsilon_{v,s}$ in the integrands of eqns.(89) and (90) to derive the expressions of the anodic and cathodic components of i^n. In this case it is the factor $\rho(\varepsilon_{v,s})f(\varepsilon_{v,s})$ in the resulting expression of the cathodic hole current, namely the number of occupied levels at the top of the valence band, which can be approximately identified with $\rho(\varepsilon_{v,s})$, since most of the electron levels there are occupied even in a *p*-type semiconductor, especially at a negative overpotential. Conversely, the factor $\rho(\varepsilon_{v,s})[1-f(\varepsilon_{v,s})]$ in the expression for the anodic hole current density i^P_a is equal to the number density $p(x=0)$ of holes at the top of the valence band. Taking eqn.(47) into account and noting that a given overpotential η alters the electric potential ϕ_{sc} at the semiconductor boundary by an amount $-\eta$ with respect to its equilibrium value, we may set $\rho(\varepsilon_{v,s})[1-f(\varepsilon_{v,s})]=p_{eq}(x=0)\exp(e\eta/kT)$, where $p_{eq}(x=0)$ denotes the $p(x=0)$ value at equilibrium. Hence, the hole current density i^P is given by:

$$i^P = i^P_a - i^P_c = i^P_0(e^{e\eta/kT}-1) \tag{98}$$

where i^P_0 is the *hole exchange current density*. Figure 23 shows schematically the strongly asymmetric i vs. η plot for a *p*-type semiconductor/solution interface with a net predominance of the hole current over the electron current. The increase in the cathodic

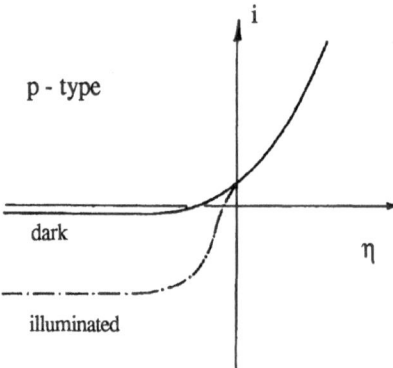

Figure 23. Current vs. overpotential curves at a *p*-type semiconductor/solution interface in the dark and under illumination.

63

current following illumination is now due to the strong electric field in the space charge region at a negative overpotential, which prevents electron-hole recombination by moving holes towards the bulk semiconductor and electrons towards the surface, where they are captured by ox molecules.

The rectifying properties of semiconductors affect in an analogous way the current vs. potential curves at *p*-type semiconductor/*n*-type semiconductor interfaces (briefly, *p-n junctions*) and metal/semiconductor interfaces. Thus these interfaces, which are used in *photovoltaic cells* for solar energy conversion, exhibit highly asymmetric current vs. potential curves which satisfy eqn.(98) (see, for instance, eqn.4 in Perfetti's contribution to this volume, for n=1). The applied overpotential η has no effect upon the current flowing in one direction, whereas it exerts its maximum influence (in that the whole electrostatic potential energy $e\eta$ appears in the Boltzmann-like factor $\exp(e\eta/kT)$) in the opposite direction.

REFERENCES

(1) N.W. Ashcroft and N.D. Mermin, *Solid State Physics*, Sauders College, Philadelphia, 1976.
(2) R.A. Marcus, J. Phys. Chem., **6 7**,853, 1963; J. Chem. Phys., **4 3**,679,1965.
(3) V.G. Levich in *Advances in Electrochemistry and Electrochemical Engineering*, P. Delahay Ed., Vol.4, p.249, Interscience, New York, 1966.
(4) R.R. Dogonadze in *Reactions of Molecules at Electrodes*, N.S. Hush Ed., p.135, Wiley Interscience, 1971.
(5) P.P. Schmidt in *Specialist Periodical Report on Electrochemistry*, Vol.5, p.1, Chemical Society, 1975.
(6) V.A. Myamlin and Yu.V. Pleskov, *Electrochemistry of Semiconductors*, Plenum, New York, 1967.
(7) Yu.V. Pleskov and Yu.Ya. Gurevich, *Semiconductor Electrochemistry*, Consultants Bureau, New York, 1986.
(8) R. Memming in *Electroanalytical Chemistry*, A.J. Bard Ed., Vol.11, p.1, Marcel Dekker, New York, 1979.
(9) A. Zangwill, *Physics at Surfaces*, Cambridge University Press, Cambridge, 1988.
(10) F. García-Moliner and F. Flores, *Introduction to the Theory of Solid Surfaces*, Cambridge University Press, Cambridge, 1979.
(11) D. M. Mohilner in *Electroanalytical Chemistry*, A.J. Bard Ed., Vol.1, p.241, Marcel Dekker, New York, 1966.
(12) B.H. Bijsterbosch and J. Lyklema, Adv. Colloid Interface Sci., **9**,147,1978.

(13) H.P. van Leeuwen and J. Lyklema in *Modern Aspects of Electrochemistry*,
        J.O'M. Bockris, B.E. Conway and R.E. White Eds., Vol.17, p.411, Plenum,
        New York, 1986.

(14) H. Ti Tien, *Bilayer Lipid Membranes (BLM)*, Marcel Dekker, New York, 1974.

(15) H. Rottenberg in *Bioelectrochemistry III. Charge Separation Across
        Biomembranes*,G. Milazzo and M. Blank Eds.,p.55,Plenum, New York, 1990.

(16) D. Walz, Biochim. Biophys. Acta, **505**,279,1979.

(17) P. Schoch, D.F. Sargent and R. Schwyzer, J. Membrane Biol., **46**,71,1979.

(18) S. McLaughlin, Annu. Rev. Biophys. Biophys. Chem.,**18**,113,1989.

(19) H. Gerischer in *Solar Energy Conversion*, B.O. Seraphin Ed., Springer Verlag,
        Berlin, 1979.

STRUCTURE AND ELECTRONIC PROPERTIES OF METAL SURFACES

K. WANDELT

Institut für Physikalische und Theoretische Chemie
Universität Bonn
Wegelerstr. 12
D-5300 Bonn 1
F.R.G.

ABSTRACT. Due to the asymmetric environment of surface atoms and, hence, the existence of unsaturated, "dangling" bonds at crystal surfaces the properties at the surface are different in almost every respect from those in the bulk of the same material. Crystallographically surfaces are relaxed or even reconstructed. As a consequence of this geometrical rearrangement also the electronical properties at the surface are different from the bulk properties. For two- or multicomponent solids the surface composition is often found to deviate from the bulk composition due to segregation effects.

This chapter concentrates on structural and electronical properties of metal surfaces. After the Introduction some general aspects of surfaces and surface analytical methods are addressed. Then a discussion of structural peculiarities of perfect single crystal surfaces, namely surface relaxation and surface reconstruction, follows.

In reality, however, surfaces are never perfect but always contain a variety of structural (steps, kinks, vacancies etc.) and chemical defects (heteroatoms), which, in fact, may have a decisive effect on the properties of the surface as a whole. Therefore, the second part of this chapter is devoted to the identification of atomic scale surface irregularities and to the characterization of their local physical properties. Finally, a brief correlation between these localized properties and their influence on adsorbed molecules is presented.

R. Guidelli (ed.), Electrified Interfaces in Physics, Chemistry and Biology, 65–115.
© 1992 *Kluwer Academic Publishers.*

1.   Introduction

Surface and interface physics and chemistry provide the basic research for many modern technologies and processes such as Thin Film Technologies (as e.g. applied in microelectronics, inorganic and organic coatings), Materials Science ("Surface - a new state of matter" /1/), Energy Conversion (new energy sources like electro-chemical fuel cells, solar cells, catalytic refinement of fossil fuels), Chemical Production (heterogeneous catalysis) and Micro- and Nanometer Technologies, to name only a few. Obviously the success of these technologies rests on an ever im-proving understanding of the physical and chemical properties of surfaces and in-terfaces as well as of molecular surface processes such as sticking, chemical bonding and interfacial charge transfer, diffusion, structural ordering or disordering etc. This chapter will mainly concentrate on physical properties of metal surfaces.

"Surface science" of the past thirty years has concentrated on the phenomenon "surface" per se, that is on the question: How do the properties of the solid change along the z-coordinate (perpendicular to the surface) approaching the surface from the bulk? In order to answer this question much work was done with single crystal surfaces, which were assumed to be homogeneous and periodic within the surface (xy-plane). As a result it is clear by now that in most cases the properties of the very surface differ widely from those of a bulk lattice plane parallel to the surface /2/. Crystallographically surfaces are relaxed or even reconstructed. This is accompanied by a surface specific electronic (surface core level shifts, surface states, two-dimensional band structure) and vibrational (surface phonons) structure. For two- or multicomponent solids the surface composition is often found to deviate from the bulk composition due to segregation effects.

However, in reality surfaces are never perfect but always show a distribution of structural and chemical defects (within the xy-plane of the surface) such as steps, kinks, adatoms, vacancies, heteroatoms, etc. as sketched in Fig. 1. There is abundant evidence, that physical and chemical processes at surfaces are strongly influenced (if not dominated) by these surface defects. Among these processes are, for instance, the scattering of light and particles; adsorption, chemical reactions and hetero-geneous catalysis; crystal growth, epitaxy, and many more. Thus, much is known about the **influence** of surface defects, but little is known about the local **properties** of different kinds of surface defects as well as about the range of these properties. As with the z-dependence of the physical properties near a solid surface the variation of the surface properties within the xy-plane must be known on the atomic scale to arrive at a full understanding of the various surface processes mentioned above. This is possible nowadays since the techniques and methods of the necessary resolution are available now.

This chapter will present results on both perfect single crystal surfaces and surfaces with atomic scale structural and chemical heterogeneity. In particular, the influence of surface defects on adsorbed species will also be addressed.

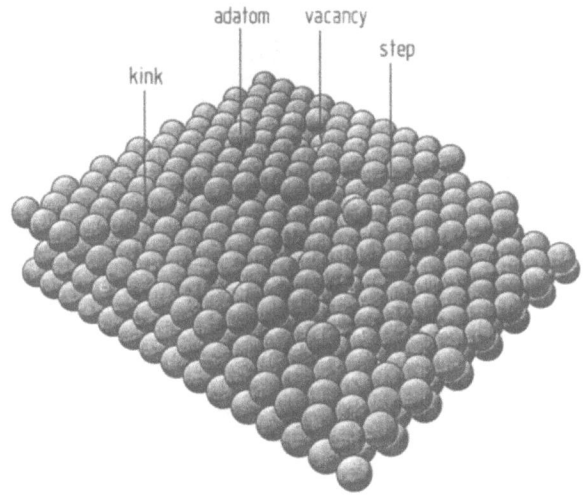

kink   adatom   vacancy   step

Fig. 1   Schematic representation of a surface with structural defects such as kinks, adatoms, vacancies, steps etc.

2. Fundamental Aspects of Surfaces

Solids are limited by surfaces. Regular crystals form planar surfaces as depicted schematically in Fig. 2a. The peculiarities of a surface become immediately clear if we form new surface by cutting the crystal in half (see Fig. 2b). We realize that it may cost quite some energy, for instance, in the case of hard metal crystals, to perform the cut. Some of this energy will be dissipated as heat, but most of the energy is simply required to break up the chemical bonds between the atoms on both sides of the cut. This energy remains stored in the unsaturated, "dangling" bonds which now "point out" of the surface. The total excess energy stored in all these "dangling" bonds of a newly created surface makes up the surface free energy or surface tension σ. The existence of this surface excess energy is one of the important physical parameters which explains the deviation between surface and bulk properties.

The surface free energy is face specific. This face specificity of σ is easily conceivable in terms of the number of broken bonds per atom at surfaces of different crystallographic orientation, as will be discussed in more detail in section 4.1.2. A rough estimate of the energy per broken bond can be obtained from the heat of sublimation of the respective material: Sublimation involves breaking of all bonds per atom and thus, to a first approximation, provides a measure of the energy per half-bond per atom, that is per "dangling" bond per surface atom. Surface relaxation,

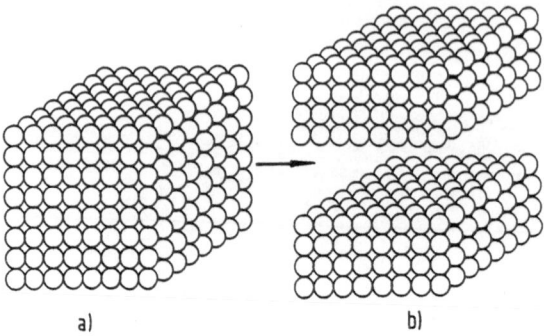

a)   b)

Fig. 2   a) Schematic representation of a single crystal. b) Creation of new surface requires energy which remains stored in the broken surface bonds.

surface reconstruction (see section 4.1.1. and 4.1.2), surface segregation as well as adsorption of particles at surfaces are driven by the attempt of the system to lower the energy per broken bond, to lower the number of broken bonds or to involve the surface "dangling" bonds in new bonds (with the adsorbate).

The second important consequence of the cut through the crystal is the fact that the new surface atoms have lost their neighbors on one side. As a result surface atoms are held in an asymmetric potential, which is expected to alter their lattice position as well as their electronic and dynamic properties compared to those of an isotropically surrounded bulk atom.

Both the surface excess energy as well as the broken symmetry perpendicular to the surface (z-direction) are the fundamental reasons for the deviating, two-dimensional physics of surfaces compared to the threedimensional physics of the bulk of the same material.

3.   Surface Analysis

An abundance of experimental techniques has been developed in order to investigate the structural, electronic and dynamic properties as well as the composition of solid surfaces. Several books and articles exist which provide excellent introductions to the various methods /e.g. 2a, 2c/. These methods need to be both very sensitive and surface specific. Surface specificity is required to obtain information only about the first few outermost atomic layers, say up to three or five atomic layers, isolated from the bulk background. This implies at the same time that the amount of material within the detected volume is of the order of some 10^{12}-10^{13} atoms or 10^{-10} mole only. Hence, the techniques must be highly sensitive. Finally, the experiments need to be carried out under ultrahigh-vacuum conditions (UHV) in order to keep a surface, once cleaned by cycles of heating, ion bombardment and/or chemical treatments, clean for the duration of the experiment. In modern commercial UHV-equipment pressures of $\leq 1 \times 10^{-10}$ Torr are routinely achieved which provides stable surface conditions for maybe one to two hours before adsorption from the residual gas begins to affect the surface properties.

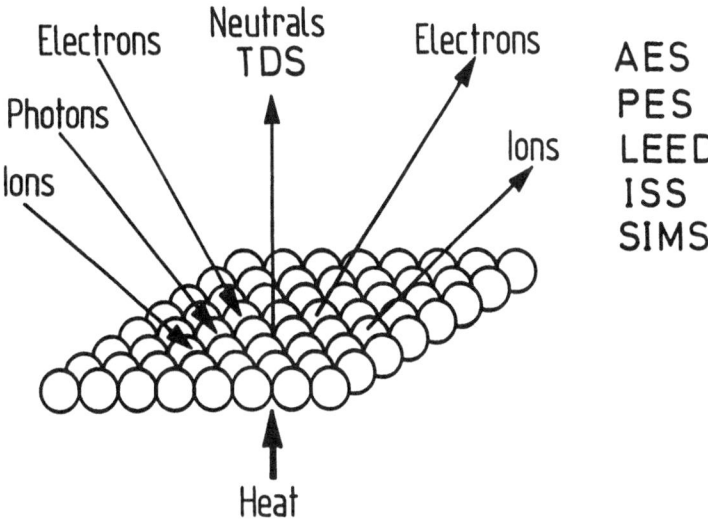

Fig. 3 Illustration of the operation of surface techniques. The acronyms are explained in the text.

Most standard surface analytical techniques involve beams of low energy electrons, ions, photons or neutral particles which are used to exit or to scatter from the layer of surface atoms (see Fig. 3). Thermal energy atom beams are scattered from the outermost atomic layer. Low energy ion and electron beams (0-1000eV) penetrate only a few atomic layers into the solid and may result in the reemission/emission of primary/secondary ions or electrons which by virtue of their energy and momentum carry information about the surface only. Low energy photons, though penetrating deeply into the solid, may excite low energy photoelectrons which, due to their low energy, can leave the solid only if they originate again only from the first few outer surface layers. Fig. 3 gives the acronyms of some of the most frequently applied surface techniques: Auger Electron Spectroscopy (AES), Photoelectron Spectroscopy (PES) with UV-light (UPS) or X-rays (XPS), High Resolution Electron Energy Loss Spectroscopy (HREELS), Low Energy Electron Diffraction (LEED), Ion Scattering Spectroscopy (ISS) and Secondary Ion Mass Spectrometrie (SIMS). Roughly speaking AES and PES provide information about the surface composition as well as about the surface electronic structure, while LEED and ISS yield information about the surface crystallography. Since ISS is also mass specific, the scattered beam of primary ions carries also information about the surface composition. In SIMS primary ions knock surface atoms or molecules away (sputter erosion) which are detected by a mass spectrometer. Finally, in HREELS scattered low energy electrons (0-5eV) incur energy losses (10-1000meV) due to characteristic surface vibrational modes.

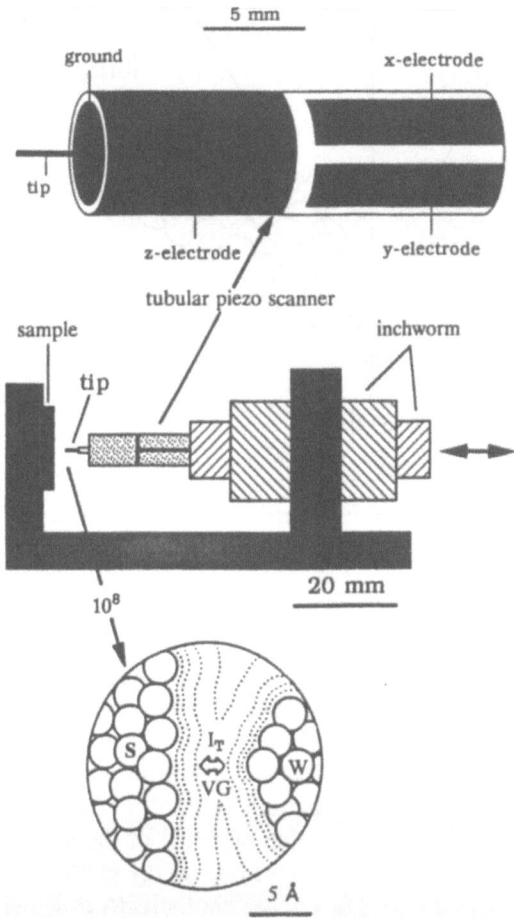

Fig. 4 The principle of a scanning tunneling microscope (STM) on the basis of a tubular piezo scanner. The "inchworm" realizes the coarse approach /4/.

Most fascinating of course, is the realization of Scanning Tunneling Microscopy (STM): A fine metal tip (e.g. of tungsten, W in Fig. 4) pointing to the sample surface is mounted, for instance, at the end of a tube of piezo-ceramics which is coated with metal film electrodes on the inside as well as on the outside. In Fig. 4 it is easily recognized that the outer electrode is divided into five parts, the +x, -x, +y, -y-electrodes and the z-electrode ring. By applying an appropriate voltage between the inner electrode and the x-electrodes (y-electrodes) the piezo-tube bends into x-(y-) direction and moves the tip W in x-(y-)direction across the surface. Applying a voltage between the inner electrode and the outer z-ring expands or contracts the piezo-tube and thereby allows to control the width d_z of the vacuum gap V_G between sample (S) and tip (W) of a few Ångstroms with an accuracy of better than 0.1 Å /4/. A close-up (Fig. 4) shows the individual atoms on both sides of the vacuum gap (V_G); under these conditions the electronic wave functions of atoms of both sides begin to overlap and a tunneling current I_T can be drawn through the gap when a small voltage is applied. Since the tunneling depends exponentially on both the width of the vacuum gap and the height of the tunneling barrier, small changes in either one will lead to drastic changes of the tunneling current. Upon scanning the tip across the surface in x- and y-direction one may either keep the tip at constant height and measure the $I_T(x,y)$ or keep I_T constant by adjusting d_z by means of an electronic feedback mechanism. Twodimensional representations of $I_T(x,y)_{Z=const}$ or $d_z(x,y)_{I=const}$ contain the (coupled) information about the real-space surface topography and the variation of the tunneling barrier (which depends on the local charge density) on an atomic scale. Examples of STM images are shown here in Figs. 6 and 23. For excellent reviews about Scanning Tunneling Microscopy see e.g. Refs. 3.

4.   Properties of Single Crystal Surfaces

4.1.   SURFACE CRYSTALLOGRAPHY

The aim of surface crystallography is to provide full information about the lattice positions of surface atoms. These positions may deviate both perpendicular and parallel to the surface from those of the extrapolated bulk lattice /5,6/.

The in-plane surface lattice positions are generally given with respect to those of the bulk. Within a bulk plane parallel to the surface indentical lattice positions are connected by translation vectors:

$$T_B = h' \vec{a}_1 + k' \vec{a}_2 \tag{1}$$

where a_1 and a_2 are the unit vectors and h' and k' are the two-dimensional (2D) Miller indices. The periodicity within the actual surface layer or within the overlayer of an adsorbed species may be described by the different 2D lattice vector:

$$T_S = h' \, \vec{b}_1 + k' \, \vec{b}_2 \quad . \tag{2}$$

Two different procedures are generally applied to express the surface periodicity with respect to the bulk structure. The Wood-nomenclature uses the ratios between the unit vectors of overlayer and bulk:

$$\frac{|\vec{b}_1|}{|\vec{a}_1|}, \quad \frac{|\vec{b}_2|}{|\vec{a}_2|}, \quad \vartheta \tag{3}$$

and ϑ describes a possible angular rotation between both lattices. This Wood-nomenclature requires some common periodicity between the substrate and the overlayer in order to arrive at simple rational numbers (see e.g. Fig. 5). Universally applicable, however, is the socalled matrix-notation:

$$
\begin{aligned}
b_1 &= m_{11} \, \vec{a}_1 + m_{12} \, \vec{a}_2 \\
b_2 &= m_{21} \, \vec{a}_1 + m_{22} \, \vec{a}_2
\end{aligned}
\tag{4}
$$

with the transformation matrix:

$$M = \begin{vmatrix} m_{11} & m_{12} \\ m_{21} & m_{22} \end{vmatrix} \tag{5}$$

Examples for both terminologies are found throughout this chapter.

Changes of the lattice spacing perpendicular to the surface, that is between the first and the second layer, the second and the third layer etc., may be expressed by the percentage of the change in bond lengths between atoms of neighboring layers.

Deviations between the in-plane periodicity from that of the bulk are called "surface reconstruction". Changes of the interlayer spacings perpendicular to the surface are termed "surface relaxation".

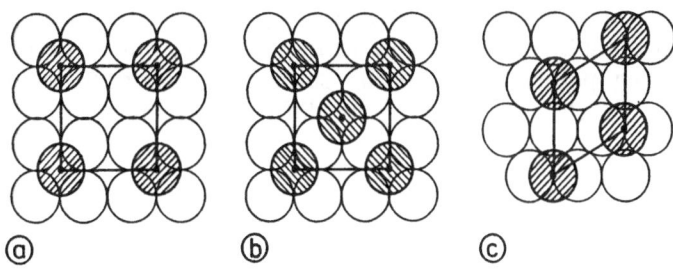

Fig. 5 Examples of adsorbate overlayer structures denoted in the Wood-nomenclature a) (2x2), b)√2 x √2/R45° ≙ c(2x2), c) √3 x √3/R30°

Nowadays it appears most natural to visualize the structure of a metal surface by direct STM imaging. As an example Fig. 6 reproduces the image of an Al(111) surface /7/. This image is, indeed, most impressive because the individual Al-atoms are clearly resolved even though the (111) surface is the most densely packed one and Al is an sp-metal. Both facts together should give rise to a very small corrugation of this surface. Yet, the atoms are clearly visible which underlines the fascinating power of the STM-technique.

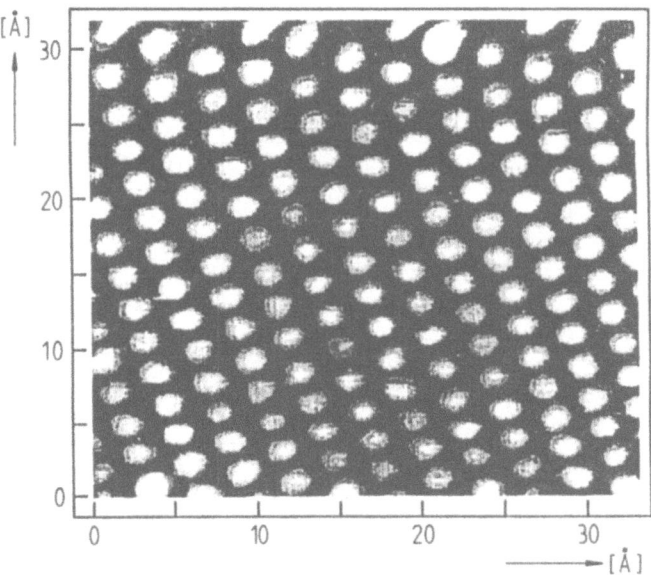

Fig. 6 Scanning Tunneling Microscopy (STM) relief of an Al(111) surface showing the individual Al atoms /7/.

74

However, one must realize that an STM image provides only information about the atomic topography parallel to the surface; nothing can be derived about the interlayer-spacing perpendicular to the surface. This information can only be obtained from the "classical" scattering or diffraction techniques such as LEED or ISS.

4.1.1. Surface Relaxation. The simple LEED pattern of a twodimensional surface lattice (x,y-plane) contains information about the periodicity parallel to the surface. In fact, the LEED pattern corresponds to the reciprocal surface lattice /2a/. The intensity within the diffraction spots, however, varies with the energy of the primary electrons due to constructive or non-constructive interference between contributions of scattered electron waves from the first, the second and deeper layers, respectively. Constructive interference, as described by the Laue condition in z-direction, is determined by the interlayer spacing perpendicular to the surface. Thus LEED-intensity measurements as a function of the primary electron energy, socalled I/V-

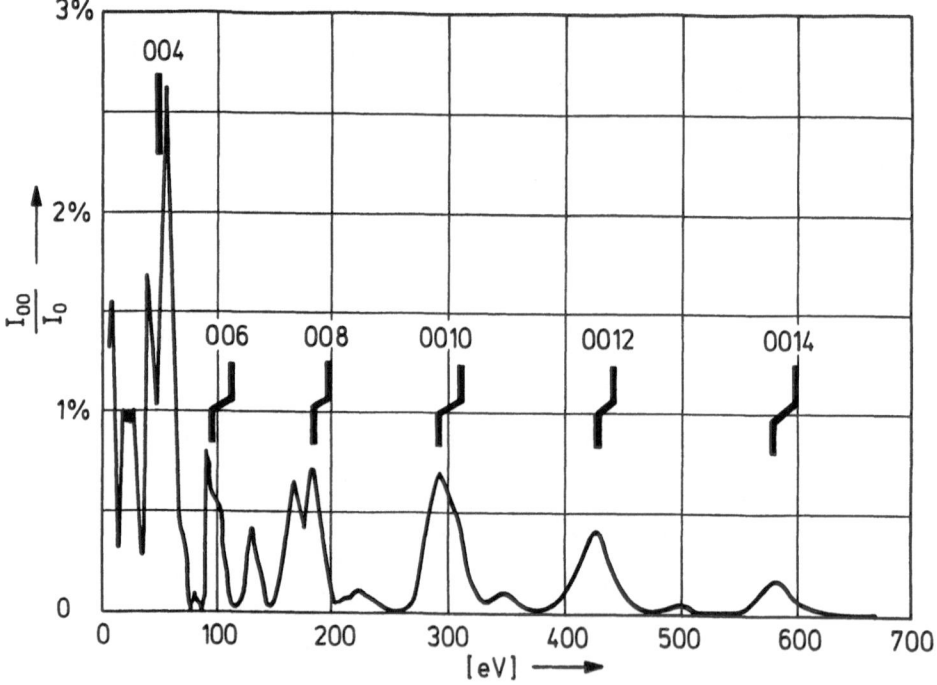

Fig. 7 Intensity/voltage (I/V)-curve for the (0,0)-beam of a clean Ni(100) surface at normal incidence. The intensity of the diffracted beam, I_{00}, is normalized with respect to the intensity of the primary beam, I_0. (From Christmann et al., Ref.8)

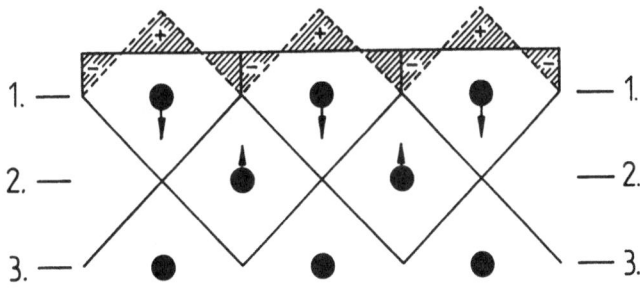

Fig. 8 Schematic representation of the socalled Smoluchowski electron smoothing effect at surfaces, which provides an explanation for the face-specificity of the surface relaxation effect (see text).

curves, yield interlayer spacings or, therefrom, bond lengths between successive layers perpendicular to the surface. As an example Fig. 7 reproduced the variation of the intensity I_{oo} of the (0,0) beam (normalized to the primary electron current I_o) from a Ni(100) surface /8/. The overall shift of the Bragg-maxima to lower energies compared to the expected bulk positions (black cranked bars) is due to the inner potential felt by the electrons inside the crystal. The relative separation between the Bragg peaks contains the information about any surface relaxation.

Many surfaces have been shown to be relaxed /5/ as evidenced by Table 1. Almost exclusively a **contraction** is found between the first and the second layer; the percentages in Table 1 correspond to contractions of bond lengths between atoms of the first and the second layer. A clear dependence of this effect on the crystallographic orientation of the surface can also be noticed. The more open, that is the less densely packed, the surface is the more pronounced is the contraction effect.

Face	Metal		
fcc (110)	Ag 2%, Al 3-4%, Ni 1.5%, Cu 3-4%		
bcc (100)	Fe 1.5%, Mo 4%, W 2-4%		
bcc (111)	Fe 1.5%		
fcc(100)	Al, Co, Ni, Cu, Rh, Ag, Pt	⎫	
bcc (110)	Na, Fe, W	⎬ < 1%	
fcc(111)	Al, Co, Ni, Cu, Ag, Ir, Pt, Au	⎪	
hcp(0001)	Be, Ti, Co, Zu, Cd	⎭	

Table 1 Bond length changes between the first and the second layer (compared to the bulk value) due to surface relaxation at different metal surfaces as derived from LEED I/V-measurement.

The general observation of an interlayer contraction is easy to conceive in view of the asymmetric environment of the surface atoms. There are only neighbors on one side attracting the surface atoms toward the bulk.

The more subtle effect of the face specificity of the surface relaxation can be explained on the basis of the Smoluchowski electron smoothing effect /9/ as illustrated in the cross-section shown in Fig. 8. There the black dots indicate the nuclei of the atoms on their lattice positions. The squares around the dots represent (a 2D cross-section through) the corresponding Wigner-Seitz cells, which in the bulk are units of charge neutrality. At the surface the Wigner-Seitz cells protrude out of the surface (dashed lines). This is an energetically unfavorable charge distribution: the "electron sea" tends to form a flat surface (oversimplified by the horizontal full line in Fig. 8). As a consequence electronic charge flows from the protruding edges of the Wigner-Seitz cells into the valleys between them, thereby leading to an excess of negative charge in the valleys and a corresponding deficiency on the edges. In simple terms this results in a repulsion between the nuclei and the positive edges and, hence, in an inward-relaxation of the first surface layer of atoms as indicated by the arrows. More precisely the charge redistribution at the surface leads to an increase of the self-energy of each Wigner-Seitz cell at the surface, which can be lowered by an inward shift of the first plane of nuclei. Both the charge redistribution and, as a consequence, the inward-relaxation become more pronounced with increasing surface roughness, that is the more the Wigner-Seitz cells protrude out of the (average) surface plane. This is in full accord with the message of Table 1.

This model can even be extended to a prediction of a geometrical relaxation of the second atomic layer. The nuclei of the second layer are just below the negative excess charge in the valleys. Attractive interaction between both would, in fact, suggest some **outward**-relaxation of the second layer. This is, indeed, also in agreement with very detailed LEED-I/V measurements as shown at the end of the next section.

Here we summarize, that as a consequence of the broken symmetry in z-direction surface-nearest atomic layers exhibit the phenomenon of geometrical relaxation; the first layer is generally relaxed inward while the second layer relaxes outward (see section 4.1.2.). As a consequence these outermost layers spatially form a new compact entity which is somewhat decoupled from the bulk. This fact makes it, indeed, rather suggestive to expect this "surface skin" to exhibit its own 2D physics.

4.1.2 Surface Reconstruction. While surface relaxation describes only changes in crystallographic distances in z-direction, surface reconstruction denotes changes in atomic positions parallel to the surface. Indeed, the crystallographic faces of the metals listed in Table 2 are known for quite some time to exhibit a different surface periodicity than expected from a simple truncation of the bulk lattice. According to the reconstruction mechanism irreversible, diffusive surface reconstructions and reversible, displacive reconstructions are to be distinguished.

Face	fcc metals	bcc metals
(100)	Ir, Pt, Au	V, Cr, Mo, W
(110)	Ir, Pt, Au, Pb	
(111)	Au	

Table 2 Metal surfaces exhibiting surface reconstruction

4.1.2.1. Irreversible, diffusive surface reconstructions. As an example Fig. 9 reproduces two LEED pattern from an Ir(100) surface, denoted by (1x1)Ir(100) and (5x1)Ir(100). The (1x1) pattern shows a square arrangement of the LEED spots as expected from an face-centered-cubic (fcc)(100) face. This pattern, however, is not easy to obtain; it requires some rather involved preparation procedure of the Ir(100), which will be described in the section 4.1.2.3. Here we realize, that the stable LEED picture of the **clean** Ir(100) surface is the (5x1) pattern. If, for instance, the (1x1) pattern has been obtained after the special surface treatment explained in the next section, simple heating to around 800 K leads to an irreversible (1x1) → (5x1) transformation /10a/. Obviously the thermodynamically stable

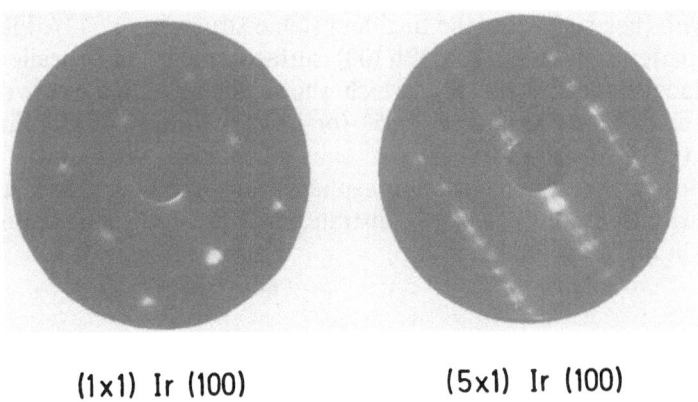

(1x1) Ir (100)   (5x1) Ir (100)

Fig. 9 LEED pattern of a metastable (1x1)Ir(100) and a stable (5x1) Ir(100) surface.

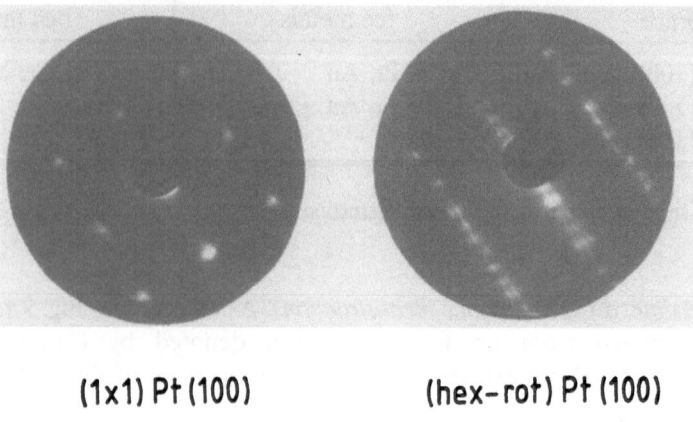

(1x1) Pt (100)   (hex-rot) Pt (100)

Fig.10   LEED pattern of a metastable (1x1) Pt(100) and a stable (hex-rot) Pt(100) surface (see text).

structure of the Ir(100) surface is not a simple square arrangement of the surface atoms. Similar as well as even more complicated pattern have been observed for the other metal surfaces listed in Table 2. For instance, the Pt(100) surface undergoes a whole sequence of irreversible structure transformations: (1x1) → (1x5) → (5x25) → "(hex-rot)" with (hex-rot) being the final and stable structure /11-14/. Fig. 10 shows the LEED pattern of a) the (1x1)Pt(100) surface and b) the socalled "hex-rot" Pt(100) surface modification /15/, which shows slightly tilted rows of multiple diffraction spots. The denotations "1x5" (or "5x1", which is only rotated by 90° compared to "1x5" on the (100) square substrate lattice, see below) and "5x25" correspond to the Wood-nomenclature explained in section 4.1. Their unit vectors are aligned with those of the square substrate. The (hex-rot) modification is better represented in matrix-notation

$$\begin{pmatrix} 12 & 1 \\ -1 & 5 \end{pmatrix}$$

because of the small rotation angle.

These unexpected LEED pattern from the Ir(100) and the Pt(100) surface (as well as similar observations from the Au(100) surface /16/) can be explained by assuming that the first atomic layer of these samples are **reconstructed** to form a **hexa-**

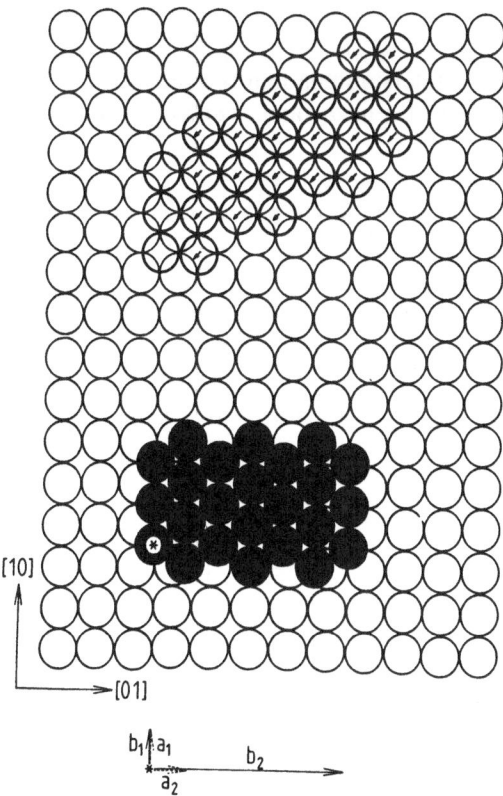

Fig.11 Illustration of the hexagonal reconstruction of the (100) surface of iridium, platinum, and gold (see text). Below the structure model the unit vectors of the square substrate (a_1, a_2) as well as of the (5x1) coincidence mesh of the reconstructed surface layer (b_1, b_2) are shown (the star at the origin of these vectors has to be placed on the star in the black surface atom).

gonal arrangement of the atoms. This is illustrated in Fig. 11. The open and empty circles represent the square arrangement of an ordinary fcc(100) surface. In the upper half of the figure an island of 24 atoms is placed on the substrate, with the atoms residing on fourfold hollow-sites, thus, forming again a square structure. If these atoms are squeezed together along the direction of the little arrows until they form the island shown in the lower half of the figure (black atoms), we notice that their arrangement has changed to a hexagonal one. Double scattering of electrons from such an arrangement, namely a hexagonal monolayer on a square substrate, can reproduce the (5x1) as well as the (1x5) and (5x25) LEED pattern from the Ir(100) and Pt(100) surfaces, respectively. In order to explain the more complex pattern of the (hex-rot)Pt(100) the hexagonal overlayer needs only to be rotated by ~ 0.8° compared to the alignment shown in Fig. 11 (lower half) /6/.

It is, indeed, the present understanding that the reconstructed surfaces of Ir(100), Pt(100) and Au(100) consist of a hexagonal overlayer of monoatomic thickness on the square structure of the second layer. This conclusion from LEED observations has, in the meantime, also been ascertained by other techniques such as Thermal Energy Atom Scattering (TEAS) with He-beams /17/ and Scanning Tunneling Microscopy (STM) /18/ (see also Fig. 23). Even a Au(100) electrode surface in solution shows the hexagonal reconstruction as evidenced by in-situ STM studies /19/.

Obviously the question arises as to why these surfaces prefer to expose a hexagonal surface arrangement rather than the natural square structure. What is the driving force for this reconstruction of the surface? To a first approximation this can be explained in terms of the surface free energy, that is in terms of the broken bond

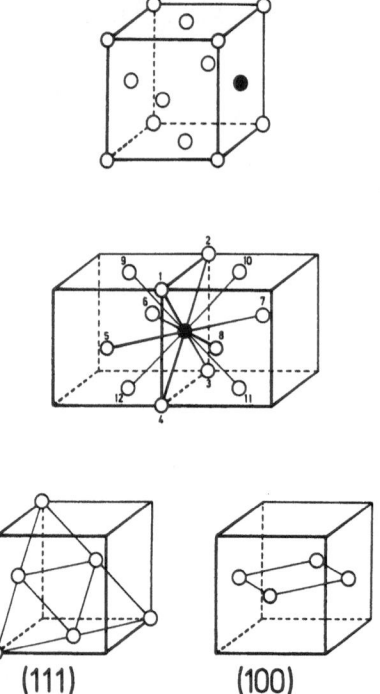

Fig.12 Atomic positions in an face-centered-cubic (fcc) metal lattice. Top: fcc-unit cell; center: 12 nearest neigbors of an atom; bottom: orientation of a (111) and a (100) crystal plane.

model introduced in section 2. Fig. 12 shows at the top the unit cell of an fcc lattice, containing atoms at the corners as well as at the centers of the faces of the cubic cell. At the bottom Fig. 12 indicates the atoms which belong to a (111) and (100) plane put through the unit cell. In the center of Fig. 12 the black test atom is surrounded by the nearest atoms of its two neighboring unit cells. A (100) surface plane through the black atom contains the atoms 5,6,7 and 8, and the atoms 1,2,9 and 10 are cut away. Hence, an atom of a (100) surface has 4 "dangling" bonds. In turn, a (111) surface plane through the black atom contains the atoms 2,7,11,4,5 and 9, and the atoms 1,8 and 10 have been cut away. An atom within a hexagonal (111) surface has only three "dangling" bonds. Consequently, the fcc(111) surface has a lower, in fact, the lowest surface free energy (see Au(111) below). The atoms in the (111) face are most densely packed and, consequently, suffer from the least number of broken bonds. The hexagonal reconstruction of the Ir(100), Pt(100) and Au(100) surface is driven by this lowering of the surface free energy. Note that the atomic density of a hexagonal layer is ~20% larger than that of a square (1x1) structure. Hence, the reconstruction illustrated in Fig. 11 involves a substantial surface diffusion of atoms.

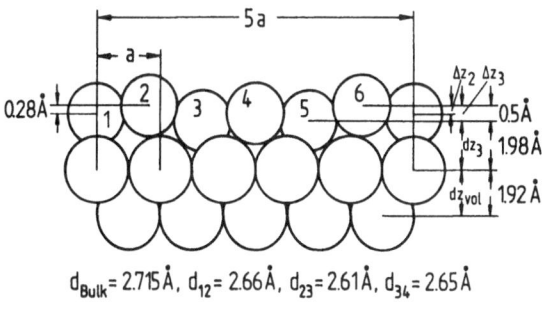

$d_{Bulk} = 2.715 \text{Å}$, $d_{12} = 2.66 \text{Å}$, $d_{23} = 2.61 \text{Å}$, $d_{34} = 2.65 \text{Å}$

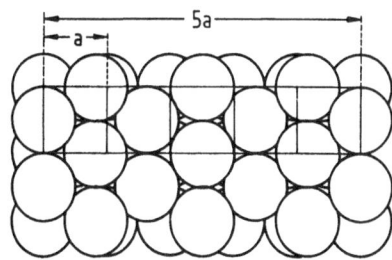

Fig.13 Top: Cross section, and bottom: Top-view of the (5x1) coincidence mesh of the structure model of the reconstructed (5x1) Ir(100) surface as derived from LEED I/V-measurements. d_{12}, d_{23}, d_{34} are interatomic distances between the respective surface atoms. d_{Bulk} is the interatomic distance in the bulk. Δz_2, Δz_3, dz_3, dz_{vol} refer to the corresponding vertical spacings (Ref.20).

It is obvious, that the mismatch between the hexagonal overlayer and the square structure of the substrate will result in a **non-planar** surface. This has also been confirmed by LEED, TEAS, STM etc. The most complete structure model for the (5x1)Ir(100) surface as derived from LEED is summarized in Fig. 13 /20/. The top shows a cross section through the surface and the bottom the corresponding top view. The (5x1) coincidence mesh is clearly seen; 6 surface atoms fall onto 5 atoms in the second layer. The side view shows the buckling of the surface layer and gives all the relevant distances. All in all the contraction within the surface amounts to 2,1-2,4% while the bond-lengths between first and second layer atoms are contracted by 4-5%. The latter result corresponds to a surface relaxation, which is in any case energetically favorable as discussed in the previous section. The slight compression of the first layer (compared to a regular (111) face) is obviously energetically overcompensated by the reduced number of "dangling surface bonds".

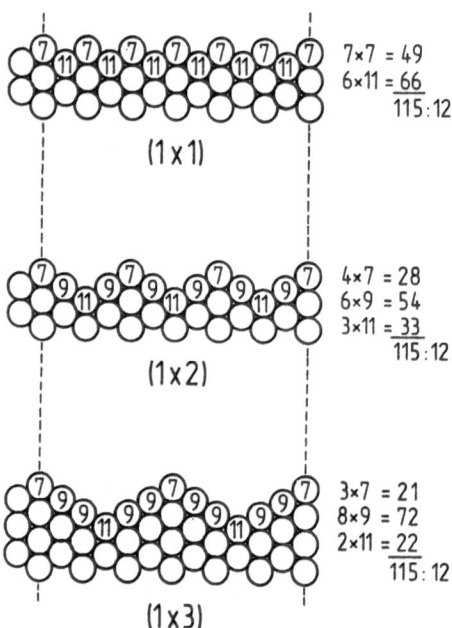

Fig.14 Missing row models of the (1x2) and (1x3) reconstructed (1x1) fcc (110) surface. The average number of bonds per surface atom, however, is the same in all three cases, namely 115 : 12 = 9,6 bonds/atom, and, hence, the number of unsaturated "dangling" bonds as well.

Table 1 indicates that also the (110) surfaces of the same fcc-metals are reconstructed /5/. Most often (1x2) and sometimes (1x3) pattern are observed instead for
these surfaces. The present understanding of these reconstructions is based on
socalled "missing row" models as illustrated in Fig. 14. The (1x2) structure is created
by taking out every second row of surface atoms of a regular (1x1)fcc(110) surface.
A (1x3) structure is formed by taking out 3 rows. These reconstructions obviously
involve again a massiv mass-transport. It is generally believed that step defects,
which are always present at surfaces (see section 5.2.1.) act as sinks for the removed
atoms and as nucleation centers for the growth of the (likewise reconstructed) up-
step layer. The existence of the deep grooves on the (1x2) and (1x3) reconstructed
surfaces is consistent with TEAS /21/, ISS /22/ and STM /23,24/ data. The most
complete structure model for the (1x2)Au(110) surface is displayed in Fig. 15 /25/;
all relevant interatomic distances are listed in the table below. The model involves
the first four atomic layers.

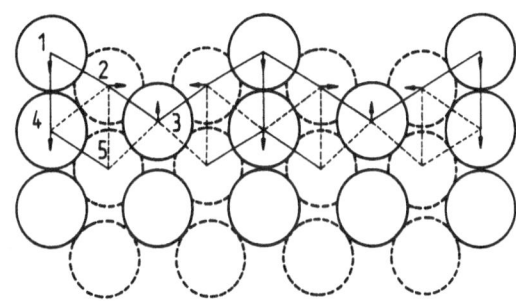

	Interatomic distance [Å]	Deviation from bulk [%]
r_{12}	2.74 Å	-2.8 %
r_{23}	2.79 Å	-5.0 %
r_{24}	3.01 Å	-3.3 %
r_{25}	2.94 Å	+4.4 %
r_{35}	2.96 Å	+2.0 %
r_{45}	2.84 Å	+2.7 %
r_{Bulk}	2.884 Å	-1.5 %

Fig.15 Cross section through the structure model of the (1x2) reconstructed Au(110) surface as
derived from LEED I/V-measurements. The table gives the interatomic distances between the
respective atoms in the model. Full lines represent interatomic distances which are contracted, while
dashed lines represent interatomic distances which are expanded compared to the interatomic distance
in the bulk. The small arrows indicate the displacement of the respective atoms from their original bulk
positions (Ref.25).

A close inspection of the different interatomic distances and of their deviations from the bulk value (in percent), indicated that those drawn in full in Fig. 15 are reduced, while those drawn in dashed line are slightly expanded. In particular, all bonds within the surface are contracted as well as those from the highest atoms into the bulk. This results in a more rounded surface topography and, in the end, in a lowered surface area.

In a first attempt in order to explain the driving force for these fcc(110) reconstructions it was believed that again the formation of the (111)-microfacets, which form the slopes of the deep grooves, is the reason for the reconstruction. This argument, however, can be easily disproved by Fig. 14. The numbers assigned in Fig. 14 to the surface atoms for all three surface modifications, namely the (1x1), (1x2) and (1x3) structure, correspond to the number of **remaining** next-nearest neighbors of the respective surface atom. For the surface regions between the two vertical dashed lines, which all three comprise 12 surface atoms, the **average number of next-nearest neighbors is the same**, as evidenced by the calculations on the right-hand side. Consequently, also the **average** number of broken bonds is the same, which does not provide a driving force for the (1x1) → (1x2) or (1x1) → (1x3) reconstruction.

A further indication for the fact that the simple formation of close-packed (111)-like surface layers or facets can **not** be the only explanation for the occurence of surface reconstructions results also from the observation, that the Au(111) surface itself is again reconstructed /26/. The outermost layer of atoms is hexagonally close-packed, but rotated slightly with respect to the second (hexagonal) layer. The reason for this rotation must be of a more subtle nature than a simple close-packing of the surface-atoms.

Obviously, the driving force for surface reconstruction does not only depend on the **number** of broken bonds but also on the **nature** of the surface bonds involved, that is, in other words, on the detailed electronic band structure at the surface. In particular, surface states may play an important role. Socalled "surface charge density waves" have been invoked /27/, but a complete theory of surface reconstructions is by far not available yet.

4.1.2.2. Reversible, displacive surface reconstructions. All the surface reconstructions described so far are the result of an **irreversible** surface transformation from the metastable unreconstructed (1x1) to the stable reconstructed surface modification. A brief discussion of the mechanism of this structure transition will be given in section 4.1.2.3. Quite a different kind of surface reconstruction is encountered with e.g. the W(100) or Mo(100) surface. These surfaces undergo a **reversible** surface structure transition below room temperature. For instance, Fig. 16 shows the LEED pattern of the W(100) surface at ~ 370 K and at ~ 150 K /28/. At 370 K the pattern corresponds to a normal (1x1) bcc(100) structure. At 150 K the extra spots suggest a c(2x2) surface reconstruction; hence, reconstruction occurs upon **cooling** in contrast to the fcc(100) surfaces discussed above. Warming the surface again above room temperature restores the (1x1) modification. The intensity profile of one of the

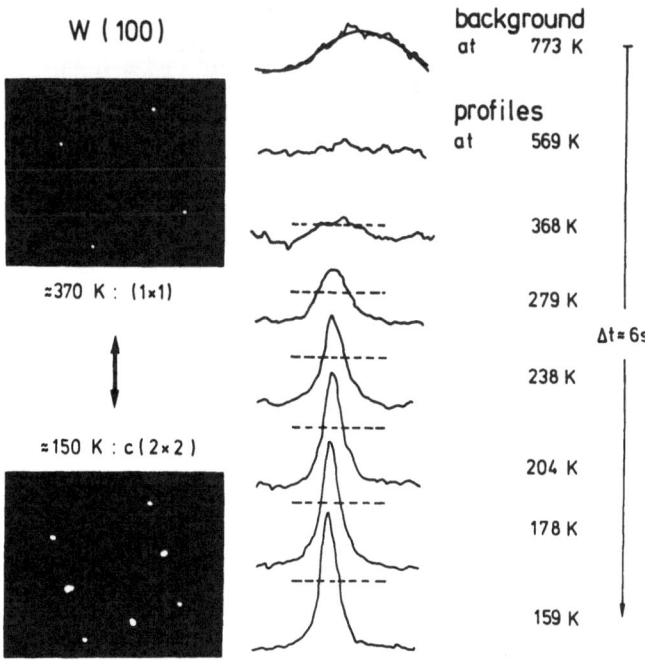

Fig.16 LEED pattern of the (1x1)W(100) and c(2x2)W(100) surface modifications together with the intensity profile of the c(2x2) extra spot as a function of temperature. (From Ref.28)

extra LEED-spots is shown on the right-hand-side of Fig. 16 as a function of temperature. The transformation is completed within a few seconds. This seems to exclude a massiv transport over long distances across the surface. In fact, it is believed that the surface atoms are only slightly displaced from their regular (room temperature) lattice positions. Two possible models are displayed in Fig. 17. In model a) the surface atoms are alternatingly displaced inward and outward of the surface /28/. In model b) the surface atoms are displaced parallel to the surface to form zig-zag rows as shown /29/. In both cases the displacements are only a few tens of an Ångstrom (< 0.3 Å) and, thus, easily reversible. Therefore these reconstructions are also termed "displacive reconstructions".

4.1.2.3. Mechanism of surface reconstruction. The mechanisms of the irreversible reconstructions of the fcc(100) and fcc(110) surfaces must be rather complex because they involve massive mass transport due to surface diffusion. In order to elucidate these mechanisms it is necessary to prepare the metastable (1x1) confi-

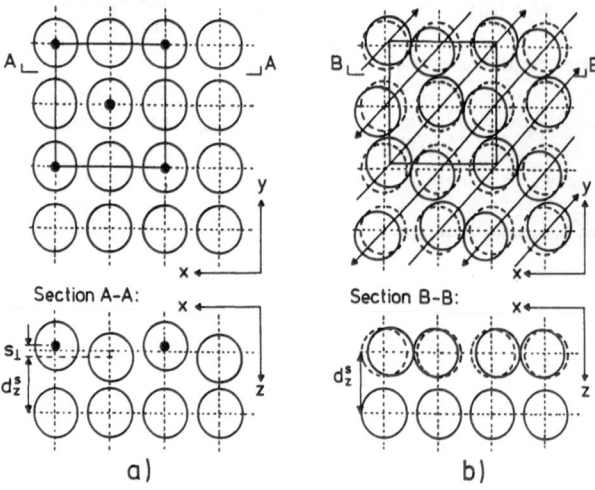

Fig.17 Structure models of the c(2x2)W(100) surface: a) Model of vertical displacement of alternate atom, b) model of lateral displacement (zig-zag-model) of alternate rows of tungsten atoms. (From Ref.29)

gurations in the first place.

All recipes for the preparation of the metastable (1x1) modifications are based on the validity of an energy triagram as illustrated in Fig. 18, which displays the (surface) free energy of the crystal under consideration as a function of both the surface configuration (R = reconstructed, B = bulk like) and the coverage θ of an adsorbate, e.g. CO, O_2, H_2. The experiment proves that the most stable configuration of the clean surface ($\theta = 0$) is characterized by the energy minimum R(0) corresponding to the reconstructed surface. The energy B(0) of the clean metastable (1x1) surface must be higher, because heating (thermal activation) causes reconstruction. In turn, gas adsorption and, if necessary, some heating lifts the reconstruction. Consequently, the energy B(1) of the adsorbate covered unreconstructed surface is lower than the energy R(1) of the adsorbate covered reconstructed configuration. As a function of coverage θ the energy of the reconstructed and unreconstructed (bulk like) surface may vary along the lines R(0)-R(1) and B(1)-B(0), respectively. B(0) and R(0), and likewise R(1) and B(1) are separated by respective activation barriers, because at sufficiently low temperatures the configurations R(1) and B(0) can be isolated experimentally (see e.g. Refs.10 and 15). Based on these energy considerations clean unreconstructed (metastable) surface configurations (characterized by B(0)) can be prepared as follows. The clean surface is covered with an adsorbate (e.g. H_2, CO, O_2, NO) and warmed up without desorbing the

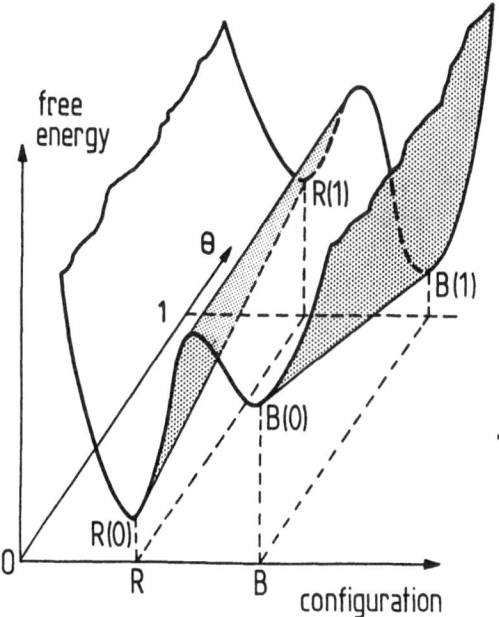

Fig.18 Free energy of a surface as a function of configuration (R = reconstructed, B = bulk-like) and coverage Θ. Note the reversal of the minimum energies at Θ = 0 and Θ = 1 between the reconstructed and the bulk-like surface modification.

adsorbed molecules. This will produce state B(1). In the next step the adsorbate must be removed in a way so that the energy of the system remains near the bottom of the "valley" B(1)-B(0). For instance, CO reacts with O_2 from the gas phase at mild temperatures to form CO_2, which desorbs. Only care must be excercised not to overcome thermally the activation barrier between the "valleys" B(1)-B(0) and R(0)-R(1) at any point along B(1)-B(0). Following this principle, the (1x1)Ir(100) and the (1x1)Pt(100) surfaces shown in Figs. 9 and 10 have been prepared /10,15/. It should be mentioned, however, that there is an ever ongoing discussion as to how clean the obtained surfaces at B(0) really are, or whether there is always a small (possibly undetectable) residue of adsorbate necessary to stabilize the configuration B(0).

In any case, having prepared the metastable "clean" configuration B(0) it is possible to promote reconstruction by thermal activation into minimum R(0) and to investigate this transformation by following any physical property which enables distinction between the unreconstructed (B(0)) and reconstructed (R(0)) surface. One possibility is the LEED pattern itself and the intensity of the extra-spots characteristic for the reconstructed surface, as for instance displayed in Fig. 16. The change in the intensity profile, namely the width of the profile contains information about the microscopic mechanism of the structure transformation /30/.

88

As a consequence of the different atomic structure of the unreconstructed and reconstructed surface also the electronic properties of both surface configurations are expected to be different. This is verified by Fig. 19 which shows the HeII excited UPS valence band spectra of three different Ir surfaces, namely (1x1)Ir(100), (5x1)Ir(100) and Ir(111). The (1x1) and (5x1) spectra differ widely. Conversely, the (5x1)Ir(100) and Ir(111) show much similarity, strongly supporting the notion that the (5x1)Ir(100) surface corresponds to a hexagonally close-packed layer similar to the Ir(111) surface /10/.

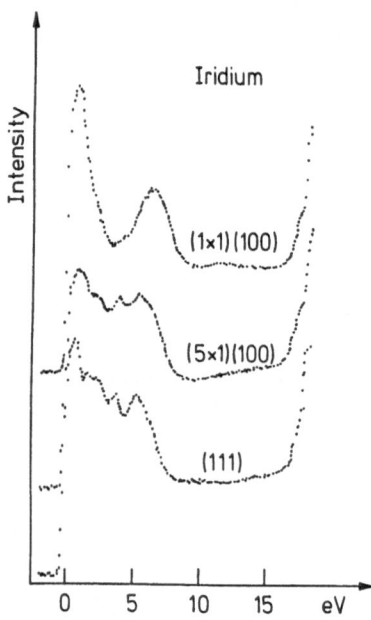

Fig.19 UV-Photoemission valence band spectra (He II excited) of the three indicated surfaces. Note the similarity between the hexagonally reconstructed (5x1)Ir(100) - and the Ir(111)-surface. (From Ref.10a)

Another physical quantity which enables distinction between the unreconstructed and the reconstructed surface modification and which, thus, can be used in order to investigate the mechanism of surface reconstructions is the work function φ. For example, Fig. 20 shows the change in work function when heating the (1x1)Ir(100) surface to ~ 1200 K with a heating rate of 5 K/s (full line); the work function of the final (5x1) surface configuration is ~ 160 meV lower than that of the (1x1) structure /10a/.

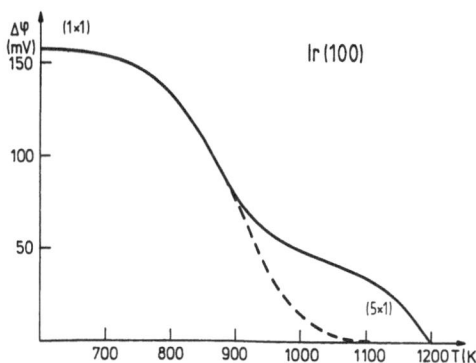

Fig.20 Irreversible work function change between the metastable (1x1)Ir(100) and the stable reconstructed (5x1)Ir(100)-surface observed as a function of annealing temperature. Full line: Experimental data; dashed line: Fit according to a first order transformation mechanism. (After Ref.10a)

As discussed in great detail in section 5.1 work function differences between different homogeneous samples or between unlike patches at inhomogeneous surfaces are also reflected in the electron binding energies of rare gas atoms adsorbed on these surfaces or patches, respectively. As an example Fig. 21 displays HeI-excited UPS spectra from the valence band region of a (1x1)Ir(100) and a (5x1) Ir(100) surface, both surfaces being clean or covered with one monolayer of xenon, respectively. The two sharp extra peaks due to Xe adsorption in both cases correspond to the $Xe(5p_{1/2})$ (at high binding energy) and the $Xe(5p_{3/2})$ final states of the photoionized, adsorbed Xe atoms. It is clearly noted that the electron binding energy (with respect to the Fermi level $E_F = 0$) of e.g. the $Xe(5p_{1/2})$ signal, namely $E_B^F(5p_{1/2})$, is larger on the (5x1) surface than on the (1x1) surface. This is precisely in line with the general observation $\Delta E_B^F(5p_{1/2}) = -\Delta \varphi$ discussed at length in section 5.1, and which can also be used to examine the mechanism of the (1x1)→(5x1) structure transformation. This is demonstrated by Fig. 22. This figure shows expanded $Xe(5p_{3/2,1/2})$ spectra of Xe adsorbed on a (1x1) (bottom spectrum) and a (5x1) (top spectrum) Ir(100) surface as well as on (1x1)Ir(100) surfaces which were shortly heated (flashed) to the indicated temperatures and then quenched to 100 K /10/. In particular the $5p_{1/2}$ signals manifest the appearance and the progressive growth of a $5p_{1/2}$ signal from (5x1) patches, which grow at the expense of the (1x1) surface area. Note, for instance, the clear superposition of (1x1)- and (5x1)-$5p_{1/2}$ signals after the flash to 900 K. This series of spectra is a direct indication of a nucleation and growth mechanism underlying the irreversible (1x1) → (5x1) structure transition. A similar conclusion was drawn from the temperature dependent intensity profiles in Fig. 16 for the reversible reconstruction of the W(100) surface as well as from the

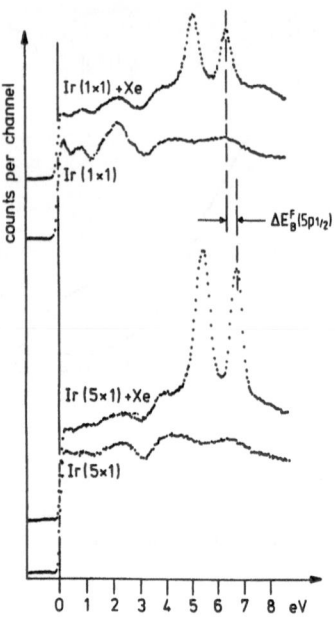

Fig.21 $5p_{3/2,\,1/2}$ photoemission signals from Xe adsorbed on an (5x1)Ir(100)- and (1x1)Ir(100)-surface respectively. $\Delta E_B^{\ F}(5p_{1/2})$ corresponds to the work function difference between both surface modifications (From Ref.10b)

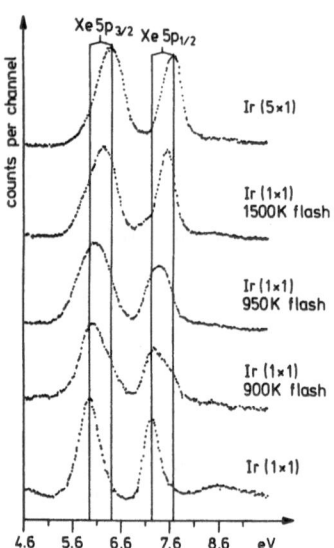

Fig.22 $5p_{3/2,1/2}$ photoemission spectra of xenon adsorbed on Ir(100) surfaces exhibiting (1x1) and (5x1) as well as partially reconstructed surface structures. On the partially reconstructed surfaces the Xe photoemission indicates a superposition of (1x1) and (5x1) patches.

temperature dependence of the work function change between (1x1) and (5x1)Ir(100) shown in Fig. 20. In the latter case the first part of the $\Delta\varphi$(T)-curve (that is below 900 K) is compatible with a first order transformation rate which leads to the dashed line in Fig. 20 and which also corresponds to a mechanism of nucleation and growth of (5x1) patches within the initial (1x1) surface /10a/.

This latter example no longer deals with perfect single crystal surfaces. The partially reconstructed Ir surfaces underlying Fig. 22 are highly imperfect containing (5x1) and (1x1) patches, steps and kinks around the (5x1) islands on the (1x1) substrate due to the shrinkage illustrated in Fig. 11, probably misfitboundaries between coalescing (5x1) islands, etc. This example provides a natural transition to the discussion of imperfect surfaces in the following sections.

5.   Properties of Imperfect Surfaces

As mentioned in the Introduction, in reality surfaces are never perfect, but exhibit a variety of structural and chemical defects as illustrated in Fig. 1. As a consequence the physical and chemical properties of these surfaces, in particular the electron density, vary within the (x,y) surface plane. This was already suggested by the partially reconstructed Ir(100) surfaces and Fig. 22 discussed in the previous section. A complete understanding of surfaces ultimately requires the characterization of the lateral variation of their properties with atomic scale resolution. The necessary techniques are available now. However, the only reasonable way to understand these

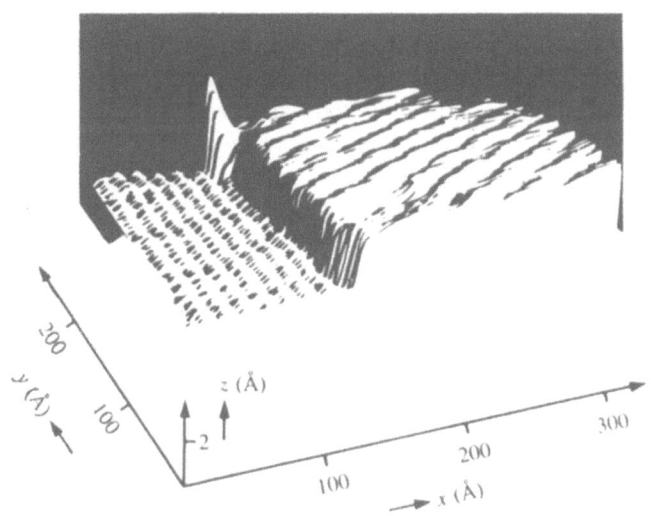

Fig.23   STM image from a monoatomic step on a reconstructed Pt(100) surface (From Ref.18).

complex surfaces appears to be the use of model surfaces which are prepared to contain first predominantly one kind of defect only. In the next step two kinds of defects may be introduced, and so on. In this way largely defect surface structures can be synthesized stepwise. Each kind of defect can be prepared separately and studied in isolation. Or different kinds of defects can be combined, which also reveals possible synergetic effects between them.

Obviously Scanning Tunneling Microscopy (STM) appears best suited to study the atomic scale heterogeneity of realistic surfaces. As one of the early examples Fig. 23 shows the STM picture of a step edge of monoatomic height on a reconstructed Pt(100) surface /31/. The step edge is clearly observed. Its height was the lowest one observed on the surface and was therefore assigned to be monoatomic. Very interesting is the wavy structure on both sides of the step. The periodicity of these waves is 14 Å and reflects the mismatch between the (1x1) Pt(100) underlayer and the reconstructed hexagonal top layer. The corrugation was estimated to be ~0.4 Å in close agreement with the data from Fig. 13. Interesting is the different direction of the wavy structure on either side of the step, that is on the down-step terrace compared to the up-step terrace. The step acts as a boundary between reconstructed domains of different orientation with respect to the (1x1) underlayer.

In the meantime the capabilities of STM have much expanded and improved as is, for instance, manifested by Fig. 6 and a wealth of interesting and important results has been accumulated by now (see e.g Ref.32). In this chapter another technique, namely Photoemission of Adsorbed Xenon (PAX), will be described which also has the capability to provide some information about atomic-scale surface disorder and, in particular, about the variation of the surface potential, that is the "local work function", with nearly atomic "resolution". Variations of the local surface potential may be correlated with variations of the surface charge density. While the STM is an imaging technique which provides real space images from the atomic structure of surfaces, the PAX-technique is a kind of decoration method which enables the titration of specific surface sites which are characterized firstly by their adsorption energy for the Xe probe atoms and secondly by their local surface potential. The next section contains a detailed description of the PAX-technique. Then various examples will be presented in the following sections which demonstrate the capability of the PAX-technique to add to the understanding of the properties of atomic scale structural and chemical defects on surfaces. The examples will include the characterization of step-defects, of submonolayer films of metal adsorbates, as well as of individual adsorbed ions, namely potassium ions. In section 6 the importance of short-range variations of the local surface potential will be discussed briefly.

5.1 THE PAX-METHOD

Below ~ 80 K the rare gas xenon can be "physisorbed" on any solid surface. The UV (HeI) excited photoemission spectrum of Xe adsorbed on a Ru(001) surface is displayed in Fig. 24. The two sharp extra peaks (above the Ru background, dashed

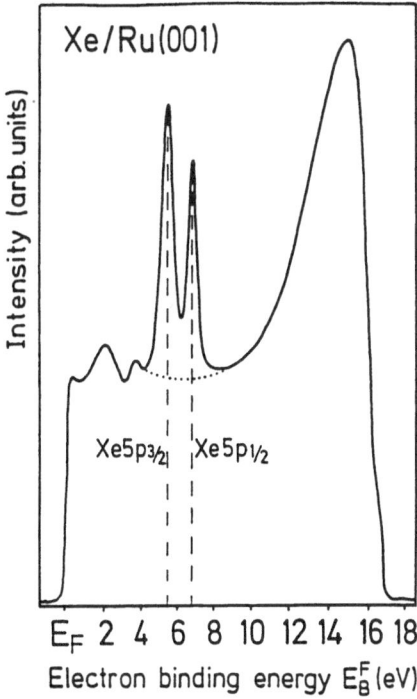

Fig.24 UV photoemission spectrum of a xenon covered Ru(001) surface excited with HeI radiation of 21.22 eV photon energy. The two sharp and intense signals between 5 and 7 eV below E_F arise from the $5p_{3/2}$ and the $5p_{1/2}$ photoemission final states of adsorbed Xe. The dashed line indicates the substrate background emission.

line) between 5 and 7 eV below E_F arise from the $5p_{3/2}$ and $5p_{1/2}$ photoemission final states of adsorbed Xe atoms. For the sake of clarity and ease we will concentrate on the $5p_{1/2}$ signal only, because its physical structure is somewhat simpler than that of the $5p_{3/2}$ signal, which becomes evident from Fig. 25. The $5p_{1/2}$-signal consists of only one Lorentzian line while the $5p_{3/2}$ consists of two Lorentzian lines, all three arising from the $5p_{1/2}$ ($m_j = \pm 1/2$), the $5p_{3/2}$ ($m_j = \pm 1/2$) and the $5p_{3/2}$ ($m_j = \pm 3/2$) final states of photoionized Xe atoms, respectively. Due to the presence of the surface as well as of interactions between the adsorbed Xe atoms and twodimensional band structure formation within the Xe overlayer the two $5p_{3/2}$ final states are no longer degenerate as in the gas phase /33-36/. In the following, the important surface structural information will be retrieved from the $5p_{1/2}$ signal only, namely from its energy position, its intensity as well as its shape. But all the arguments and conclusions drawn hereof hold likewise for the $5p_{3/2}$ signal.

94

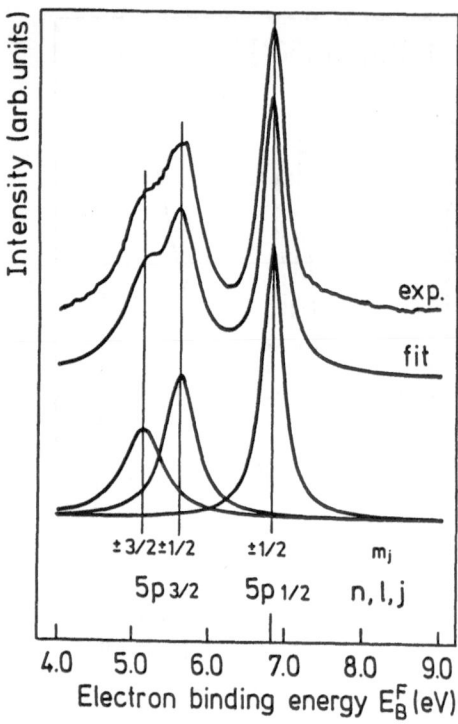

Fig.25 Experimental $5p_{3/2,1/2}$, UPS spectrum (upper trace) of one complete monolayer of Xe adsorbed on a complete monolayer of Au on Ru(001). The fit curve corresponds to the best fit with three Lorentzian functions representing the $5p_{3/2}$ ($m_j = \pm 3/2$), $5p_{3/2}$ ($m_j = \pm 1/2$) and $5p_{1/2}$ ($m_j = \pm 1/2$) final states of photoionized adsorbed Xe. n, l, j, m_j = quantum numbers. The sharp $5p_{1/2}$ peak is best used for the distinction of different coexisting Xe adsorption states on a heterogeneous surface.

In a number of publications /e.g. 10b,37-39/ it has been shown, that the ionization potential of adsorbed Xe atoms (with respect to the vacuum level V) is rather independent of the substrat:

$$E_B^V(5p_{1/2}) = E_B^F(5p_{1/2}) + \varphi \approx constant . \qquad (6)$$

This correlation could, in fact, be verified by now for more than 25 single crystal substrates of quite different nature, including transition metals (Pd, Pt, Ir, Ni etc.),

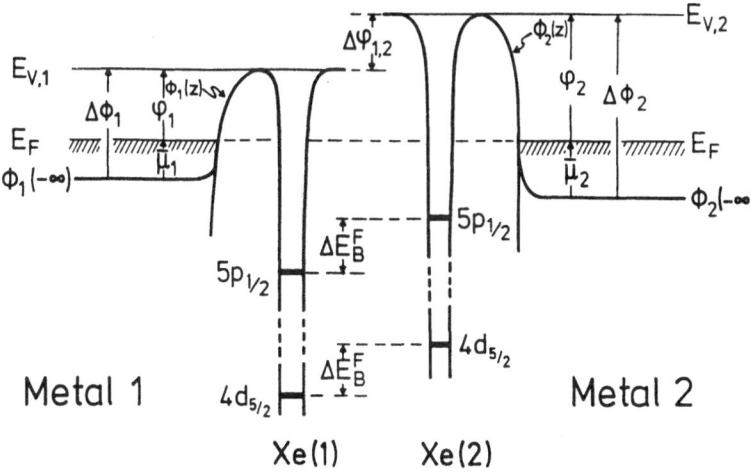

Fig.26 Potential energy diagram for Xe atoms adsorbed on two homogeneous and infinitely extended metal surfaces with different work functions φ_1 and φ_2. $\bar{\mu}$ = chemical potential, $\Delta\phi$ = surface dipole barrier. ϕ (- ∞) is the electrostatic potential inside the metal, $E_V \equiv \phi$ (+ ∞) is the vacuum potential outside of either surface. Note that the Xe potential well floats with the respective surface potential.

noble metals (Cu, Ag, Au), alkali metals (K, Cs), semiconductor surfaces (Si, ZnO) as well as oxides (TiO_2, ZnO). There are two physical reasons for this surprising invariance of $E_B^V(5p_{1/2})$. Firstly, the Xe-substrate interaction is very weak in all cases; typical Xe physisorption energies are smaller than ~ 8 kcal/mole. As a consequence initial state bonding shifts, and in particular their changes between different substrates, are negligibly small (< 0.2 eV). Secondly, a Xe atom is rather big (diameter 4,5 Å). Therefore the center of an adsorbed Xe atom is located mainly outside of the steeply varying electrostatic surface potential ϕ (Fig. 26), which arises from the surface dipole layer. As a consequence the potential well of an adsorbed Xe atom is "pinned" to the vacuum potential outside the surface dipole barrier and "floats" up and down when the work function φ of the substrate is changed, for whatever reason, e.g. when changing the crystallographic orientation of the substrate surface, when changing the substrate material or when preadsorbing some other adsorbate. This "floating" model is illustrated in Fig. 26 and has received strong theoretical /40/ as well as experimental support. For instance, the same invariance as expressed in equ. (6) could be verified for the Xe(4d) and Xe(3d) core levels /41/.

96

An immediate consequence of equ. (6) is:

$$\Delta \ E_B^F(5p_{1/2}) = -\Delta \varphi \qquad (7)$$

as becomes clear from Fig. 26. The important point now is, that this equation holds also for heterogeneous surfaces: Xe atoms being adsorbed on two "patches" of the

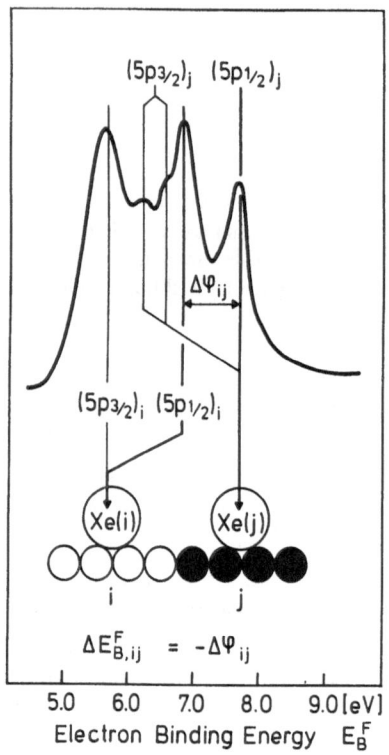

Fig.27 Superposition of $5p_{3/2, 1/2}$ spectra from two coexisting Xe adsorption states on two kinds of surface patches i and j on a heterogeneous surface. The $(5p_{3/2})_i$ signal is only broadened, while the $(5p_{3/2})_j$ peak is split (see e.g. Fig. 24). The shift $\Delta E_B^F (5p_{1/2})_{i,j}$ is a measure of the "local work function" difference $\Delta \varphi_{i,j}$ between both patches.

same surface, which have different **local** work functions, differ in their $E_B^F(5p_{1/2})$ binding energy values (with respect to the Fermi level E_F) accordingly, and two Xe(5p) spectra appear simultaneously and shifted by

$$\Delta E_B^F(5p_{1/2}) = -\Delta\varphi_{local} \qquad (8)$$

with respect to each other. This is illustrated in Fig. 27. Equation (8) is the basis of PAX. Considering further, that the adsorption energy E_{ad} of Xe also depends on the chemical nature and coordination of the specific adsorption site, different surface sites can be populated successively in the sequence of decreasing E_{ad} and can thus be characterized by PAX separately on an atomic scale /37-39,42/. The role of the adsorbed Xe atoms is nothing more than to deposit a "test electron" (bound e.g. in the 5p level) at a particular surface site (controlled by E_{ad}) which then via photo-emission provides information about the surface potential at this site. Beyond the distinction of different kinds of surface sites by their $5p_{1/2}$ electron binding energy (qualitative analysis) evaluation of the corresponding partial intensities yields the surface concentration of each kind of site (quantitative analysis). The following sections present selected examples.

5.2   CHARACTERIZATION OF SURFACE IMPERFECTIONS WITH PAX: CASE STUDIES

5.2.1. Surface Steps. Steps and kinks are ubiquitous surface structure defects as has become clear from Scanning Tunneling Microscopy. The influence and the proper-ties of these step defects at surfaces are studied best with well prepared vicinal surfaces as model systems /43/, because they enable a control of the average step density and the crystallographic step direction. Vicinal surfaces are prepared by cutting a single crystal under a small angle ($\alpha < 10°$) off a low index crystal plane. After polishing and cleaning in ultrahigh vacuum such an equilibrated vicinal surface assumes a rather regular step structure which can be characterized very accurately by STM /e.g. Ref.31/ and LEED /43/ with respect to the step height (which is mostly monoatomic), the crystallographic direction of the step ledges as well as the mean width W of the terraces between the steps. The terraces between two adjacent steps have the same structure as the parent low-index plane ($\alpha = 0°$), even if this surface tends to reconstruct as evidenced by Fig. 23. Fig. 28 shows schematically an fcc(100) surface with three monoatomic steps of different orientations [hkl]. Steps with [110] direction are formed from close-packed rows of atoms resulting in smooth ledges. Steps with directions γ off the [110] direction include kinks. In the [100] direction ($\gamma = 45°$) the step should be fully kinked.

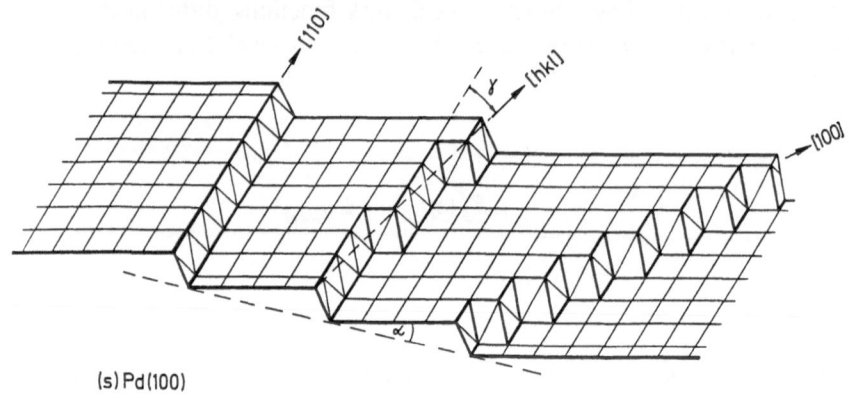

(s) Pd(100)

Fig.28 Schematic representation of stepped (s)Pd(100) vicinal surfaces with monoatomic steps of different orientation (hkl).
α = vicinal angle, γ = step direction off the (110)-direction.

Step wedges (including kinks) provide surface sites with high coordination to the substrate, and therefore physisorbed xenon atoms will preferably occupy (down-) step and kink sites before the terrace sites are populated /e.g.44/. This assumption is particularly justified in the case of physisorbed gases which are held at the surface by isotropic van der Waals forces, but has also been proved by direct STM-observations at low temperature. Eigler and Schweizer /45/ "decorated" a Pt(111) surface with a small coverage of xenon atoms at 4K and imaged their distribution be means of an STM. Most of the Xe atoms were found along step edges (on the lower terrace). A few Xe atoms were found on the terraces. In order to show why the latter Xe atoms were held on the terraces, Eigler and Schweizer did the most "obvious" experiment: They kicked a Xe atom with the STM tip away and found - no wonder - a single vacancy site underneath. The Xe atom itself jumped to another high-coordination vacancy site the position of which was known from the previous image.

$Xe(5p_{3/2, 1/2})$ UPS spectra from a stepped Pd(100) surface are displayed in Fig. 29. It is clearly seen, that at low Xe coverages obtained after Xe exposures < 4L Xe (spectra a-c) the $5p_{1/2}$ peak is located at somewhat higher electron binding energy than a second $5p_{1/2}$ peak which first becomes visible in spectrum d (4L) and which finally overgrows the low coverage signal. Spectrum g corresponds to a complete monolayer of Xe on this surface; its intensity serves as reference for the assignment of the lower coverages. A careful decomposition of the $5p_{1/2}$ intensity of the spectra from Fig. 29 using three Lorentzian lines as shown in Fig. 25 per Xe adsorption state yields two $5p_{1/2}$ lines which are separated by ~ 350 meV. At Xe monolayer saturation the intensities of both $5p_{1/2}$ peaks reflect the relative abundances of step-

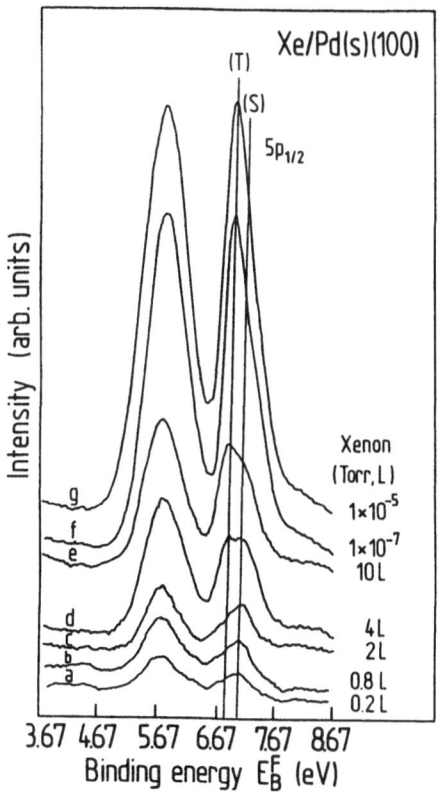

Fig.29 Xe ($5p_{3/2,1/2}$) spectra for Xe adsorption on a stepped Pd(100) vicinal surface. Note the sequential population of the step (S) and terrace sites (T) with xenon. The highest spectrum corresponds to a complete Xe monolayer on this surface.

and terrace-sites on this surface which are known from a LEED-analysis. Of course, the initial selective population of the step sites clearly determines the assignment of the two $5p_{1/2}$ photoemission peaks seen in Fig. 29, in that the **higher** binding energy peak (at low coverage) corresponds to Xe atoms at step sites [Xe(S)] and the lower binding energy peak to Xe atoms on terrace sites [Xe(T)]. Besides the energy shift of $E_B^F \approx 350$ meV between them, the absolute $E_B^F(5p_{1/2},S)$ value is very close to the value found on a perfect Pd(110) surface, namely 7,03 eV /46/. In turn, the $E_B^F(5p_{1/2},T)$ value is near the value found on a perfect Pd(100) surfaces ($\alpha = 0°$ in Fig. 28). Very similar results have been obtained with other Pd(100) vicinal surfaces of different step density and step orientation /39,47/, and in all cases the experimentally determined intensity ratios $(Xe(T):Xe(S))_{exp}$ agreed within $\sim$ 15 % with the $(Xe(T):Xe(S))_{model}$ value estimated on the basis of an ideal terrace-ledge-kink model of the respective vicinal surface as derived from LEED and populated with hard-sphere Xe atoms.

Using equ. (8) the local surface potential at the step sites is $\Delta\varphi_{S,T} = \Delta E_B^F(S,T) \approx$ 350 meV lower than at the terrace sites. This is a consequence of the so-called Smoluchowski smoothing effect /9/. The electronic charge near the surface does not follow the abrupt step geometry but flows from the upper step edge into the lower step wedge (Fig. 30). This creates a local extra dipole which is antiparallel to the normal surface dipole layer. Indeed, systematic studies with regularly stepped surfaces /47-49/ revealed a linear decrease of the (macroscopic) work function of

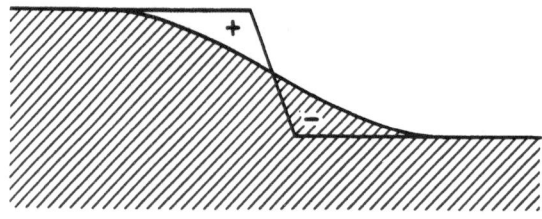

Fig.30 Schematic representation of the Smoluchowski electron smoothing effect near a surface step. Charge flows from the upper step edge into the lower step wedge, thereby creating a localized extra dipole which counteracts the normal surface dipole.

these surfaces as a function of step density. This linearity originally led to the **indirect** conclusion, that the step induced charge redistribution and, hence, the associated extra dipoles which counteract the normal surface dipole is really confined to the immediate vicinity of the steps. Each step adds an incremental dipole which leads to a linear decrease of the macroscopic work function with increasing step density. The present PAX results provide **direct** evidence that the surface potential is, indeed, **locally** lowered by ~ 350 meV at a step site compared to a terrace site. Moreover, from the fact, that the $(Xe(T) : Xe(S))_{exp}$ intensity ratio determined at Xe monolayer saturation always agrees with the expectation based on the model derived numbers of available step and terrace sites, it follows conclusively that only those Xe atoms "feel" the lowered surface potential which are in **immediate** contact with the step; already the next nearest row of Xe atoms away from the step does not contribute to the Xe(S) signal. Consequently, PAX proves directly that the work function decrease is strictly **localized** along the steps. This, in turn, supports the notion, that the lateral "resolution" of the PAX technique is of the order of one Xe atom diameter, namely ~ 5 Å.

These selected results demonstrate the capability of the PAX-method to contribute to a characterization of atomic surface steps. Of course, both the number of step sites and the local surface potential at a step position are basic ingredients for an understanding of e.g. the catalytic activity of structural surface defects. In fact, the local surface potential decrease at step sites may be enormous as has been shown with PAX measurements on stepped surfaces of other metals. Table 3 summarizes PAX derived local surface potential decreases from stepped (s) and defected (d) metal surfaces; they reach the value of ~ 1 eV at steps on a Pt(111) surface. A consequence of this enormous surface potential difference, which decays over a distance of a few Ångstroms only, and of the associated local fields is addressed in section 6 briefly.

Surface	$\Delta \varphi_{loc}$[mev]
(s)Pt(111)	980
(s)Ru(001)	700
(d)Ru(001)	~ 500
(s)Pd(111)	500
(s)Pd(100)	300
(d)Pd(110)	~ 0
(s)Rh(110)	400
(d)Ag(111)	150
(d)Cu(111)	~ 150

Table 3 Local surface potential differences between step (s) and (sputter-induced) defect (d) sites and perfect sites on the various metal surfaces.

5.2.2. Metal Adsorbates. Epitaxial metal monolayers have attracted increasing attention recently because of their potential technological importance. For instance, epitaxial monolayers of a magnetic material on a non-magnetic substrate like Fe/Cu(100) /50/, Fe/Au(100)/51/ may exhibit new magnetic behavior due to an altered geometrical structure and their electronic interaction with the substrate.

The nucleation and growth (condensation) of thin metal films is governed by atomistic processes. Individual atoms are accommodated on the substrate. They may, thermally activated, diffuse across the surface, recombine with other atoms forming stable homogeneous nuclei of a few atoms, or may be incorporated into the atomic edges of already existing immobile islands of larger size. Eventually the growing islands may coalesce giving rise to atomic mismatch and defect sites along the island (grain) boundaries. The case of heterogeneous nucleation per se is controlled by atomic scale defects like steps, kinks and point defects in the substrate surface. Any such structural (and chemical) surface disorder is accompanied by a corresponding variation of the surface electronic structure and energy, and results in a distribution of surface sites of different adsorption and reaction behavior. Hence, the ultimate goal of understanding the growth and the properties of thin films again calls for techniques which allow the characterization of the initial stages of nucleation and of surface morphology on an atomic scale.

The PAX-technique can also provide quantitative information about the concentration and the lateral distribution of heteroatoms on a metal surface. In part the PAX results complement STM reliefs in a unique way, in that 1.) the PAX intensities are a quick and direct measure ("titration") of the density of specific surface sites, and 2.) the $E_B^F(5p_{1/2})$ values of the adsorbed Xe probe atoms yield "local work functions" (as described in section 5.1), which do not seem to be obtainable by STM.

As an example, in this section the evaluation of PAX spectra from Ru surfaces which were covered with submonolayer amounts of Cu will be described /52/. The Cu/Ru system is known as a **bimetallic** catalyst, in which the Cu is present only **on** the surface of the Ru, because both metals are completely immiscible in the bulk. Earlier model studies with submonolayer deposits of Cu on a Ru(001) single crystal surface pointed to the formation of Cu islands of monoatomic thickness, because Cu-TDS spectra from the Ru(001) substrate indicated a quasi-zeroth order desorption behavior as well as stronger Cu-Ru bonds than Cu-Cu bonds. Here we discuss PAX spectra from submonolayer deposits of Cu on a perfectly flat Ru(001) single crystal surface as well as a Ru(001) surface which prior to Cu deposition was slightly sputter-roughened with Ar^+ ion bombardment. The results provide a rather direct insight into the completely different growth mode of the Cu layers on both substrates.

Fig. 31 shows a series of $Xe(5p_{3/2, 1/2})$ spectra from the perfect Ru(001), which prior to Xe adsorption, was covered with ~ 0.4 monolayers of Cu by vapor deposition and was well annealed at 520 K. Up to an exposure of 3L Xe the spectra exhibit two $5p_{1/2}$ states (A and C) at $E_B^F = 6.7$ eV and $E_B^F \approx 7.3$ eV, respectively. This becomes particularly evident from Fig. 32 which displays difference curves between consecutive spectra from Fig. 31. These difference curves accentuate the intensity added by each new Xe dose and are therefore termed "incremental spectra". The dominant peak A at 6.7 eV which appears first is close to the position as on clean Ru, and is therefore assigned to Xe atoms on bare Ru patches [Xe(Ru)]. Above 3L Xe exposure a new $5p_{1/2}$ peak B emerges at $E_B^F \approx 7.4$ eV (Fig. 31). Its $5p_{3/2}$ counterpart grows between the $5p_{3/2}$ and $5p_{1/2}$ signals of the Xe(Ru) state. This is again seen best in the incremental spectra of Fig. 32. Signal B is very close to the (high-coverage) position on Cu(111) and is therefore assigned to Xe atoms **on** the deposited Cu [Xe(Cu)]. The fact, that these Cu sites are populated with Xe after the Ru sites is in agreement with the lower Xe adsorption energy on Cu compared to Ru. The $Xe(Cu)5p_{3/2}$ signal (between the Xe(Ru) peaks) is clearly split into two peaks (see Figs. 31 and 32). As stated in section 5.1 it is generally accepted that this splitting is mainly a consequence of the formation of a two-dimensional (2D)

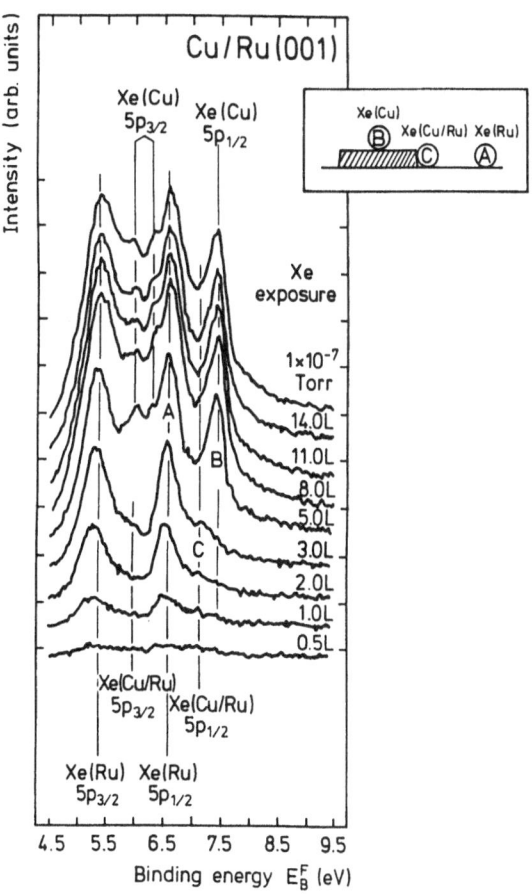

Fig.31 $Xe(5p_{3/2,1/2})$ spectra from a Ru(001) surface covered with ~0.4 monolayers of Cu in the form of well equilibrated 2D Cu islands. Three different Xe(5p) spectra (A,B,C) develop with increasing Xe coverage, which can be associated with the three different kinds of adsorption sites on this surface, namely A on free Ru sites, B on top of the Cu islands and C at the Cu island boundaries, as illustrated in the inset. The highest spectrum corresponds to a complete Xe monolayer on this bimetallic surface.

electronic band structure within the adsorbed Xe layer, which in the present case is only conseivable if the deposited and well annealed Cu layer forms flat islands, so that the Xe atoms on top can also form a densely packed overlayer of sufficient lateral extension. Hence, the structure of the Xe(Cu) spectrum itself, namely the $5p_{3/2}$ splitting, carries the unambiquous information that submonolayers of Cu on a flat Ru(001) surface form islands, which should be monoatomically thick, because the Cu-Ru interaction is stronger than the Cu-Cu interaction as mentioned above. This together with the fact, that Cu and Ru are immiscible, and that therefore the

Cu islands are **on** the Ru surface, leads to the assignment, that the peak C in Figs. 31 and 32 corresponds to Xe atoms at the Cu/Ru step sites along the Cu island boundaries as illustrated in the inset of Fig. 31. Since here the Xe atoms are in contact with both Ru and Cu atoms these sites are henceforth called "mixed" sites [Xe(Cu/Ru)]. They are populated rather early because here they have the form of high-coordination step sites (see also section 5.2.1.). Similar results have also been obtained with Ag and Au submonolayers on Ru(001). The partial intensities of all three Xe states, Xe(Ru), Xe(Cu) and Xe(Cu/Ru), are a quantitative measure of the relative abundance of these three kinds of surface sites, and the local surface potential at the mixed Cu/Ru boundary sites is intermediate between those **on** Ru and **on** Cu patches, respectively. The latter two are in agreement with the macroscopic work functions of a clean Ru(001) surface and a Ru(001) surface covered with one complete monolayer of Cu. Hence, locally the Cu islands behave like a complete Cu monolayer.

The assignment of peak C in Figs. 31 and 32 being due to Xe at the mixed Cu/Ru boundary sites is strongly supported by PAX spectra from a Ru(001) surface which prior to Cu deposition was slightly sputtered with Ar^+ ions of 2 keV. The corresponding integral and incremental PAX spectra are shown in Figs. 33 and 32. Although the overlayer of 0.3 ML Cu on the defected Ru(001) surface was again annealed for 10 min at 520 K (as on the flat Ru surface described above) prior to Xe adsorption, **no** $Xe(Cu)(5p_{1/2})$ signal can be resolved on this surface. Instead, the

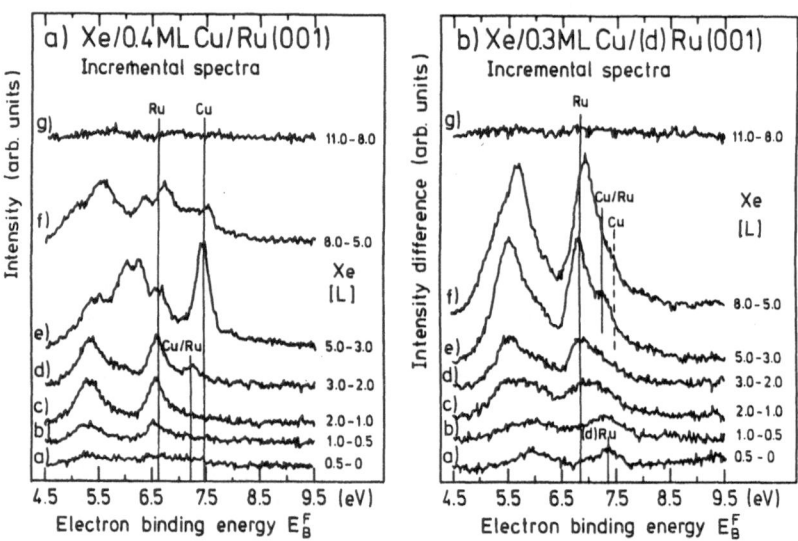

Fig.32 Difference curves between subsequent spectra from Fig. 31 and from Fig. 33. These socalled "incremental spectra" accentuate the sequential population of surface sites in the order of decreasing Xe adsorption energy E_{ad}.

Xe(Cu/Ru) intensity is much more pronounced indicating a higher concentration of mixed Cu/Ru sites as a consequence of a more uniform dispersion of the Cu particles across the whole surface. Obviously, in this case the formation of 2D Cu islands is prevented due to the high concentration of sputter induced defect sites which act as heterogeneous nucleation centers. Note also that at no stage the $5p_{3/2}$ signal from this surface shows any obvious splitting suggesting Cu clusters on the surface too small for a 2D electronic band structure to form within the Xe on top. This Cu/Ru example shows that PAX cannot only distinguish qualitatively between the three possible surface sites for Xe adsorption, namely Ru, Cu and Cu/Ru sites, but also provides a quantitative measure of their relative surface concentration, of their local surface potential by virtue of their $5p_{1/2}$ electron binding energies using equ.(8), as well as of the distribution of the Cu deposit, namely 2D island formation on the flat Ru(001) substrate versus nearly atomic dispersion on a sputtered Ru substrate. In particular, these latter results have served as reference data for PAX measurements with Cu/Ru-powder samples /52/. Even with such highly dispersed samples PAX-measurements are possible - in contrast to meaningful STM-studies.

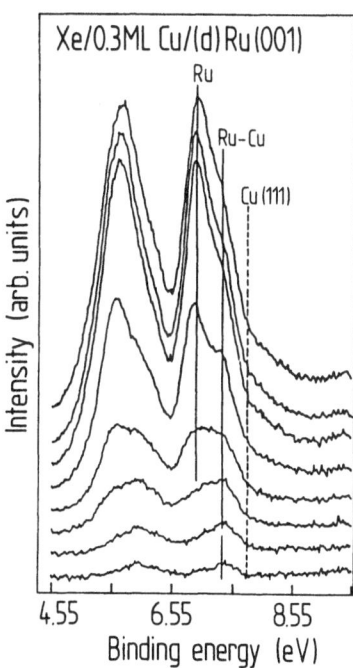

Fig.33 Xe($5p_{3/2,1/2}$) spectra of Xe adsorbed on a bimetallic Cu/Ru surface. Before Cu deposition, the Ru surface was subjected to ion bombardment. Note the lack of any resolved Xe(Cu) peak compared to Fig. 31, suggesting a highly dispersed distribution of the Cu due to the Ru surface defects.

5.2.3. Adsorbed Alkali Ions. It is well established by many different experimental studies, that alkali metals adsorbed on both metal and semiconductor surfaces may **reduce** the work function by up to more than 4 eV /e.g.53/ depending on the alkali metal/substrate system and the adsorbate coverage. As an example Fig. 34 shows the deposition of potassium on Ru(001). Typically at low coverage ($\Theta_{ALK} \leq 0.1$ ML) the work function decreases linearly with increasing alkali metal coverage. Around $\Theta_{Alk} \sim 0.5$ ML the work function change curve passes through a shallow minimum, and at monolayer completion ($\Theta_{ALK} = 1$ ML) the work function φ approaches already the value of the respective bulk alkali metal. This general behavior, which is not restricted to metal substrates, but has been observed for semiconductor surfaces likewise (e.g. Refs.54,55), is explained, since Langmuir /56-58/, in terms of a strong, coverage dependent charge transfer from the alkali atoms to the substrate surface, because of the low ionization potential of alkali metals. As a result of this charge

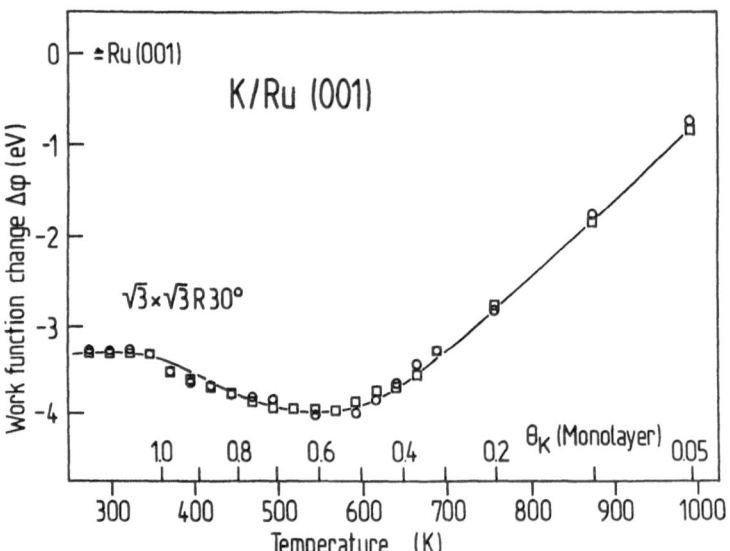

Fig.34 Potassium induced work function change $\Delta\varphi$ on a Ru(001) surface as a function of potassium coverage Θ_K. In contrast to most other presentations of this kind the Ru(001) surface was first covered with a thick, bulk like potassium film of work function $\varphi = 2.26$ eV, which was then briefly heated to successively higher temperature in order to desorb the potassium stepwise. The remaining Θ_K are measured by Auger electron spectroscopy and calibrated against a saturated monolayer which manifests itself by a sharp $\sqrt{3} \times \sqrt{3}$ R30° LEED structure.

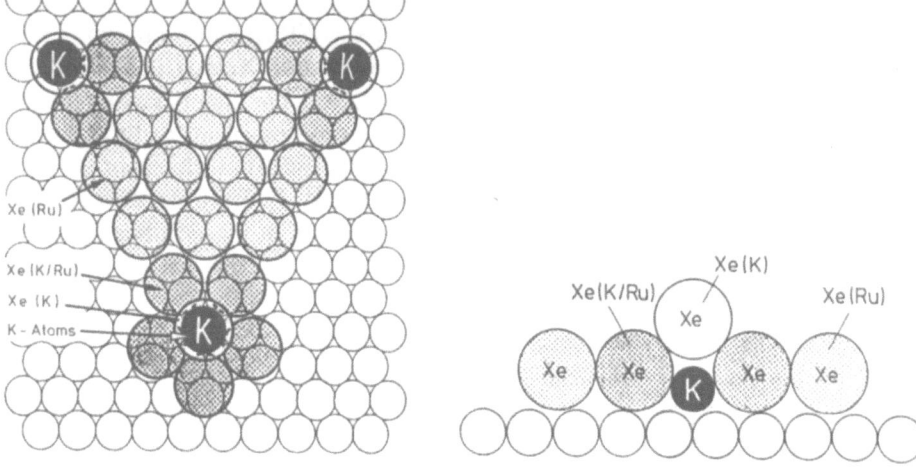

Fig.35 Structure model of a Xe covered 0.05 ML K/Ru(001) surface. The small white circles represent the Ru(001) plane. Note that at this K-coverage 4 Xe atoms (dotted atoms) may be accommodated between two nearest K-atoms (black atoms); 5 Xe atoms fit around on K-atom and 1 Xe atom (large white atoms) on top of one K-atom. These three Xe states are denoted Xe(Ru), Xe(K/Ru) and Xe(K), respectively.

donation each adsorbed alkali atom (partial ion)leads to the creation of a dipole, the moment μ of which is antiparallel to that of the substrate surface dipole layer, thereby causing the observed work function **decrease.** At low alkali metal coverage the partially positive alkali ions repel each other electrostatically and it is safe to assume that individual ions are uniformly distributed across the surface. At these low coverages each ion donates the same amount of charge to the substrate as must be concluded from the **linear** initial decrease of the $\Delta\varphi(\Theta)$-curves. An important question concerns the spatial distribution of the negative extra charge which is donated to the substrate. Since this extra surface charge is expected to influence the surface potential of the substrate, locally resolved work function change measurements by means of PAX again appear to be an appropriate way to probe the range of the charge redistribution around individual alkali metal atoms adsorbed on a metal or semiconductor surface.

Fig. 35 shows a model of the Ru(001) basal plane (white atoms) covered with 0.05 ML of potassium (black atoms) and with an overlayer of Xenon at 50 K (dotted atoms on Ru-atoms and big white atoms on K-atoms). This model is to illustrate two points which are important for the discussion of the experimental data shown below. Firstly, Xe atoms tend to form a hexagonally close-packed overlayer of monoatomic thickness nearly independent from the substrate structure. Secondly, five Xe atoms fit around one K-atom on the Ru(001) substrate. All three Xe-states, namely Xe on free Ru(001) sites [Xe(Ru)], Xe next to a K-atom [Xe(K/Ru)] and Xe ontop of a K-atom [Xe(K)], can be distinguished in the corresponding Xe(5p) photoemission spectrum excited with HeI-radiation (see Fig. 36a-d). Both dominant emission peaks, Xe($5p_{3/2}$) and Xe($5p_{1/2}$), actually consist of a superposition of up to three peaks. This is clearly seen with the $5p_{1/2}$ signal of the highest spectrum which corresponds to a complete Xe monolayer over the whole 0.05 ML K/Ru(001) surface. The assignment of the three corresponding Xe-states as given in Fig. 36a has been dis-

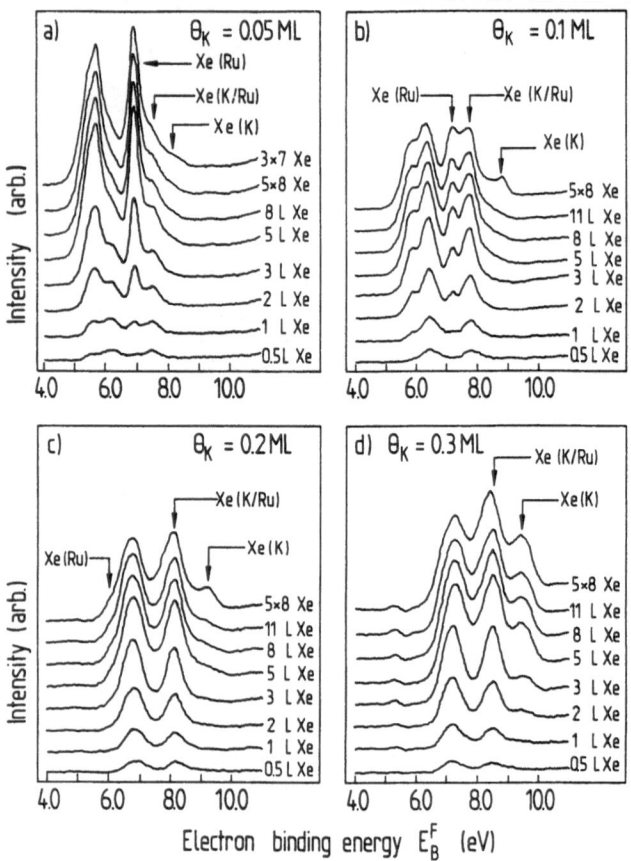

Fig.36 Xe($5p_{3/2,1/2}$) photoemission spectra for different Xe coverages on four Ru(001) surfaces covered with 0.05 ML, 0.1 ML, 0.2 ML and 0.3 ML potassium, respectively. In each panel the highest spectrum corresponds to a saturated Xe monolayer on the respective surface. Note the successive growth of up to three Xe states as the Xe coverage increases as can be seen most clearly from the $5p_{1/2}$ emission which in the four panels occurs at electron binding energies E_B^F greater than a) 6.7 eV, b) 7.0 eV, c) 7.5 eV, and d) 7.9 eV. In panel c) Xe(Ru) points to the $5p_{3/2}$ component of the corresponding emission.

cussed in detail elsewhere and can be supported as follows in analogy to the Cu/Ru(001) system described in the previous section. The Xe(Ru) peak is most intense and has an electron binding energy very close to that found on pure Ru(001). The Xe(K) peak is smallest, it is shifted most from the Xe(Ru) peak position and it appears only near monolayer completion (because the adsorption energy of Xe on potassium is much smaller than on Ru). The Xe(K/Ru)-state populates first (because the sites next to a K-atom on Ru(001) provide the highest coordination for the physisorbed Xe probe atoms) and has an electron binding energy between those of the Xe(Ru)- and the Xe(K)-states. A quantitative decomposition of the Xe monolayer spectrum from Fig. 36a (using sets of three Lorentzians per Xe state as shown in Fig. 25) into the three contributions is shown in Fig. 37a. The three vertical arrows mark the energy E_B^F of the three $Xe(5p_{1/2})$ signals, from which the local work functions φ_{loc} have been calculated using equ. (8) (see Fig. 37a). The percentages denote the relative intensity of each Xe-state. Note that the Xe(K/Ru)-state is five times more intense than the Xe(K)-state in excellent agreement with the expectation based on the structure model in Fig. 35.

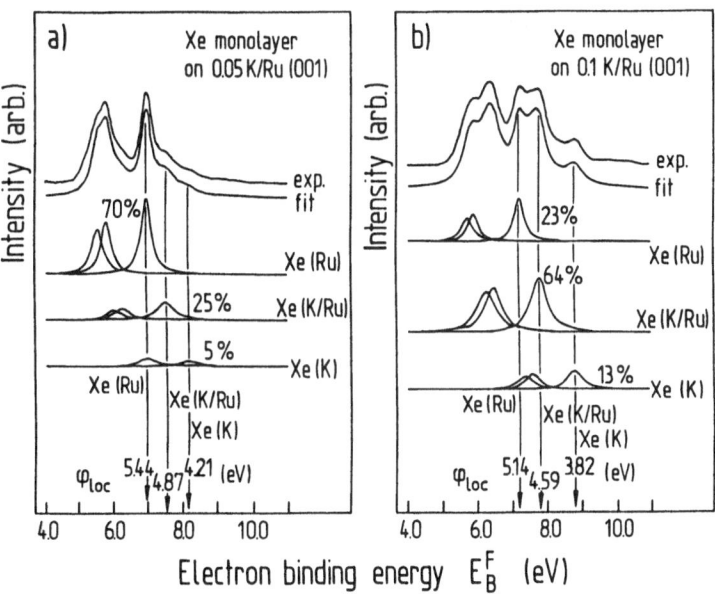

Fig.37 Decomposition of the Xe monolayer PAX spectra from a) the 0.05ML K/Ru(001) and b) the 0.1ML K/Ru(001) surface into the three component spectra arising from Xe-atoms on free Ru(001) sites [Xe(Ru)], next to K-atoms [Xe(K/Ru)] and on top of K-atoms [Xe(K)]. The percentages give their relative intensities; the φ_{loc}-values are calculated using equs.(8) and correspond to the local work function at the respective adsorption site.

A decomposition of the PAX-monolayer spectrum from the 0.1 ML K/Ru(001) surface (Fig. 36b) is shown in Fig. 37b. Again the intensity ratio between the Xe(K/Ru) and the Xe(K) state is 5 as expected, but the electron binding energies of all three Xe states and, hence, the corresponding local work functions are slightly changed with respect to those from the 0.05 ML K/Ru(001) surface. This trend continues for Θ_K > 0.1 ML (Figs. 36c and d), and also the intensity ratio of the Xe(K/Ru) and Xe(K) states is no longer near five. Both deviations are easy to conceive in terms of the average distance between two K-atoms on the surface at different coverages. At Θ_K = 0.05 ML this interatomic distance permits to place 4 Xe atoms between two K-atoms (see Fig. 35), two of which are Xe(K/Ru)-atoms, while the other two are Xe(Ru)-atoms. Obviously these latter two experience a local work function which is hardly changed (80 meV) from that of a K-free Ru(001) surface. At coverages Θ_K > 0.05 ML, the spheres of electronic perturbation around the individual K-atoms begin to overlap thereby causing a shift of the **whole** PAX spectrum including the Xe(Ru) state (Fig. 37b). From this it can be concluded that the range of electronic perturbation is mainly confined to within a radius of r ≈ 8 Å around a K-atom, namely $1/2\ r_K + 3/2\ r_{Xe}$; this is rather short ranged. For coverages Θ_K ≥ 0.2 ML Xe(K/Ru)-atoms begin to belong to two K-atoms simultaneously. As a consequence the intensity ratio I(Xe(K/Ru))/I(Xe(K)) decreases below 5.

The strongly localized work function decrease measured with the 0.05 ML K/Ru(001) surface next to a K-atom (by means of Xe(K/Ru)($5p_{1/2}$)) and ontop of a K-atom (by means of Xe(K)($5p_{1/2}$)) are in very good agreement with theoretical predictions by Lang and coworkers /61/. At the respective locations of the centers of Xe(K/Ru)- and Xe(K)-atoms they predict local surface potential decreases of $\varphi_{loc(K/Ru)}$ = -550 meV and $\varphi_{loc(K)}$ = -1300 meV, respectively. The corresponding experimental values are -570 meV and -1230 meV as calculated from Fig. 37a.

By the way, a rather similar behavior was also observed for K-adsorption on a semiconductor surface, namely Si(100)2x1 /53,59/. But while on the 0.05 ML K/Ru(001) surface the Xe(Ru) state was hardly shifted (~ 80 meV) from the position found on bare Ru(001), on the 0.04 ML K/Si(100)2x1 surface the Xe(Si) state was found to be shifted by ~ 400 meV compared to the clean Si(100)2x1 surface. This seems to be in accord with the different screening lengths typical for metals versus semiconductors.

5.2.4. Comparison between PAX and STM. As demonstrated in this part it is possible to determine local work functions by means of PAX with a lateral "resolution" of 5-10 Å. Whenever available, theoretical predictions are in good agreement with the experimental findings. As a result detailed information can be obtained about the concentration and distribution of structural and chemical defects at metal surfaces as well as about their local surface potential on an atomic scale. The presented examples included the characterization of surface steps; of adsorbed potassium ions on both a metal and a semiconductor surface, and the growth of evaporated metal films. In most of these cases the substrate was a Ru(001) surface, because the adsorbed metals could be completely desorbed again in order to restore the clean substrate. The described examples are case studies, which are to show the capability and the methodical procedure of PAX measurements. But it is obvious, that similar experiments may provide valuable microscopic information also about other structurally and chemically heterogeneous surfaces likewise. It is also believed that the verification of strong local surface potential variations and, hence, the existence of strong localized fields arising hereof is relevant for an understanding of processes at both solid/gas and solid/liquid interfaces (see section 6).

Comparing PAX with Scanning Tunneling Microscopy (STM) it appears that to some extent both techniques provide complementary information. STM reliefs provide a real image of the atomic topography of a surface. At metal surfaces the spatial resolution is ~ 1 Å laterally (x,y) and 0.1 Å vertically (z). Because of this high resolution the STM is uniquely suited to characterize the location and the geometry of individual defects. But also because of its very high resolution the STM should not be utilized to scan large surface areas in order to determine defect densities. This information may be easier obtained by means of PAX. The partial PAX (Xe $5p_{1/2}$) intensities are a direct measure of the relative surface concentrations of specific kinds of surface sites, e.g. structural defects or hetero-atoms. In many respects it is enough to know the number of certain defects rather than their exact location, for example, when evaluating the reactivity and catalytic activity of a surface. This argument may be extended to the characterization of powder samples, where the application of the STM does not seem to be of obvious help. In turn, PAX studies of a bimetallic Cu/Ru powder catalyst have provided interesting information about purity and surface roughness of the particles /52/

6.   Influence of Surface Imperfections

Even though the strong influence of atomic scale surface defects is known and well documented through the macroscopic parameters of many surface processes such as sticking, heterogeneous catalysis, film growth and surface reactions, little is known as yet about the local physical properties of the ubiquitous structural and chemical defects at surfaces. However, these local properties are ultimately responsible for the influence of the imperfections. Only their characterization on the atomic scale will lead to the key insights. For instance, the dominantly very short ranged surface potential variations near steps, adatoms and heteroatoms were not considered until recently.

As was shown in sections 5.2.1 and 5.2.3 by means of PAX there are rather strong surface potential differences over short distances (~5 Å) within the surface (xy-plane), e.g. between defect sites and their immediate regular surrounding. The same is also true between nearest and next-nearest neighbor sites around a heteroatom. For instance, the first ring of Xe atoms around an individual (electropositive) potassium atom on Ru(001) senses a local surface potential which is ~500 meV lower than that felt by the second ring of Xe atoms around the same K atom /53,60/ in rather good agreement with theoretical calculations /61/. The opposite

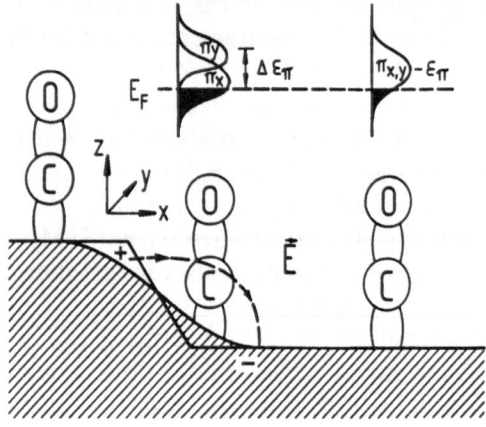

Fig.38 Schematic representation of the surface Stark effect on adsorbed CO molecules due to the strong lateral field near a step defect. Near the step the degeneracy of the antibonding $2\pi^*$-derived molecular orbital is lifted leading to an increased $2\pi^*$-electron occupation and, hence, to a C-O-bond weakening (From Ref.62)

is expected for the change in local surface potential around an electronegative heteroatom, e.g. chlorine or oxygen.

The experimental verification of this rather strong and short-ranged difference of the surface potential near atomic scale structure defects or heteroatoms may, indeed, provide a clue for an understanding of the influence of these surface irregularities on the properties of adsorbed molecules. These strong potential gradients are expected to give rise to a rather strong localized electric field **parallel** to the surface and, hence, to a lateral surface Stark effect acting on adsorbed molecules near these defects. As illustrated in Fig. 38 with CO this lateral surface Stark effect will lead to a shifting and splitting of the $2\pi^*$ derived resonance of adsorbed CO molecules and, as a result, to an increased electronic "back donation" into the $2\pi^*$ derived orbital. This, in turn, will weaken the intramolecular CO bond, because the $2\pi^*$ orbital is of anti-bonding nature, and will modify the reactivity of the CO molecules accordingly. A strong lowering of the intramolecular vibration frequency for CO adsorbed at the steps of a Ni[5(111)x(110)] surface as a precursor state for dissociation was actually observed with HREELS /63/. Full dissociation even of adsorbed O_2 molecules on a stepped Pt surface at 100 K could only be assigned to the step sites but not to the terrace sites /64/.

References

/1/ A conference of this title was held in Japan in 1990

/2/ a) G.Ertl, J.Küppers, "Low Energy Electrons and Surface Chemistry (VCH-Verlag, Weinheim, 1985)

b) G.A.Somorjai, "Chemistry in Two Dimensions Surfaces" (Cornell University Press, Ithaca, 1981)

c) D.P.Woodruff, T.A.Delchar, "Modern Techniques of Surface Science" (Cambridge University Press, Cambridge, 1986)

d) A.Zangwill, "Physics at Surfaces" (Cambridge University Press, Cambridge, 1988)

/3/ a) R.J.Behm, W.Hösler, in: Physics and Chemistry of Solid Surface VI, Eds. R. Vanselow, R.Howe, Springer Series in Surface Sciences 5, Eds. G.Ertl, R. Gomer (Springer, Berlin, 1986), p. 361

b) Y.Kuk, P.J.Silverman, Rev.Sci.Instruments 60 (1989) 165

/4/ a) Th.Fries, C.Becker, M.Böhmer, K.Wandelt, Fesenius Z.Analyt.Chem., (1991) in press

b) C.Becker, Th.Fries, K.Wandelt, U.Kreibig, G.Schmid, J.Vac.Sci.Technol. 89 (1991) 810

/5/ M.A.van Hove, in: The Nature of the Surface Chemical Bond, T.N.Rhodin, G.Ertl (Eds.), North-Holland, Amsterdam, 1979

/6/ K.Heinz, K.Müller "Structural Studies of Surfaces", Springer Tracts in Modern Physics, Vol. 91 (Springer, Berlin, 1982)

/7/ J.Wintterlin, J.Wiechers, H.Brune, T.Gritsch, H.Höfer, R.J.Behm, Phys.Rev.Lett. 62 (1989) 59

/8/ K.Christmann, G.Ertl, O.Schober, Surf.Sci. 40 (1973) 61

/9/ R.Smoluchowski, Phys.Rev. 60 (1941) 661

/10/ a) J.Küppers, H.Michel, Appl.Surf.Sci. 3 (1979) 179

b) J.Küppers, H.Michel, F.Nitschke, K.Wandelt, G.Ertl, Surf.Sci. 89 (1979) 361

/11/ P.Heilmann, K.Heinz, K.Müller, Surf.Sci. 83 (1979) 487

/12/ C.R.Helms, H.P.Bonzel, S.Kelemen, J.Chem. Phys. 65 (1976) 1773

/13/ K.Heinz, E.Lang, K.Strauss, K.Müller, Appl.Surf.Sci. 11/12 (1982) 611

/14/ M.A. van Hove, R.J.Koestner, J.P.Bibérian, L.L.Kesmodel, I.Bartos, G.A.Somorjai, Surf.Sci. 103 (1981) 189, 218

/15/ B.Pennemann, K.Oster, K.Wandelt, Surf.Sci. (1991)

/16/ J.F.Wendelken, D.M.Zehner, Surf.Sci. 71 (1978) 178

/17/ a) K.Kern, P.Zeppenfeld, R.David, G.Comsa, Phys.Rev.Lett.59 (1987)79

b) K.E.Kuhnke, PhD-Thesis, Universität Bonn, 1991

/18/ W.Hösler, R.J.Behm, E.Ritter, IBM J.Res.Develop. 30 (1986) 403

/19/ a) R.J.Behm, private communication

b) D.M.Kolb, Ber.Bunsengesellschaft Phys.Chem. 92 (1988) 1125

/20/ W.Moritz, Habilitations-Thesis, Universität München 1984

/21/ T.Engel, K.H.Rieder, in: "Structural Studies of Surfaces", Springer Tracts in Modern Physics, Vol. 91 (Springer, Berlin, 1982)

114

/22/ H.Niehus, J.Vac.Sci.Technol. A5 (1987) 751, and references therein

/23/ G.Binnig, H.Rohrer, Ch.Gerber, E.Weibel, Surf.Sci. 131 (1983) L379

/24/ N.Garcia, C.Ocal, F.Flores, Phys.Rev.Lett. 50 (1983) 2002

/25/ W.Moritz, D.Wolf, Surf.Sci. 163 (1985)

/26/ a) U.Harten, A.M.Lahee, J.P.Toennies, Ch.Wöll, Phys.Rev. Lett. 54 (1985) 2619

   b) J.V.Barth, H.Brune, G.Ertl, R.J.Behm, Phys.Rev.B42 (1990) 9307

/27/ A.Bartolini, F.Ercolessi, E.Tosatti, in: The Structure of Surfaces II, Eds. J.F.van der Veen, M.A. van Hove, Springer Series in Surface Sciences, Vol. 11 (Springer, Heidelberg, 1988) p.132

/28/ P.Heilmann, PhD-Thesis (Erlangen, 1979), see also Ref. 6.

/29/ M.K.Debe, D.A.King, Surf.Sci. 81 (1979) 193

/30/ M.Henzler, "Electron Diffraction and Surface Defect Structure", in: Electron Spectroscopy for Surface Analysis, Topics in Current Physics 4, Ed. H.Ibach (Springer, Heidelberg, 1977) p. 116

/31/ a) W.Hösler, R.J.Behm, E.Ritter, IBM J.Res.Develop., 30 (1986) 403
   b) R.J.Behm, W.Hösler, E.Ritter, G.Binnig, Phys.Rev.Lett. 56 (1986) 228

/32/ Proceedings of the STM'86, Surf.Sci. 181 (1987); Proceedings of the STM'87, J.Vac.Sci. Technol. A6 (1988); Proceedings of the STM'88, J.Microsc. 152 (1989); Proceedings of the STM'89, J.Vac.Sci.Technol A8 (1990); Proceedings of the STM'90, J.Vac.Sci.Technol. B9 (1991)

/33/ M.Scheffler, K.Horn, A.M.Bradshaw, K.Kambe, Surf.Sci. 80 (1979) 69

/34/ A.Cassuto, J.J.Erhardt, J.Cousty, R.Riwan, Surf.Sci. 194 (1988) 579

/35/ B.J.Waclawski, J.F.Herbst, Phys.Rev.Lett. 35 (1975) 1594

/36/ R.P.Antoniewicz, Phys.Rev.Lett. 38 (1977) 374

/37/ K.Wandelt, J.Vac.Sci.Technol. A2 (1984) 802

/38/ K.Wandelt, in: Studies in Surface Science and Catalysis, Eds. B.Delmont, J.T.Yates, Vol. 32, Thin Metal Films and Gas Chemisorption, Ed.P.Wissmann (Elsevier, Amsterdam, 1987) 280-368

/39/ K.Wandelt, in: Chemistry and Physics of Solid Surfaces VIII, Eds. R.Vanselow, R.Howe, Springer Series in Surface Sciences, Vol. 22 (Springer, Berlin, 1990) 289-346, and references therein

/40/ N.D.Lang, A.R.Williams, Phys.Rev.B25 (1982) 2940

/41/ J.Behm, C.R.Brundle, K.Wandelt, J.Chem.Phys. 85 (1986) 1061

/42/ A.Jablonski, S.Eder, K.Wandelt, Appl.Surf.Sci 22/23 (1985) 309

/43/ H.Wagner, in: Springer Tracts in Modern Physics, Vol. 85 (Springer, Berlin, 1979)

/44/ K.Wandelt, J.Hulse, J.Küppers, Surf.Sci. 104 (1981) 212

/45/ D.Eigler, E.K.Schweizer, presented at the German Physical Society Meeting, Regensburg, March 1990; at the 5th Conf. on STM/STS, Baltimore, July 1990; and J.Vac.Sci.Technol. (1991) in press

/46/ K.Wandelt, J.Hulse, J.Chem.Phys.80 (1984) 1340

/47/ S.Daiser, Diplom-Thesis, Universität München, 1981

/48/ K.Besocke, H.Wagner, Surf.Sci. 52 (1975) 653

/49/ K.Besocke, H.Wagner, Phys.Rev. B8 (1973) 4597

/50/ A.Clarke, P.J.Rous, M.Arnott, G.Hennings, R.F.Willis, Surf.Sci. <u>192</u> (1987) L843

/51/ W.Dürr, M.Taborelli, O.Paul, R.Germar, W. Gudat, D.Pescia, M.Landolt; Phys.Rev.Lett. <u>62</u> (1989) 206

/52/ K.S.Kim, J.H.Sinfelt, S.Eder, K.Markert, K.Wandelt, J.Chem.Phys. <u>91</u> (1987) 2337

/53/ K.Wandelt, in: Physics and Chemistry of Alkali Metal Adsorption, Eds. H.P.Bonzel, A.M.Bradshaw, G.Ertl (Elsevier, Amsterdam, 1989) 25-44

/54/ E.M.Oellig, R.Miranda, Surf.Sci. <u>177</u> (1986) L947

/55/ G.Surnev, Surf.Sci. <u>110</u> (1981) 45

/56/ I.Langmuir, J.Amer.Chem.Soc. <u>54</u> (1932) 2798

/57/ I.Langmuir, J.B.Taylor, Phys.Rev. <u>40</u> (1932) 463

/58/ J.B.Taylor, I.Langmuir, Phys.Rev. <u>B44</u> (1933) 423

/59/ P.Pervan, E.Michel, G.R.Castro, R.Miranda, K.Wandelt, J.Vac.Sci.Technol. <u>A7</u> (1989) 1885

/60/ K.Markert, K.Wandelt, Surf.Sci. <u>159</u> (1985) 24

/61/ J.K.Norskov, S.Holloway, N.D.Lang, Surf.Sci. <u>150</u> (1985) 65

/62/ a) B.Gumhalter, K.Hermann, K.Wandelt, Vacuum <u>41</u> (1990) 192
b) K.Hermann, B.Gumhalter, K.Wandelt, Surf.Sci.(1991) Proc.ECOSS-11, in press

/63/ W.Erley, H.Ibach, S.Lehwald, H.Wagner, Surf.Sci. <u>83</u> (1979) 585

/64/ S.Daiser, K.Wandelt, Surf. Sci. <u>128</u> (1983) L213

PHYSICS OF SURFACES

R. Del Sole
Dipartimento di Fisica, II Università di Roma "Tor Vergata"
v. E. Carnevale, 00173 Roma, Italy

ABSTRACT. A brief account of the most important experimental and theoretical methods employed in surface physics is given. The use of optical spectroscopy in the study of clean semiconductor surfaces is described in detail.

1. Introduction

In the last twenty years surface science has widely grown. The preparation of crystal surfaces under ultra high-vacuum conditions made possible the study of the atomically clean surfaces with structural techniques sensitive to the geometry of the first atomic layer. Moreover the developments of surface-sensitive spectroscopic techniques allowed the study of the electronic structure of the surfaces.

In these lectures I will give a short description of semiconductor surfaces. I hope that it will be clear also to people not directly involved in surface physics. In Section 2 I will briefly recall the more common experimental methods. The treatment will be very short because they have been described in other lectures of this school. In Section 3 I will give an idea of the modern theoretical techniques, which allow *ab initio* calculation of the total energy and of the electronic structure.

Section 4 is devoted to the optical spectroscopy of surfaces. Here I will give a rather complete account of the purposes of this technique, and of the experimental and theoretical methods involved.

2. Experimental Techniques

We list below the most widely used experimental techniques, together with the information we can get from each of them.

2.1. STRUCTURAL PROBES

2.1.1. *Low energy electron diffraction (LEED) [1].* Electrons of kinetic energy of the order of 50 eV impinge normally on the surface. They are backscattered by the crystal potential at different angles. The surface periodicity is directly observed via the diffraction pattern, which gives an image of the reciprocal mesh. Very often, in addition to the reciprocal-lattice vectors expected to appear because of bulk periodicity,

R. Guidelli (ed.), Electrified Interfaces in Physics, Chemistry and Biology, 117–132.
© 1992 *Kluwer Academic Publishers.*

fractional-order spots are observed. For instance, in the case of the freshly cleaved Si(111) surface, half-integer spots are observed along the direction of one of the principal vectors of the reciprocal lattice. It means that the periodicity has doubled along the real-space direction corresponding to that vector: people say that a 2x1 reconstruction has occurred.

To determine the geometry underlying a given reconstruction is difficult, because LEED kinematics (i.e. the positions of the spots) gives information on the periodicity, but not on the real space geometry. One has to guess a model, and compare the predictions based on it with experiments. Some information on the geometry can be obtained from the study of the intensity of each diffraction spot as a function of the energy of incoming electrons (dynamical LEED). However it is not as clear-cut as that given by kinematical LEED.

2.1.2. *Light-ion or atom backscattering.* The first method [2] uses beams of H^+ or He^{++} ions with energies between one and hundreds of keV. The ions are scattered by the nuclei of the crystal following the dynamics of Rutherford scattering and also loose energy along straight lines due to the interaction with the electrons. By suitably choosing the direction of the incoming beam and detecting the number of backscattered ions as a function of the energy and of the outcoming directions, information on the atomic positions in the first few layers can be obtained. The second method [3] samples the corrugation of the last atomic plane and in particular that of the tail of electron charge-density.

2.1.3. *Scanning Tunnel Microscopy (STM) [4].* A metal tip is positioned very near the sample, so that electrons can tunnel, according to the bias, to or from the sample. While the tip scans the surface laterally, its distance from the sample is adjusted in such a way to keep the current constant. This is roughly equivalent to keep constant the distance between the electron clouds and the tip. STM gives information on the corrugation of the electron density, and therefore on atomic positions. The piezoelectric devices emploied allow a lateral resolution of the order of 1 Å and a vertical resolution of the order of 0.1 Å.

2.2. PROBES OF THE ELECTRONIC STRUCTURE

The electronic structure of the surface can strongly depend on the atomic geometry. The relaxation or the reconstruction can deeply change the electronic structure with respect to that of the ideally terminated crystal. Therefore many techniques devised to test the electronic structure give information also on the atomic structure.

2.2.1. *Photoelectron Spectroscopy [5].* Choosing photon energy in such a way to produce emitted electrons in the energy range where the escape depth has a minimum value (a few Å), one gets a picture of the density of states of the last atomic layers. Bulk and surface states contribute to it: however, the surface features can be identified by modifying them through the adsorption of different atoms at the surface. Photoemission with *angular resolution* allows to get information about the k-dispersion of the bands. In this case surface states show up clearly, since their energy does not depend on the k-component normal to the surface, say k_z (that is on photon energy, if the parallel component of **k** is fixed).

Empty surface bands can be studied by *inverse photoemission,* which detects the energy of the photon emitted when an electron of an incoming beam of given E and **k** falls into an empty conduction band.

2.2.2. Optical spectroscopy [6]. The surface contribution to reflectance or absorption of light is small, of the order of a few percent. Appropriate conditions apt to enhance surface sensitivity must be chosen. This spectroscopy gives information on the *near surface* joint density-of-states, which not always coincides with the joint density of surface states. A detailed treatment will be given in Section 4.

2.2.3. Electron energy loss spectroscopy (EELS) [7]. This is a natural complement to the optical spectroscopy. Measuring energy and momentum of backscattered electrons, one gets information on the probability for an electron of loosing a certain amount of energy and momentum. This is related to the dielectric function of the crystal [8]. Despite some complications in the interpretation of the data, it turns out to be a surface-sensitive probe, particularly in the more recent version of high-resolution EEL spectroscopy, using a low-energy incoming beam [7].

3. Theoretical Methods

3.1. SLAB GEOMETRY

To calculate the electronic states of a semi-infinite crystal is a difficult task, because of the large extension and lack of periodicity along the direction perpendicular to the surface, say z. Therefore calculations are usually carried out for finite slabs of ten or twenty atomic layers. The idea underlying this method is that such slabs are thick enough to allow the presence of a bulk-like region in the middle, so that each surface has the same electronic structure as the surface of a semi-infinite crystal.

The single slab is used in connection with the methods which use localized orbitals as basis set. These are the semi-empirical tight-binding method, discussed in the next Subsection, and the local-orbital versions of *ab initio* methods. If, on the other hand, plane waves are used, it is convenient to retain a periodicity also along z. In this case the *repeated slab geometry* is used. Attention must be paid to the vacuum region separating the slabs: it must be thick enough to avoid the unwanted interaction of surfaces belonging to different slabs. A number of missing layers between 5 and 10 is usually sufficient.

3.2. THE SEMI-EMPIRICAL TIGHT-BINDING METHOD

The eigenfunctions of the Schrödinger equation can be written as combinations of the orbitals $\phi_{\alpha n}(r)$ centered at atomic sites (α labels the orbital and n the atom in the unit cell):

$$\psi_k(r) = N^{-1/2} \Sigma_{\alpha nm} \, c_{\alpha nm}(k) \Sigma_R e^{ik \cdot (R+R_{mn})} \, \phi_{\alpha n}(r\text{-}R\text{-}R_{mn}) \qquad (1)$$

where **R** are the vectors of the 2D lattice, **k** is the 2D wave vector in the surface plane, and the 3D-vectors R_{mn} give the atomic positions inside the unit cell of the slab at each plane m. The coefficiens $c_{\alpha nm}(k)$ are obtained by solving the secular problem

$$\Sigma_{\alpha'n'm'} \{ H_{\alpha nm,\alpha'n'm'}(k) - E(k) \, \delta_{\alpha,\alpha'} \, \delta_{n,n'} \, \delta_{m,m'} \} \, c_{\alpha'n'm'}(k) = 0 \; , \qquad (2)$$

where the matrix elements of the Hamiltonian are

$$H_{\alpha nm,\alpha'n'm'}(k) = (1/N) \, \Sigma_R \, e^{ik \cdot (R+R_{mn}-R_{m'n'})} \, x$$
$$\int \phi^*_{\alpha n}(r\text{-}R\text{-}R_{mn}) \, H \, \phi_{\alpha'n'}(r\text{-}R_{m'n'}) \, dr \qquad (3)$$

In the semi-empirical tight binding method the integrals in (3) are not computed, but used as fitting parameters in order to reproduce the bulk band structure. Only interactions with few neighbor shells are retained, while the others are put equal to zero. If the geometry of the surface is modified, the interaction parameters can be varied with the distance according to an inverse power law [9].

The main problem of the semi-empirical tight-binding method is that the choice of the parameters is not unique. It depends on the number of orbitals and of neighbors retained, and, even with these numbers fixed, it is not fully determined. The very reason of this is that the localized orbitals spanning a given set of bands, the Wannier functions, are themselves not uniquely defined. The use of these parameters for surface calculations relies on the assumption that they are related to really localized orbitals: in fact their energies must not be perturbed by the missing layers. It is clear from this that the choice of parameters well suited for surface calculations is a difficult task, not to be pursued by a definite recipe, but rather on a try-and-test basis. A popular choice for covalent solids is that of Vögl and coworkers [10], who use 5 orbitals per atom: s, p_x, p_y, p_z, and s*. The last orbital mimics d orbitals, which are essential in order to give a good description of the lowest conduction bands. This approach, which will be recalled in Section 4 for the calculation of the optical properties, has been emploied in the cases of vacancies, surfaces and defects, with good results [11].

The main advantage of the tight-binding method for surface calculations is that the secular matrix results to be relatively small. For instance, in the case of Si(111)2x1, one can make calculations for a slab of 18 layers (36 atoms in the unit cell of the 2D lattice)) using 180 basis functions. Pseudopotential calculations for the same slab should involve about 2000 plane waves.

3.3. AB INITIO METHODS

In the following we will present the ingredients of a modern surface band-structure calculation. We will describe the use of the self-consistent pseudopotential method in the framework of the density functional theory (DFT).

We consider the many-electron problem in a crystal. We replace the true potential of each ion by a pseudopotential, which is shallow in the region near the atomic core [12]. The pseudopotential is chosen in such a way to reproduce the atomic wave functions outside the core region. In this way we will not get from our calculation the core states, but we deal with shallow potentials, with the advantage of using a few plane waves.

The DFT ensures that the total electron energy is a unique functional of the electron density $n(r)$, and has a minimum when the density is the true one [13]. This variational principle can be put in the form of one-particle Schrödinger-like equations, from whose solutions the electron density and total energy can be derived [14].

The eigenvalues of these one-particle Kohn-Sham equations are generally considered as band energies. This is not rigorously justified, since these equations have been derived only to the aim of minimizing the total energy. In fact the band gaps calculated in this way are smaller than experimental ones by 0.5 or 1 eV in semiconductors [15]. In spite of this, the error in band positions, ~ 0.5 eV, is the smallest error given by any method involving a local potential.

The Kohn-Sham equations are:

$$\{-\hbar^2 \nabla^2/2m + V(r) + e^2\int dr'\, n(r')/|r-r'| + V_{xc}(r)\}\psi_i(r) = \varepsilon_i\, \psi_i(r), \qquad (4)$$

where $V(r)$ is the ionic pseudopotential, the integral is the potential generated by the electron charge distribution, and $V_{xc}(r)$ is the exchange-correlation potential. In principle it can be calculated if the exchange-correlation part of the *universal* energy functional is known. Unfortunately this is not the case: it is this ignorance which hampers the development of an exact DFT calculation.

The exchange-correlation potential is well known for the homogeneous electron gas: numerical (Monte Carlo) calculations [16] have been parametrized in order to give a tractable and accurate expression for $E_{xc}^{HEG}(n)$, the exchange-correlation energy per electron as a function of the density n. In the case of real solids, one resorts to the Local Density Approximation (LDA): we divide the solid in small volumes, of constant electronic density. We assume that the contribution of each small volume to E_{xc} is found according to the law valid for the homogeneous electron gas of the same density. By summing all the contributions we get

$$E_{xc} = \int dr\, n(r)\, E_{xc}^{HEG}(n(r)) \qquad (5)$$

Most of the total energy calculations are carried out within the LDA DFT. The calculated structural properties, for instance the equilibrium lattice constant, bulk modulus, etc. are in very good agreement with experiment [17]. The atomic geometry of surfaces is also very well described within this approximation [18].

As anticipated above, significant discrepancies are present between calculated and measured excitation energies. In particular, the calculated gap between filled and empty states is always too small, by 0.5 or 1 eV in semiconductors [19]. This is due to the fact that DFT is a *ground state* theory, from which we cannot, in principle, get information about excited states. Their energies should be found as the poles of the E-dependent

Green's functions, computed according to many-body techniques [15]. If one carries out this program, which involves the calculation of the Green's function G and of the screened interaction W (therefore called the GW method), one meets again a one-electron Schrödinger like equation similar to (4), where the exchange-correlation potential is replaced by an energy-dependent non-local self-energy operator [15,20]. The resulting wave functions are practically equal to those of Kohn-Sham, but the energies are different: they agree with experimental excitation energies within 0.1 or 0.2 eV. The GW method is the best way presently available for the calculation of electron energies in extended systems. The price for the increased precision, with respect to LDA-DFT, is an increase of the computing time of two orders of magnitude!

An empirical method which allows to find correct gaps (in bulk) without carrying out the cumbersome Green's function calculation, is to use Slater's exchange in (4), instead of V_{xc}. Before the formulation of DFT, Slater proposed a local approximation to exchange [21]:

$$V_x(\mathbf{r}) = -(3e^2/2\pi) \{3\pi^2 n(\mathbf{r})\}^{1/3} \tag{6}$$

After the development of DFT, it became clear that the real exchange potential in Kohn-Sham equations should be 2/3 of this. Nevertheless, if one uses (6) rather the right exchange potential, and neglects correlation, one gets the right bulk gaps! A different version of this method, called the X_α method, uses a value of α, comprised between 0.8 and 1, in front of Slater's exchange, adjusted to fit experimental spectra.

4. Optical Spectroscopy

4.1. EXPERIMENTAL METHODS

The optical probe has the advantage, with respect to electron or photoelectron spectroscopy, of not charging or damaging the sample, and of a better energy resolution. On the other hand, since light penetration and wave length are much larger than surface thicknesses (a few Angstroms), it is poorly sensitive to surfaces. Some tricks must be arranged in order to increase its surface sensitivity.

In some experiments [22-28] the reflectance of the clean surface is measured first, and then the reflectance of the surface after chemisorption (often of oxygen). Their percentage difference, the differential reflectance (DR), is related to the surface structure. However, to what extent it is related to the clean or to the chemisorbed surface, whether or not it is sensitive to the spectrum of surface states, and/or to the atomic structure of the surface are in general open questions. In spite of this, in some cases this technique has yielded important contributions to the understanding of the atomic and electronic structure of surfaces [22,23,25,29].

Reflectance anisotropy (RA) spectroscopy (sometimes called reflectance-difference spectroscopy) consists of measuring the relative reflectance difference RA of two orthogonal light polarizations (x and y) in the surface plane [30-34]. Since the bulk optical properties of cubic materials are isotropic, any observed RA must be related to

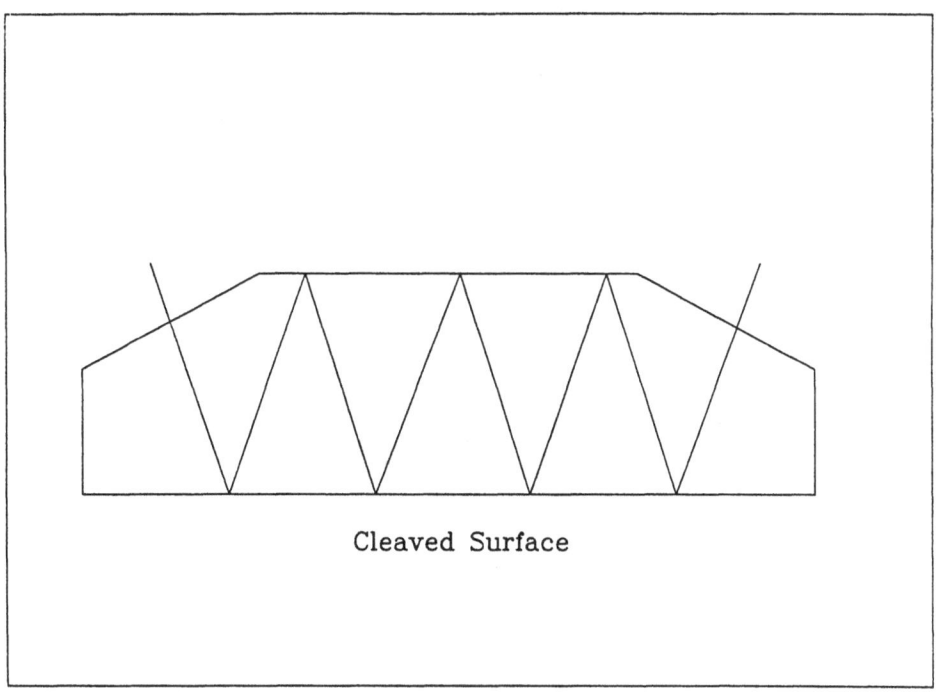

Figure 1. Sketch of the geometry used in the experimental arrangement allowing multiple internal reflections. Light undergoes four internal reflections at the cleaved surface.

the reduced symmetry of the surface. The great advantage of this technique with respect to DR is that it does not involve chemisorbed surfaces, so that it can also be used to monitor the growth of epitaxial superlattice structures. The theoretical interpretation is also in principle simpler, since an important and poorly known information, i.e. the atomic structure of the chemisorbed surface, is not necessary. This technique seems however to be less sensitive than DR to the spectrum of surface states.

It is clear from the above that theoretical work is needed in order to fully exploit the potentiality of surface optical spectroscopy. In the following we will illustrate the use of optical spectroscopy in two cases, namely Si(111)2x1 and GaAs(110). We will discuss what information can be extracted, the methods of calculation, and their results and limits.

4.2. Si(111)2x1

The most important achievement of surface optical spectroscopy has been its contribution to understanding the atomic and electronic structure of freshly cleaved clean Si and Ge (111) surfaces. Low-energy electron diffraction yields evidence of 2x1 reconstruction, but does not inform on the underlying atomic distortions. Understanding these has taken more than twenty years of work. Several structural models have been proposed to explain the 2x1 reconstruction, but no conclusive evidence for any of them was found until a few years ago. The buckling model [35], for long time generally accepted, has been strongly questioned since it does not agree with angle-resolved photoemission measurements [36,37]. Among the other models, Pandey's chain model [38], formulated in 1981, is in good agreement with many experiments, and is now believed to be correct. A big piece of evidence in its favor is its ability to explain the DR optical anisotropy, which cannot be explained by any of the other models.

DR experiments have been carried out for the first time on Ge(111)2x1 by Chiarotti and coworkers in 1968 [22]. In order to increase the surface sensitivity, an experimental set-up allowing multiple total internal reflections was used (see Fig. 1). At frequencies below the bulk gap, light absorption can only occur at the surface, causing a reduction of the reflected light. A DR peak around 0.45 eV was found (also for Si(111)2x1 [23]), of intensity of about 5%, which yielded some of the first evidence of surface states. After oxygen chemisorption, dangling bonds are saturated, and therefore optical transitions across them occur at higher energies: in this case the DR signal is related only to the surface states of the clean surface. The relatively large intensity of such transitions have allowed their observation also by means of external normal-incidence DR [24], and therefore the extension to frequencies above the bulk gap.

The breakthrough in DR spectroscopy occurred in 1984 with the use of polarized light on Si(111) samples with single-domain 2x1 reconstruction [25,26] (Fig. 2). The measured 100% anisotropy of the Si 0.45 eV peak yielded a big piece of evidence in favour of Pandey's chain model [38], as will be discussed below.

The surface contribution to reflection of normally incident light, polarized along the α direction, by a semi-infinite crystal is computed according to [39]:

$$(\Delta R_\alpha/R_0) = 4(\omega/c)\text{Im}\{\Delta\varepsilon_{\alpha\alpha}(\omega)/[\varepsilon_b(\omega)-1]\} \ . \tag{7}$$

Here $\Delta R_\alpha = R_\alpha - R_0$ is the difference between the actual reflection coefficient R_α, including surface effects, and R_0 given by Fresnel formula; $\varepsilon_b(\omega)$ is the bulk dielectric constant; and

$$\Delta\varepsilon_{\alpha\alpha}(\omega) = \int_{-\infty}^{+\infty}dz \int_{-\infty}^{+\infty} dz'[\varepsilon_{\alpha\alpha}(z,z';\omega)-\delta(z-z')] \ , \tag{8}$$

where $\varepsilon_{\alpha\beta}(z,z';\omega)$ is the microscopic dielectric tensor of the vacuum-crystal interface. Its off-diagonal ($\alpha\neq\beta$) components, which are very small in all cases considered here, have been neglected in (8).

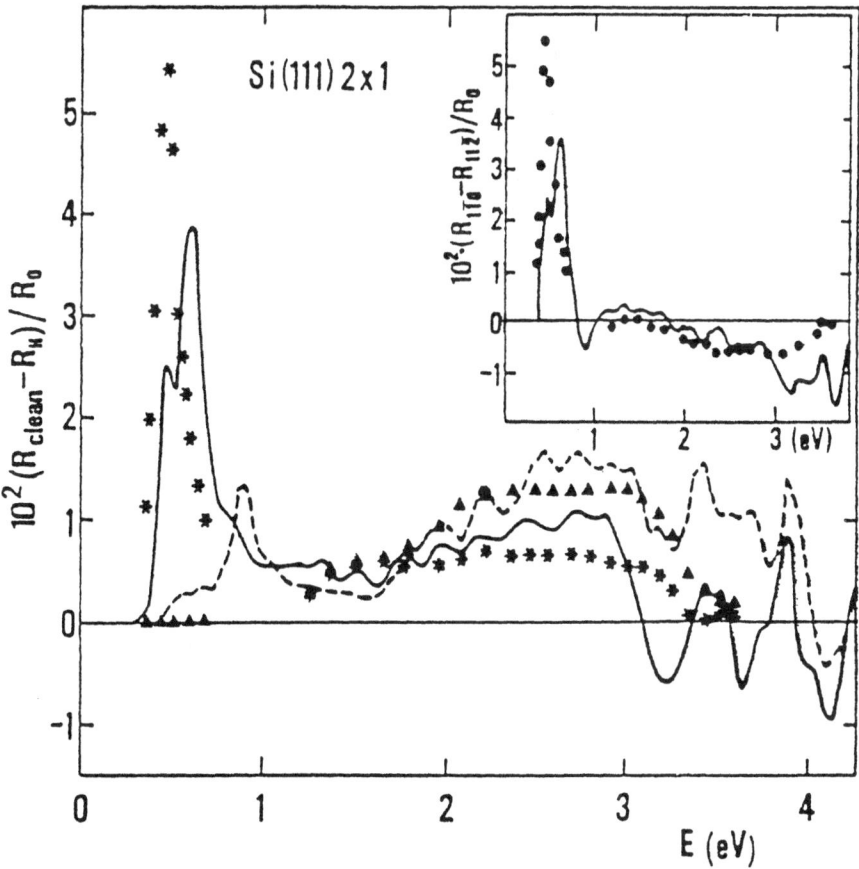

Figure 2. Differential reflectivity of Si(111)2x1, calculated as the difference between the clean 2x1 and hydrogen-covered 1x1 surface reflectivities [41]. Full line, y polarization (parallel to the chains); dashed line, x polarization (perpendicular to the chains); stars and triangles, experimental results [26] for y and x polarizations respectively. Inset - reflectance anisotropy: full line, calculation; dots, experimental results.

The geometry of the chain model for the (2x1)-reconstructed (111) surface of silicon is characterized by the surface atoms being close as bulk nearest neighbors ($d \approx 2.35$ Å) and forming zig-zag chains along the [1$\bar{1}$0] direction, while interchain distances are rather large (see Fig. 3). Each surface atom has a dangling bond (DB), from which near-gap surface states mainly originate. The relevant features of these surface bands can be reasonably well reproduced by a second-neighbor tight-binding model including DB orbitals only [29]. For $\hbar\omega < 1$eV, it is reasonable to restrict our attention to these states, since transitions involving other surface or bulk states are expected to occur at higher energies. The surface bands are shown in the inset of Fig. 4, together with the angle-resolved photoemission data of [36] and [37].

The calculated reflectivity for polarization parallel (y) and perpendicular (x) to the chain direction is shown in Fig. 4. The main peak at 0.45 eV is present only for y-polarization, as a consequence of the chain structure. The polarization dependence of

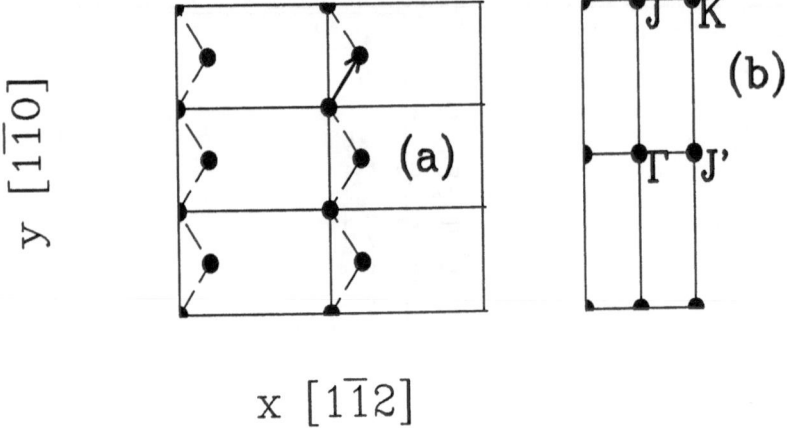

y [1̄10]

x [11̄2̄]

Figure 3. Atomic arrangement in the surface plane (a) and Surface Brillouin Zone (b) for the chain model of Si(111)2x1. Solid circles denote atomic positions, and dashed lines represent bonds in the surface plane.

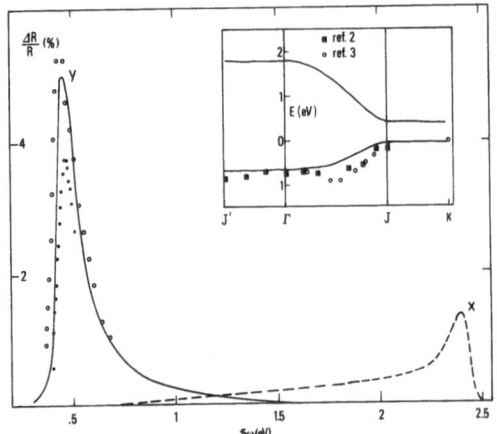

Figure 4. Differential reflectivity of the chain model of Si(111)2x1 for light polarized parallel (y) and perpendicular (x) to the chains [29]. Circles and solid dots are experimental data for y polarization from [25] and [40] respectively. In the inset, the calculated DB bands are shown (full lines). Squares and circles are angle-resolved photoemission data from [37] and [36] respectively.

the reflectivity at the peak energy is shown in Fig. 5 (solid line). This, as well the asymmetrical lineshape and the absolute value of the peak height in Fig. 4, well agree with experiment [25,40]. The polarization dependence predicted for the buckling model is also shown in Fig. 5 (dashed line). It drastically differs from the experiment. We thus conclude that the buckling model [35] must be ruled out, since it cannot account

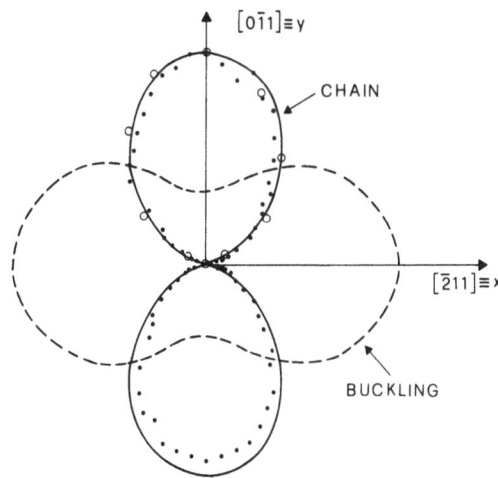

Figure 5. Polar plot of the DR peak at 0.45 eV of Si(111)2x1 vs. polarization direction in the (111) plane. Solid curve, chain model; dashed curve, buckling model; circles and dots, experimental data from [25] and [40] respectively.

for the observed polarization dependence of the reflectivity.

The evaluation of the optical properties described so far, based on a two-dimensional tight-binding model of the surface electronic structure, must be replaced by a more realistic calculation, if the above-gap frequency range, where transitions involving bulk as well surface states can coexist, is of interest. A method for the realistic evaluation of surface reflectivity within a tight-binding scheme has been developed and applied to Si(111)2x1 and to Si(110):H surfaces [41]. We start from bulk bands calculated using the tight-binding Hamiltonian of Vögl et al. [10]. We calculate the electronic states and the optical properties of a Si slab of 18 atomic layers along the [111] direction, with the surfaces reconstructed according to Pandey's chain model.

DR can be simulated by subtracting from the clean-surface reflectivity that of the hydrogenated surface. (It is simpler to carry out calculations for hydrogen rather than for oxygen covered surfaces.) This difference is shown in Fig. 2 for x and y polarizations. A strong peak at 0.6 eV is present for y and not for x polarization, in agreement with experiment (where the peak occurs at 0.45 eV). All features of DB->DB transitions are found to be in agreement with the predictions of the simple two-dimensional calculation shown in Fig. 4. The agreement with experiment is a little bit worse than in Fig. 4 for the ~0.5 eV peak, which can be understood recalling that no additional parameter has been introduced in the realistic calculation to reproduce surface features. A quite good agreement is found for the structures in the range 2-4 eV. If one repeats the calculation for the hydrogenated surface by retaining the 2x1 reconstruction, these structures become about three times weaker [41]. A substantial contribution to them arises therefore from the reflectance difference between

H-covered (111)1x1 and (111)2x1 surfaces. This difference is not related to surface states but is of structural origin.

4.3. GaAs(110)

We consider this surface as an example of the cleavage surfaces of III-V compounds. These surfaces do not show any reconstruction, but dynamical-LEED analyses yield evidence of a relaxation of the atoms in the first layer, consisting of a rotation of the Ga-As bond, with the Ga and As atoms moving inward and outward, respectively [42]. Surface states are degenerate with bulk bands, so that no clear-cut surface structure originates below the bulk gap, at variance with the cases of Si and Ge (111)2x1 surfaces. It is interesting to understand which kind of information can be extracted from optical spectrocopy in this case.

DR experiments using polarized light have been carried out by Selci et al. [28] on GaAs(110) (see Fig. 6). RA experiments have been carried out by Berkovits et al. [33] (see Fig. 7). All spectral structures occur at frequencies above the bulk gap. The main problem for all above-gap structures is their more or less direct relation to surface states, or, in other words, how large the contribution from bulk states is. In order to answer this question, we have applied to GaAs(110) a method for the realistic calculation of surface optical properties based on self-consistent pseudopotentials.

We improve on the tight-binding method by using the selfconsistent electronic structure determined with local pseudopotentials and a plane-wave basis set, within the X_α method, with $\alpha=1$ (in other words, using Slater's exchange in order to get the right gaps) [43-46]. Not only we find more realistic electronic states, but also reduce the computing time. The reason of this reduction is that the momentum operator, whose matrix elements must be computed to find the optical properties, is diagonal in the plane-wave, but not in the local-orbital, representation. As a consequence, shorter computer time is needed to calculate *optical properties* using plane waves in the case of a thin slab. We consider repeated GaAs slabs of 15 (110) layers, separated by 9 missing layers, with the clean surfaces in the relaxed configuration with a 27.5° rotation angle [47]. We repeat the calculation for the unrelaxed H-covered surfaces.

The calculated RA of clean GaAs(110) is compared with experiment in Fig. 7. The agreement of peak positions and strengths is reasonable. The sensitivity of this technique to the spectrum of surface states can be determined by decomposing the calculated RA into various components, namely surface to surface, surface to bulk, bulk to surface and bulk to bulk state transitions. When this is done, it is apparent that the contribution of bulk to bulk state transitions is dominant [46]. Such large contribution of bulk states can be understood as follows. The energy positions of bulk states is not altered by the presence of the surface; however, this does not imply that the surface does not perturb bulk wave functions as well. If a surface sensitive experimental technique is used, the reduced symmetry of bulk states in the surface region cannot be neglected. In particular, bulk wave functions of the truncated crystal have cubic symmetry well inside the solid, where the atoms are in cubic environment, but they have the same reduced symmetry of surface-state wave functions near the surface.

The calculated DR spectrum of GaAs(110) is compared with experiment in Fig. 6. It can be seen that the agreement is not very good. The reason is probably the inadequacy

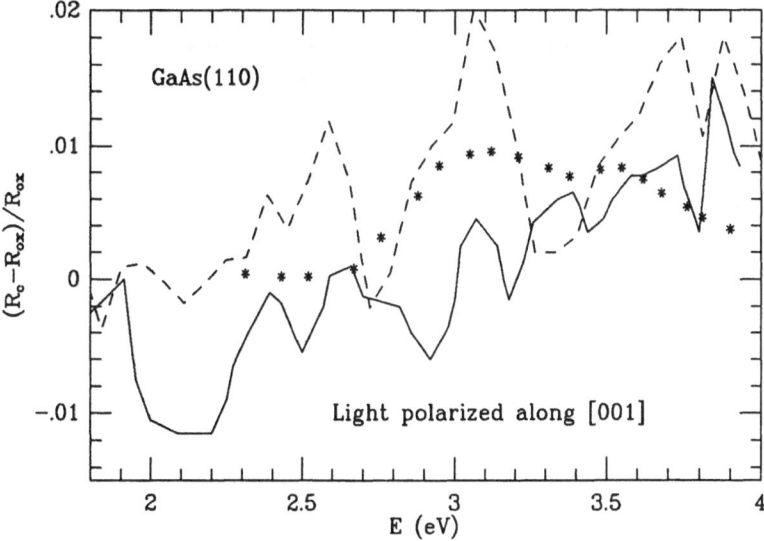

Figure 6. Differential reflectivity of GaAs(110): stars, experiment [28]; solid line, calculation [46] for hydrogen atoms sitting atop first-layer gallium and arsenic atoms; dashed line, the same for hydrogen atoms aligned along the missing bonds.

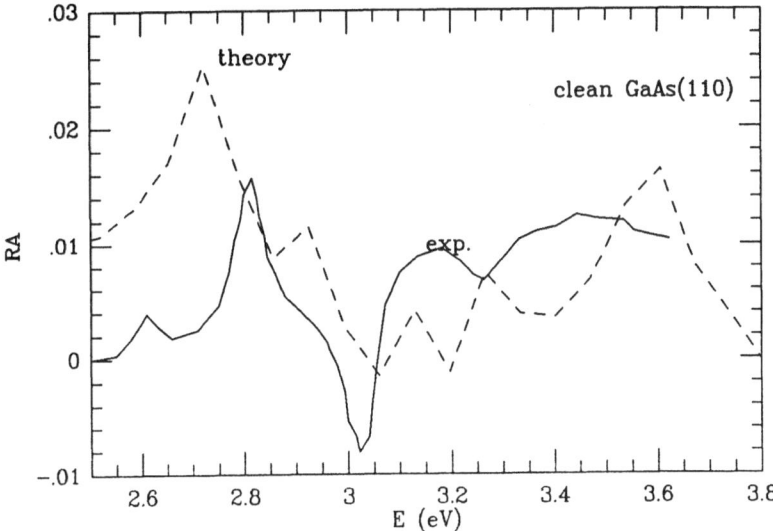

Figure 7. Reflectance anisotropy of clean GaAs(110). Experiment and theory from [33] and [46] respectively.

of hydrogenated surfaces to mimic the oxidized surfaces involved in experiments. On the other hand, the great sensitivity of DR to the surface structure is demostrated by the comparison of the solid curve (calculated for H-atoms on top of first-layer atoms) with the dashed curve (calculated for H-atoms aligned along the missing bond directions) in Fig. 6: such a small structural difference leads to qualitatively different curves. We cannot expect therefore similar lineshapes for H- and O-covered surfaces, unless in special cases. A decomposition of DR as it was made for RA shows that transitions across bulk states give a significant contribution, of about 50%.

4.4 CONCLUSION

When good agreement between theory and experiment occurs, optical spectroscopy can yield important information on surface structure, as in the case of Si(111)2x1. However not always theory gives a good description of experiments. This is partially due to the low degree of sophistication of the theory, which does not yet account properly for many-body corrections to the band structure [48] and to optical properties [8], and partially to the difficulty of describing the real oxidized surfaces. Spurious effects present in the experiments, as the Franz-Keldysh effect [33,34], may also be sources of discrepancies. In spite of this, some conclusions can be drawn. DR is strictly related to the spectrum of surface states below the first bulk absorption peak. Around this, a strong contribution of transitions across bulk states, depending on the atomic structure of the surface, is present. Such contribution is dominant in the case of RA. DR and RA are strongly sensitive to the atomic structure of the clean and chemisorbed surfaces; however an improvement of calculation ability is necessary in order to fully exploit this as a tool of surface investigation.

REFERENCES

[1] Pendry J. B. (1974) Low Energy Electron Diffraction, Academic Press, New York

[2] Van der Veen J. F. (1985) Surface Sci. Reports 5, 199

[3[Cardillo M. J. (1981) Phys. Rev. B23, 4279

[4] Binning G., Rohrer H., Gerber Ch., and Weibel E. (1983) Phys. Rev. Lett. 50, 120

[5] Chiang T. C., Knapp J. A., Aono M. and Eastmann D. E. (1980), Phys. Rev. B21, 3513; Williams G. P., Smith R. J. and Lapeyre G. J. (1978), J. Vac. Sci. Technol. 15, 1249

[6] Del Sole R. (1990), Mat. Sci. and Engin. B5, 177

[7] Math R., Lüth H. and Ritz H. (1983), Solid State Commun. 46, 343; Di Nardo N. J., Thompson W. A., Schell Sorokin A. J. and Demuth J. E. (1986), Phys. Rev. B34, 3007

[8] Del Sole R. and Fiorino E. (1984), Phys. Rev. B29, 4631

[9] Harrison W. A. (1980), Electronic structure and the properties of solids, Freeman, San Francisco

[10] Vögl P., Hjalmarson H. P. and Dow J. D. (1983), J. Phys. Chem. Solids 44, 365

[11] Mailhiot C., Duke C. B. and Chadi D. J. (1985), Surface Sci. 149, 366; and references therein

[12] Cohen M. L. and Chelikowsky J. R. (1988), Electronic structure and optical

properties of solids, Springer-Verlag, New York
[13] Hohenberg P. and Kohn W. (1964), Phys. Rev. 136 B, 864
[14] Kohn W. and Sham L. J. (1965), Phys. Rev. 140 A, 1133
[15] Hybertsen M. S. and Louie S. G. (1986), Phys. Rev. B 34, 5390; Godby R. W., Schlüter M. and Sham L. J. (1986), Phys. Rev. Lett. 56, 2415
[16] Ceperley D. M. and Alder B. J. (1980), Phys. Rev. Lett. 45, 566
[17] Van Camp P. E., Van Doren V. E. and Devreese J. T. (1988), Phys. Rev. B 38, 12675
[18] Northrup J. E. and Cohen M. L. (1982), Phys. Rev. Lett. 49, 1349
[19] Bechstedt F. and Del Sole R. (1988), Phys. Rev. B 38, 7710
[20] Godby R. W., Schlüter M. and Sham L. J. (1988), Phys. Rev. B 37, 10159
[21] Slater J. (1963) Quantum Theory of Molecules and Solids, McGraw-Hill, New York
[22] Chiarotti G., Del Signore G., and Nannarone S. (1968), Phys. Rev. Lett. 21, 1170
[23] Chiarotti G., Nannarone S., Pastore R. and Chiaradia P. (1971), Phys. Rev. B 4, 3398
[24] Chiaradia P., Chiarotti G., Nannarone S. and Sassaroli P. (1978), Solid State Commun. 26, 813; Nannarone S., Chiaradia P., Ciccacci F., Memeo R., Sassaroli P., Selci S. and Chiarotti G. (1980), Solid State Commun. 33, 593
[25] Chiaradia P., Cricenti A., Selci S. and Chiarotti G. (1984), Phys. Rev. Lett. 52, 1145
[26] Selci S., Chiaradia P., Ciccacci F., Cricenti A., Sparvieri N. and Chiarotti G. (1985), Phys. Rev. B 31, 4096
[27] Wierenga P. E., van Silfhout A. and Spaarnay M. J. (1979), Surface Sci. 87, 43; (1980) ibidem 99, 59
[28] Selci S., Ciccacci F., Cricenti A., Felici A. C., Goletti C. and Chiaradia P. (1987), Solid State Commun. 62, 833
[29] Del Sole R. and Selloni A. (1984), Solid State Commun. 50, 825
[30] Aspnes D. E. and Studna A. A. (1985), Phys. Rev. Lett. 54, 1956
[31] Aspnes D. E. (1985), J. Vac. Sci. Technol. B 3, 1138; (1985) ibidem B 3, 1498
[32] Acosta-Ortiz S. E. and Lastras-Martinez A. (1987), Solid State Commun. 64, 809
[33] Berkovits V. L., Makarenko I. V., Minashvili T. A., and Safarov V. I. (1985), Solid State Commun. 56, 449
[34] Berkovits V. L., Ivantsov L. F., Makarenko I. V., Minashvili T. A., and Safarov V. I. (1987), Solid State Commun. 64, 767
[35] Haneman D. (1961), Phys. Rev. 121, 1093
[36] Uhrberg R. I. G., Hansson G. V., Nicholls J. M. and Flödstrom S. A. (1982), Phys. Rev. Lett. 48, 1032
[37] Himpsel F. J., Heimann P. and Eastman D. E. (1981), Phys. Rev. B 24, 2003
[38] Pandey K. C. (1981), Phys. Rev. Lett. 47, 1913; (1982) ibidem 49, 223
[39] Del Sole R. (1981), Solid State Commun. 37, 537
[40] Olmstead M. A. and Amer N. A. (1984), Phys. Rev. Lett. 52, 1148
[41] Selloni A., Marsella P. and Del Sole R. (1986), Phys. Rev. B 33, 8885
[42] Bechstedt F. and Enderlein R. (1988) Semiconductor surfaces and interfaces, Academic Verlag, Berlin
[43] Manghi F., Molinari E., Del Sole R. and Selloni A. (1987), Surface Sci. 189/190,

132

1028

[44] Manghi F., Del Sole R., Molinari E., and Selloni A. (1989), Surface Sci.
211/212, 518

[45] Manghi F., Molinari E., Del Sole R. and Selloni A. (1989), Phys. Rev. B 39,
13005

[46] Manghi F., Del Sole R., Selloni A. and Molinari E. (1990), Phys. Rev. B 41, 9935

[47] Kahn A. (1983), Surface Sci. Rep. 3, 193

[48] Bechstedt F., Del Sole R., and Manghi F. (1990), J. Phys.: Cond. Matter 1, SB 75

AB-INITIO MOLECULAR DYNAMICS: SELECTED APPLICATIONS TO DISORDERED SYSTEMS AND SURFACES

Giulia Galli*, Francesco Ancilotto† and Annabella Selloni‡

*IBM Research Division, Zurich Research Laboratory, CH-8803 Rüschlikon, Switzerland

†Università degli Studi di Padova, Dipartimento di Fisica, via Marzolo 8, 35131 Padova, Italy

‡International School for Advanced Studies, SISSA, 34014 Trieste, Italy

ABSTRACT: The novel approach to the study of electronic and thermodynamical properties of condensed matter systems recently proposed by Car and Parrinello has made it possible to perform molecular dynamics simulations with parameter-free potentials: Interatomic forces due to the quantum electronic system are derived from first-principles within the framework of density functional theory. This new scheme has permitted the solution of problems for real materials which had been inaccessible until a few years ago. Selected applications of the Car-Parrinello method to liquids and surfaces are presented. In particular, some properties of liquid carbon, the high-pressure region of the carbon phase diagram and some properties of the (111) surface of silicon are discussed in detail.

1. Introduction

A few years ago, Car and Parrinello [1] proposed a novel approach to the study of electronic and thermodynamical properties of condensed matter systems, which has stimulated new interest and activity in both electronic structure calculations and molecular dynamics (MD) simulations. This new scheme combines the MD technique [2] for the computation of statistical properties of classical systems with the first-principles treatment of interatomic forces due to the quantum electronic system, as provided by density functional theory (DFT) [3,4].

There is a growing list of accomplishments in solving problems for real materials with the Car-Parrinello (CP) method which until a few years ago had been inaccessible. These include physical properties of disordered systems (such as carbon [5,6,7,8], silicon [9,10], gallium arsenide [11], arsenic [12], I-IV alloys [13], selenium [14]) in their liquid and amorphous states; the investigation of processes relevant to semiconductor technology, e.g. the diffusion of hydrogen in Si [15], as well as the study of surfaces reconstruction [16,17,18,19] and atomic clusters [20,21,22,23].

R. Guidelli (ed.), Electrified Interfaces in Physics, Chemistry and Biology, 133–151.
© 1992 Kluwer Academic Publishers.

Molecular dynamics [2] is a well-established methodology for the calculation of statistical properties of condensed matter systems. The essence of the method is the numerical solution of Newton's equations of motion for ensembles of atoms; these equations are solved with iterative techniques for appropriately long time intervals, and equilibrium statistical averages are computed as temporal averages over the observation time. In principle, explicit knowledge of the electronic ground state at each atomic position is necessary for a correct description of the interatomic forces. In conventional MD simulations, no attempt is made to solve this complex many-body problem, and interactions between atoms are modelled with empirical potentials. Although many systems have been investigated with model potentials, it is difficult to find empirical interactions that work for different states of matter aggregation and for a wide range of materials. In complex cases, such as those involving covalent bonds, there is no general agreement, even on such basic questions as to whether the interactions need to encompass 2-, 3-, 4- or higher n-body terms. Furthermore, the empirical MD approach suffers an important conceptual limitation: The correlation between local atomic structure, e.g. bonding properties, and atomic dynamics as well as the effect of atomic dynamics on the electronic properties are missed.

Density functional theory [3,4], in contrast, deals with the ground state of the quantum mechanical many-body system of electrons. It can provide the *exact* charge density and thus, by using the Hellman-Feynman theorem [24], exact forces on atoms for a chosen ionic configuration. In practice, the local density approximation (LDA) [4,25] to treat exchange and correlation has been shown to give very accurate energies and atomic forces for many systems of interest in materials science. Thus, there is a well-developed way to calculate forces from a fundamental electronic theory in an *ab-initio* manner. Until recently, however, LDA energies and forces have been considered prohibitively complex — from a computational point of view — for direct application to statistical mechanics simulations which usually require the study of a number of atoms ranging from several tens up to a few thousands and evolving through 10^4–10^6 configurations. DFT-LDA has mainly been used to investigate zero-temperature properties of ordered systems, such as crystals and superlattices.

The unified approach for DFT and MD [1] has made it possible to use forces derived within DFT-LDA in MD simulations: Interatomic forces are derived *ab-initio* (with no input from experiment or assumptions on the physical characteristics of the system) from the electronic ground state. This has been accomplished by (i) devising a way to generate *at the same time* ionic trajectories *and* the corresponding electronic ground state, and (ii) developing very efficient techniques for the solution of the Schrödinger equation for the single particle electronic orbitals. The first-principles MD technique can also be used within a simulated annealing approach [26] to search for the global minimum of complex energy surfaces, depending upon a large number of degrees of freedom. This approach has been adopted, for example, to investigate the structural properties of different kinds of clusters [20,21,22,23] (S, Si, Na etc.) with up to 60 atoms.

The remainder of the paper is organized as follows: In Section 2 a brief description of the CP method is given; Sections 3 and 4 are devoted to the discussion of some applications, in particular *ab-initio* simulations of liquid carbon and silicon surfaces. Finally, Section 5 contains our conclusions.

2. The Car-Parrinello Method

If the Born-Oppenheimer (BO) approximation holds, the potential energy (Φ) of an ensemble of interacting ions of coordinates $\{R_I\}$ is given by:

$$\Phi\left[\{\vec{R}_I\}\right] = \min_{\{\psi_i\}} E\left[\{\psi_i\}, \{\vec{R}_I\}\right].\tag{1}$$

where the functional $E[\{\psi_i\}, \{R_I\}]$ can be expressed within DFT [3,4] (atomic units are used: $e = \hbar = m_e = 1$):

$$
\begin{aligned}
E\left[\{\psi_i\}, \{\vec{R}_I\}\right] &= \sum_i^{occ} \int d\vec{r}\psi_i^*(\vec{r})\left(-\tfrac{1}{2}\nabla^2\right)\psi_i(\vec{r}) + \int d\vec{r}V^{ext}(\vec{r})\rho_e(\vec{r}) \\
&\quad + \tfrac{1}{2}\int d\vec{r}d\vec{r}\,'\frac{\rho_e(\vec{r})\rho_e(\vec{r}\,')}{|\vec{r}-\vec{r}\,'|} + E^{xc}[\rho_e] + \sum_{I\neq J}\frac{Z_I Z_J}{|\vec{R}_I-\vec{R}_J|}.
\end{aligned}\tag{2}
$$

$\rho_e(\vec{r}) = \sum_i^{occ}|\psi_i(\vec{r})|^2$ denotes the electronic charge density, $\{\psi_i\}$ being single particle orbitals associated with the electrons. $V^{ext}(\vec{r})$ is the total ionic potential acting on the electrons, Z_I are ionic charges and the state sum extends over the occupied states. $E^{xc}[\rho_e]$ is the exchange-correlation energy functional [3] expressed within the LDA [25].

In principle, one could perform *ab-initio* simulations by following three separate steps at each MD move: (i) solve Eq. (1) for a given ionic configuration $\{R_I\}$; (ii) compute forces acting on ions $(-\nabla\Phi)$; (iii) solve Newton equations of motion: $M_I\ddot{R}_I = -\nabla_{R_I}\Phi$. However, for most systems, this procedure is too demanding from a computational point of view and impractical.

The basic idea introduced by Car and Parrinello [1,27] to overcome this problem was to define a fictitious dynamical system whose potential energy surface is E (Eq. 2). This is regarded as a functional of two sets of *classical* degrees of freedom: the ionic (or ionic cores) coordinates ($\{R_I\}$) and the single particle orbitals of DFT ($\{\psi_i\}$). A generalized classical Lagrangian is introduced:

$$\mathcal{L}_{el} = K_e + K_I - E\left[\{\psi_i\}, \{R_I\}\right] + \sum_{ij}\Lambda_{ij}\left(\int dr\,\psi_i^*(r)\psi_j(r) - \delta_{ij}\right)\tag{3}$$

which yields the following coupled equations of motion for $\{R_I\}$ and $\{\psi_i\}$:

$$\mu\ddot{\psi}_i = -\frac{\delta E}{\delta\psi_i^*} + \sum_j \lambda_{ij}\psi_j\tag{4}$$

$$M_I\ddot{\mathbf{R}}_I = -\nabla_{R_I}E.\tag{5}$$

$K_e = \mu \int d\mathbf{r} \, |\dot{\psi}|^2$ is a classical kinetic energy term associated with the $\{\psi_i(\mathbf{r})\}$, where μ is a parameter of dimension (mass)·(length)^2; $K_I = \sum_I \frac{1}{2} M_I \dot{R}_I^2$ is the kinetic energy of the ions of masses M_I. Λ_{ij} are Lagrange multipliers that impose orthonormality constraints between the occupied single particle orbitals. In general, the trajectories generated by the coupled Eqs. (4) and (5) and those obtained with the procedure (i)-(iii) do not coincide. However, the parameter μ and the initial conditions for the electronic states $[\, \{\psi_i\}_0 \, \{\dot{\psi}_i\}_0 \,]$ can be chosen appropriately, so that the trajectories of the fictitious systems described by the Lagrangian (3) reproduce the *true* ionic trajectories very closely, e.g. those of the physical system described by the Lagrangian $\mathcal{L} = K_I - \Phi$. Therefore Eqs. (4) and (5) allow us to move ions and simultaneously to find their electronic ground state without having to repeat the full self-consistent procedure needed to solve Eq. (1) at each MD step, which is computationally very expensive.

We note that for arbitrary values of the parameter μ and initial conditions, thermal equilibration between the classical degrees of freedom $\{R_I\}$ and $\{\psi_i\}$ will be achieved after some time by equipartion. Therefore μ and $\{\psi_i\}_0 \, \{\dot{\psi}_i\}_0$ must be chosen such that the two sets of degrees of freedom are only weakly coupled; in this way the transfer of energy between them is small enough to allow the electrons to follow the ionic motion adiabatically, and thus remain close to their instantaneous BO surface. In such a metastable situation, the time required for thermal equilibration is considerably larger than the characteristic ionic relaxation times, and meaningful temporal averages can be computed.

In the case of metallic systems, the time scales of the electronic and ionic motion may become comparable, and thermal equilibration may occur in fairly short time intervals [6,7,10]. Metals at high temperature present special problems owing to the broad spectrum of their characteristic ionic frequencies. A correct computation of statistical averages can still be accomplished: The average temperature of the ionic system needs to be kept constant during the MD runs (by using, e.g. a Nose thermostat [28,29]); in addition, the electronic coordinates must be periodically quenched to keep them near their proper BO states.

In addition to the basic idea outlined above, *ab-initio* MD simulations have been made possible by the development of efficient computational techniques for the solution of the Schrödinger equation for the electrons [1,27]: Standard matrix diagonalization of the Hamiltonian is replaced by iterative matrix multiplication. The nature of the Hamiltonian (H, consisting of a kinetic part, diagonal in reciprocal space, and a potential part, diagonal in real space) is explicitly used to simplify the evaluation of $H\psi$.

In most applications of the CP method that have appeared in the literature so far and in those described in Sections 3 and 4, the electronic orbitals are expanded in plane waves, and the interaction between valence and core electrons (assumed not to contribute to the chemical bond and then frozen in their atomic arrangement) is described with *ab-initio* pseudopotentials [30,31].

In the next sections, we will discuss applications of the CP method to the simulation of disordered systems and surfaces. The examples presented are liquid carbon and Si(111).

3. First-Principles Simulations of Diamond Melting and Liquid Carbon

The phase diagram of carbon has been the subject of active research for several decades in different fields, such as condensed matter physics [32,33], geology [34,35,36] and astrophysics [37]. Unfortunately, progress has been slow: The key experimental problem is the extreme temperature required to melt both diamond and graphite, while on the theoretical side, a comprehensive model that could describe the complex bonding properties of carbon as a function of temperature (T) and pressure (P) has long been lacking.

Figure 1a shows the phase diagram of carbon as it was assumed to be until the early 1980s. The graphite-diamond phase boundary and the graphite melting curve are the only lines that seem to be well established [32], whereas the high-pressure and high-temperature regions of the phase diagram are largely unknown: The location of the diamond-liquid phase boundary is in question, and the existence of solid phases other than diamond at $P \geq 1$ Mbar has been investigated only partially at zero temperature [38,39,40]. By analogy with the melting behavior of Si and Ge, many authors [32] assumed that the diamond melting curve has a negative dP/dT slope, as shown in Fig. 1a. However, recent experiments [41,42,43,44] seem to indicate that the melting temperature of diamond increases with pressure.

Whereas the properties of solid C at low T are known quite accurately, the knowledge of the liquid-state (l-C) properties has until recently been very poor. There had been no experimental determination of the structural characteristics of l-C, and its electronic properties, in particular the question whether the system is metallic or insulating, have been the subject of many controversies [33,45,46]. Recently, Heremans et al. [33] have shown that carbon undergoes a solid-liquid transition at $T = 4450$ K and $P \leq 4$ bar, leading to a metallic system. However, earlier optical measurements of graphite melted by laser pulses supported the proposal that at low-P l-C is a semi-insulator [45].

We have determined the sign of the slope dP/dT of the diamond-liquid phase boundary in the Mbar region [8], and investigated the properties of l-C at low ($P \simeq 0$) and high ($P \geq 1$ Mbar) pressure [6,7,8] by performing first-principles MD simulations with the method described in Section 2. Since DFT has proven to be a valuable tool for the investigation of different forms of C, such as diamond [38,39,40] graphite [47], high pressure metallic phases [38] and amorphous carbon [5], our calculation of thermodynamic properties is expected to treat the complex bonding characteristics of carbon accurately, and allow realistic predictions of unknown properties in the solid and liquid states.

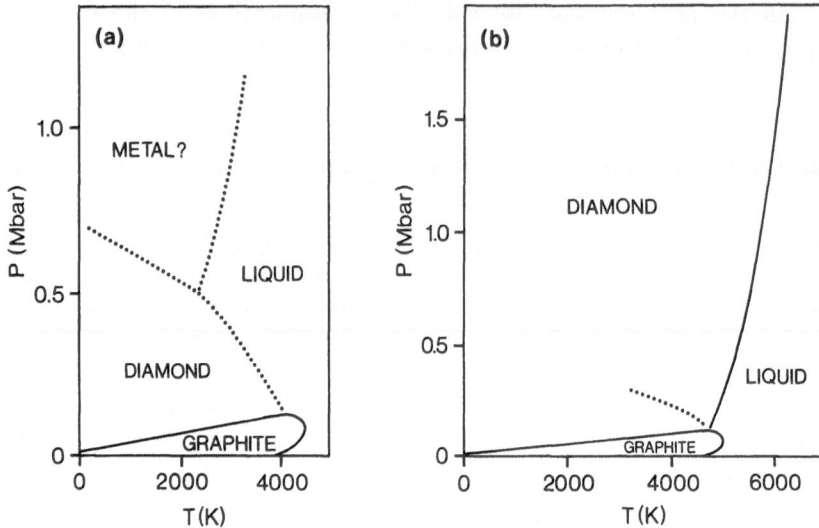

Figure 1. (a) Phase diagram of carbon as it was assumed to be until the early 1980s [32] and (b) according to the results of recent experiments [41,42,43,44] and of our computer simulations [8]. We note that only the sign and not the value of the slope of the diamond-liquid phase boundary (b) has been determined.

In our simulations, we used MD cells with 54 and 64 atoms, and a plane-wave cutoff of 35 Ry for the expansion of the electronic orbitals. Details of our calculations are given in Refs. [6,7,8].

We have established the sign of the slope dP/dT of the diamond-liquid boundary using the Clapeyron equation and direct calculations of the change in pressure upon melting at constant volume (Ω). Figure 2 shows P as a function of T computed for the solid and liquid phases. We defined our system to be a liquid when the atomic displacement as a function of the simulation time showed diffusive behavior. As is to be expected in classical systems, P varies almost linearly with T in both the solid and the liquid. The crossing point of the two curves displayed in Fig. 2 is found to be at a very high temperature: 17000 K. Therefore, provided that melting occurs at $T \leq 17000$ K, the difference in pressure between the liquid and the unstable solid is expected to be positive at the melting point. 17000 K is much higher than any reasonable estimate for the melting temperature. In fact, we can unambiguously establish that already at $T \geq 8000$ K the system is unstable towards melting. We will demonstrate this in the following.

At 8000 K, we have measured the formation energy of a vacancy, which turned out to be very close to zero, indicating a tendency of the system towards spontaneous creation of vacancies: At this temperature a melting transition is expected to occur, nucleated by the presence of the vacancy. However, to expedite matters, we further increased T to 9000 K. At this temperature, a spontaneous melting was observed

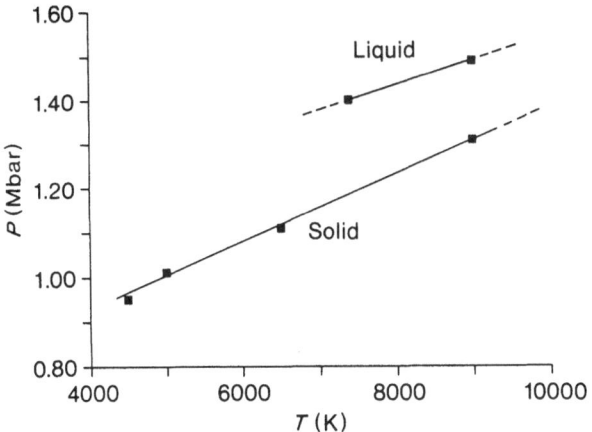

Figure 2. Pressure (P) as a function of temperature (T) for solid and liquid carbon. We obtained the liquid at 9000 K by heating the 8000 K sample (see text), whereas we generated the liquid at 7500 K by cooling the 9000 K system. From Ref. 8. Copyright 1990 by the AAAS.

only after 3000 MD steps: Melting occurred very rapidly, in a first-order fashion, and the onset of the diffusive behavior was accompanied by a sudden rise in the pressure.

This indicates that the pressure increases in going from solid to molten diamond, consistently with the results of Fig. 2. Therefore, in a constant pressure transition the system must expand $(\Omega_l > \Omega_s)$. If we assume that entropy (S) increases upon melting, from the Clapeyron equation

$$\frac{dP}{dT} = \frac{S_l - S_s}{\Omega_l - \Omega_s} \tag{6}$$

we can conclude that $dP/dT > 0$. Thus, unlike the cases of Si and Ge, liquid C is found to be less dense than diamond. Our findings are in agreement with shock wave data reported by Shaner et al. [41,42], with the results of laser-heated diamond anvil cells presented by Bassett [44], and with the results of transmission electron microscopy [35] and resistance measurements [43] of diamond melts that recently appeared in the literature (see Fig. 1b). They are in disagreement with the *old* picture of the carbon phase diagram [32], displayed in Fig. 1a, which predicted a negative slope for the diamond-liquid phase line. Our results imply that in the range of P and T thought to occur in the Earth's mantle and in outer planets, any free carbon is solid, and very likely in the diamond structure.

The two liquids generated at high P and $T = 7500$ and 9000 K, respectively, have nearly identical properties but are remarkably different from those of low-P l-C. Table 1 summarizes the characteristics of high and low-P carbon. A liquid at

Table 1. Structural and electronic properties of carbon in the solid and liquid phases. The average coordination numbers, the type of bonds and the electronic properties of the high and low-pressure liquids are the results of our computer simulations [6,7,8].

	Graphite	Diamond	Low-P liquid	High-P liquid
Average coordination	3	4	2.9	$\simeq 4$
2-fold sites 3-fold sites 4-fold sites 5-fold sites	100	100	32 52 16	25 50 25
Type of bonds	sp^2	sp^3	sp, sp^2 $(+\, sp^3)$	sp^2, sp^3 $(+\, sp)$
Electronic properties	semi-metal	insulator	metal	metal

$P \simeq 0$ and $T = 5000$ K was simulated by melting an amorphous carbon structure that had been obtained [5] with *ab-initio* MD runs at a fixed macroscopic density (ρ_d) of 2 gr cm^{-3}. We find that at low P, l-C has an average coordination of 2.9, slightly smaller than that of graphite, whereas at high P its average coordination is 4, the same as in solid diamond. In particular, the liquid state at $T = 5000$ K and $P \simeq 0$ is found to be composed of 32% 2-fold, 52% 3-fold and the remainder 4-fold coordinated atoms. A snapshot of the system is shown in Fig. 3. An analysis of partial correlation functions and angular distributions shows that 2-fold sites have the shortest bond lengths, with bond angles between 180 and 110°; 3-fold atoms can instead been regarded as distorted graphitic sites, and 4-fold ones have a wide range of preferred bond lengths and angles, quite unlike sp^3 diamond sites. In the two high-P liquids, no 2-fold coordinated atom is present; 40 to 50% of the sites are 4-fold coordinated; the remaining atoms are 3-fold and 5-fold coordinated and occur in approximately equal proportions. 3-fold atoms have the shortest bond lengths, whereas the 5-fold ones have the longest nearest-neighbor (N-N) distances. Furthermore, no 5-fold site has five bonds of approximately equal length: This suggests that the presence of atoms more than 4-fold coordinated is basically due to high temperature fluctuations, and is consistent with the results of LDA calculations [38,39,40] at $T = 0$, showing no tendency for C to have coordination higher than 4 for $P \leq 10$ Mbar.

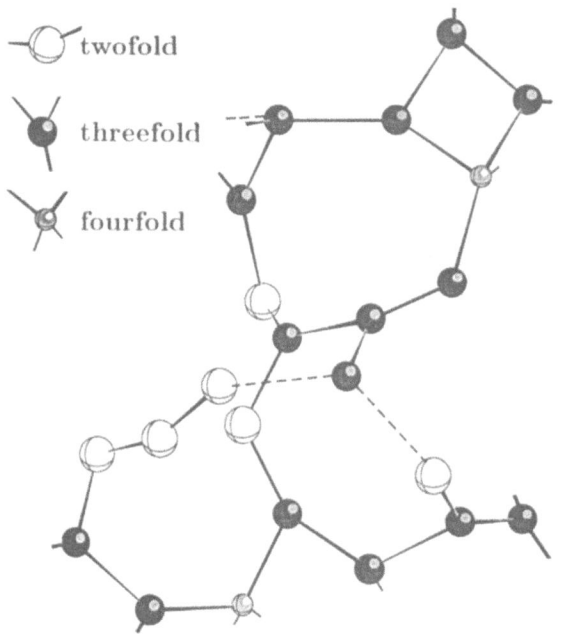

Figure 3. Several atoms of a representative configuration of low-P l-C are shown. The instantaneous configuration has been chosen such that the atomic coordination is close to the average one. White, black and grey spheres indicate 2-, 3- and 4-fold coordinate atoms, respectively. Bonds whose length is close to the first minimum of the particle-particle correlation function are represented as dashed lines. From Ref. 7.

The wide range of N-N distances (from those of the C_2 molecule to those of the diamond lattice) found in both the low and high-P liquids indicates that a variety of bonds, from sp to sp^3, are present in the two systems. However, the different kind of coordination of the two melts suggests that sp^x bonds with $1 \leq x \leq 2$ are the majority in the low-P liquid, whereas sp^x bonds with $2 \leq x \leq 3$ are prevalent in the high-P system (see Table 1). The properties of liquid carbon are very different from those of the other group-IV elements: Si and Ge, e.g., are 6-fold coordinated liquids both at low and high pressure [10].

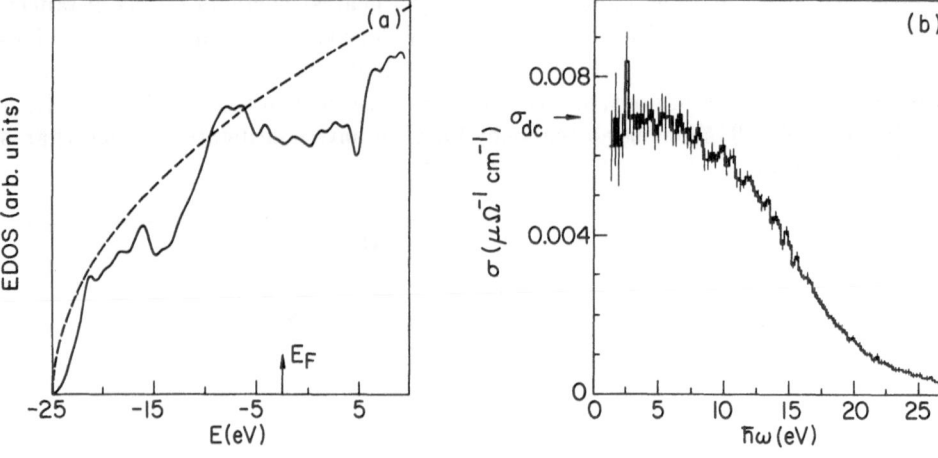

Figure 4. Computed Electronic Density of States (EDOS) and electrical conductivity (σ) of low-P l-C. EDOS and σ have been obtained as averages over ten atomic configurations, selected among the total number (10000) generated in our computer simulation, namely, one about every 1000 steps. The accuracy of the calculation has been checked by computing the density of occupied states using all 10000 configurations, which turns out to be nearly identical to that displayed in the picture. Note the similarity to the free-electron density of states, shown by the dashed line in (a). The extrapolated conductivity at zero [$\sigma(\omega \to 0)$] gives a dc value of about $0.007\mu\Omega\cdot$ cm^{-1} and a resistivity ($\rho_r = 1/\sigma_{dc}$) of $140 \pm 28\,\mu\Omega$. From Ref. 6.

Figure 4 shows the computed density of state (EDOS) and electrical conductivity (σ) of the low-P liquid. Although uncertainties in the calculated EDOS and σ are introduced by finite-size effects, and inaccuracies can possibly derive from an approximate sampling of the Brillouin Zone and from the LDA of DFT, we can draw an important qualitative conclusion from our results: The system is a good metal. A thorough analysis of the electronic properties has been carried out only for the low-P system; however the results of our simulations indicate that the high-P liquid is metallic as well. The results shown in Fig. 4 are in good agreement with reflectivity and resistivity measurements reported very recently [33,43]. We note, however, the apparent disagreement with the conclusions of earlier reflectivity studies, supporting the idea that l-C is a semi-insulator [45].

4. Surface Properties of Silicon

In this section we describe two recent applications of the *ab-initio* MD approach to semiconductor surfaces. The first concerns the 2×1 reconstruction of the Si(111) surface [16], and in particular the transformation process from the bulk-terminated 1×1 to the reconstructed 2×1 surface, including the details of the resulting structure. The second deals with the structural and electronic properties of an isolated surface point-defect, namely a neutral vacancy on Si(111)2×1 [18].

4.1 2×1 RECONSTRUCTION OF Si(111)

A central issue of semiconductor surface physics is the reconstruction that the surface layers often undergo with respect to the bulk structure. In this context, important questions concern the details of the resulting surface structure, its formation process and the changes induced in its electronic and vibrational properties. In particular, the Si(111)-2×1 surface, which appears after cleavage at low temperature, has been the subject of intense theoretical and experimental research in the past twenty years [48]. The key for the understanding of this surface was provided about ten years ago by Pandey's suggestion of a π-bonded chain model [49]. However, in spite of the success of Pandey's model, several issues concerning the static and dynamic structure of this surface have long remained unclear [16,17].

We have studied the reconstruction of the bulk-terminated Si(111) surface using the *ab-initio* MD scheme outlined in Section 2. The surface was modelled using periodically repeated slabs of eight (111) layers with eight Si atoms per layer plus a vacuum region to decouple each slab from its repeated images. The lower (ideal) surface of each slab was stabilized by a monolayer of H. Six layers of Si atoms were allowed to move in our calculations.

The search for the optimal surface geometry of the slab was undertaken with a series of different calculations. In a typical case, we started from a surface with small random displacements of the atoms from their *ideal* positions, simulating the thermal disorder induced by a temperature of ~ 50 K. If the system was then allowed to evolve spontaneously under the action of the interionic forces, a sudden increase of the ionic temperature up to 300–400 K was observed. Since the total energy is conserved, this increase is due to a gain in ionic potential energy, indicating that a structural transformation is taking place. The surface was then slowly cooled down to 0 K by rescaling the ionic velocities. The entire duration of the MD run was typically ~ 5000 time steps, i.e. ~ 1 ps.

A few of the structures characterizing the transformation process are illustrated in Fig. 5. Characteristic features are an initial 2×1 buckling distortion of the topmost layer (see (b)) and the bond-switching process between (b) and (c), which leads to the formation of 5-fold and 7-fold rings of atoms near the surface. The intermediate structure ((c), "chain high") is already chain-like, but the direction of the tilt angle of the topmost chains relative to the substrate is reversed. In the final phase (d), which corresponds to the lowest energy, the sign of the tilt is changed in agreement with experiment [50]. When we performed steepest descent (SD) calculations [1,27] to search for the local minimum closest to the ideal surface geometry, we ended up with the "chain high" arrangement, irrespective of the initial displacement amplitudes given to the atoms. The "chain high" arrangement turns

144

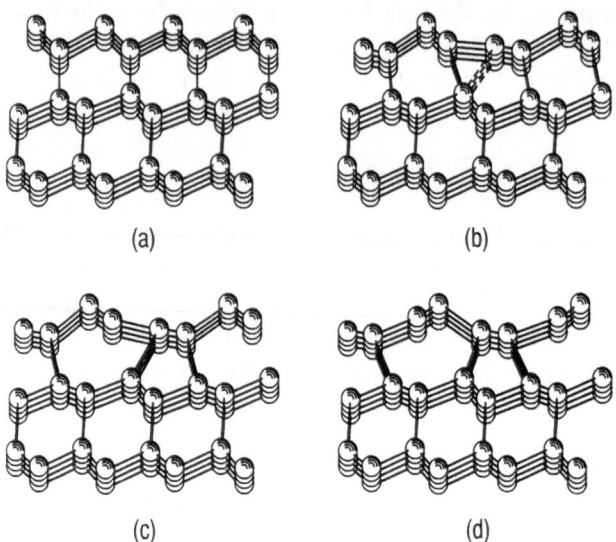

Figure 5. Atomic structures of Si (2×1) surface in the $(\bar{1}10)$ plane corresponding to selected instants in the reconstruction path: (a) ideal (initial), (b) one intermediate, (c) "chain high" and (d) "chain low" (final). From Ref. 16.

out to be a shallow local minimum that is only few meV/surface atom higher in energy than the (d) structure. Our results show that the ideal surface is unstable and that at least two relevant minima exist with 2×1 periodicity and a chain-like structure. Coexistence of the (c) and (d) structures may result in domains, which could be detected by large-scan scanning tunneling microscopy (STM) imaging.

At variance with Pandey's original suggestion, the resulting structure is characterized by a substantial buckling of the surface chains ($b \sim 0.49$ Å), i.e. the chains run along alternating "up" and "down" atoms. A nonzero buckling ($b = 0.3$–0.4 Å) is essential for a quantitative interpretation of the low energy electron diffraction data [50,51]. We found that this buckling is accompanied by a net charge transfer from the "down" to the "up" atoms of the surface chains, with respect to the case $b = 0$, which is likely to have a further stabilizing effect on the surface.

4.2 VACANCY STRUCTURE ON Si(111) 2×1

The identification and characterization of defects on semiconductor surfaces as well as the understanding of their electronic and thermodynamic properties are important issues, due to the influence of surface defects on the electrical properties of semiconductors. Moreover, surface defects often play an important role in reconstruction patterns, such as the (2×2) reconstruction of the GaAs(111) surface [52]. Limited experimental information on localized defects on Si(111) 2×1 is available. Optical [53,54], electron energy loss (EEL) [55] and STM [56] spectroscopy measurements have revealed the existence of electronic states, lying within the surface band gap, which have tentatively been associated with either step and/or point defects.

Our calculations were performed using a periodically repeated slab of six Si layers, with 12 atoms per layer. To check for finite lateral size effects some calculations with slabs of 16 atomic sites per layer were also performed. A fictitious SD dynamics [1,27] for ions and electrons was used to compute structural relaxations, in which the first five layers were allowed to move. (Although no full temperature-dependent study was carried out, we have performed short microcanonical MD runs and thus checked that the resulting structural configurations are locally stable.) We calculated the formation energy E_f for a single, fully relaxed vacancy on both the "up" and "down" atoms of the surface π-bonded chains as well as for a vacancy on the second layer, using

$$E_f = E_V[N-1] - E[N] + E_{bulk}[N]/N.$$

Here $E_V[N-1]$ is the total energy of the defect system with $N-1$ atoms, $E[N]$ is the total energy of the perfect slab of N atoms, and $E_{bulk}[N]$ is the energy of N Si atoms in the bulk phase. The resulting values are $E_f \sim 1.8\,\mathrm{eV}$ and $\sim 1.9\,\mathrm{eV}$ for the "up" and "down" atom vacancy, respectively, whereas the vacancy on the second layer requires $\sim 2.2\,\mathrm{eV}$ to be formed. For comparison, we also calculated the bulk vacancy formation energy, E_f^{bulk}, using a bulk supercell with 72 Si atoms arranged into six (111) layers, very similar to that used for the surface vacancy calculations. We found $E_f^{bulk} = 2.9\,\mathrm{eV}$, to be compared to the experimental value of $3.6 \pm 0.2\,\mathrm{eV}$ [57]. Thus the formation of a vacancy on the "up" surface atom is energetically favored with respect to other positions. No experimental values exist for E_f on Si(111) 2×1.

Figure 6 displays a top view of the fully relaxed "up" vacancy structure. This shows that the relaxation of the atoms around the vacancy is large and asymmetric. In particular, the two initially equivalent neighboring surface atoms H and L become inequivalent, with H rising by $\sim$0.9 a.u. and L lowering by $\sim$0.5 a.u.; atom C on the second layer is displaced towards L. The bonds connecting L and H to their neighbors are stretched, whereas the L–C distance changes from 7.22 (perfect surface) to 5.03 a.u. A large energy gain is associated with this relaxation: For the unrelaxed "up" vacancy $E_f \sim 3$ eV, implying an energy gain due to relaxation of as much as 1.2 eV. (The convergence of our results with respect to the number of atoms in the MD cell has been checked with additional calculations, using 16 atomic sites per layer. The resulting relaxed vacancy structure is very similar to that shown in Fig. 6; differences in the individual atomic positions are at most of $\simeq$0.1 a.u.)

Figure 7 shows the energy levels of the defect states for the unrelaxed and the fully relaxed vacancy. Four vacancy states are observed: Two are (doubly) occupied (v_1 and v_2 in Fig. 7) and two (c_1 and c_2) are empty. v_2 is a σ-bonding combination of $(L + H)$, and its energy is degenerate with the valence band continuum. v_1 is a dangling-bond state on the C atom, c_1 is the $(L - H)$ σ-antibonding state, and c_2

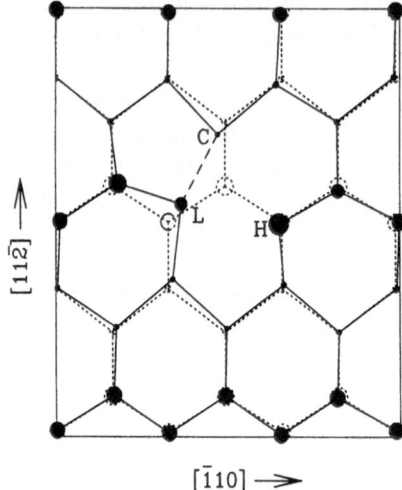

Figure 6. Top view of the first two layers of the Si (2×1) surface with a vacancy on the "up" atom of the surface chain. Circles and lines represent atoms and bonds. Dotted lines and circles: perfect surface. Solid lines and filled circles: fully relaxed vacancy. The circle size reflects the distance of an atom from the surface: the larger circles indicate surface atoms, the smaller ones second-layer atoms. The dashed line represents a weak bond between atoms L and C.

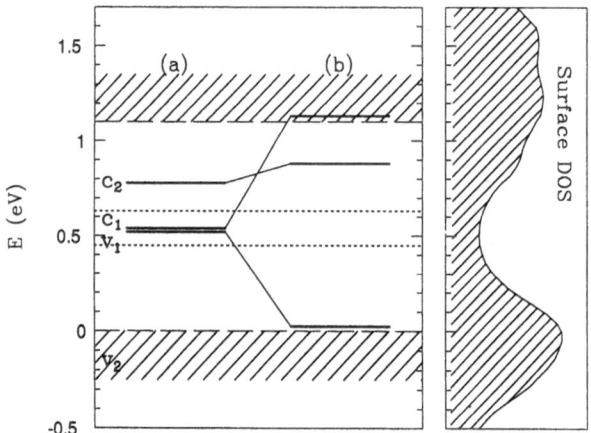

Figure 7. Schematic energy diagram of the vacancy levels for the (a) unrelaxed and (b) fully relaxed vacancy in Si(111) (2×1). The dotted lines show the position of the energy gap between occupied and empty surface states for the perfect surface. The horizontal dashed lines indicate the bulk energy gap, as obtained from the "slab-like" bulk calculation (see text). In the right part of the figure, on the same energy scale, we show the difference in the SDOS computed with and without vacancy.

is a more complicated combination between C and (L + H) π-bonding states. For the unrelaxed vacancy, v_1 and c_1 are accidentally close in energy and fall in the middle of the surface gap. After a full relaxation, they split and move out of the gap. Unlike the case of a vacancy in bulk-Si, this splitting is not driven by a lowering of symmetry: Indeed it takes place even when atoms L and H are allowed to relax around the vacancy, with the constraint of maintaining the mirror-plane symmetry of the perfect surface [18]. In other words, the symmetry-breaking relaxation of the surface vacancy is not driven by a Jahn-Teller effect. Instead, the inequivalent relaxation of H and L is accompanied by an ionic charge transfer from L to H; simultaneously the formation of a weak bond between L and C takes place [18] (dashed line in Fig. 6). Interestingly, the L to H charge transfer is qualitatively similar to that occurring between adatoms and rest atoms on Ge(111)c(2×8) [58,59] or on Si(111) 7×7 [59].

To check whether the vacancy introduces states in the surface gap, we have evaluated the difference between the surface density of states (SDOS) computed with and without the "up" atom vacancy. This is also displayed in Fig. 7, which clearly shows that most vacancy states lie well outside the surface gap.

The above features could be detected experimentally with STM techniques. STM has recently been used to determine the local electronic properties of clean Si(111) 2×1 [60]. In particular, the ratio of tip-to-sample differential to total conductivity, $(\mathrm{d}I/\mathrm{d}V)(I/V)^{-1}$, which has a relatively direct connection to the SDOS [61], has been measured.

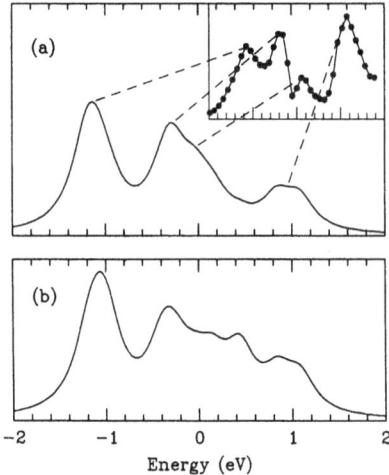

Figure 8. Calculated SDOS for the (2×1) surface (a) in the half of the supercell without defect as a function of energy (relative to the Fermi level) and (b) in the half of the cell containing the vacancy. In the inset the ratio of differential to total conductivity as measured with STM is shown on the same energy scale (from Ref. [60]). The dashed lines connect comparable peaks in the experimental and theoretical spectra.

148

In Fig. 8 we compare our calculated SDOS for the perfect surface (panel (a)) with the experimental results of Ref. [60] (shown in the inset) and the calculated SDOS of the defect system (panel (b)). The SDOS in the vacancy region has additional well-defined features with respect to the case without defects, which could possibly be detected experimentally.

6. Conclusions

We have described the *ab-initio* molecular dynamics method proposed a few years ago by Car and Parrinello [1,27], and presented applications of this novel technique to disordered systems and surfaces. In particular, we have discussed some properties of liquid carbon at high and low pressure, the slope of the diamond-liquid phase boundary and the structure of the defect-free and defective Si(111)2×1 surfaces. Our results, together with those for a variety of other systems reported in the literature over the past five years, show that the CP method has achieved two important objectives in the field of computational condensed matter physics: (i) to perform MD simulations with no input from experiment and/or assumptions on the physical characteristics of the system; the only assumptions are the validity of classical mechanics to describe ionic motion and of the BO approximation to separate nuclear and electronic coordinates; and (ii) to compute ground-state electronic properties of disordered and large systems (i.e., considerably larger than those handled by standard techniques for the diagonalization of the electronic Hamiltonian), at the level of state-of-the-art electronic structure calculations. However, the present implementations of the CP method are computationally very demanding both in terms of CPU time and memory occupation. With supercomputers, feasible *ab-initio* simulations involve cells containing $\simeq 150$ bulk atoms for systems like silicon, and $\simeq 100$ atoms for systems like carbon (for which a much larger number of plane waves is required for a proper description of its electronic orbitals than for Si). Furthermore, elements with very localized charge densities, such as transition metals or oxygen, cannot be treated efficiently, except for special cases [62,63]. Novel formulations of the equations used in density functional theory to describe the electronic wave functions as well as new pseudopotentials may help to solve these problems.

References

[1] R. Car and M. Parrinello, Phys. Rev. Lett. **55**, 2471 (1985).

[2] *Molecular Dynamics Simulation of Statistical Mechanical Systems*, edited by G. Ciccotti and W. Hoover (Plenum Press, New York, 1986).

[3] P. Hohenberg and W. Kohn, Phys. Rev. **136**, B864 (1964).

[4] W. Kohn and P. Vashishta, *Theory of the Inhomogeneous Electron Gas*, edited by S. Lundqvist and N. H. March (Plenum Press, New York, 1983) p. 79.

[5] G. Galli, R. Martin, R.M. Car, and M. Parrinello, Phys. Rev. Lett. **62**, 555 (1989).

[6] G. Galli, R. Martin, R.M. Car, and M. Parrinello, Phys. Rev. Lett. **63**, 988 (1989).

[7] G. Galli, R. Martin, R.M. Car, and M. Parrinello, Phys. Rev. B **42**, 7470 (1990).

[8] G. Galli, R. Martin, R.M. Car, and M. Parrinello, Science (1990) in press.

[9] R. Car and M. Parrinello, Phys. Rev. Lett. **60**, 204 (1988).

[10] I. Stich, R. Car, and M. Parrinello, Phys. Rev. Lett. **63**, 2240 (1989).

[11] Q.M. Zhang, G. Chiarotti, A. Selloni, R. Car, and M. Parrinello, Phys. Rev. B **42**, 5071 (1990).

[12] X.P. Li, P.B. Allen, J.Q. Broughton, R. Car, and M. Parrinello, Phys. Rev. B **41**, 3260 (1990).

[13] G. Galli and M. Parrinello, J. Phys. Cond. Matter (1990) in press.

[14] D. Hohl, R.O. Jones, R. Car, and M. Parrinello, J. Am. Chem. Soc. **111**, 826 (1989).

[15] F. Buda, G. Chiarotti, R. Car, and M. Parrinello, Phys. Rev. Lett. **63**, 294 (1989).

[16] F. Ancilotto, W. Andreoni, A. Selloni, R. Car, and M. Parrinello, Phys. Rev. Lett. (1990) in press.

[17] F. Ancilotto, A. Selloni, W. Andreoni, S. Baroni, R. Car, and M. Parrinello, submitted for publication (1990).

[18] A. Selloni, F. Ancilotto, and E. Tosatti, to be published.

[19] I. Moullett, Andreoni W., and M. Parrinello, to be published (1990).

[20] P. Ballone, W. Andreoni, R. Car, and M. Parrinello, Phys. Rev. Lett. **60**, 271 (1988).

[21] D. Hohl, R.O. Jones, R. Car, and M. Parrinello, Chem. Phys. Lett. **139**, 540 (1987).

[22] D. Hohl, R.O. Jones, R. Car, and M. Parrinello, J. Chem. Phys. **89**, 6823 (1988).

150

[23] P. Ballone, W. Andreoni, R. Car, and M. Parrinello, Europhys. Lett. **8**, 73 (1989).

[24] R. Feynman, Phys. Rev. **56**, 340 (1939).

[25] W. Kohn and L. Sham, Phys. Rev. **140**, A1133 (1965).

[26] S. Kirkpatrick, C.D. Gelatt, and M.P. Vecchi, Science **220**, 671 (1983).

[27] R. Car and P. Parrinello, *Simple Molecular Systems and Very High Density*, edited by A. Polian, P. Loubeyre and N. Boccara (Plenum Press, New York, 1988) p. 455.

[28] S. Nose, Mol. Phys. **52**, 255 (1984).

[29] S. Nose, J. Chem. Phys. **81**, 511 (1984).

[30] D.R. Hamann, M. Schlüter, and C. Chiang, Phys. Rev. Lett. **43**, 1494 (1979).

[31] G.B. Bachelet, D.R. Hamann, and M. Schlüter, Phys. Rev. B **26**, 4199 (1982).

[32] F.P. Bundy, Physica A **156**, 169 (1989).

[33] J. Heremans, C.H. Olk, G.L. Eeseley, J. Steinbeck, and G. Dresselhaus, Phys. Rev. Lett. **60**, 453 (1988).

[34] J.S. Dickey, W.A. Bassett, J.M. Bird, and M.S. Weathers, Geology **11**, 219 (1983).

[35] M.S. Weathers and W.A. Bassett, Phys. Chem. Minerals **15**, 105 (1987).

[36] J.S. Gold, W.A. Bassett, M.S. Weathers, and J.M. Bird, Science **225**, 921 (1984).

[37] M. Ross, Nature **292**, 435 (1981).

[38] M.T. Yin and M.L. Cohen, Phys. Rev. Lett. **50**, 2006 (1983).

[39] R. Biswas, R.M. Martin, R.J. Needs, and O.H. Nielsen, Phys. Rev. B **30**, 3210 (1983).

[40] R. Biswas, R.M. Martin, R.J. Needs, and O.H. Nielsen, Phys. Rev. B **35**, 9559 (1987).

[41] J.W. Shaner, J.M. Brown, C.A. Swenson, and R.G. McQueen, J. Phys. C **8**, 235 (1984).

[42] A.C. Mitchell, J.W. Shaner, and R.N. Keeler, Physica **139**, 386 (1986).

[43] M. Togaya, *Proceedings of the First International Conference on the New Diamond Science and Technology*, Tokyo, Japan (1988).

[44] W.A. Basset, Bull. Am. Phys. Soc. **35**, 465 (1990).

[45] A.M. Malvezzi, N. Bloenbergen, and C.Y. Huang, Phys. Rev. Lett. **57**, 146 (1986).

[46] D.H. Reitze, X. Wang, H. Ahn, and M.C. Downer, Phys. Rev. B **40**, 11986 (1990).

[47] S. Fahy, S. Louie, and M.L. Cohen, Phys. Rev. B **34**, 1191 (1986).

[48] D. Hanemann, Rep. Progr. Phys. **50**, 1045 (1987).

[49] K. Pandey, Phys. Rev. Lett. **47**, 1913 (1981).

[50] F.J. Himpsel, P.M. Marcus, R. Tromp, I.P. Batra, M.R. Cook, F. Jona, and H. Liu, Phys. Rev. B **30**, 2257 (1984).

[51] H. Sakama, A. Kawazu, and K. Ueda, Phys. Rev. B **34**, 1367 (1986).

[52] S.Y. Tong, G. Xu, and W.N. Mei, Phys. Rev. Lett. **52**, 1693 (1984).

[53] P. Chiaradia, G. Chiarotti, S. Selci, and Z. Zhi-Ji, Surf. Sci. **132**, 62 (1983).

[54] J. Bokor, R. Storz, R.R. Freeman, and P.H. Bucksbaum, Phys. Rev. Lett. **57**, 881 (1986).

[55] N.J. Di Nardo, J.E. Demuth, W.A. Thompson, and P. Avouris, Phys. Rev. B **31**, 4077 (1985).

[56] R.M. Feenstra, W.A. Thompson, and A.P. Fein, Phys. Rev. Lett. **56**, 608 (1986).

[57] S. Dannefaer, M. Mascher, and D. Kerr, Phys. Rev. Lett. **56**, 2195 (1986).

[58] R. van Silfhout, J.F. van der Veen, C. Norris, and J.E. Macdonald, J. Chem. Soc. Faraday Discussion **89** (1990) in press.

[59] R.D. Meade and D. Vanderbilt, Phys. Rev. B **40**, 3905 (1989).

[60] J.A. Stroscio, R.M. Feenstra, and A.P. Fein, Phys. Rev. Lett. **57**, 2579 (1986).

[61] J. Tersoff and D.R. Hamann, Phys. Rev. B **31**, 805 (1985).

[62] P. Ballone and G. Galli, Phys. Rev. B **40**, 8563 (1989).

[63] P. Ballone and G. Galli, Phys. Rev. B **42**, 1112 (1990).

THE PROBLEM OF SCHOTTKY BARRIER

P. PERFETTI

Istituto di Struttura della Materia
via E. Fermi 38
00046 Frascati, Italy

ABSTRACT. A short historical survey will be made of metal-semiconductor research and of the most widely used theoretical models that have tried to explain the physical reasons for the formation of a Schottky barrier. Among them the Schottky model, the interface states model, the metal-induced gap states (MIGS) model and the defect model will be reviewed. Experimental methods for the determination of the Schottky barrier will be presented, including a short review of methods based on transport properties and on photoelectron spectroscopy. Experiments performed at low temperatures and the main current conclusions of the scientists working in the field will be the subject of the final remarks.

1. Introduction

When a metal and a semiconductor are brought into contact to form a junction, as shown in Fig. 1, a potential barrier $q\Phi_B$ is formed, which is responsible for the rectifying properties of the system. A deviation from Ohm's law is observed when a voltage is applied across the junction; the electrical resistance is much lower for a negative voltage applied to the semiconductor of Fig. 1 than for a positive voltage. This system is called a Schottky diode and $q\Phi_B$ is the Schottky barrier.

How is $q\Phi_B$ formed? What is the physical mechanism which determines the barrier height? This is perhaps the most important problem in metal-semiconductor research (1). Even if Schottky diodes are widely employed in electronics, and there is a large variety of metal-semiconductor systems used for different purposes, the above fundamental problem remains to be solved.

From a historical point of view, rectifying properties on metal-sulfide junctions were first observed in 1874 (2), but only in 1938 (3) Schottky postulated the existence of a depletion layer on the semiconductor side of the junction to explain the observed phenomena. Even though the Schottky model was able to explain the rectifying properties of metal-semiconductor junctions, it failed to predict the Schottky barrier heights in different systems. The main reason for the failure of the Schottky model consists in the neglect of the possible role played by the interface during the formation of the junction. A first approach in this direction was made in 1947 by Bardeen (4), who proposed that the barrier height is determined by the intrinsic surface states of the semiconductor. A limitation

R. Guidelli (ed.), Electrified Interfaces in Physics, Chemistry and Biology, 153–177.
© 1992 Kluwer Academic Publishers.

154

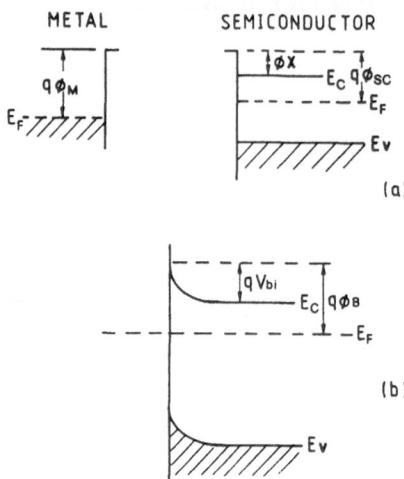

Figure 1. General band scheme of a Schottky barrier. In (a) metal and semiconductor are separated; in (b) they are brought into contact to form a junction.

to this model comes from the hypothesis that the intrinsic surface states are still present at the interface once the junction is formed. The model might still retain its validity if one considers interface states induced by the metal rather than intrinsic surface states present at the junction, and this hypothesis has been advanced by different authors in the subsequent years (5-10). The next important step in the theory was made in 1965 by Heine (11), who introduced the so called metal-induced gap states (MIGS) model where the pinning of the Fermi level E_F at the junction is determined by the tailing of the metallic wavefunction into the semiconductor. This model was subjected to different reformulations (12-13), and at present the pinning of E_F due to MIGS states is regarded (14) as the primary mechanism determining $q\Phi_B$.

From an experimental point of view we can distinguish two main approaches to the problem of Schottky barrier, namely the "device approach" and the "surface science approach". The first approach consists in evaluating barrier heights in metal-semiconductor contacts from current-voltage (I/V) and capacitance-voltage (C/V) techniques on devices, i.e. systems where the junction is formed between two bulk materials. In this approach the metal-semiconductor junction is considered as a whole system including the interfacial region, the depletion layer and the two bulk materials. Insight into the primary mechanism determining $q\Phi_B$, as obtained by transport properties, is gained from a natural average of the interface parameters originating from the different regions which contribute to the transport properties. Another important technique, the "internal photoemission" (15), is usually applied to devices, but the photoexcitation arises directly at the interface, and the technique itself is a more local probe than the techniques based on transport properties.

The aim of the "surface science approach" is to shed some light on those phenomena, arising during interface formation, which are important in establishing $q\Phi_B$. These could be substrate disruption, nucleation of reacted species, outdiffusion of substrate atoms, and adatoms clustering. This approach must be based on techniques which are very surface sensitive and which permit us to study Schottky barriers step by step during the formation of the interface. In a following paragraph we will focus our attention on photoemission techniques, although other methods, e.g. Auger Electron Spectroscopy (AES), Low Energy Electron Diffraction (LEED) and Low Energy Electron Loss Spectroscopy (LEELS), are usually employed as complementary techniques. In general these studies are performed on a semiconductor covered successively with very thin metal overlayers, ranging from a submonolayer to a few monolayers. Due to the extreme surface sensitivity of the above techniques, they become useless for an interface analysis after a few Angstroms of metal deposition. The "surface science approach" is then complementary to the "device approach", and the majority of scientists working in this field are aware that both are needed to solve the problem of Schottky barrier.

2. The Schottky Model

The Schottky model can be easily derived from Fig.1, for the case of an n-doped semiconductor, taking the vacuum level as the zero energy reference for both materials and leaving the Fermi levels to line up when the junction is formed. The current flow, due to a transfer of electrons from the semiconductor to the metal (i.e. from the material with lower work function to the other), while causing the alignment of E_F, also produces a space charge region on the semiconductor side (depletion layer). At equilibrium, the established electrical dipole opposes any further current flow. This thermodynamic equilibrium is then characterized by a Schottky barrier $q\Phi_{Bn}$ given by:

$$q\Phi_{Bn} = q (\Phi_m - \chi). \tag{1}$$

where Φ_m and χ are the metal work function and the semiconductor electron affinity.

In a very similar way the Schottky barrier height $q\Phi_{Bp}$ for a p-doped semiconductor can be derived:

$$q\Phi_{Bp} = E_G - q (\Phi_m - \chi). \tag{2}$$

For a given semiconductor, whatever the metal overlayer may be, we should have:

156

$$q\,(\Phi_{Bn} + \Phi_{Bp}) = E_G. \tag{3}$$

Two main consequences can be derived from the above model. Firstly, the Schottky barrier height depends on the metal work function (eqns. 1 and 2), and different Schottky barrier heights should result upon joining metals with different work functions to the same semiconductor. Secondly, the final position of E_F in the forbidden gap in a p-type or n-type semiconductor is the same for the same deposited metal, and the two Schottky barrier heights should add up to the energy gap.

After the formulation of the Schottky model it was clear that the formation of a depletion layer, by inducing an interface dipole and a potential step, was responsible for the observed rectifying properties. However, the model failed to explain experimental results quantitatively because a great number of junctions formed on silicon (16) (see Fig. 2) as

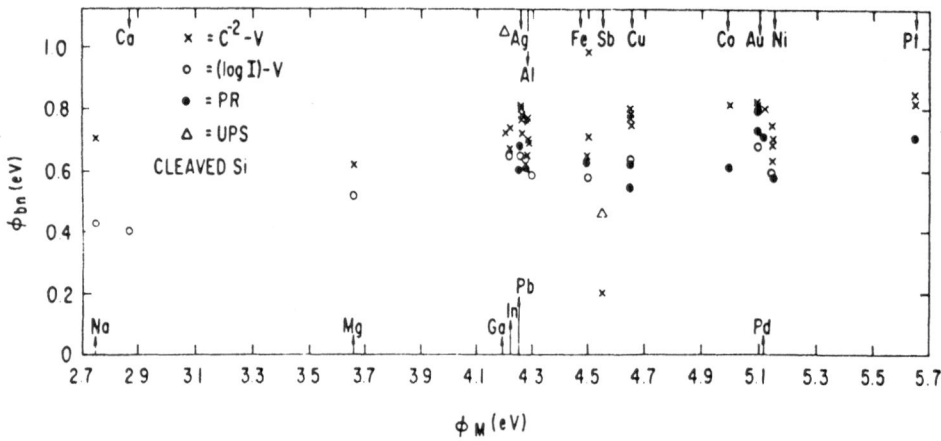

Figure 2. Comparison between measured Schottky barrier heights on vacuum cleaved silicon and the measured work functions (from ref. 16).

well as on GaAs (17) (see Fig 3) and other semiconductors, using different metals, showed no correlation at all between Schottky barrier heights and metal work functions. Instead, Eqn.(3) showed a more general validity for thick metal overlayers (18). Failures to this rule are often attributed (14) to the presence of interface defects induced by the particular Schottky barrier preparation.

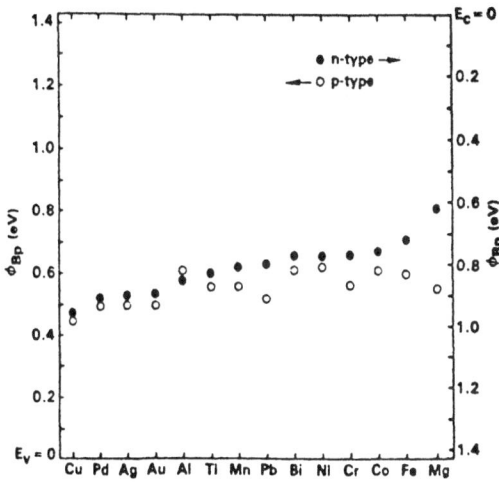

Figure 3. Schottky barrier height of metal-GaAs (100) as measured by the I-V method; Solid circles: n-type contacts (right-hand scale); open circles: p-type contacts (left hand scale) (from ref. 17).

3. The Interface States Model

The experimental evidence that the Schottky barrier height is practically independent of metal work function, led Bardeen (4) to regard the existence of surface states as the possible mechanism inducing the pinning of the Fermi level. A high density of localized surface states in the energy gap could be the necessary tank for the charge transfer at the interface leading to thermodynamic equilibrium. The intrinsic origin of these surface states, due to the breaking of the translational symmetry at the semiconductor-vacuum interface, was soon questioned, since there is no reason why intrinsic surface states should still be present at the interface once the junction is formed. Scientists following Bardeen's idea focused their attention on the extrinsic nature of interface states which are formed when the junction is made. The nature of these extrinsic states (the term "extrinsic" will be used herein to mean "induced by the metal overlayer") is still controversial, and different models, e.g. the substitutional metal impurity defects states model (19), the native point defects model (5) and the metal-induced gap states model (MIGS) (11-13), have been suggested.

3.1. THE BARDEEN MODEL

The model is illustrated in Fig 4 (20), where an experiment is simulated in which a metal and a semiconductor, separated in Fig. 4a, are connected by a wire on their back sides (not shown, Fig. 4b) forcing their Fermi levels to coincide; a potential difference $\Delta = q(\Phi_m - \chi - \Phi_{Bn})$ is established in the region between the separated faces. The existence of localized surface states in the gap already creates a depletion layer and a band bending at the clean semiconductor surface, and the Fermi level is pinned by the high density of surface states (Fig. 4a). When the two systems approach each other the potential drop across the

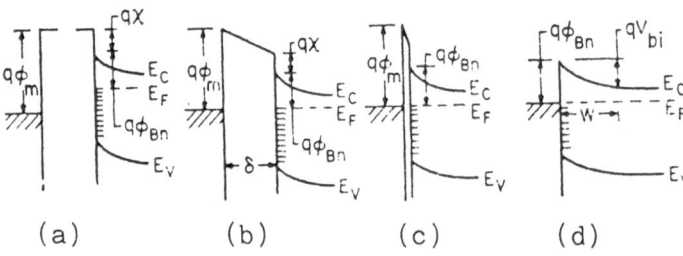

Figure 4. Band scheme in the Bardeen model (from ref. 20).

semiconductor side is controlled by the surface charge density with a small movement of the Fermi level. When metal and semiconductor are joined together the Schottky barrier height $q\Phi_{Bn}$ is practically the same as the one existing at the clean surface. The clear consequence of the model is the independence of $q\Phi_{Bn}$ from the metal work function. It should be stressed that the existence of surface states at free semiconductor surfaces is not a general case. In fact, some of the III-V semiconductors (21) do not show any surface state contribution in the energy gap, and a flat band scheme at the vacuum-semiconductor interface occurs. The last consideration rules out the possibility, at least for III-V semiconductors, that the intrinsic surface states remain at the interface after contact with the metal and are responsible for the pinning of E_F. The previous model still retains its validity if one substitutes the words "surface states" with the words "interface states". These are electronic states in the energy gap which arise from the interaction of metal adatoms with the semiconductor surface. Among the various models so far proposed, the unified defect model (5) and the metal-induced gap states model (MIGS) (11-13) are the most successful.

3.2. THE UNIFIED DEFECT MODEL

This model, developed for III-V semiconductors and proposed by the Stanford group (5), suggests that the origin of interface states is due to native defects formed during the first steps of metal deposition. These native defects could be anion antisite defects (i.e. the anion takes the place of the cation, e.g. As on a Ga site in GaAs), or cation antisite defects, to take into account the pinning on p- and n-doped substrates. Vacancies of group III or group V elements as well as more complicated complexes are also considered as possible interface defects. Despite the great efforts from both an experimental and a theoretical point of view, the electronic nature and density of these defects have not been well established. The model was suggested by the experimental findings reported in Figs. 5 and 6. Figure 5 shows the evolution of the Fermi level position as a function of different metal and oxygen coverages on n- and p-type GaSb. Two main results support the defect model: the very low coverage inducing the pinning of E_F and the independence of Schottky barrier height from the type of adatoms. Similar results, summarized in Fig. 6, were obtained for GaAs and InP. The first finding was regarded by the authors as evidence that metal overlayers need not be thick to produce the pinning of E_F; the second result was used to discard the hypothesis that the Schottky barrier height is determined by energy levels directly induced by the orbitals of the adatoms; in fact these adatoms are so different that it is difficult to believe that they may induce states having approximately the same energy.

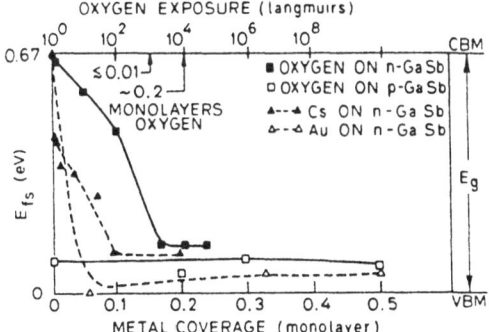

Figure 5. E_F behaviour vs. oxygen, Cs and Au coverage on GaSb (from Ref.5).

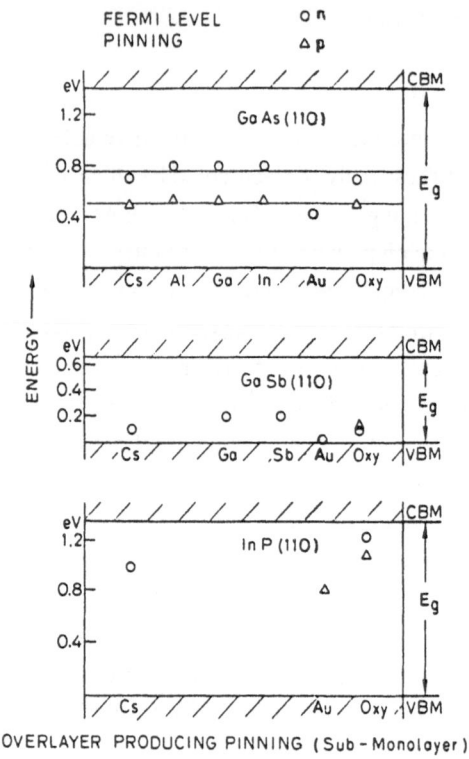

Figure 6. Pinning position of the Fermi level in the energy gap of GaAs 110, GaSb 110 and InP 110 for different metal overlayers (from Ref.5).

It is important to note that the small amount of deposited metal required to pin E_F implies that the screening charge comes from the semiconductor depletion layer, which is of the order of 10^3 Å depending on the doping. A very few native defects of density 10^{12} cm^{-2}, a surface density equivalent to the bulk doping density, can pin the Fermi level in this case. A calculation of the Fermi level position as a function of the interface state density for two different defects levels was carried out in Ref. 22 and reported in Fig. 7. The density of interface defects which is required to pin E_F is so small that it is well below the limits of detectability of many local surface probes used to measure the density of states. This intrinsic experimental difficulty makes it hard to verify the validity of the unified defect model, and many experimental efforts have been made in the last years to study Schottky barriers at low temperatures in order to try to identify the energy levels of the above defects (see Ref. 14 and references therein).

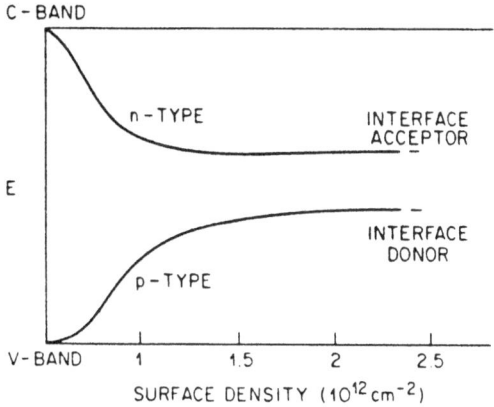

Figure 7. Position of the Fermi level as a function of the interface state density for two different defects levels (from Ref. 22).

3.3. THE METAL-INDUCED GAP STATES MODEL (MIGS)

E_F pinning by metal-induced interface states was considered by Heine (11) and Pellegrini (23), and was further elaborated by Yndurain (24), Tejedor *et al.* (12), Tersoff (13), and Louie and Cohen (12). This model, which was first developed for thick metal layers (>1 monolayer), is based on the concept of metal wavefunctions tunneling into the semiconductor energy gap. The first few layers of a semiconductor near the interface are metal-like, and a new density of interface states is formed which is compensated for by a decrease of the density of states in the conduction and valence bands of the semiconductor. This is shown in Fig. 8 for the case of a one-dimensional model of a metal-covalent

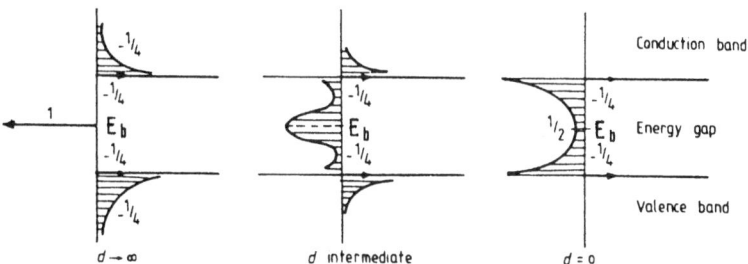

Figure 8. One dimensional model of a metal-covalent semiconductor interface (from Ref. 25).

semiconductor interface. The figure shows the case of a decoupled interface (a surface state is present in the middle of the gap having a half occupancy), the intermediate situation (the surface state is broadened) and the case of intimate contact, when a continuum of states is formed. In all the above cases a charge neutrality level E_B exists, which is the separation between conduction and valence-like states. It is important to note that, in the simple example reported in Fig. 8, E_B lies at the same position as the surface state on the clean surface. Then E_B seems to give a natural estimate of the local quasi-Fermi level or "charge neutrality level" in the semiconductor, once the interface is formed. The same arguments apply to a three-dimensional case (25), and the role of E_B at the interface is equivalent to that of E_F in a metal. When a metal-semiconductor junction is formed E_F and E_B must be aligned, otherwise an electrostatic dipole would form restoring the bands line-up as required by the charge neutrality. Examples of calculations of the above models and a comparison between experimental and theoretical Schottky barrier heights are reported in table 1 for different semiconductors, using gold as the metal overlayer (26). Agreement is good except for the case of GaAs. Even in the latter case, a better fitting of the conduction band gave a value of 0.5 eV for E_B, in agreement with the experimental value. Since in the above calculations the band structures are an important ingredient in the determination of E_B, the accuracy of the band structure used gives an indication of the magnitude of the error introduced in the theoretical estimates.

The concept that the charge-neutrality level position can be dependent on the interface conditions was recently introduced by Platero *et al.* (27), who studied the variation of E_B for the case of Al-Si interface using different sites for the aluminum adatoms. They found variations up to 0.2 eV with respect to the position of E_B on the free semiconductor surface. These results also extend the validity of the MIGS model down to one monolayer of deposited metal.

Table 1

Band gap E_G, experimental barriers $q\Phi_B$ (Au on p-type material) and calculated effective charge-neutrality level E_B (from Ref. 26).

	E_G	$q\Phi_B$	E_B
Si	1.11	0.32	0.36
Ge	0.66	0.07	0.18
GaP	2.27	0.94	0.81
InP	1.34	0.77	0.76
AlAs	2.15	0.96	1.05
GaAs	1.43	0.52	0.70
InAs	0.36	0.47	0.50
GaSb	0.70	0.07	0.07

Despite these theoretical efforts the MIGS model bases its validity on the metallicity of the overlayer, and it is not yet clear how it could explain the Fermi level pinning positions in the submonolayer regime as reported in Fig.5.

4. Schottky Barrier Height as Measured by Electrical Properties of Junction

The band scheme of Fig. 4 tells us that there is an easy flow of current when the semiconductor is negatively biased and the electrons from the semiconductor side see a lower potential barrier (direct polarization); the opposite is true for inverse polarization when the semiconductor is positively biased. The current-voltage characteristics (I-V) can be expressed by the familiar thermionic emission formula (20):

$$I = I_s \exp(qV/nkT) \, [1 - \exp(-qV/kT)]. \tag{4}$$

Here n is an "ideality" factor (the best diodes have n very close to unity), V is the applied voltage, T is the temperature and I_s is the saturation current density obtained in reverse bias conditions. It is easily seen from Eqn. (4) that a plot of $\ln\{I/[1-\exp(-qV/kT)]\}$ vs. V is linear for all V values; the corresponding plot permits us to measure I_s and n from the intercept and slope of the best fit of experimental data to a straight line. The Schottky barrier height $q\Phi_B$ is related to I_s by:

$$I_s = A^* \, T^2 \, \exp[-q(\Phi_B - \Delta\Phi)/kT] \; \text{A cm}^{-2}, \tag{5}$$

where A^* is the Richardson constant and $\Delta\Phi$ is the image force correction (20). The effective work function $q(\Phi_B - \Delta\Phi)$ as measured by the I-V characteristics must be corrected for the image force contribution to obtain the real $q\Phi_B$ metal-semiconductor Schottky barrier height.

The other widely employed technique for the measurement of $q\Phi_B$ is the capacitance-voltage method. In this case a plot of $1/C^2$ vs. V can be analyzed using the following expression, valid for the case of an ideal abrupt junction and for an n-type sample:

$$1/C^2 = 2(\varepsilon \, q \, N_D)^{-1} \, (V_{bi} - V - kT/q) \; \text{F}^{-2} \, \text{cm}^2, \tag{6}$$

where N_D is the donor concentration and V_{bi} is the built-in potential (see Fig. 4). The intercept on the voltage axis V_i is related to the built-in potential by $V_i = V_{bi} - kT/q$ and the Schottky barrier height equals $q\Phi_B = q(V_i + \delta + kT/q)$, where δ is the energy separation between the Fermi level and the conduction band in the bulk. Some results obtained with the above methods are reported in Fig. 3.

The above electrical methods are strictly valid only if the doping level is constant up to the interface, giving rise to an abrupt junction and without any interface charges present therein. Interface charges and impurity gradients tend to modify the intercept, e.g., in the C-V plot. This is the reason why measurements of Schottky barrier heights may often depend on the preparation of samples which may induce inhomogeneity in the interface region. These limitations opened the way to other more local techniques, such as photoelectron spectroscopy, with the aim of performing a direct measure of $q\Phi_B$.

5. Photoemission Measurements of Schottky Barrier Height

This approach requires the growth *in situ* of a thin metal overlayer on top of a semiconductor. Ultraviolet and soft X-ray photons impinging on the the sample create photoelectrons. Due to the short electron mean free path inside the two materials, the

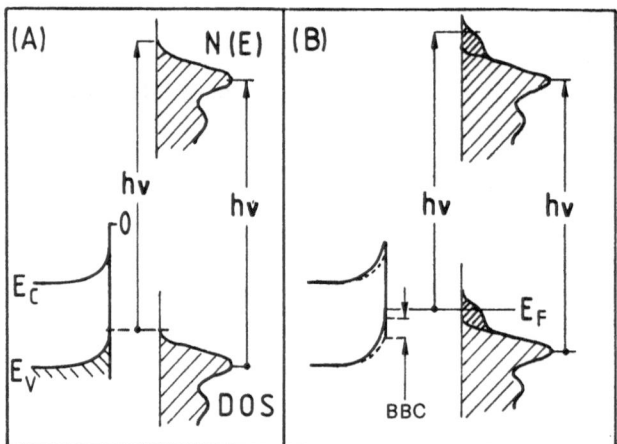

Figure 9. Photoemission measurement of a Schottky barrier height. (A) The first part of the experiment is performed on the clean surface of the substrate. Photons of energy h v create photoelectrons whose energy distribution reflects the density of electronic states (DOS), as a first approximation. (B) When a thin metal overlayer is deposited and the Fermi edge is clearly visible, the Schottky barrier height can be measured on the basis of the knowledge of the distance of E_F from the top of the valence band once the band bending is taken into account, for example from the movement of the core lines during metal deposition. Since both valence band and core lines are important in this analysis, these experiments need Synchrotron radiation as the photon source providing photons with an energy range large enough to excite shallow (valence) and deep (core) electrons .

photoelectrons originate from the overlayer and from the substrate region immediately close to the interface. Their measured distribution in energy is an image of the local density of states at the interface, at least as a first approximation.

The principle of this method is illustrated in Fig. 9. The first part of the experiments is performed on the clean surface of the substrate (Fig. 9A). As a first approximation, the density of the electronic states (DOS) of the substrate is given by the energy distribution of photoelectrons, $N(E)$. Notice that the position in energy of E_v on the surface can be measured by subtracting $h\nu$ from the position of the leading edge of $N(E)$. Figure 9B shows the situation when a thin metal overlayer is deposited. If the overlayer has a thickness comparable with the escape depth (a few Angstroms depending on the electron energy), it is possible to have an electron contribution from both the substrate and the overlayer. Photoemission measurements obtained step by step at different metal coverages allow us to follow the evolution of the density of interface states until the Fermi edge onset is established. The deposition of metal overlayers induces band bending changes which can be followed by the movement of the Fermi level onset, if it is visible in the spectra, or by the energy position of the core levels. An example of the above analysis is reported in Fig. 10 (see W. A. Spicer, Ref. 28), which shows a series of photoemission valence bands for

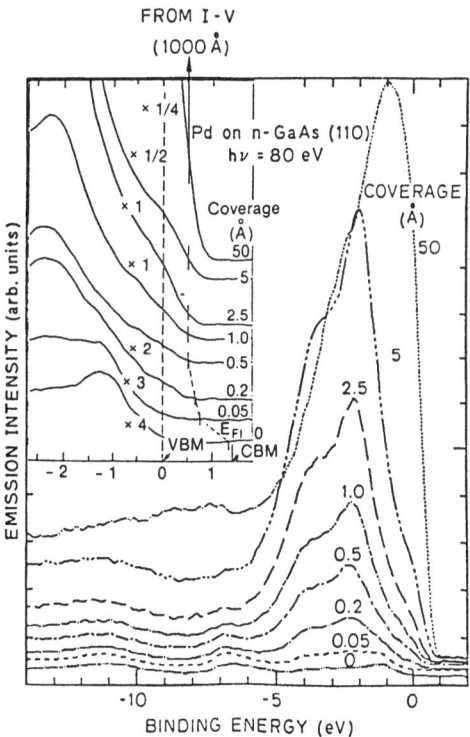

Figure 10. Photoemission valence band of GaAs covered with Pd (from Ref. 28).

166

the Pd-GaAs junction at different Pd coverages. The position of the Fermi level on the clean surface, which coincides with the bulk energy position of the n-type GaAs (110), moves towards the middle of the gap and is stabilized when a film of Pd, 0.5 Å thick, is deposited (see the inset in the figure). The E_F position does not change for higher Pd depositions and, in this particular case, the final position is nearly coincident with that obtained by I-V measurements on metal layers 1000 Å thick. From Fig. 10 it is also clear

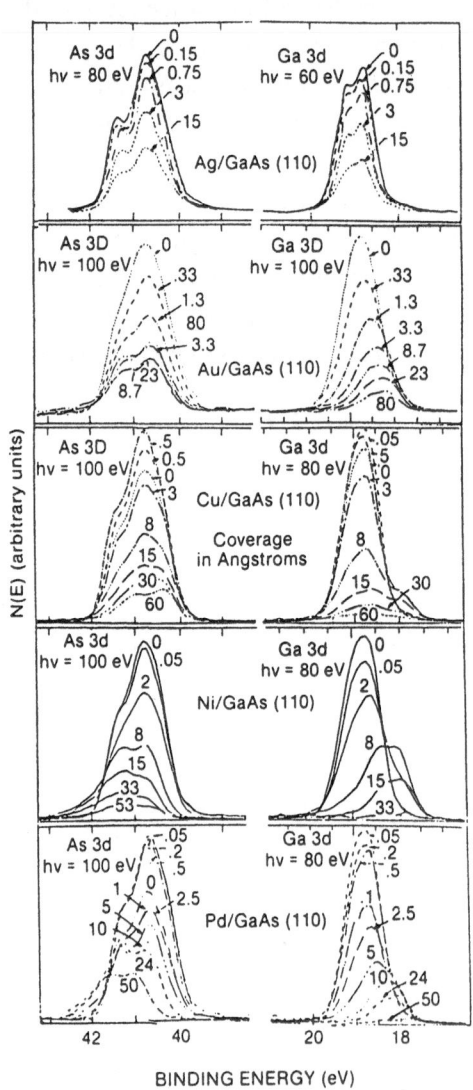

Figure 11. Examples of As and Ga core levels evolution as a function of different metal overlayers (from Ref.28).

that photoemission data allow us to obtain some information on the "chemistry" occurring at the interface. The series of new valence band structures induced by Pd deposition, which stabilizes into a Pd-like valence band for 50 Å, indicates a certain interaction between palladium atoms and the substrate. In general this interaction is also checked by the core levels evolution as a function of metal coverage, where the chemical interaction may result in core levels shifts changing the overall lineshape of core lines. Variations in the energy position of core levels not affected by chemical shifts give the band bending changes. Examples of core level analysis for different metals on GaAs are reported in Fig. 11 (28), where shifts due to band bending have been subtracted and only shifts due to chemical reactions are present.

The results of Fig. 11 were regarded by the authors as evidence for the validity of the native defect model, since E_F pinning is reached in the submonolayer regime and at a coverage stage where the chemistry does not appear to be a dominant factor.

It is important to note that the morphology of the grown film depends strongly on the temperature of the substrate during growth. During the last years a number of works trying to investigate the importance of interfacial reactivity on the basis of experiments at liquid nitrogen temperatures (LNT) and room temperature (RT) have appeared (29-32). A good example is provided by the photoemission study of the Na-GaAs(110) system (30).

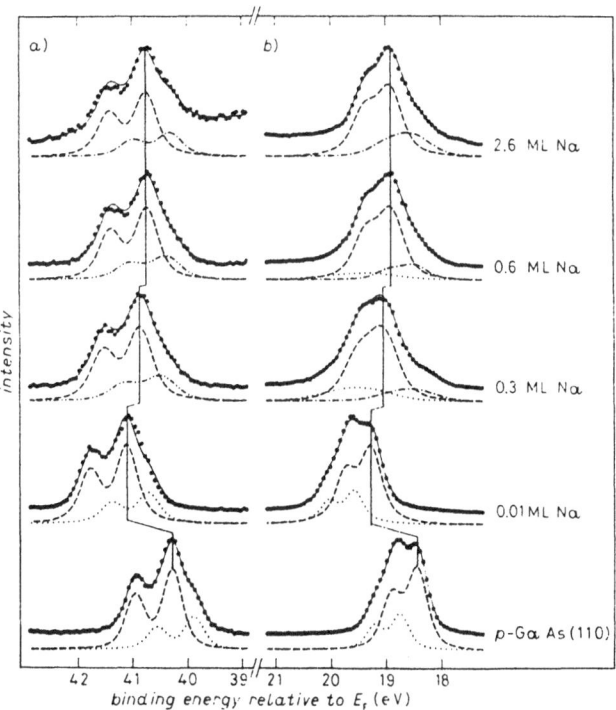

Figure 12. LNT Ga and As core lines vs. Na coverage in GaAs (from Ref. 30).

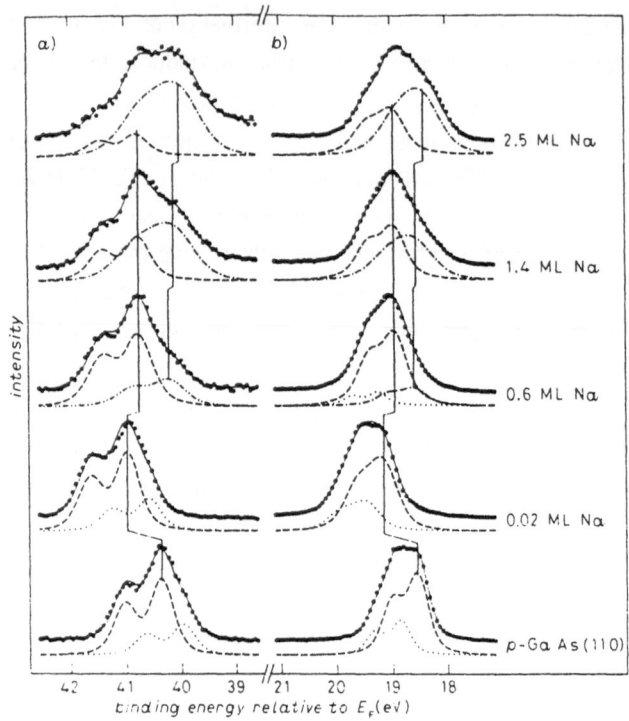

Figure 13 . Ga and As core lines vs. Na coverage taken at room temperature (RT) on GaAs (from Ref.30).

Figures 12 and 13 show the gallium and arsenic core lines for different sodium coverages at LNT and RT respectively. For the clean surface the two spin-orbit splitted doublets related to bulk and surface contributions are evidenced by least squares fitting of experimental data to two Lorentian doublets convoluted by a Gaussian to account for instrumental resolution. The contribution of chemically shifted components is denoted by a dashed dotted line. The latter contribution is much more evident at RT than at LNT, thus pointing out the importance of using different temperatures to select different degrees of reactivity. The common binding energy variations of As-3d and Ga-3d bulk substrate core levels with varying Na coverage, as observed for reactive and non reactive interfaces, provide an indication of the band bending occurring at the interface. These variations, corrected for finite sampling depth (see Ref. 30) and referred to the conduction band minimum CBM and to the valence band maximum VBM, are reported in Fig. 14 for n-type and p-type GaAs(110). In Fig. 14 CBM and VBM are kept fixed and the Fermi level is

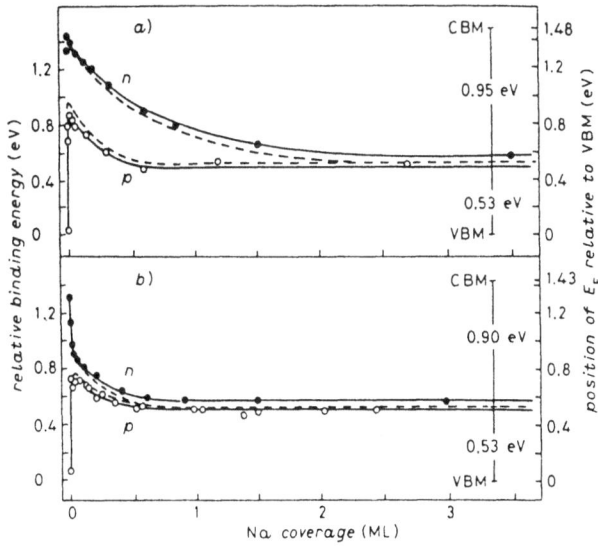

Figure 14. E_F behaviour vs. Na coverage in n-type and p-type GaAs substrates at LNT (a) and RT (b) (from Ref. 30).

allowed to move correspondingly. First of all we notice a drastic difference between n- and p-type samples at very low coverages (0.01 ML). For non-reactive interfaces (LNT) the shift by 1 eV in the p-type sample is not observed in the n-type sample, and this is considered by the authors to be due to electron transfer from Na to p-GaAs holes, while the opposite process from n-GaAs to Na is inhibited by the electronegativity of sodium. For reactive interfaces (RT) the E_F variation is almost symmetrical and the common position of E_F has been attributed to the presence of defects levels. The E_F behaviour in the intermediate coverage range has been regarded as due to the presence of metal-induced gap states, which become the dominant E_F pinning factor for Na coverages greater than two monolayers. It is important to point out that the data taken at LNT are affected by the occurrence of surface photovoltage, as will be apparent from the next example, and some of the conclusions by the authors which are based on these data may be in error.

Other photoemission measurements performed at LNT and RT give similar results, except for the sudden variation of E_F on p-type samples as observed for very small amounts of Na. Indeed, this behaviour may be strictly related to the very low electronegativity of sodium. The general behaviour is reported in Fig. 15, which clearly shows the symmetric

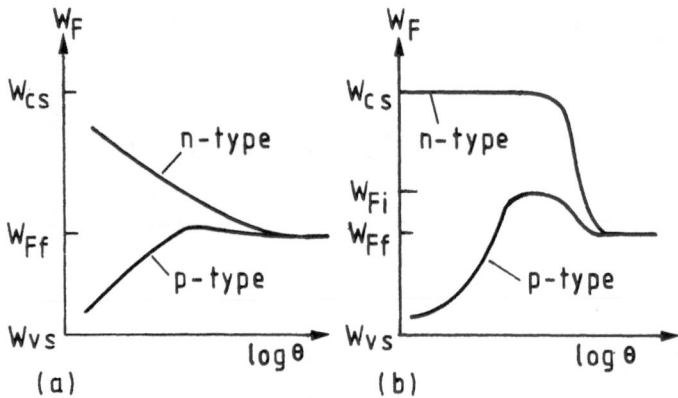

Figure 15. General behaviour of the movement of E_F inside the gap at RT (a) and LNT (b) as a function of log θ , where θ is the coverage.

vs. asymmetric behaviour of E_F at LNT and RT on p- and n-type samples. Over the coverage range from 1 to 10 monalayers E_F reaches a common value at low and high temperatures. This occurs when the overlayer thickness ensures the growth of a metal-like film. In this case the Fermi level pinning is dominated by the MIGS states. At intermediate coverages the behaviour of E_F has been explained (14) in terms of differences in surface mobility of atoms arriving on the substrate at different temperatures. It has been estimated that an aluminum atom arriving on a surface makes an order of 10^4 jumps per second at room temperature but only one jump in 10^4 days at 100 K. It is highly probable that at RT the atoms of the overlayers tend to form clusters and islands even in the submonolayer regime. At LNT the atoms are more likely to deposit to form an homogeneous layer without any strong interactions between them. In this case it might be possible that native defects play an important role in pinning E_F, and this is the explanation provided (14) to justify the intermediate flat part of E_F behaviour in p-type samples at LNT. Donor native defects have been claimed in the latter case.

Measurements at low temperatures seem to define the coverage regions where the native defect model and MIGS model are valid, the first one being valid at very low metal coverages and the second at thicknesses where the metal-like behaviour of the overlayer is reached. Recently, the interpretation of the LNT results has been questioned by different authors (31,32). The presence of a surface photovoltage effect can give rise to a non-equilibrium charge distribution at the interface, changing the position of the Fermi level in the energy gap with respect to its thermodynamic equilibrium value. If we consider Fig. 1 and refer to a very thin metal-overlayer, photons of sufficient energy impinging on the metal side create electron-hole pairs which are separated by the electric field present therein. While some of the electrons are swiped out, holes accumulate into the notch creating a non-equilibrium charge distribution and a quasi-Fermi level at a lower energy than the equilibrium E_F position. The influence of this effect depends on photon flux, semiconductor energy gap and electron-hole recombination rate, which is strongly dependent on temperature. No doubt, all the Schottky barrier heights derived from photoemission data at low temperatures must be analyzed taking into account this phenomenon. Measurements performed ad room temperature are differently affected depending on the semiconductor energy gap. While for GaAs-metal systems (GaAs has an energy gap of 1.4 eV) the influence seems to be small (32), of the order of 0.1 eV for high photon intensities, for GaP-metal interfaces (GaP has an energy gap of 2.2 eV) it can be higher than 0.5 eV (31). Figure 16 shows the valence band during the first stages of formation of the GaP-Ag junction for n- and p-type (110) GaP substrates. The vertical dashed line corresponds to the position of the Fermi level of a bulk gold sample in close contact with the GaP substrate. The presence of surface photovoltage is evidenced by the different position of E_F at small coverages with respect to the bulk value (reference line). This difference tends to zero with increasing coverage. When a thick metal overlayer is

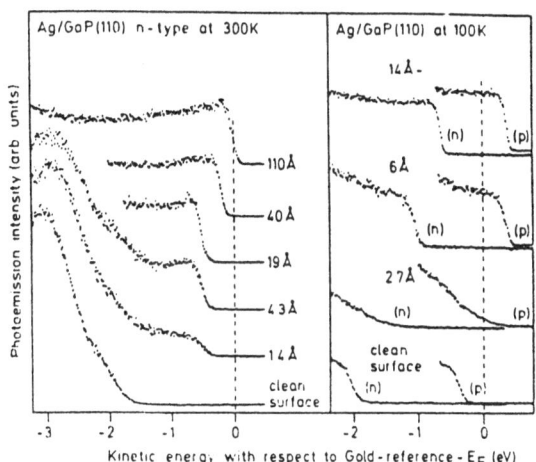

Figure 16. Position of the quasi Fermi level at the GaP-Ag interface vs. Ag coverage as referred to the position of E_F in bulk Au. The behaviour of the quasi Fermi levels is opposite in p-type and n-type samples, indicating the presence of a surface photovoltage (from Ref. 31).

formed, photon penetration through the metal film drops practically to zero and the quasi-Fermi level coincides with the E_F bulk position. The shift is opposite for n- and p-type substrates, thus providing clear evidence for the presence of a surface photovoltage *vs.* a charging effect. Once the positions of E_F in the energy gap are corrected for this effect, the results reported in Fig. 17 are obtained. The position of E_F for differently doped materials is quite similar in the intermediate coverage range, and is not so far from that of the bulk value. These results seem to exclude that differences in the morphology of the grown film at submonolayer coverages are important in determining the E_F behaviour.

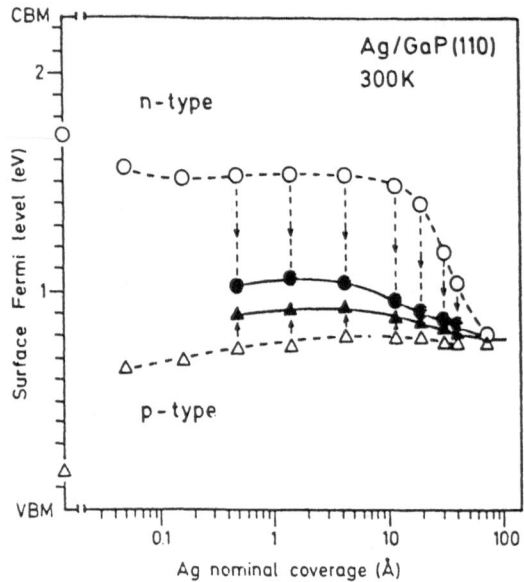

Figure 17. Position of E_F in n- and p-type GaP as corrected for the surface photovoltage (from Ref. 31).

6. Scanning Tunneling Spectroscopy Measurements of Schottky Barrier Height

The scanning tunneling microscope (STM) has proved to be a powerful tool for studing the spectroscopy of surface states (33). The electrons leaving the tip at different voltages have different energies and can tunnel to different empty states of the sample surface. The opposite is true for electrons leaving the filled electronic states of the substrate towards the tip. The spectroscopy mode of the STM is then obtained by measuring the tunneling current

I and by varying the positive or negative potential of the tip with respect to the sample. The spectrum of surface state density is given directly by the derivative of the I/V curves (34). The great advantage of this type of spectroscopy stems from the atomic resolution of the instrument, which allows the local density of states to be probed. This technique has been recently applied to the problem of Schottky barrier formation. The results obtained on GaAs-Sb (34) and GaAs-Fe (35) can be regarded as a real milestone towards the final solution of this problem. Here the results obtained on Fe clusters deposited on p-GaAs (110) will be described. Figure 18 shows a 100x100 Å STM image of two Fe clusters

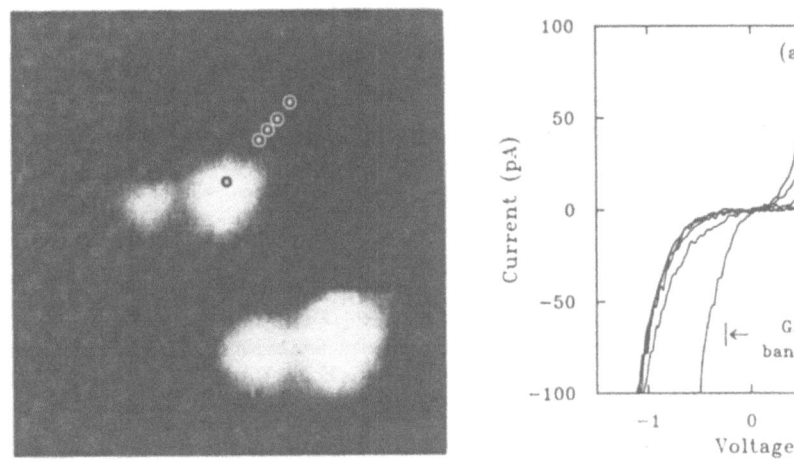

Figure 18 . STM images of Fe clusters on GaAs and corresponding I/V curves (from Ref. 35).

(white regions in the figure). The GaAs substrate appears in the form of rows of As atoms along the 110 direction. The spatial distribution of the electronic states can be evidenced by the selected I-V curves taken with the tip at the different positions marked by circles in Fig. 18. The corresponding I-V characteristics are also reported in the same figure. Curve (a) with the tip on top of the Fe cluster shows a finite differential conductance dI/dV for V=0. This provides clear evidence in favour of the metallicity of clusters. Curve (e), which was obtained with the tip well inside the GaAs substrate, shows no current contribution from electrons tunneling in the gap region, 1.5 eV wide, where there is a lack of states. The intermediate curves, also recorded on the GaAs substrate, show some tunneling current contribution in the gap region. These states, present all around the cluster regions, are the metal-induced gap states due to the electron wavefunctions tunneling into the semiconductor gap. This is clearly apparent from Fig. 19, which shows the logarithm of

174

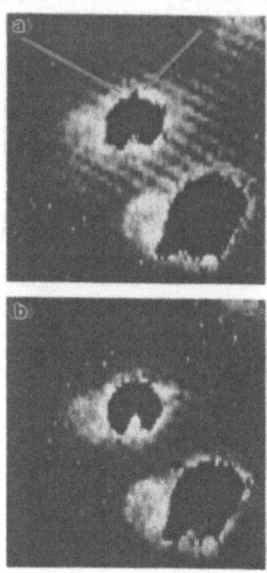

Figure 19. Logarithm of the differential conductance, for a tip voltage of 1.1 eV and 1.0 eV (from Ref. 35).

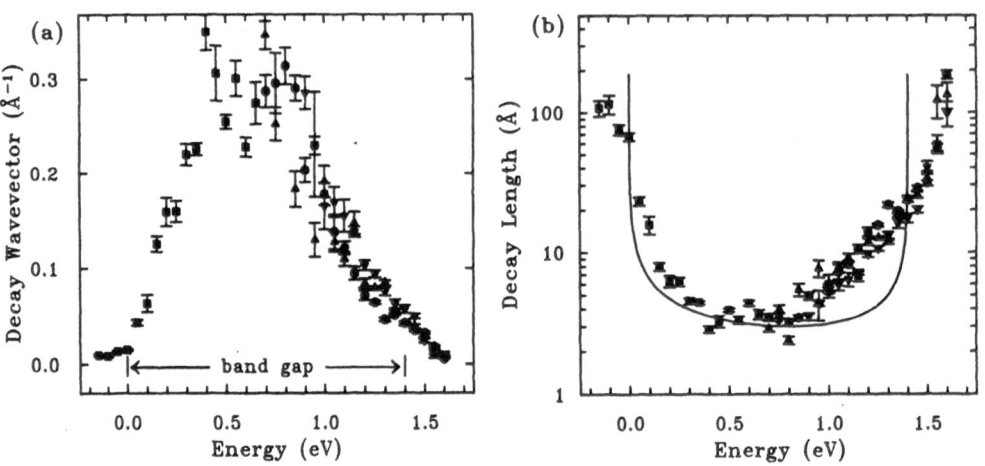

Figure 20. Decay wavevector and decay length as obtained upon analysing a complete set of data similar to those in Fig. 19 (from Ref. 35).

the differential conductance for a tip voltage of 1.1 and 1.0 eV. The MIGS states at V=1.1 eV extend more into the GaAs side (white regions in the figure) and the decay length diverges for tip voltages near the band edges (the GaAs energy gap is 1.5 eV as measured by the I-V relationships of Fig. 18). The decay wavevector and decay length obtained by a complete analysis of the I-V curves are reported in Fig 20. This is what is expected from the MIGS model, which is schematically represented in Fig. 21. The metal electron wave

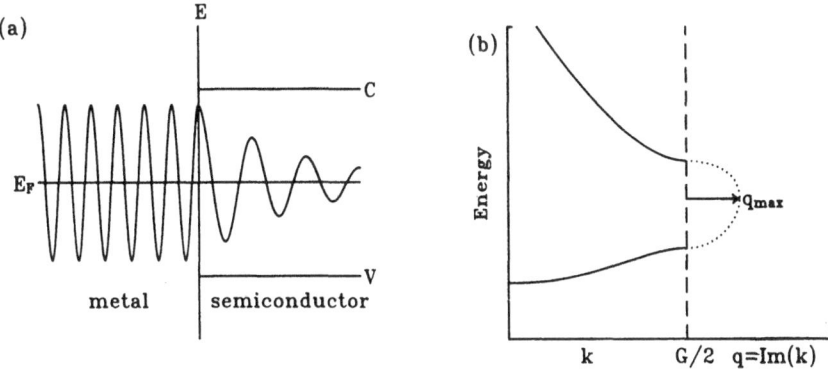

Figure 21. Metal electron wavefunction tunneling in the semiconductor energy gap (a). E-k dispersion relationship of a two band model of a semiconductor. The dashed line denotes a continuation of the k axis for imaginary wave vectors at q=0.

functions tunneling in the gap have the form $\psi(x) = \exp(ikx) = \exp(ik_r x) \exp(-qx)$, where $k_r = G/2$, $q = \mathrm{Im}(k)$, and G is a reciprocal lattice vector. The wave vector q is a maximum near the middle of the gap (see Fig. 21) and vanishes at the band edges, and the decay length diverges as the wave functions became plane waves-like. These results are completely accounted for by the experimental findings of Fig. 20.

7. Conclusion

This short review of different approaches to the understanding of Schottky barrier does not cover the great amount of work done by other scientists on the subject, but, in the author's opinion, it outlines the fundamental lines of research which have generated all other branch lines. From the above analysis, and having in mind Figs. 15 and 17, the pinning of E_F beyond one or two monolayers of deposited metal is essentially determined by the metal-induced gap states. The very rapid variation of E_F at submonolayer coverages, as shown by Figs. 5 and 14 for GaAs, remains to be explained. If in this coverage regime the adatoms coalesce to form clusters, their local metallicity is still the dominant factor affecting the Fermi level position. In the case in which the adatoms do not interact with each other, we do not have a continuum of interface induced states, but rather localized states which could be acceptor- or donor-like. In this case the adatoms may happen not to give the charge to these states, and the nonequilibrium conditions determining the E_F behaviour are then described by a charge transfer from the semiconductor depletion layer to the adatoms induced states (22) (for the case of an n-type semiconductor, acceptor-like states would take the effect into account). If, instead, the adatoms are able to furnish the charge to the tunneling states in the gap, as in the MIGS model, the position of the Fermi level would then be determined by this extra charge. One may also postulate an interplay of different mechanisms in the submonolayer regime, or possibly other mechanisms not taken into account in this review. This point will be certainly the subject of further experimental and theoretical investigation in the next years.

References

(1) Milnes A. G. and Feucht D. L. (1972) in *Heterojunction and Metal-Semiconductor Junctions*, Academic Press, New York.

(2) Braun F. (1874), Pogg. Ann. **153**, 556.

(3) Schottky W. (1938), Naturwissenschaften **26**, 843.

(4) Bardeen J. (1947), Phys. Rev. **71**, 717.

(5) Spicer W. E., Lindau I., Skeath P., and Yu C. Y.(1986), J. Vac. Sci. Technol. **17**,1019.

(6) Williams R. H.(1981), J. Vac. Sci. Technol. **18**, 929.

(7) Williams R. H.(1983), Surf. Sci. **132**, 122.

(8) Monch W. (1983), Surf. Sci. **132**, 92.

(9) Allen R. E. and Dow. J. D. (1979), J; Vac. Sci. Technol. **19**, 383.

(10) Daw M. S. and Smith D. L.(1979), Phys. Rev. **B20**, 5150.

(11) Heine V. (1965), Phys. Rev. **138**, 1689.

(12) Flores F. and Tejedor C. (1979), J. Phys. C: Solid State Phys. **12**, 731; Flores Tejedor C. and Louis E. (1977), Phys. Rev. **B16**, 4695; Louie J. G. and Cohen M. L. (1976), Phys. Rev. **B13**, 2461.

(13) Tersoff J. (1984), Phys. Rev. Lett. **52**, 465; Tersoff J.(1984), Phys. Rev. **B30**, 4874.

(14) Monch W.(1989), Appl. Surf. Sci. **41/42**, 128.

(15) Coluzza C., Lama F., Frova A., Perfetti P., Quaresima C., Capozi M.(1988), J. Appl. Phys.**64**, 3304.

(16) Freeouf J. L.(1983), Surf. Sci. **132**, 233.

(17) Waldrop J. L. (1984), J. Vac. Sci. Technol. **B2**, 445.

(18) Newman N., Spicer W. E., and Weber E. R. (1987), J. Vac. Sci.Technol.**B4**, 1020.

(19) Huges G., Ludeke R., Shaffler F., Rieger D. (1986), J. Vac. Sci. Technol. **B4**, 924.

(20) Sze S .M.(1969), in *Physics of Semiconductor Devices*, John Wiley & Sons Publishers, New York.

(21)Van Laar J., Huijser H. and Van Roy T. L. (1977), J. Vac. Sci. Technol. **14**, 894.

(22) Zur A., Mc Gill T. C. and Smith D. L.(1983), Phys. Rev. **B28**, 2060.

(23) Pellegrini B.(1973), Phys. Rev. **B7**, 5299.

(24) Yndurain F. (1977), J. Phys **C4**, 2849.

(25) Flores F. and Tejedor C. (1987), J. Phys. **C20**, 145.

(26) Tersoff J. (1986), Surf. Sci. **168**, 275.

(27) Platero G., Vergas J. A. and Flores F. (1986), Surf. Sci. **168**, 100.

(28) Spicer W. E., Kendelewicz T., Newman N., Chin K. K. and Lindau I. (1986), Surf. Sci. **168**, 240.

(29) Stiles K., Horn S. F., Kahn A, McKinley J., Kilday D.G. and Margaritondo G.(1988), J. Vac. Sci. Technol. **B6**, 1392.

(30) Prietsch M., Lanbschat C., Domke M., and Kaindl G., Europhys. Lett. **6**, 451.

(31) Alonso M., Cimino R. and Horn K. (1990), Phys. Rev . Lett. **64**, 1947.

(32) Waddill G. D., Aldao C. M., Capsso C., Benning P. J., Yongiun Hu, Wagener T. J., Jost M. B. and Weaver J. H. (1990), Phys Rev. **B41**, 5960.

(33) Stroscio J. A., Feenstra R. M. and Fein A. P. (1987), Phys. Rev. Lett. **58**, 1668.

(34) Fenstra R. M. and Martensson (1988), Phys. Rev. Lett. **61**, 447.

(35) Stroscio J. A., First P. M., Dragoset R. A., Whitman L.J., Pierce D.T. and Celotta R. J. (1990), J. Vac. Sci.Technol. **A8**, 284.

THE SEMICONDUCTOR/ELECTROLYTE INTERFACE

Laurence M Peter
Department of Chemistry
The University
Southampton SO9 5NH
United Kingdom

ABSTRACT.

This chapter presents a brief survey of fundamental aspects of the semiconductor/electrolyte interface and introduces some of the experimental techniques that are used to characterise single crystal semiconductors and oxide coated metals. These methods include photocurrent spectroscopy, non steady-state photocurrent techniques and *in-situ* infrared spectroscopy.

1. INTRODUCTION.

The interfacial properties of semiconductors are important in many applications. This chapter deals with the properties of the junction formed when a semiconducting or insulating phase is contacted by an electrolyte. This situation is encountered more often than might be supposed. It includes not only the obvious case where a bulk (usually single crystal) semiconductor is in contact with an electrolyte, but also the circumstance where a metal electrode is covered with a thin film of oxide or other compound which is an insulator or semiconductor.

Single crystal semiconductor electrodes have played an important role in the search for solar energy conversion devices, and much of the impetus for research in semiconductor electrochemistry has been provided by the need to develop alternative sources of energy. Although systems based on photo-electrochemical processes have not been developed into practical devices, semiconductor electrochemistry continues to be important in various steps in electronic device fabrication such as etching, micro-machining and contacting.

Semiconductor electrochemistry also overlaps strongly with more conventional electrochemistry. Thin semiconducting or insulating films on metals are common, for example, in battery and corrosion electrochemistry, and the solid-state properties of different surface phases may determine the electrochemical behaviour of the complete system. Often, the techniques of semiconductor electrochemistry can be applied to these more complex systems in order to provide information that is not accessible from conventional methods.

The discussion of fundamentals in this chapter is necessarily very brief; the objective is to provide a minimum level of understanding so that some interesting aspects of semiconductor electrochemistry can be considered in more detail. Background information can be found in several detailed treatments of semiconductor electrochemistry [1-4].

Some of the techniques and applications of semiconductor electrochemistry are considered in the second half of the chapter. The main

R. Guidelli (ed.), Electrified Interfaces in Physics, Chemistry and Biology, 179–199.
© 1992 *Kluwer Academic Publishers*.

objective here is to illustrate a few of the powerful methods currently
available and to examine the prospects for future developments in this
exciting field. The extension of the methodology of semiconductor
electrochemistry to develop *in-situ* spectroscopic techniques to study metal
electrodes is also considered and illustrated with examples.

2. ELECTRONS IN SOLIDS.

The discussion of the junction between a semiconductor and a second
phase starts from the bulk electronic properties. The electron energy
levels in a solid with a regular periodic crystal lattice are grouped
together in *energy bands* that are separated by *energy gaps*. A proper
discussion of the band model is outside the scope of this chapter; for the
present purposes we need only to know that the electron energy depends on
the reciprocal lattice vector, giving rise to the kind of band diagram
shown in **fig 1** for GaAs.

In the case of an intrinsic (undoped) semiconductor at 0 K, the
valence band is completely filled, whereas the higher energy conduction
band is vacant. Transitions from the valence to the conduction band can be
brought about by absorption of a photon (see figure 1) or, at higher
temperatures, by thermal excitation of electrons across the forbidden
energy gap.

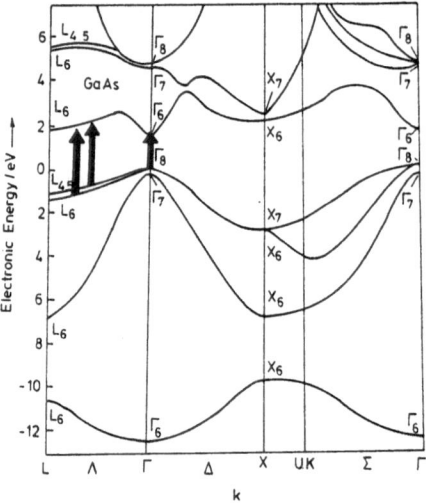

FIGURE 1. Band diagram for GaAs. The arrows show optical transitions.

The equilibrium occupation of electron energy levels is determined by
the *Fermi Dirac* distribution function, $f_{FD}(E)$, which describes the
probability that an electron state of energy E is occupied;

$$f_{FD}(E) = 1/[1 + \exp\{(E-E_F)/kT\}] \qquad\qquad 1)$$

where E_F is the *Fermi energy*. The probability that the same state is
unoccupied is $(1 - f_{FD})$. In the particular case of an *intrinsic* (i.e.

undoped) semiconductor, n, the density of electrons in the conduction band, is equal to p, the density of electron vacancies or *holes* in the valence band. The values of n and p depend in turn on the products of the occupation probabilities f_{FD} and $(1 - f_{FD})$ and the density of states functions near the conduction and valence band edges. It can be shown that the product of the electron and hole densities is given by

$$n.p = N_c.N_v \exp(-E_g/kT) = n_i^2 \qquad\qquad 2)$$

where E_g is the energy gap, N_c and N_v are the effective density of states in the conduction and valence bands respectively and n_i is the *intrinsic* carrier density.

Eqn 2 also applies to *extrinsic (doped)* semiconductors. In this case, however, the densities of electrons and holes are no longer equal. The accidental or deliberate inclusion of ionisable electron donors or electron acceptors in the semiconductor lattice gives rise to n- or *p-type* semiconductivity respectively. At ambient temperatures, where the dopant species are completely ionised, the density of the majority carriers (electrons for n-type and holes for p-type) is determined by the concentration of dopant species, i.e. either $n \gg n_i$ or $p \gg n_i$. Provided that electronic equilibrium is maintained, the density of minority carriers can be obtained from eqn 2 if the majority carrier density is known.

It is convenient to be able to represent the occupation of the conduction and valence bands schematically, and this can be done by indicating the position of the Fermi energy relative to the valence and conduction bands. For an intrinsic semiconductor, eqn 2 indicates that the Fermi energy must be located close to the centre of the energy gap in order for the condition $n = p$ to be fulfilled. On the other hand, the condition $n \gg n_i \gg p$ for an n-type semiconductor must correspond to the Fermi energy being located close to the conduction band edge. Similarly, for a p-type semiconductor, the Fermi energy is located close to the valence band edge.

3. ELECTRONS IN THE SOLUTION PHASE.

The conduction electrons in solids can be considered as delocalised, whereas by contrast electrons on ionic or molecular species in solution are localised on orbitals. Each ion is surrounded by fluctuating solvent dipoles that modify its ionisation energy and electron affinity. The fluctuations of the electron energy levels are large, typically of the order of eV, even though the thermal fluctuations of the dipoles are of the order of kT. In the theory of electron transfer developed by Marcus and others[2,5-7], these fluctuations are responsible for the formation of the transition state in the electron transfer process between the species O and R which are interconverted by the reaction

$$O + ne \rightleftharpoons R \qquad\qquad 3)$$

The mean or most probable energies for the oxidised and reduced species (E_0 and E_R) do not coincide since the charge on 0 and R differs by ne. The probability distribution functions, $W(E)$, about the mean energies for 0 and R are usually assumed to be Gaussian as shown in **fig 2**.

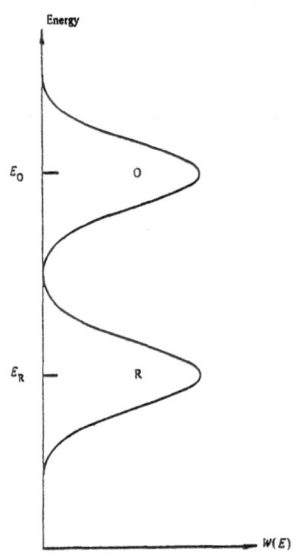

FIGURE 2. Probability distribution functions for solution redox species

4. ELECTRONIC EQUILIBRIUM AT THE SEMICONDUCTOR/ELECTROLYTE INTERFACE.

Electronic equilibrium can be established across the junction between a semiconductor and a second phase provided that electron transfer is possible. If the contacting phase is an electrolyte solution containing redox species **0** and **R**, then equilibrium is achieved by partial oxidation or reduction of solution species. The system is at equilibrium when the free energy or *electrochemical potential* of electrons is the same on both sides of the interface. In semiconductor physics, the electrochemical potential of electrons is equivalent to the Fermi energy which must be the same on both sides of the junction. The electrochemical potential of electrons in the electrolyte is not easy to visualise since we are dealing with discrete ions, but nevertheless we can define a *redox Fermi level* which is related to the *redox potential* of the 0/R couple on the electrochemical scale (the standard hydrogen electrode is taken as zero by convention). If the formation of the junction results in the extraction of majority carriers from the semiconductor, a *space charge*, consisting of ionised dopant atoms, will be formed in the solid. If, on the other hand, junction formation results in the injection of majority carriers into the solid, the excess charge will be mobile and concentrated close to the interface. These two situations correspond to the formation of *blocking* and *ohmic* contacts respectively.

The existence of a space charge in the semiconductor leads to a characteristic distortion of the energy bands which arises from the potential distribution across the interface. Typically, this *band bending* may extend over a distance of 10^{-5} cm or more. The establishment of equilibrium at the semiconductor/electrolyte interface is illustrated in **fig** 3. Note that a blocking contact is formed in this case because electrons are removed from the n-type semiconductor.

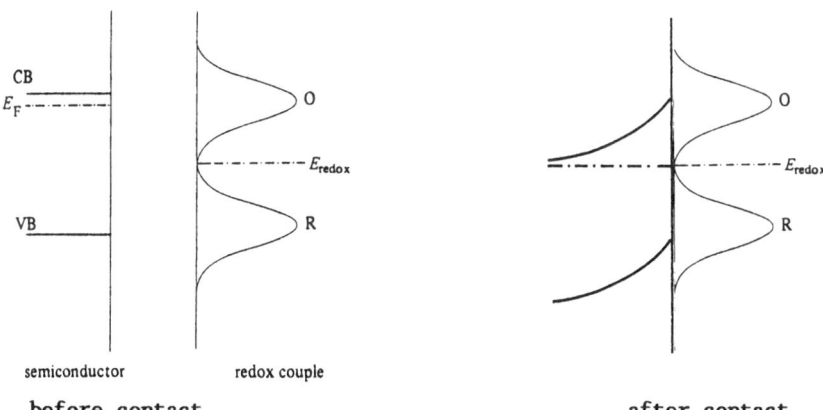

before contact   **after contact**

FIGURE 3. Formation of a blocking semiconductor/electrolyte contact .

5. POTENTIAL DISTRIBUTION AT THE SEMICONDUCTOR/ELECTROLYTE JUNCTION.

Let us consider the special case where there are no redox species in solution (other than those associated with the solvent). The potential difference across the interface can be changed over a wide range by application of an external potential source. The distribution of charge and potential will depend on the applied potential, and at a particular value of the electrode potential, there will be no excess charge in the semiconductor. This potential is referred to as the *flatband potential*, E_{fb}, since the energy bands are no longer distorted by the presence of excess charge. If we consider an n-type semiconductor, a positive space charge (consisting of ionised donor states) develops when the potential is made more positive than E_{fb}, whereas a negative electron charge accumulates at the surface when the potential is negative of E_{fb}. The potential and charge distribution in the case of the blocking contact are shown in **fig 4.**

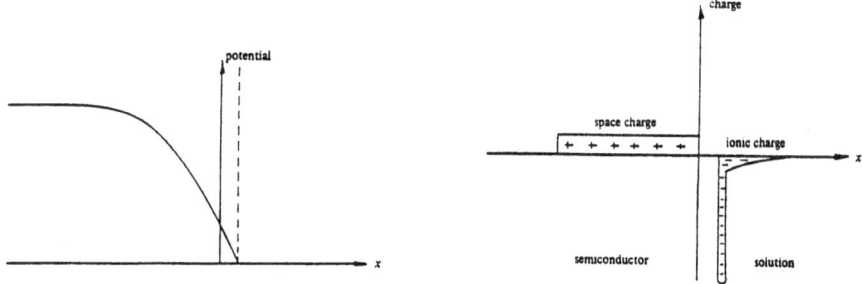

FIGURE 4. Potential and charge distributions at blocking contact.

6. *THE SPACE CHARGE CAPACITANCE.*

The preceding discussion showed that the space charge developed under depletion conditions depends on electrode potential; the derivative of charge with respect to potential corresponds to a *space charge capacitance*, $C_{sc} = dQ_{sc}/dE$. For an n-type semiconductor, C_{sc} is given by the *Mott-Schottky* equation

$$C_{sc}^{-2} = (2/qN_d\epsilon\epsilon_o).(\Delta\phi - kT/q) \qquad\qquad 4)$$

where N_d is the donor density, ϵ is the relative permittivity and $\Delta\phi$ is the potential drop across the space charge region. If the doping density is small, $\Delta\phi$ is almost identical to $E - E_{fb}$, but for highly doped semiconductors, changes in the potential drop on the Helmholtz double layer become significant [3].

In principle, plots of C_{sc}^{-2} *vs* electrode potential can be used to determine values of the flatband potential and dopant densities. The 'capacitance' is often measured as the out of phase component of the electrode admittance at a single frequency. This implies that the electrode impedance can be represented as a parallel RC network and that the series resistance is negligible. Furthermore, the capacitance of the Helmholtz double layer is assumed to be much larger than the space charge capacitance provided that the semiconductor doping is not too high. In practice, however, the impedance of semiconductor electrodes is usually more complicated, so that Mott-Schottky plots often exhibit frequency dependent slopes and intercepts. There are many reasons for such behaviour, and generally it is preferable to carry out a proper frequency response analysis and to use a more complex equivalent circuit to fit the data (non-linear fitting programs are available commercially to ease this task). Examples of Mott-Schottky plots that show frequency dispersion are shown in **fig 5**.

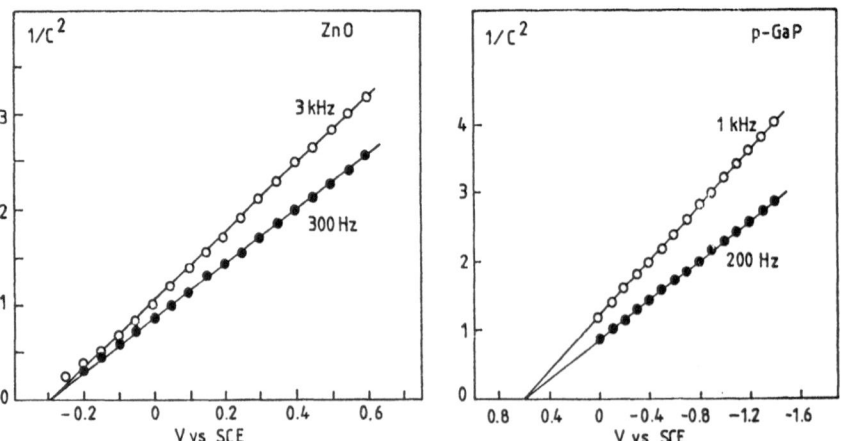

FIGURE 5. Mott Schottky plots for n-ZnO and p-GaP.

7. THE ILLUMINATED SEMICONDUCTOR/ELECTROLYTE JUNCTION.

Under depletion conditions, the potential distribution in the semiconductor/electrolyte junction assists the separation of electron-hole pairs created by illumination as shown in **fig 6a**.

The fate of the minority carrier when it reaches the surface depends on *thermodynamic* and *kinetic* factors. Many n-type semiconductors are thermodynamically unstable with respect to *photoanodic corrosion*. This means that the energetically most favourable reaction involves progressive rupture of the semiconductor bonds by the trapping of holes (i.e. electron vacancies in the valence band). The dissolution of the lattice is a multistep electron transfer reaction. For example, the photoanodic dissolution of n-GaAs involves the capture of six holes :

$$GaAs + 6 \; h^+ \rightarrow Ga(III) + As(III) \qquad\qquad 5)$$

The semiconductor can be stabilised kinetically if a suitable redox system is chosen which either scavenges holes rapidly before the lattice is ruptures or which effectively 'repairs' partially ruptured bond by fast electron donation.

The generation and collection of minority carriers (i.e. holes for an n-type semiconductor) is described quantitatively using the three characteristic lengths shown in **fig 6b**.

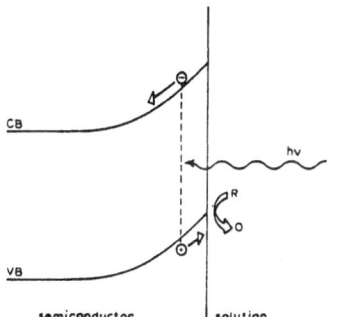

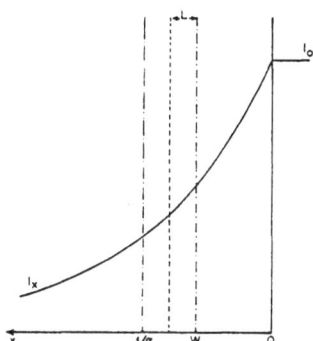

    (a) photoexcitation of carriers   (b) characteristic lengths

FIGURE 6. Photogeneration and collection of electron-hole pairs.

The width, **W**, of the space charge region is given by

$$W = (2\Delta\phi\epsilon\epsilon_o/eN_d)^{1/2} \qquad\qquad 6)$$

The minority carrier diffusion length, L_p, is determined by the lifetime, τ_p and diffusion coefficient, D_p, of minority carriers;

$$L_p = (\pi D\tau_p)^{1/2} \qquad\qquad 7)$$

The penetration depth of the light, $1/\alpha$, is determined by the absorption coefficient α. The photocurrent is a linear function of the incident photon flux (corrected for reflection), I_o, and is given by the *Gärtner equation* [2,4,8].

$$\Phi = j_{photo}/I_o = 1 - \{\exp(-\alpha W)/(1 + \alpha L_p)\} \qquad 8)$$

where Φ is the *photocurrent conversion efficiency*. Eqn 8 forms the basis of a useful method of determining minority carrier diffusion lengths. Plots of $-\ln(1-\Phi)$ vs $(E-E_{fb})^{1/2}$ are expected to give straight lines from which α and L_p can be determined independently by using the slopes and intercepts. An example of the application of this method to determination of the electron diffusion length in p-GaP [9] is shown in **fig 7**. The photocurrent-voltage curves were measured at different wavelengths and at low intensities to avoid problems associated with changes in the surface properties of the semiconductor.

The diffusion length of electrons in p-Gap derived from the data shown in fig 7 is 0.07 μm. The analysis based on eqn 8 requires precise measurement of the photocurrent conversion efficiency and proper correction for reflection losses. The validity of the analysis should always be tested by checking that measurements at different wavelengths give identical values of L_p and that the values of α are consistent with available data.

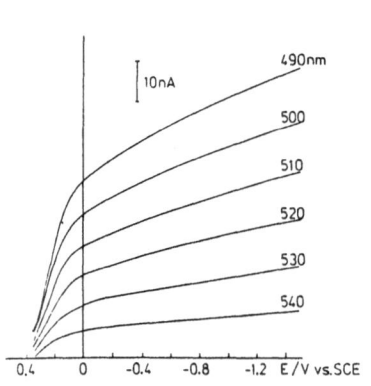

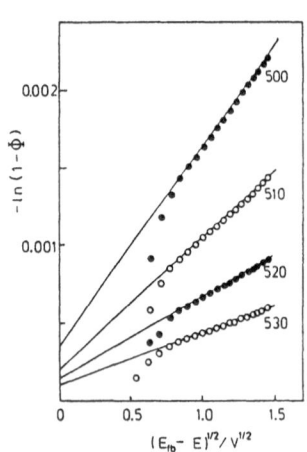

(a) photocurrent-potential plots   (b) Gärtner equation plots

FIGURE 7. Determination of electron diffusion length in p-GaP.

8. DEVIATIONS FROM IDEALITY; RECOMBINATION.

At large values of band bending, most real semiconductor/electrolyte junctions follow the behaviour predicted by the Gärtner equation, but closer to E_{fb}, the photocurrent is usually smaller than expected (this can be seen, for example, in fig 8). There are several possible reasons for this. The most likely explanation is that bulk or surface energy levels are present which assist the *recombination* of the photogenerated electron-hole pairs as illustrated in **fig 8** [10].

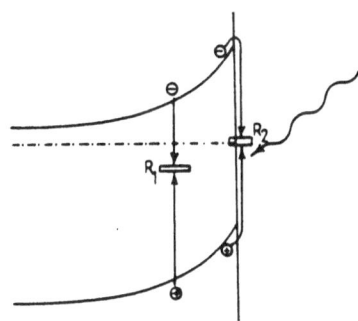

FIGURE 8. Bulk (R_1) and surface (R_2) recombination.

An alternative explanation of deviations from the Gärtner equation is that illumination results in the development of a surface charge on the semiconductor which alters the potential distribution across the semiconductor so as to lower the band bending. Theories of space charge and surface recombination have been developed for the steady state, but they usually fail to explain fully the observed behaviour. More recent developments, concerned with the non steady-state response of the illuminated junction, are discussed in section 11.

9. PHOTOCURRENT MULTIPLICATION PROCESSES.

The previous section considered why the experimental photocurrent should be smaller than the value predicted by the Gärtner equation. However, there are several examples of photoelectrochemical processes in which the photocurrent is *larger* than predicted by the Gärtner equation. This implies that the quantum efficiency must be greater than unity in some circumstances. An example of *photocurrent doubling* has been observed during the photoreduction of oxygen at p-GaP[11]. **Fig 9** shows that the photocurrent in the presence of oxygen is almost double the value observed with deoxygenated solutions.

The mechanism involves the following scheme

$$h\nu \quad \rightarrow \quad h^+ + e^- \qquad\qquad 9a)$$

$$O_2 + H^+ + e^- \rightarrow \quad HO_2^{\cdot} \qquad\qquad 9b)$$

$$HO_2^{\cdot} + H^+ \quad \rightarrow \quad H_2O_2 + h^+ \qquad\qquad 9c)$$

The first step involves the capture of a photogenerated electron (minority carrier), whereas the second step involves *injection of a hole (majority carrier)*. The overall one photon process results in the two electron reduction of oxygen to hydrogen peroxide, giving a maximum quantum efficiency of 2. However, reaction 6c has to compete with the alterative capture of a second photogenerated electron

$$HO_2 + H^+ + e^- \rightarrow H_2O_2 \qquad \qquad 9d)$$

As a consequence, the quantum efficiency falls towards unity as the light intensity is increased and the photogenerated electrons become more readily available. The competition between routes 9c and 9d has been used to estimate the first order rate constant for hole injection by the HO_2 intermediate [11]. The rather low value of 2.5×10^4 s^{-1} suggests that injection occurs via a surface bound state that is located above the valence band edge.

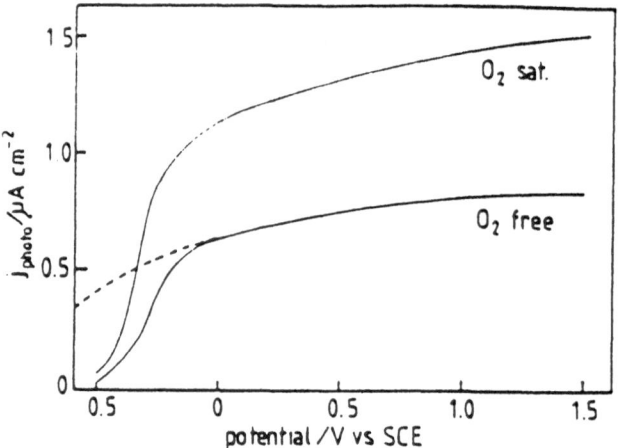

FIGURE 9. Current doubling reduction of O_2 at p-GaP in 0.5 M H_2SO_4.

There are many examples of current doubling at n-type semiconductor electrodes (ZnO, CdS, TiO$_2$)[2]. The photooxidation of formic acid, for example, is thought to involve the following sequence

$$h\nu \rightarrow h^+ + e^- \qquad \qquad 10a)$$
$$HCOOH + h^+ \rightarrow HCOO^- + H^+ \qquad \qquad 10b)$$
$$HCOO^- \rightarrow CO_2 + H^+ + e^- \qquad \qquad 10c)$$

In this case, the capture of a hole is followed by injection of an electron. Recent work[12], however, suggests that reactions 10a to 10c probably oversimplify the current multiplication process. It appears that electron injection involves a surface state located below the conduction band edge rather than the highly reducing HCOO$^-$ species.

Current multiplication processes are also considered in section 13, which is concerned with non steady-state photocurrent response.

10. PHOTOCURRENT SPECTROSCOPY.

One of the most important techniques in semiconductor electrochemistry is photocurrent spectroscopy. The method is most powerful when it involves quantitative determination of the photocurrent conversion efficiency, Φ (see eqn 9), as a function of photon energy. The required sensitivity is achieved by using chopped illumination and lock-in detection, and the incident photon flux is usually measured with a calibrated photodiode. Spectra are corrected to constant incident photon flux, and less commonly for reflection losses. It is important to check that the photocurrent is a linear function of intensity before correction to constant photon flux is carried out.

Fig 10a shows an example of a high resolution photocurrent spectrum of p-GaP. The indirect and direct band transitions can be seen clearly. The low energy photoresponse is particularly important since it can reveal the presence of interband states. Electrons can be excited from these interband states at energies smaller than E_g, allowing density and position of the states in the gap to be estimated. *Sub-bandgap photocurrent spectroscopy* is therefore a useful extension of the method, and **fig 10b** illustrates the detection of hydrogen which diffuses into the GaP lattice during the photocathodic evolution of hydrogen. The incorporated hydrogen acts as an efficient recombination centre, and this system is therefore considered in more detail later in this chapter.

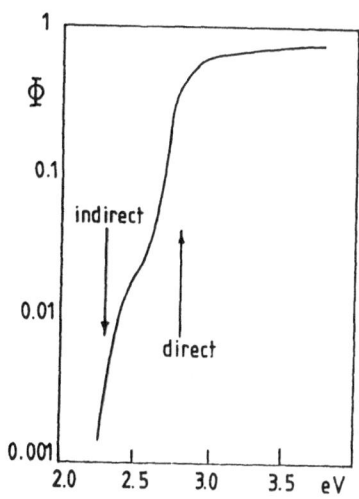

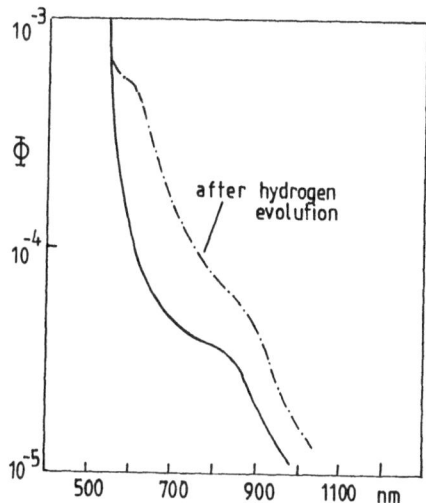

(a) photocurrent spectrum of p-GaP   (b) sub bandgap response of p-GaP

FIGURE 10. Photocurrent spectra obtained for p-GaP in 0.5 M H_2SO_4.

Photocurrent spectroscopy can also be used as an *in-situ*
spectroscopic technique to investigate thin corrosion layers on metals. In
this case, the film is usually thinner than the penetration depth $1/\alpha$ of
the illumination, so that it is necessary to take reflection at the
film/metal interface into account. There are at least two different
mechanisms of photocurrent generation at oxide films on metals. The first,
which involves photoexcitation of an electron-hole pair, is analogous to
the process discussed above for bulk semiconductors. The second process
involves absorption of light by the *metal* and internal photoemission of an
electron into the oxide film. These two processes are contrasted in **fig 11**.

Photocurrent spectroscopy can be used as a quantitative tool to
identify surface films on metals and to estimate their thickness [13]. The
technique is sensitive down to almost monolayer levels, and it has been
used, for example, to derive bandgap data for thin films of Bi_2S_3 under
conditions where conventional absorbance measurements are unreliable [14].

Bismuth can be anodised in alkaline solution to form an insulating
layer of amorphous Bi_2O_3. The anodised electrode exhibits both anodic and
cathodic photocurrent responses [15] (this kind of behaviour is expected if
the oxide film is an insulator). The photocurrent spectra measured for
oxide films on bismuth are shown in **fig 12**. They provide a particularly
good example of the two mechanisms. The anodic photocurrent spectrum is due
to band-band excitation, whereas the cathodic photocurrent spectrum
exhibits a well-defined additional response at low energies due to internal
photoemission. The threshold for internal photoemission obtained from a
Fowler plot of $\Phi^{1/2}$ v. hν is exactly half the bandgap energy of Bi_2O_3,
indicating that the Fermi energy is indeed located midway in the gap as
expected for an insulator (*cf*. fig 11).

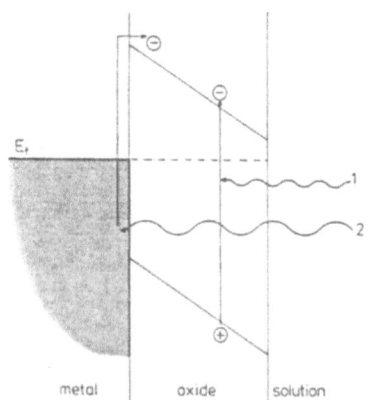

FIGURE 11. Photocurrent generation in oxide films on metals.

Similar results have been obtained with oxide films on lead [16]. Photoemission at the metal/solution interface has also been extensively investigated[17]. The quantum efficiencies for the process are usually much smaller than for internal photoemission at the metal/oxide interface, and the information available is more limited since the threshold energy is determined by the electrode potential.

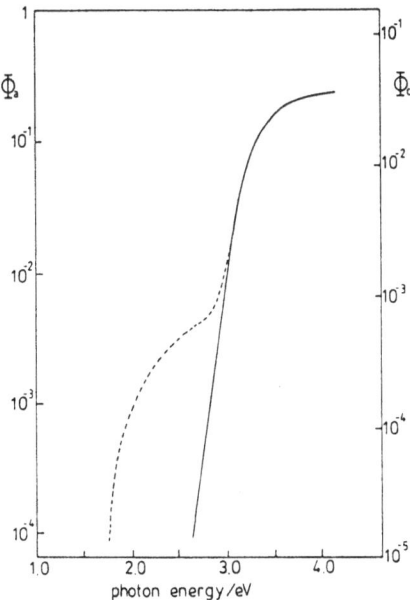

FIGURE 12. Photocurrent spectra for anodised bismuth .
(anodic ___, cathodic ....)

11. NON STEADY-STATE PHOTOCURRENT TECHNIQUES.

Since the rate constants of electron transfer reactions at metal electrodes are strongly potential dependent, the kinetics of electrode processes are widely studied using potential pulse or sweep profiles. This approach is not useful in the case of semiconductor electrodes, since most of the potential drop is located in the space charge region rather than at the interface. However, it is possible to perturb photoelectrochemical processes at semiconductor electrodes by using pulsed, chopped or modulated illumination[18]. Dynamic aspects of semiconductor photoelectrochemistry have been reviewed recently[19], and only a few topics will be considered here.

The response of a p-GaP electrode to chopped illumination is shown in **fig 13**. The relaxation and overshoot in the photocurrent response are typical of a recombination process, and in this particular case it has been established that photogenerated electrons recombine via proton states created during the diffusion of hydrogen into the semiconductor lattice (*cf*. fig 11b). The reaction sequence involves minority carrier capture by the recombination centre followed by majority carrier capture [9]:

$$h\nu \quad \rightarrow \quad h^+ + e^- \tag{11a}$$

$$H^+ + e^- \quad \rightarrow \quad H^\cdot \tag{11b}$$

$$H^\cdot + h^+ \quad \rightarrow \quad H^+ \tag{11c}$$

When the light is switched on, the concentration of $H^\cdot$ begins to increase and as a consequence an anodic current begins to flow as holes are consumed in reaction 11c. The total current therefore falls from its initial instantaneous value towards a lower steady state value. When the light is interrupted, the electron flux falls almost instantly to zero, but reaction 11c continues until excess $H^\cdot$ atoms have been consumed. It is this process which is responsible for the anodic overshoot in the transients in fig 13.

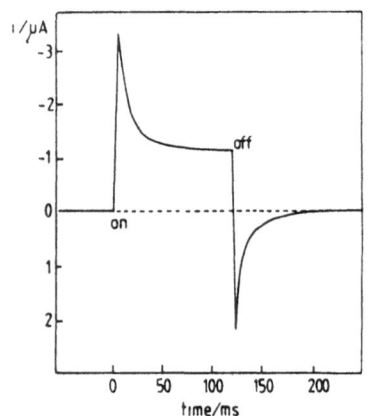

FIGURE 13. Photocurrent transient observed for p-GaP in 0.5 M H_2SO_4.

The decay transients are first order, and the pseudo first order rate constant, k_{rec}, for reaction 11c can be expressed as

$$k_{rec} = \sigma_h \cdot \upsilon_h \cdot p_{surf} \tag{12}$$

where σ_h is the hole capture cross section of $H^\cdot$, υ_h is the thermal velocity of holes and p_{surf} is the surface density of holes. If the potential distribution across the semiconductor/electrolyte junction behaves ideally, the surface density of holes is given by the Fermi-Dirac function which reduces to the Boltzmann distribution under depletion conditions:

$$p_{surf} = p_{bulk} \cdot \exp(q(E - E_{fb})/kT) \tag{13}$$

Consequently, the transient decay rate is expected to show a 'Nernstian' potential dependence of 59 mV per decade. In fact, the ideal Nernstian slope is rarely observed at potentials where recombination is important. This suggests that changes in the potential drop in the Helmholtz layer due to surface charging cannot be neglected [19,20].

12. INTENSITY MODULATED PHOTOCURRENT SPECTROSCOPY (IMPS).

The most sophisticated non stationary photocurrent technique involves excitation with an intensity modulated laser beam[19-21]. The apparatus used is shown in **fig** 14. The intensity of the laser beam is modulated sinusoidally by the acousto-optic modulator and the frequency response analyzer is used to relate magnitude and phase of the photocurrent response to those of the incident illumination, and the ratio j_{photo}/I_o is displayed in the complex plane.

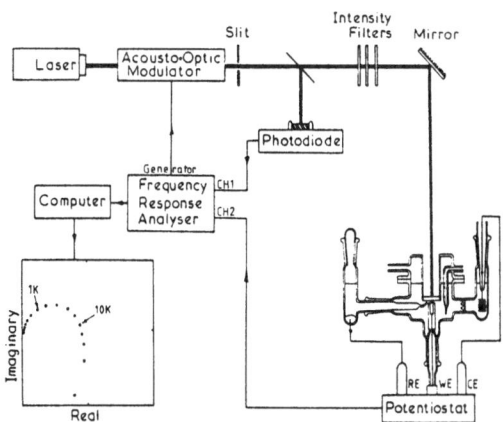

FIGURE 14. Experimental arrangement for IMPS.

In general, the ac photocurrent response contains two components. The first is the minority carrier current excited by the illumination, and since transport through the space charge is very rapid, this component remains in phase with the excitation. The second component is the majority carrier current. It is opposite in sign to the minority carrier current, and it exhibits a frequency dependent phase lag and attenuation because it arises from changes in the population of the surface recombination centres. At high frequencies, the ac component of the majority carrier flux becomes negligible, so that the effects of surface recombination are effectively 'frozen out'.

The IMPS response is also sensitive to the RC time constant made up by the combination of the space charge capacitance and the solution resistance. This leads to attenuation of the ac photocurrent signal at very high frequencies.

The overall IMPS response is illustrated by the theoretical response shown in **fig** 15. The low frequency semicircle in the *upper* complex plane is due to recombination, and the value of k_{rec} determines ω_{max}. The low frequency intercept on the real axis corresponds to the steady state photocurrent. The IMPS response then crosses the real axis and enters a semicircle in the *lower* complex plane because the signal is attenuated by the space charge capacitance in parallel with the solution resistance. The $R_{sol}C_{sc}$ time constant is equal to $1/\omega_{min}$.

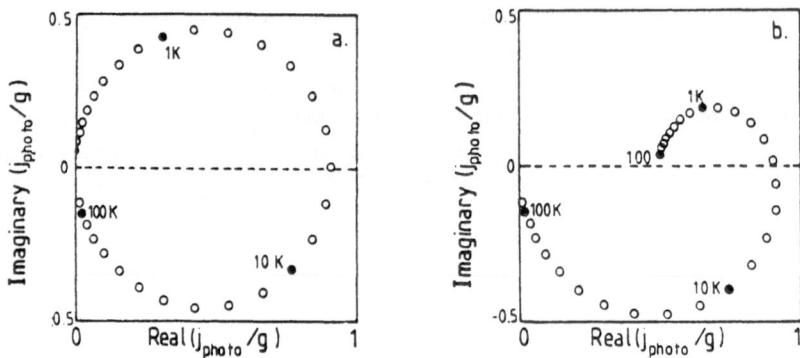

FIGURE 15. Theoretical IMPS responses.
(a) 100% recombination, (b) 50% recombination.

A typical set of experimental IMPS responses is shown in **fig 16** to illustrate the effect of band bending. The figure also shows the corresponding responses to chopped illumination. It can be seen that the semicircle in the upper complex IMPS plane is related to the recombination process which gives rise to the relaxation and overshoot in the chopped photocurrent response. The frequency response analysis is particularly useful because the rate constant k_{rec} is given directly by the value of ω_{max}. The results show that recombination becomes more rapid as the potential approaches flatband, whereas at high band bending, the recombination semicircle disappears, and the corresponding response to chopped illumination loses the decay and overshoot components.

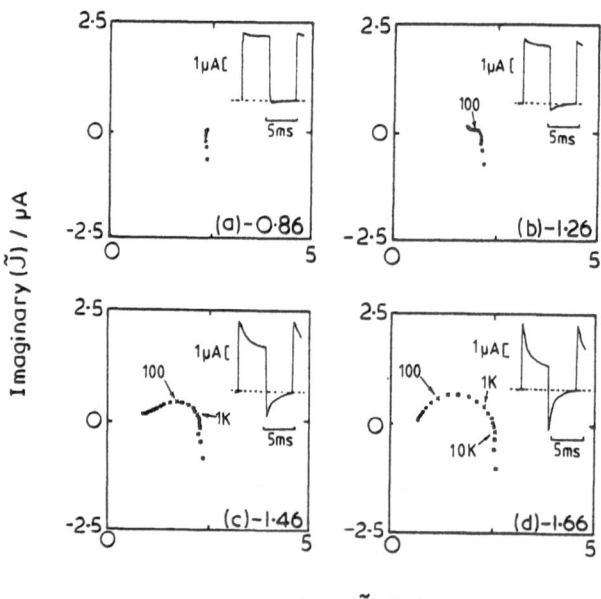

FIGURE 16. Set of IMPS responses for n-GaAs in K₂Se solution.

13. IMPS STUDIES OF PHOTOCURRENT MULTIPLICATION.

The power of the IMPS method is demonstrated by the successful analysis of several photocurrent multiplication reactions. The full theory is outside the scope of this brief overview, but in general terms it can be seen that the periodic photocurrent response will contain contributions from minority carrier capture and majority carrier injection. These contributions cannot be separated in the steady state, but under periodic conditions they exhibit quite different frequency dependencies. The minority carrier reaction is very rapid, and attenuation is negligible in the measurable frequency range. By contrast, majority carrier injection may be sufficiently slow that it gives rise to a characteristic phase lag and attenuation. The frequency dependent phase lag is again manifest as a semicircle, but now it is in the *lower* complex plane because the current due to majority carrier injection has the *same* sign as the minority carrier current (contrast this with the recombination case discussed in section 12). At high frequencies, the ac component associated with injection is attenuated and the semicircle in the lower plane tends towards unit quantum efficiency on the real axis because only the minority carrier component remains (the effect of the $R_{sol}C_{sc}$ time constant has been neglected for simplicity).

Fig 17 contrasts theoretical and experimental IMPS responses for the current doubling reduction of oxygen on p-GaP [23] (see also section 9). The rate constant for hole injection by the HO_2 intermediate obtained directly from the value of ω_{min} agrees well with the value estimated from the intensity dependence of the photocurrent conversion efficiency (see section 9).

IMPS has also been used to investigate the photocorrosion of n-Si in ammonium fluoride solutions [24,25]. In this case, photocurrent *quadrupling* has been observed at low light intensities, and the IMPS analysis shows that hole capture by a surface bond is followed by the injection of three electrons. These injection steps give rise to characteristic relaxation semicircles in the lower complex plane as shown in **fig 18**.

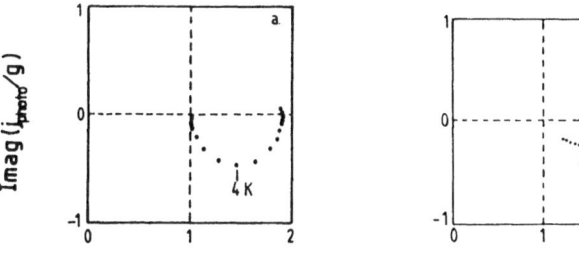

FIGURE 17. (a) Theoretical IMPS response for current doubling.
(b) Experimental IMPS response for p-GaP/O_2.

196

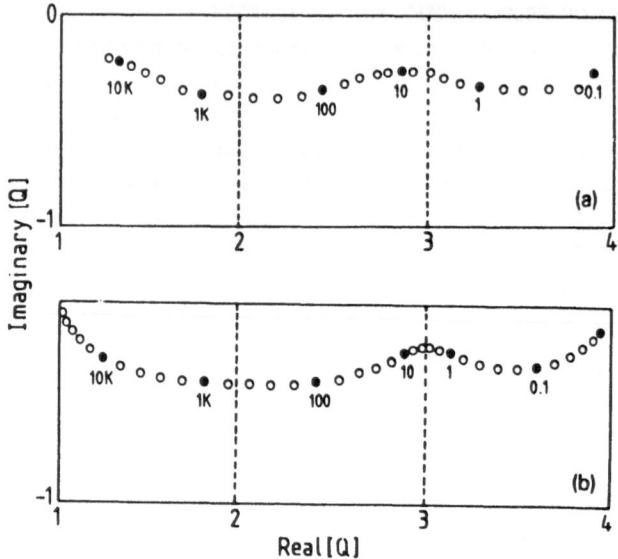

FIGURE 18. IMPS responses for n-Si in NH₄F.
(a) experimental, (b) simulated.

14. IN-SITU SPECTROSCOPIC METHODS IN SEMICONDUCTOR ELECTROCHEMISTRY.

A wide range of in-situ spectroscopies can be applied to
semiconductor electrodes. In the uv/visible, reflectance spectroscopy and
spectroscopic ellipsometry can be used to study the optical properties of
the semiconductor itself as well as to characterise the formation of
surface layers under electrochemical conditions. Modulation spectroscopies
have been widely used to study semiconductor electrodes because the optical
properties can be perturbed by an externally applied excitation [   ].

The technique of *electrolyte electroreflectance* is an example of a
modulation spectroscopy. It relies on the fact that the optical properties
of the solid are perturbed by an electric field. The corresponding change
in reflectance is small (typically less than 0.1%), but it can be measured
easily using potential modulation and lock-in detection in the
configuration shown in **fig 19**.

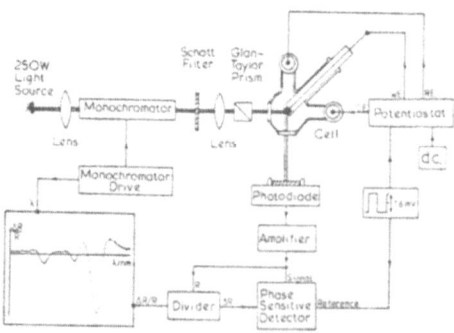

FIGURE 19. Experimental arrangement for electrolyte electroreflectance.

EER spectra normally exhibit sharp third derivative structure at the band gap energy and at other energies corresponding to critical points in the Brillouin zone. The normalised reflectance, $\Delta R/R$, depends on the square of the electric field in the so-called low-field limit[26], and under depletion conditions this in turn means that $\Delta R/R$ varies linearly with the amplitude of the potential modulation. **Fig 20** is an example of the kind of EER spectra observed for III-V semiconductors. The inset shows that $\Delta R/R$ varies linearly with modulation amplitude.

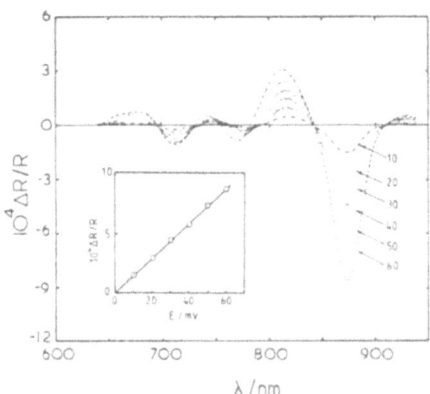

FIGURE 20. EER spectra for n-GaAs.

EER spectroscopy has been used to derive information about the potential distribution across the semiconductor/electrolyte interface. The advantage of the method is that it is sensitive only to the electric field in the space charge region, so that in principle it should be possible to deconvolute the overall potential drop into contributions associated with the space charge and Helmholtz regions. In practice, this is complicated by the fact that the field in the space charge region is non-uniform [27,28]. Another approach that we are examining is to study the frequency dependence of the EER response and to compare it with the corresponding impedance response. In principle, it should be possible to deconvolute the electrode impedance in this way.

A second technique that has proved to be a powerful tool in the investigation of the semiconductor/electrolyte interface is *in-situ infrared spectroscopy*. The method has been applied to metal electrodes where it is possible to modulate the IR absorbance by perturbing the potential. As discussed in section 11, in the case of a semiconductor, it is more convenient to perturb the system with *light*. Multiple internal reflectance FTIR spectroscopy has been used to follow changes in the surface composition of n-Si during photodissolution in ammonium fluoride solutions [29,30]. Spectra were recorded in the dark and under illumination, and the normalised difference spectrum was obtained as a function of the duration and intensity of illumination. **Fig 21** shows that a strong Si-H stretch develops after illumination and photodissolution of the silicon. This signal decays only slowly in the dark, so it is not due to an unstable intermediate. In fact, the signal arises from the transient formation of a layer of porous or amorphous silicon that is formed during the photo- dissolution process. Further work is in progress to improve the time resolution of this new method.

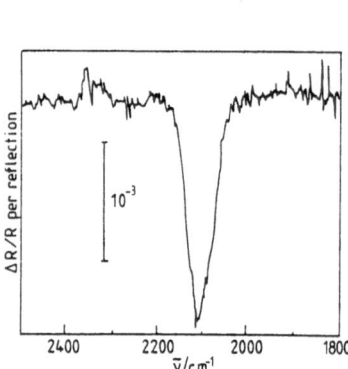
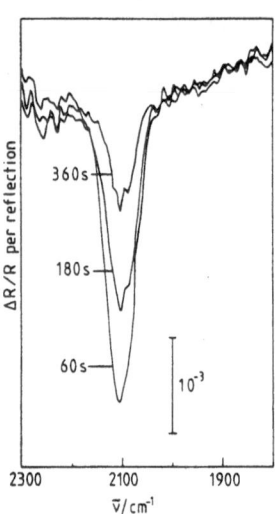

FIGURE 21. FTIR spectra of Si-H surface on photoetched n-Si.
(a) after photoetching, (b) decay of signal in dark.

15. OUTLOOK.

This survey of the semiconductor/electrolyte interface has been necessarily brief. The progress made in the last 20 years has been substantial, but there is a need for more surface sensitive techniques to characterise the chemical and physical changes that take place at the surface of a semiconductor. Some of the models that are used to describe the interface will need revision as new information becomes available. The role of the surface has certainly been underestimated, and even though the materials are single crystals, the degree of characterisation of the real surfaces has been inadequate. In the case of metal electrodes, it has been possible to characterise surfaces *ex-situ* by the techniques such as LEED and Auger spectroscopy. Little comparable work has been carried out with semiconductor electrodes. *In-situ* methods look more promising; the advent of *in-situ* infrared spectroscopy should allow semiconductor electrochemistry to enter the same phase of molecular resolution that has been achieved in the past few years in the case of metal electrodes.

ACKNOWLEDGMENTS.

The support of the UK Science and Engineering Council is gratefully acknowledged. I should also like to thank my collaborators who have worked with me in the area of semiconductor electrochemistry.

REFERENCES.

1. V.A. Myamlin and Yu.V. Pleskov, *Electrochemistry of semiconductors*; Plenum: New York, 1967.

2. S.R. Morrison, *Electrochemistry of semiconductors and oxidised metal electrodes*; Plenum: New York, 1977.

3. Yu.V. Pleskov and Yu.Ya. Gurevich, *Semiconductor Photoelectrochemistry*; Consultants Bureau: New York, 1986.

4. A. Hamnett in *Comprehensive Chemical Kinetics: Volume 27*, Editor R.G. Compton; Elsevier: Amsterdam, 1987. p 61.

5. R.A. Marcus, J. Chem. Phys., **43**, 679 (1965).

6. H. Gerischer, Z. Phys. Chem. N.F. **27**, 48 (1961).

7. for a recent review see M.J. Weaver in reference 4, p 1.

8. W.W. Gärtner, Phys. Rev., **116**, 84 (1959).

9. J. Li, R. Peat and L.M. Peter, J. Electroanal. Chem., **165**, 41 (1984).

10. L.M. Peter in *Electrochemistry*; Specialist Periodical Report, Editor D. Pletcher; Royal Society of Chemistry: London 1984. p 66.

11. J. Li and L.M. Peter, J. Electroanal. Chem., **182**, 399 (1985).

12. P. Herrasti and L.M. Peter, J. Electroanal. Chem. (*in press*).

13. L.M. Peter, Ber. Bunsenges. Phys. Chem., **91**, 419 (1987).

14. L.M. Peter, J. Electroanal. Chem., **98**, 49 (1079).

15. L.M. Castillo and L.M. Peter, J. Electroanal. Chem., **146**, 37 (1983).

16. J.S. Buchanan and L.M. Peter, Electrochim. Acta, **32**, 127 (1987).

17. Y.Y. Gurevich, Y.V. Pleskov and Z.A. Rotenberg, *Photoelectrochemistry*,; Consultants Bureau: New York (1980).

18. L.M. Peter in *Photocatalysis and the Environment*. NATO ASI Ser. C. Editor M. Schiavello; Kluwer Academic: Dordrecht (1988). p 343.

19. L.M. Peter, Chem. Rev., **99**, 753 (1990).

20. R. Peat and L.M. Peter, Ber. Bunsenges. Phys. Chem. **91**, 381 (1987).

21. L.M. Peter, J. Li, R. Peat, H.J. Lewerenz and J. Stumper, Electrochim. Acta, **35**, 1657 (1990).

22. L.M. Peter and J. Li, J. Electroanal. Chem., **193**, 27 (1985).

23. R. Peat and L.M. Peter, J. Electroanal. Chem., **209**, 307 (1986).

24. H.J. Lewerenz, J. Stumper and L.M. Peter, Phys. Rev. Lett., **61**, 1989 (1988).

25. L.M. Peter, A.M. Borazio, H.J. Lewerenz and J. Stumper, J. Electroanal. Chem. **290**, 229 (1990).

26. D.E. Aspnes in *Handbook on Semiconductors Vol 2*; Editor M. Balkanski. North Holland: New York, 1980. p 109.

27. A. Hamnett in *Comprehensive Chemical Kinetics, Vol 29*. Editor R. Compton. Elesevier: Amsterdam, 1987. p 385.

28. L.M. Abrantes, R. Peat, L.M. Peter and A. Hamnett, Ber. Bunseges. Phys. Chem., **91**, 369 (1987).

29. L.M. Peter, D.J. Blackwood and S. Pons, Phys. Rev. Lett., **62**, 308 (1988).

30. L.M. Peter, D.J. Blackwood and S. Pons, J. Electroanal. Chem, **294**, 111 (1990).

STARK EFFECT ON ADSORBATES AT ELECTRIFIED INTERFACES

A.STELLA
Dipartimento di Fisica
Universita' di Pavia
Italy

ABSTRACT. In this paper we shall report a few examples of evidence of vibrational bands due to the presence of molecular species at Si-SiO$_2$ and Si-electrolyte interfaces as well as of electric-field modulated absorption of adsorbed or of implanted complexes in these systems. For both interfaces, the multiple internal reflection configuration has been adopted, since it may allow a remarkable improvement in the detection limit of the bands investigated. The specific behaviour of SiH and SiOH vibrational bands in the Si-SiO$_2$ system and of H$_3$O$^+$, H$_2$O, D$_2$O bands of the Si-electrolyte interface will be discussed. The dependence of these bands upon the electric field will be studied and a few conclusions will be drawn.

1. Introduction

We will illustrate the importance of applying very high electric fields on molecular complexes, so as to permit the study of Stark effects, and more generally of electromodulation, up to field values exceeding those usually reached in gaseous systems by one or two orders of magnitude. Two specific cases, in which such a requirement can be met, will be studied here, namely the effect of high electric fields on molecular complexes at Si-SiO$_2$ and at Si-electrolyte interfaces.

The adoption of a multiple internal reflection optical technique in the infrared allows us to maximize the sensivity, for both detection and analysis of the lineshape.

2. Si-SiO$_2$ Interface

The molecular Stark effect for gaseous rotating-vibrating molecules has been investigated on numerous occasions (1). Electric fields are limited to approximately 7×10^4 V/cm by the breakdown of the gas. The splitting or shift of the vibration-rotation lines is predicted to be quadratic in the electric field for both polar and nonpolar molecules, with a few exceptions. Although less often discussed, the peak intensity of a vibration-rotation line also varies quadratically with the electric field; hence, if the shift cannot be resolved, the integrated absorption coefficient can still be measured.

R. Guidelli (ed.), Electrified Interfaces in Physics, Chemistry and Biology, 201–212.
© 1992 *Kluwer Academic Publishers.*

Very little work has been carried out with nonrotating molecules imbedded in solids. The vibrational absorption band due to OH$^-$ dipoles in KBr was studied by Handler and Aspnes in an electric-field-modulation experiment using peak fields of 3×10^4 V/cm (2). Analysis of the lineshape of the electroabsorption signal with light polarized both parallel and perpendicular to the applied field allowed the authors to ascertain that the dipole orientation is the physical mechanism responsible for absorption modulation. Since the applied voltage was entirely ac ($V_{dc} = 0$ in their case), the field changed from - E to + E when the voltage was modulated between - V and + V. The signal, which depends on the square of the field, was then detected at double frequency.

The case of electronic bands due to color centers in alkali halide crystals has been treated in detail, the main result for the Stark effect on the F center being a shift and a distortion of the broad line, such that there is no net change in integrated absorption. The observed line has usually one large lobe in one direction of change of absorption and two small lobes in the opposite direction (3). The peak height of a given lobe is quadratic in the electric field. These changes in absorption are typically 10^5 times smaller than the absorption itself. Applied electric fields as large as 10^5 V/cm were used.

Electroabsorption studied in several other systems shows some of the features of the molecular Stark effect. For example, Anastassakis et al. measured the electric-field induced absorption of an infrared inactive vibrational band at 1336 cm^{-1} in diamond using an electric field of 1×10^5 V/cm, and verified the quadratic dependence of absorption on the electric field (4). Gobrecht *et al.* measured the electric-field induced absorption of the water molecule in the infrared at 3300 cm^{-1} at the surface of Ge in H_2SO_4 electrolyte. As for the electromodulation of electronic bands due to impurities in semiconductors, the most detailed work was performed by Jonath *et al.*, who applied to GaAs an ac voltage superimposed on a dc bias and observed first-and second-harmonic electroabsorption spectra just below the band gap (5,6).

For a simple symmetric Stark effect, the absorption coefficient is quadratic in the electric field (applied voltage V). The two typical modulation procedures can be illustrated as follows: (I) $V = V_{dc} + V_{ac}$, (II) a total ac modulation detected at 2f (a derivative scheme). The first produces a modulated absorption which is linear in V_{dc} as illustrated, while in the second the modulated absorption is quadratic.

In addition to the molecular Stark effect, we may expect to see interface-state (IS) absorption modulated by the electric field, as studied by Harrick. This absorption is roughly uniform in wavelength, except that there is an additional long wavelength accumulation of free carriers in the space-charge region or induced inversion near the surface. Let us consider a SiO_2 film, 1 μ thick, which was thermally grown on a substrate trapezoid of a high-resistivity p-type Si ($\rho \approx 5.10^3$ Ω cm). Hydrogen was subsequently implanted in the SiO_2 layer with a fluence of 2.10^{17} cm^{-2} at 95 KV, and a 2000 Å thick Au film was then evaporated on top of the oxide.

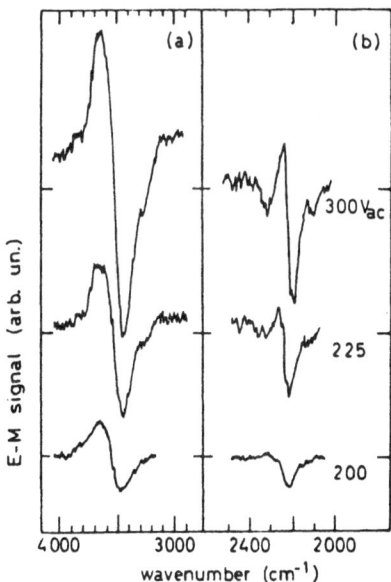

Figure 1-Experimental lineshapes of electric-field modulated MIR vibrational bands for SiOH (a) and SiH (b). V_{ac}=200, 225 and 300 V starting from the bottom in both cases.

A light beam passing through the Si trapezoid was totally reflected 18 times at the Si-SiO$_2$ interface, giving a multiple internal reflectance (MIR) spectrum which shows the presence of SiH and SiOH vibrational bands. The evanescent field of the infrared radiation decayed exponentially through the oxide layer to a small value at the SiO$_2$-Au interface. The MIR response was modulated with a.c. electric fields up to $1.3.10^6$ V/ cm, while d.c. fields were used to control the band bending at the interface.

Conventional and electric-field modulated MIR spectra were recorded using unpolarized as well as polarized light. The a.c. field dependence of $\Delta I/I$ at a given wavelength for both SiH and SiOH lines was found to be quadratic. Moreover, the lineshape modification with increasing electric field intensity turned out to change gradually its profile, initially characterized by a negative lobe, into a dispersive-like behaviour, at the highest fields applied.

Figure 1 shows the experimental electromodulation lineshapes for SiOH (left) and SiH (right) for V equal to 200, 225 and 300 V when starting from the bottom. One can observe the lineshape modification for the SiH line, which takes place at fields 4.5 times higher than for the SiOH line. Such a dispersive-like character of the lineshape at the highest applied field can be explained in terms of a combination of dipole strengthening and Stark shift. The latter, although not measurable due to the relatively poor resolution ($\approx$100 cm^{-1}), may actually contribute to determine the lineshape.

We may represent the $\alpha(v)$ function as

$$\alpha(v) = K \, |\langle\mu\rangle|^2 \, v_oN \, \frac{J}{(v-v_o)^2+J^2/4} \qquad (1)$$

where N is the number of bonds per unit volume, $\langle\mu\rangle$ is the dipole moment between the two states involved in the transition, v_o is the line frequency , J is the experimental linewidth and K is a constant which fits the experimental data. The effects of a static electric field on (1) are represented by a strengthening of the dipole moment and a shift of the line frequency v_o . If μ is written as

$$\mu = \mu_o + \varepsilon \, (r\text{-}r_o) \qquad (2)$$

where μ_o, the static component of the dipole moment , plays no role in the transition and r_o is the bond length, the strengthening can be pointed out, to first order, through ε in the following way:

$$\varepsilon = \mu_o/r_o + \mu'/r_o = \varepsilon^o + \varepsilon' \qquad (3)$$

where μ' is the dipole component induced by the electric field. Taking into account (1), (2) and (3), the absorption coefficient in the presence of the field may be represented through the harmonic approximation, as follows:

$$\overline{\alpha}(v) = K'(\varepsilon^o+\varepsilon')^2 \, N \, \frac{J}{(v-v_o)^2+J^2/4}$$

where K' incorporates all the parameters which depend neither on the field nor on the bond.

The Stark effect on v_o can be treated with a perturbative approach: we find out that no line shift is produced by the electric field for a harmonic oscillation, to second order terms included. We therefore calculated the first-order correction on a Morse oscillator. If $E=10^6$ V/cm is parallel to the dipole moment we obtain $\Delta v_o=.2$ cm^{-1} for SiH and 1.6 cm^{-1} for SiOH. As expected, the shift for SiOH is larger than for SiH and may account, within a factor of two, for the experimental observation that the lineshape switches to a dispersive-like character at field values which are lower than for the SiH line.

The effects of strengthening and shift can now be recombined, to give theoretical lineshapes to be compared with the experimental curves of the figure. In the case of the SiOH band, a theoretical lineshape which is very similar to the experimental ones is reported for specific values of ε' and ε^o .

3. Si-Electrolyte Interface

High electric fields can exist across the space-charge layer of the semiconductor surface, and across the Helmholtz and Gouy layers in the liquid. The electric fields can be large ($\geq$ 10^6V/cm) and give rise to appreciable electromodulation effects of semiconductor interband transitions, of interface states, of free-carrier absorption, and perhaps of molecules in the interface region, either as a Stark effect or as a change in population of molecular species. While electromodulation of interband transitions of semiconductors at a semiconductor/ electrolyte interface (electroreflectance) is usually performed in a "near-normal-incidence" reflection scheme for a transparent liquid in the visible spectrum, an internal-reflection configuration can be adopted when using aqueous electrolytes and operating at $E_g \geq h\nu$) in the near infrared, were E_g is the fundamental band gap. As we shall discuss, we observed signals due to free carriers, interface states and molecular species, and we will show that these effects can be individually separated and studied.

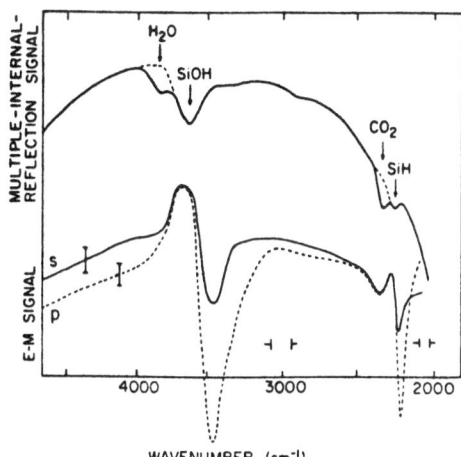

Figure 2. The upper curve is the unpolarized multiple internal reflection spectrum showing the SiOH and SiH vibration bands at 3640 and 2260 cm^{-1}. Remnants of atmospheric H_2O and CO_2 are present. The two lower curves represent s- and p-polarized electromodulation spectra. The gain is increased by a factor 10^5 for the lower spectrum. The lower abscissa is the zero signal for both spectra.

The internal-reflection spectroscopic system, covering the $5000 \div 1600$ cm^{-1} infrared region, was described in detail elsewhere. Small Si trapezoides were attached to a liquid cell, so that at the Si/electrolyte interface several internal reflections occurred. No absorption bands due to the glue were observed. A cut-on filter blocked second-order radiation at shorter wavelengths. The trapezoid was in the fore optics. One electrical lead was attached to the Si with an In-Ga eutectic and silver paint. A platinum wire served as a second electrode and the Si was grounded. No standard reference electrode was utilized. Two electrolytic solutions (0.1 M H$_2$SO$_4$ and 0.1 M KOH) were used at room temperature.

After several hours in KOH electrolyte, the Si surface began to roughen, reducing the signal substantially due to scattering. The evanescent-wave intensity decayed into the electrolyte to 1/e at a distance of 1.5 μ at 4000 cm^{-1}. The Si was n-type with a resistivity of $\approx$ 10 Ω cm (n $<10^{15}$ cm^{-3}). The large surface had a (100) orientation which was mechanically polished. Figure 2 shows an internal-reflection spectrum as obtained with a mechanical light chopper, with the cell either empty (dashed curve) or filled with 0.1M H$_2$SO$_4$ (solid curve). The main features are a water vibration band at 3400 cm^{-1} (45% absorptance) and the CH bands near 2900 cm^{-1} due to absorbed hydrocarbons. The vibration bands of gaseous H$_2$O and CO$_2$ also occurred, because the optical path was not always dry.

The 0.1 M H$_2$SO$_4$ electrolyte is noncorrosive. An anodic oxide grows on the Si under anodic bias, but does not etch off under cathodic bias. On the other hand, the 0.1 M KOH electrolyte is an etchant which dissolves Si much faster than SiO$_2$ (probably < 400 A/min for Si and < 0.6 A/min for SiO$_2$), although we did not carry out a direct measurement of etching rates at this molarity. An anodic oxide can also be grown in KOH. Since we have worked with both "wet" and "dry" samples performing both electrochemical and metal-oxide— semiconductor measurements, we will usually refer to the Pt electrode voltage with respect to the Si as to the gate voltage. This is opposite to the electrochemical convention according to which an anodic and a cathodic voltage of Si are used to denote a voltage positive and negative with respect to a standard electrode.

To probe electromodulation effects, a voltage V$_{ac}$ (peak-to-peak) was applied, superimposed on a dc bias voltage V$_{dc}$. The modulation frequency was 260 Hz and most electromodulation spectra were recorded at 1f. When a positive bias is applied to the sample, the bands are bent down, which results in an accumulation in n-type Si. The carrier concentration in the space-charge region increases as the bias voltage increases. As a result, the free-carrier absorption coefficient α(FC) will also increase (at least for low bias). When the sinusoidal modulation voltage is superimposed on the bias voltage, a concomitant modulation of α occurs, denoted by $\Delta\alpha$. Note that V$_{ac}$, the total applied voltage (V$_{dc}$ + V$_{ac}$) and the modulated one are all in phase, i.e., when V$_{ac}$ is a maximum, the same is true for (V$_{ac}$ +V$_{dc}$) and α.

At short wavelengths the free-carrier effect should be negligible. If the line in Fig.3

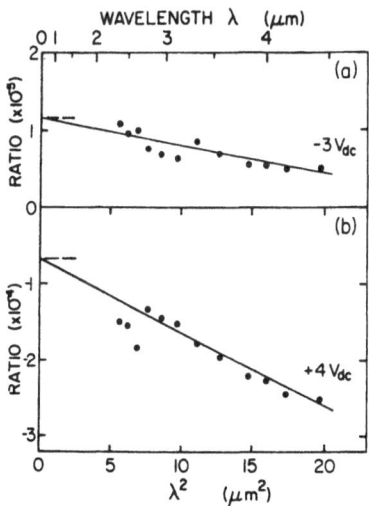

Figure 3. Ratio of electromodulation curves with the internal-reflection spectrum (solid curve). Spectrum (b) has a dependence characteristic of free-electron absorption in Si for accumulation.

is extrapolated to $\lambda^2=0$, a negative intercept, which is presumably due to interface states, is obtained. Thus, the electromodulation signal at $V_{dc}=4$ V has free-carrier as well as interface-state components, which are both negative. Therefore, the interface-state absorption coefficient $\alpha(IS)$ must have a positive slope.

A spectrum for a negative bias of -3V (spectrum 1) for the H_2SO_4 electrolyte is shown in figure 3b. In the spectrum the wavelength dependence is weak, indicating that the free-carrier effect was becoming negligible. The fact that the slope did not reverse sign indicates that the space-charge region was not inverted. Since inversion did not occur, $\alpha(FC)$ for negative bias was not significant. The electromodulation signal for negative bias was predominantly due to interface states and was positive. Thus, $\alpha(IS)$ must have a negative slope. The electromodulation signal was null when the bias voltage was close to zero, and hence $\alpha(IS)$ has a minimum near zero for a fixed wavelength.

Spectral scans near zero bias were examined, exhibiting distinct lines on a low background. Since these spectra are indicative of molecular vibrations, we shall refer to them as to molecular-electromodulation (ME) spectra. The ME spectrum consisted of two main lines of opposite sign, with a clue of a third line in between. The peak heights of the two lines turned out to be a maximum when the dc bias was adjusted to annul the electromodulation signal at higher wavenumbers. We typically made this annulling adjustement near 4010 cm^{-1}. Depending on the anodization history, a few tenths of a volt near zero were required. It may be just a coincidence that the ME signal is a maximum when the electromodulation signal is annulled, or the two may be related. Spectra recorded for both positive and negative biases of 1 V exhibited the two lines, but they were much weaker. The spectral slit width was ≈ 100 cm^{-1} at 3600 cm^{-1} and 64 cm^{-1} at 2800 cm^{-1}. Increasing the resolution by a factor of two did not reveal any structure in these lines.

On the basis of line position, we have tentatively ascribed the lines at 3610, 3350, and 3000 cm^{-1} to an OH mode, an H_2O mode and an hydrated H_3O^+ mode, respectively. The OH hydroxyl band typically falls in the 3680-3500 cm^{-1} region in most samples. Hydrated H_3O^+ has lines in the 3000-2900 cm^{-1} region. A liquid H_2O band occurs near 3400 cm^{-1}.

A spectrum obtained when replacing H_2O by D_2O (while still using H_2SO_4) is reported in Fig.4. The prominent bands at 2690 and 2290 cm^{-1} are ascribed to OD and D_3O^+ in view of the isotopic shifts. For example, OH_3 was analyzed by comparison with NH_3 to obtain the normal-mode assignments. The mode is presumably the band we observe. The ratios of the mode frequencies for gaseous $v_3(NH_3)/v_3(NDH_2)$ and

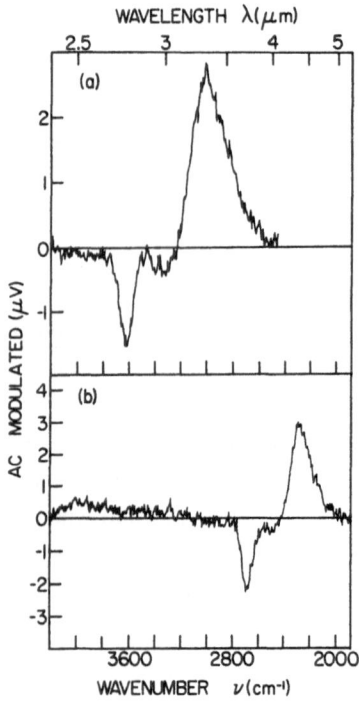

Figure 4. (a) Electroabsorption due to molecular electromodulation after the interface-state absorption signal in Fig.1(b) was annulled. (b) Electroabsorption due to molecular electromodulation using D_2O with H_2SO_4. The optical path was dried.

$v_3(NH_3)/v(ND_3)$ and the ratio H_3O^+/D_3O^+ from our experiment are 1.32±0.01. The OH mode for liquid organic molecules with single hydroxyl groups is typically 3630±10 cm^{-1}, comparable with the OH line position we measured. The ratio of modes for such organic liquids and the ratio from our measurement are 1.35±0.01. In a similar expenment using Ge, Gobrecht *et al.* observed only a H_2O band for H_2SO_4 and a D_2O band for D_2SO_4. They did not report any OH or H_3O^+ bands.

A ME spectrum for the KOH electrolyte at V_{dc} =0.3 and V_{ac}=0.6 V, together with the mechanically chopped spectrum, is shown in Fig.5. A single broad line, attributed to H_2O, appears in the ME spectrum at 3400 cm^{-1}. There is no evidence for OH or H_3O^+ bands, whose positions are marked by the arrows.

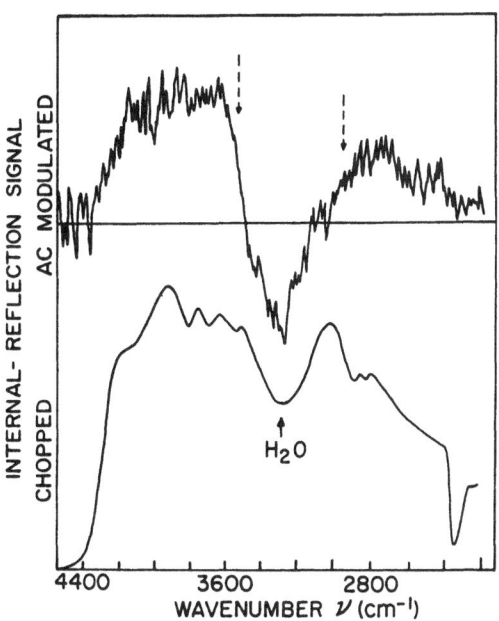

Figure 5. Modulated electroabsorption (upper curve) and internal-reflection (lower curve) spectra of Si in a 0.1M KOH solution (V_{dc}=+0.3V, V_{ac}=0.6V). The ME signal was measured at 1f. The optical path was not dried.

The electromodulation signal was not quite annulled, leading to the positive signal near 4000 cm^{-1}. The sample had an oxide of unknown thickness. However, the initial oxide film that grows as soon as anodization begins in aqueous KOH, is 20 Å thick. The dc bias was slightly cathodic so that the oxide was being etched. As a result, the dc bias had to be continuosly adjusted in the anodic direction to keep the electromodulation signal annulled at 4010 cm^{-1}. Eventually, the passivation voltage was reached, probably because the oxide had been etched off. At this point the sample reanodized and the electromodulation signal became large. Then, to annul the electromodulation signal, the dc bias had to be shifted in the cathodic direction and the whole process started again.

The dependencies of the peak line intensities upon the modulation voltage for the OH and H_3O^+ lines in H_2SO_4 and for the H_2O line in KOH are shown in Fig. 6. Recall that $V_{dc} \approx 0$. In general the ME signals are linear in ac voltage, although we do not know how the electric fields in the interfacial region depend on the applied voltage. As is known, in a "dry" $Si/SiO_2/Au$ structure the peak absorptance due to a Stark effect on Si-H and Si-OH complexes in SiO_2 varied quadratically with the applied voltage for total ac modulation. Because of the relatively low resolution, we were unable to observe any small shifts of the lines as a function of voltage.

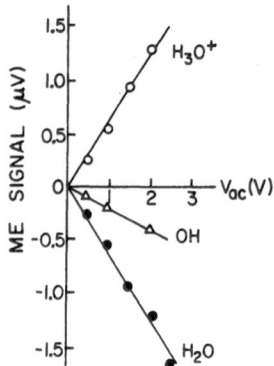

Figure 6. V_{ac} dependence of molecular electromodulation signal: $v(H_3O^+)=2980$ cm^{-1} in a 0.1M H_2SO_4 solution; $v(H_2O)=3315$ cm^{-1} in a 0.1M KOH solution.

The major fraction of the ME signal was detected at the frequency of the modulating voltage, while the Stark effect in the dry sample was mainly expected at 2f, appearing at 1f only due to the asymmetry in the applied electric field across the oxide, as a result of depletion in the Si. We are therefore inclined to conclude that the operating mechanism was not a modulation of the dipole moment but rather an electric-field induced variation of the population of molecular complexes near the interface.

In the case of the Si-electrolyte interface, we have three types of modulated infrared internal-reflection signals for n-type Si in aqueous H_2SO_4 and KOH. There is an interface-state signal which can change sign as a function of dc bias, and perhaps of wavelength. Due to lack of a calculation of the optical matrix elements both for intra-interface-state transitions and for transitions from valence band to interface state and to conduction band, it is not possible to identify the observed transitions unambiguously.

The free-carrier signal, which varies as λ^2, is due to accumulation of electrons for positive gate bias. No inversion free-carrier signal was observed for negative gate bias. When the dc bias is adjusted near zero, the interface-state signal can be annulled and new signals are revealed, which are due to a vibrational band of OH (perhaps in water clusters in or on the anodic oxide film covering the Si or in SiOH), a band of H_3O^+ (perhaps due to a population change associated with electrolysis effects), and a band of H_2O (perhaps due to flip-flop of adsorbed water molecules). The experimental phases of these vibrational-band signals along with the free-carrier and interface-state signals are used to determine the dependence of each absorption mechanism on dc bias in a qualitative way.

4. Conclusions

In the case of the Si-SiO$_2$ system, we were able to apply electric fields higher than 10^6 V cm^{-1} and to observe Stark effects on molecular complexes, immersed in a background absorption due to the presence of interface states or free carriers.

In the Si-electrolyte case, the water line observed for both electrolytes may be due to dipole orientation (flip-flop) of adsorbed H_2O molecules. For positive bias the hydrogen is preferentially oriented away from the surface. If the population of the two orientations were the same, the electromodulation signal would occur at 2f. Because of band-bending effects, the two orientations have probably different populations at the extremes of the ac modulation, which would result in a signal at 1f. The observed sign of the H_2O signal requires that the population of the hydrogen-oriented-down H_2O (positive bias) be larger than the hydrogen-oriented-up H_2O (negative bias). We were not able to carry out s- and p-polarization studies at 1f and we saw no signal at 2f because of a poor signal-to-noise ratio. These experiments were performed for the case of OH⁻ in KCl by Handler and Aspnes (2).

Electromodulation at both single and double frequency was detected, the electric field dependence was studied very accurately and the variation of the lineshape with electric field was related to a combination of shift and dipole strengthening.

References

(1) Palik E.D., Holm R.T., Stella A., and Holmes H.L., (1982) "Stark effect of SiOH and SiH molecules in SiO$_2$",Journal of Applied Physics **5 3**, 8454-8461.

(2) Handler P., Aspenes D.E., (1966) "Electric-field modulation of the vibrational absorption of OH⁻ in KBr", Physical Review Letters **17**,1095-1097.

(3) Grassano V.M., (1977) "The stark effect of localized electronic centres", Il Nuovo Cimento **39B**, 368-377.

(4) Anastassakis E., Iwasa S., and Burstein E., (1966) "Electric-field-induced infrared

absorption in diamond", Physical Review Letters **17**,1051-1054.

(**5**) Jonath A.D., Vronkov E., Bube R.H., (1975) "Extrinsic electroabsorption in High-resistivity GaAs", Journal of Applied Physics **46**,1754-1766.

(**6**) Palik E.D., Holm R.T., and Stella A., (1984) "Electric-field-modulated, infrared internal-reflection study of the silicon-electrolyte interface", Journal of Applied Physics **56**,843-849.

THERMODYNAMICS OF ADSORPTION

ROGER PARSONS
Department of Chemistry
University of Southampton
Southampton, SO9 5NH
United Kingdom

ABSTRACT.   The use of thermodynamics in obtaining information about
interfaces is outlined for both fluid/fluid and solid/fluid interfaces.
Some of the special problems which arise with interphases containing
charged species are indicated.   In particular the problem of the location
of the charge is discussed.   The use of adsorption isotherms is
considered.

1. Introduction

There is now such a variety of probes for the study of interphases that
one might reasonably ask whether the essentially macroscopic tool of
thermodynamics can compete with methods giving microscopic information
like vibrational spectroscopy or the scanning tunnelling microscope.
Nevertheless, the powerful methods introduced by J.W. Gibbs over a century
ago [1] still seem capable to providing quantitative information about the
composition of interphases and, in particular, about the charge in
electrified interphases.   Hence some discussion of the capabilities of
thermodynamics is still appropriate.   The account given here relies
substantially on the approach used by Frumkin [2, 3] which does not seem
to have gained wide acceptance outside the circle of his coworkers and yet
must be considered the most useful way of treating the behaviour of
electrified interphases [4].

2. The liquid/liquid interphase with free charge transfer.

A specific example of an interphase between two immiscible liquids will
be discussed, but the method is applicable to any fluid/fluid interphase
including, for example, the case when one fluid is a liquid metal like
mercury.

For the interphase between water (W) and an immiscible non-
aqueous solvent (O) referred to as oil, a system is considered in which

R. Guidelli (ed.), Electrified Interfaces in Physics, Chemistry and Biology, 213–227.
© 1992 *Kluwer Academic Publishers.*

the oil contains two electrolytes tetrabutylammonium tetraphenyl borate (BP) and tetramethyl ammonium tetraphenyl borate (MP) and the aqueous solution contains tetramethylammonium chloride (MC). This interphase may be represented as

$$
\begin{array}{c|c}
O & W \\
BP & MC \\
MP &
\end{array}
\tag{1}
$$

The B^+ and P^- ions are unable to cross into the water phase and the Cl^- ion is unable to cross into the oil phase, at least in a range of conditions which are accessible experimentally. On the other hand the M^+ ion can cross the interphase freely and reach an equilibrium distribution between the two phases.

The Gibbs-Duhem equation for this interphase can be written

$$
-d\gamma = S^\sigma dT - V^\sigma dp + \Gamma_{BP}d\mu_{BP} + \Gamma_{MP}d\mu_{MP} + \Gamma_{MC}d\mu_{MC}
$$

$$
+ \Gamma_o d\mu_O + \Gamma_w d\mu_w
\tag{2}
$$

where γ is the interfacial tension, S^σ is the entropy of unit area of the interphase, V^σ the volume of unit area of the interphase, Γ is the amount of the species indicated by the subscript in unit area of the interphase and μ is the chemical potential of that species. The system described here has five components; the degree of dissociation of the electrolytes need not be specified as the Γ's are concerned only with the amount of a given component.

The interphase is necessarily in contact with the neighbouring phases for which the Gibbs-Duhem equations

$$
S^o dT - V^o dp + n^o_{BP}d\mu_{BP} + n^o_{MP}d\mu_{MP} + n^o_o d\mu_o = 0
\tag{3}
$$

$$
S^w dT - V^w dp + n^w_{MC}d\mu_{MC} + n^w_w d\mu_w = 0
\tag{4}
$$

may be written.

At equilibrium the temperature, pressure and chemical potentials are uniform through the two phases and so two variables may be eliminated between the three equations (2-4). If μ_o and μ_w are eliminated the Gibbs adsorption equation is obtained:

$$
-d\gamma = s^\sigma dT - v^\sigma dp + \Gamma^{o,w}_{BP}d\mu_{BP} + \Gamma^{o,w}_{MP}d\mu_{MP} + \Gamma^{o,w}_{MC}d\mu_{MC}
\tag{5}
$$

where the coefficients of the differentials on the right-hand side are the so-called 'excess' quantities.

$$
s^\sigma = S^\sigma - S^w \frac{\Delta_w}{\Delta} - S^o \frac{\Delta_o}{\Delta}
\tag{6}
$$

$$
v^\sigma = V^\sigma - V^w \frac{\Delta_w}{\Delta} - V^o \frac{\Delta_o}{\Delta}
\tag{7}
$$

$$\Gamma_{i.}^{o,w} = \Gamma_{i-} - n_i^w \frac{\Delta_w}{\Delta} - n_{i.}^o \frac{\Delta_o}{\Delta} \tag{8}$$

and

$$\Delta = \begin{vmatrix} n_o^w & n_w^w \\ n_o^o & n_w^o \end{vmatrix} \tag{9}$$

$$\Delta_w = \begin{vmatrix} n_w^w & \Gamma_w \\ n_w^o & \Gamma_o \end{vmatrix} \tag{10}$$

$$\Delta_o = \begin{vmatrix} n_o^w & \Gamma_o \\ n_w^w & \Gamma_w \end{vmatrix} \tag{11}$$

This derivation has assumed that the interphase is a region of finite thickness which may be treated as having properties somewhat similar to a bulk phase. The extensive properties of equation (2) depend on the thickness assumed for the interphase while the 'excess' quantities of equation (5) do not. This is the importance of considering excess quantities which are accessible from experiment.

This treatment differs from that used originally by Gibbs who represented the real system by one in which the two bulk phases were uniform up to an interface of infinitessimal thickness. The differences of properties between the real system and this model system were attributed to the interface. With equivalent assumptions these two methods yield identical results in terms of excess quantities. It should also be noted that the form of equation (5) is the same as that for an interphase between two non-ionic phases.

The surface excesses of the three solute components can be obtained by measuring the interfacial tension as a function of composition:

$$\Gamma_{BP}^{o,w} = - (\partial\gamma/\partial\mu_{BP})_{T,P,\mu_{MP},\mu_{MC}} \tag{12}$$

$$\Gamma_{MP}^{o,w} = - (\partial\gamma/\partial\mu_{MP})_{T,P,\mu_{BP},\mu_{MC}} \tag{13}$$

$$\Gamma_{MC}^{o,w} = - (\partial\gamma/\partial\mu_{MC})_{T,P,\mu_{BP},\mu_{MP}} \tag{14}$$

The excess entropy and volume can be obtained from measurements of γ as a function of temperature and pressure.

In an electrochemical system, it is often convenient to measure chemical potentials by measuring the emf of appropriate cells. Thus if an electrode reversible to anion P^- is put into the oil phase and an electrode reversible to cation M^+ in the water phase, the cell may be written

$$(1) \mid P \mid BP, MP \mid MC \mid M \mid (2) \tag{15}$$

where (1) and (2) are the terminals of the same metal. The variation of the cell emf is then

$$dE_{MP} = d(\phi^2 - \phi^1) \tag{16}$$

where ϕ^α is the inner potential of phase α. The electrochemical potential of electrons in phase α, $\tilde{\mu}_e^\alpha$, may be written

$$\tilde{\mu}_e^a = \mu_e^\alpha - F\phi^\alpha \tag{17}$$

so that

$$d(\phi^2 - \phi^1) = d(\tilde{\mu}_e^1 - \tilde{\mu}_e^2)/F \tag{18}$$

because the chemical potential μ_e^α is the same in the two terminals of the same metal.

Consideration of equilibrium in the two electrodes leads to the relation

$$d\tilde{\mu}_e^1 = d\tilde{\mu}_p^o \tag{19}$$

and

$$d\tilde{\mu}_2^2 = -d\tilde{\mu}_M^w = -d\tilde{\mu}_M^o \tag{20}$$

since M^+ is in equilibrium across the oil/water interphase. Thus

$$FdE_{MP} = d\tilde{\mu}_P^o + d\tilde{\mu}_M^o = d\tilde{\mu}_{MP} \tag{21}$$

Consequently equation (13) may be written

$$F \; \Gamma_{MP}^{ow} = - \; (\partial\gamma/\partial E_{MP})_{T,P,\mu_{BP},\mu_{MC}} \tag{22}$$

The left-hand side of (22) is formally a charge and may be described as the charge

$$Q_{MP} = F\Gamma_{MP}^{o,w} = F\Gamma_{M+}^{o,w} \tag{23}$$

due to the cation M^+ because the amount of M^+ in the interphase is identical to the amount of MP; however, nothing can be said about the location of this charge (i.e. of M^+) within the interphase under these conditions. Equation (22) is known as a Lippmann equation and the relation between γ and E at constant T, P, μ_{BP}, μ_{MC} is an electrocapillary curve.

An alternative experiment could use an electrode reversible to M^+ in the oil phase and an electrode reversible to C^- in the water phase. Then, with similar arguments to those used above, it can be shown that

$$FdE_{MC} = -d(\tilde{\mu}_{M+}^w + \tilde{\mu}_{c-}^w) = -d\mu_{MC}^w \tag{24}$$

and equation (14) becomes

$$F\Gamma_{C^-}^{o,w} = -Q_{MC} = F\Gamma_{MC}^{o,w} = (\partial\gamma/\partial E_{MC})_{T,P,\mu_{BP},\mu_{MP}} \tag{25}$$

This is a second Lippmann equation related to a second electrocapillary curve and the charge Q_{MC} is related to the charge due to the anion C^- which might reasonably be expected to be located on the aqueous side of the interphase.

A third possibility is that of using an electrode reversible to the cation B^+ in the oil phase and an electrode reversible to the cation M^+ in the water phase. Then

$$FdE_{MB} = d\bar{\mu}_{M^+}^o - d\bar{\mu}_{B^+}^o \tag{26}$$

If the concentration of MP is much less than that of BP in the oil phase and it is kept constant, then $d\bar{\mu}_M^o$ may be neglected and provided that the concentration of BP is not too high

$$FdE_{MB} = -d\bar{\mu}_{B^+}^o \simeq -d\mu_{BP}^o/2 \tag{27}$$

Then to a good approximation

$$Q_{BP} = F\Gamma_{DA}^{o,w} = F\Gamma_{DA}^{o,w} \simeq 2(\partial\gamma/\partial E_{MB})_{T,P,\mu_{MP},\mu_{MC}} \tag{28}$$

This third Lippmann equation gives the charge in the interphase due to the cation B^+ which may be supposed to be located largely on the oil side of the interphase.

It is evident that three electrocapillary curves may be obtained for this interphase and correspondingly three charges, two of which can be located on either side of the interphase, but because the third has a distribution which is not accessible, the total charge distribution cannot be obtained. There are three potentials of zero charge also, depending on the conditions of measurement. It should be noted that the use of electrochemical measurements is not necessary to obtain this information although it may be convenient.

The analysis of a metal/electrolyte interphase like that of thallium amalgam in contact with an aqueous solution of $TlNO_3$ and KNO_3 is completely analogous to that of the system described here.

3. The liquid/liquid interphase with no charge transfer.

This condition can be derived from the previously discussed interphase by allowing the concentration of tetramethylammonium tetraphenylborate (MP) in the oil phase to become vanishingly small. Then the chemical potential μ_{MP} cannot be controlled by determining the composition of the system, but it may be controllable by the cell potential E_{MP} if the cell is furnished with an electrode reversible to P^- in the oil phase and one reversible to M^+ in the water phase.

The charge Q_{MP} can then be found from equations (22) and (23).

However, it should be noted that the use of (22) requires μ_{BP} and μ_{MC} to be kept constant. Since BP is now the only solute in the oil phase, this means that the composition of both phases is to be kept constant. Under these conditions the nature of the electrodes used to make contact with the two phases is no longer important; all types of electrode will yield the same change of cell emf. Since no other chemical potential in this cell can be expressed as a cell emf, (22) is the unique Lippmann equation for this system. Further the charge Q_{MP} obtained from the slope of the unique electrocapillary curve is also unique and gives the charge on the aqueous side of the interphase. The interphase may thus be regarded as a molecular condenser with charge $-Q_{MP}$ on the oil side and Q_{MP} on the water side.

The surface excesses of the other components can be obtained from equations (12) and (14) in the form

$$\Gamma_{BP}^{o,w} = -(\partial\gamma/\partial\mu_{BP})_{T,P,E_{MP},\mu_{MC}} \tag{29}$$

and

$$\Gamma_{MC}^{o,w} = -(\partial\gamma/\partial\mu_{MC})_{T,P,E_{MP},\mu_{BP}} \tag{30}$$

This information can be expressed in an alternative way in terms of the ionic surface excesses

$$\Gamma_{B+}^{o,w} = \Gamma_{BP}^{o,w} \tag{31}$$

$$\Gamma_{P-}^{o,w} = Q_{MP}/F + \Gamma_{BP}^{o,w} \tag{32}$$

$$\Gamma_{M+}^{o,w} = Q_{MP}/F + \Gamma_{MC}^{o,w} \tag{33}$$

$$\Gamma_{C-}^{o,w} = \Gamma_{MC}^{o,w} \tag{34}$$

This system has a precise analogue in the classical ideal polarized system of mercury/electrolyte.

It should be noted that when there is no charge transfer across the interphase the electrocapillary equation (i.e. the Gibbs adsorption equation) at constant T and p must take the form

$$-d\gamma = Q_{MP}dE_{MP} + \Gamma_{BP}^{o,w} d\mu_{BP} + \Gamma_{MC}^{o,w} d\mu_{MC} \tag{35}$$

whereas when there is an equilibrium of a charged species across the interphase several alternative versions are possible. Experimentally the existence of this equilibrium can often be controlled by making measurements on different time scales, e.g. at short times or high frequencies the charge transfer can be 'frozen'. The appropriate time scale depends on the system.

4. Solid/Fluid Interphases - Basic Problems

The analysis given so far would be applicable to the solid/fluid interphase if a quantity equivalent to the interfacial tension could be measured for this interphase. Surface tensions of solids are difficult to measure and have been measured only under special circumstances, for example, at temperatures just below the melting point. These measurements are not particularly accurate and at present are not useful for the application of the Gibbs equation.

The problem of the equivalent for a solid phase of the liquid interfacial tension arises because of the very low mobility of atoms or molecules in a solid phase. It is therefore difficult to ensure that the formation of a new phase is an equilibrium process. Even if a state of partial equilibrium may be acceptable, there may be different ways of achieving this and different "equilibrium" states may be achieved using these different routes in the same system. The surface tension of a liquid exists because the stress, which is isotropic in the bulk of the liquid, becomes anisotropic at the surface. The same is probably true in a solid, but the stress in the bulk of a solid is not necessarily isotropic. Also the quantity analogous to the surface tension of a liquid may itself not be isotropic.

To overcome this type of problem attempts have been made [5] to use quantities which must be isotropic such as the surface free energy. However, this depends on the choice of reference component, i.e. the components eliminated between equations (2), (3) and (4) and so its value is somewhat arbitrary. A closer approach to a measurable quantity is the free energy of formation of the interface, which might be measured in terms of the external work done by the system in the formation of the interface by an isothermal reversible process in which no "volume work" is done. Such conditions are difficult to achieve in practice. In effect, this method is equivalent, in the case of liquids, to defining the surface tension by the equation

$$\gamma = (\partial A_e / \partial A_s)_{T, V^\sigma, E, \mu} \tag{36}$$

where A_e is the electrochemical Helmholtz energy of the interphase and A_s is the area of the interphase. The state of the liquid-liquid interphase is completely defined by the variables appearing in this equation, but this is no longer true when the system contains a solid whose state of internal strain varies from point to point. The use of (36) for such a system requires a knowledge of which variables must be maintained constant during the differentiation, and in general such knowledge is not available.

Defay et al. [5] point out that these problems can be avoided by defining the interfacial tension using a relation which contains only integral quantities. In the metal/electrolyte system this definition takes the form

$$\gamma = (A_e / A_s) - \sum_{j=2}^{j=J} \Gamma_{j,e}^{(1)} \mu_{j,e} - \sum_{j=J^\alpha+2}^{j=J_0-1} \sum_{k=0}^{k=K-1} \Gamma_{j,k}^{(0)} \mu_{j,k} \tag{37}$$

in which A_e is defined with respect to the same reference components as the surface excesses $\Gamma_{j,e}$ and $\Gamma_{j,k}$. The advantage of this formula, is that the chemical potentials in the solid phase may be replaced by their values in a fluid phase in equilibrium with the solid, is of practical use only in rather simple systems like that of a crystal in equilibrium with its saturated solution. It would be difficult to apply to many electrochemical systems. Nevertheless, the definition is perhaps useful in a formal sense.

The quantity defined in (37) is perhaps better not called by the terms "surface" or "interfacial tension", and several other terms have been proposed. The situation has been summarized by Linford [6] who suggests the term "specific surface work" and the symbol γ_π for this quantity, which may be thought of as the work of forming unit area of new surface by cleavage under ideal conditions. He suggests that the work required to form unit area of new surface by stretching under equilibrium conditions should be called the "surface stress" g_{ij}. This is, in general, an anistropic quantity and is therefore represented by a tensor. It is related to γ_π by

$$g_{ij} = \delta_{ij}\gamma_\pi + (\partial\gamma_\pi/\partial\epsilon_{ij})_{\text{all other strains}} \tag{38}$$

where $\delta_{ij} = 0$ if $i \neq j$, $\delta_{ij} = 1$ if $i = j$, and ϵ_{ij} is the natural strain (i.e. increase is length per unit length. g_{ij} may be defined as numerically equal to the force acting in the jth direction per unit length of exposed edge, the edge being normal to the ith direction, that must be applied to a terminating surface to keep it in equilibrium, the ith and jth directions lying in the plane of the surface. For an isotropic surface the shear stresses, i.e. the values of g_{ij} for which $i \neq j$, become zero and the g_{ij} may be replaced by the mean surface stress g, where

$$g = \tfrac{1}{2}(g_{ii} + g_{jj}) \tag{39}$$

and the strain ϵ_{ij} may be replaced by the mean strain dA_s/A_s, so that (38) becomes

$$g = \gamma_\pi + A_s d\gamma_\pi/dA_s \tag{40}$$

For liquids the second term on the right-hand side of (40) is zero and $g = \gamma_\pi = \gamma$ the interfacial tension.

Couchman and Davidson [7] have recently suggested that $d\gamma$ in the Gibbs equation for a liquid should, for a solid, be replaced by $(\gamma_\pi - g_{ij})d\epsilon_e + d\gamma_\pi$, where ϵ_e is the elastic surface strain. They suggest that, even with unrealistically large values of $\gamma_\pi - g_{ij}$ and the electrostriction coefficient (which gives the effect of field on ϵ_e) the first of these terms contributes negligibly in comparison with $d\gamma_\pi$. Since the field arises from a change in composition of the interphase, it seems reasonable to extend this argument to the effect of composition changes on ϵ_e. It this is valid then it is probable that the specific surface work γ_π may be taken as the appropriate variable, replacing the surface or interfacial tension in solid systems. This is consistent with (37) if that equation is taken as defining γ_π.

Mohilner and Beck [8] following Linford [9] point out that surface area changes will be neither wholly plastic (as in cleavage) or wholly elastic (as in equilibrium stretching). To deal with this situation, a generalized surface parameter may be defined which is conjugate to the general (part plastic, part elastic) surface area change. For an *isotropic* solid this will be the sum of two contributions taken in proportion to the fractions of the two types of strain, i.e.

$$\gamma^s = (d\epsilon_p/d\epsilon_{tot})\ \gamma_\pi + (d\epsilon_e/d\epsilon_{tot})\ g \qquad (41)$$

where ϵ_p and ϵ_e are the plastic and elastic contributions to the total strain ϵ_{tot}. In this case the interfacial tension term in equation (2) must be replaced by $d\gamma_\pi + (\gamma_\pi + g)\ d\epsilon_e$. This means that to recover the equations deriving from (2) in the above, the elastic strain should be kept constant. Although this is not usually possible, the effect of this additional term is probably small [8].

It may be concluded that provided an appropriate property exists for a solid/fluid interface to replace the interfacial tension in a fluid/fluid interface then the Gibbs equation may be used. This can be done in practice by using the cross-differential relationships which are then valid. Provided that no charge transfer occurs the charge Q_{MP} can be measured or quantities related to it such as the interfacial capacity. This will be illustrated in the next section. However, it is important to point out that this route is not applicable when there is equilibrium of a charged species across the interphase. The only way of applying the thermodynamic route to such interphases when an interfacial tension or equivalent is not measurable is to use the technique of 'freezing out' the charge transfer mentioned in the previous section. However, it may be remarked that if a method exists to determine the surface excess of one component then the variation of this with composition of the bulk phase could be used with the aid of cross-differential relationships to determine the surface excesses of other components. In principle, radiotracer experiments or optical experiments could be used for this, but a high precision would be required and it is doubtful whether present accuracy in the experiments would be sufficient.

5. Solid/Fluid Interphases - measurement and interpretation.

In the case of ideally polarized or "blocked" interphases where no charged transfer across the interphase is possible, direct measurement of charge as, for example, in chronocoulometric experiment is possible. The related, differential capacitance, C, is also a precise route as measured directly with a bridge or a phase-sensitive detector as in a frequency response analyser. Linear sweep voltammetry can provide similar information since the current density is given by

$$j = d\sigma/dt = (d\sigma/dE)(dE/dt) \qquad (42)$$

and dE/dt is the sweep rate v which is constant. Hence

$$j = Cv \qquad (43)$$

Here σ is the physical charge on either side of the interphase which is equal to the Q_{MP} of section (3). The solid/fluid interphase then has a precise analogue of equation (35) and the usual cross-differential relations can be obtained from this complete differential equation, e.g.

$$(\partial Q_{MP}/\partial \mu_{BP})_{T,P,\mu_{MC},E_{MP}} = (\partial \Gamma_{BP}^{o,w}/\partial E_{MP})_{T,P,\mu_{MC},\mu_{BP}} \qquad (44)$$

Thus the surface excess may be obtained from measurements of the charge as a function of composition, using the integral of (44):

$$\Gamma_{BP}^{o,w} = \int_{E_{MP(ref)}}^{E_{MP}} (\partial Q_{MP}/\partial \mu_{BP})_{T,P,\mu_{MC},E_{MP}} \, dE_{MP} + \Gamma_{BP}^{o,w}(E_{ref}) \qquad (45)$$

provided that a value of the surface excess is known at some potential: $\Gamma_{BP}^{o,w}(E_{ref})$. This method has been used for much of the information about interfacial composition derived for the solid metal/electrolyte interphase.

A particular type of metal/electrolyte interphase deserves special comment although it can be treated in the way described above. This is the case when adsorption results in charge transfer. The interface remains blocked since the charge remains in the interphase and does not cross it. Nevertheless, there is no unambiguous way of locating it.

An example of this type of interphase is that in which a sub-monolayer of a metal or hydrogen is deposited, e.g. in the cell

$$Tl \mid TlCl + KCl \mid Tl \mid Ag \qquad (46)$$

At constant T and p, the Gibbs equation may be written

$$-d\gamma = \Gamma_{TlCl}d\mu_{TlCl} + \Gamma_{KCl}d\mu_{KCl} + \Gamma_{Tl}d\mu_{Tl} \qquad (47)$$

Where the superscripts indicating surface excess quantities have been omitted; these are in fact excesses with respect to the solvent water, since the metal electrode is pure Ag, i.e. the solubility of Tl in Ag may be neglected.

The emf of cell (46) may be expressed

$$dE_{Tl} = d(\tilde{\mu}_e^{Tl} - \tilde{\mu}_e^{Ag})/F \qquad (48)$$

The equilibrium at the left-hand electrode may be expressed

$$d\mu_{Tl} = d\tilde{\mu}_{Tl+}^{S} + d\tilde{\mu}_e^{Tl} \qquad (49)$$

and that at the right-hand electrode by

$$d\mu_{Tl}^{M} = d\tilde{\mu}_{Tl+}^{S} + d\tilde{\mu}_e^{Ag} \qquad (50)$$

where the superscript S denotes the solution and M the monolayer. With these two equations (48) becomes

$$FdE_{Tl} = -d\mu_{Tl}^M \tag{51}$$

which shows that the emf of this cell measures the chemical potential of Tl in the monolayer. Since Tl is present only in the monolayer, this is the only way in which its chemical potential can be controlled. From (47) and (51) it follows directly that

$$F \ \Gamma_{Tl} = (\partial\gamma/\partial E_{Tl})_{T,P,\mu_{TlCl},\mu_{KCl}} \tag{52}$$

or expressed as a charge

$$Q_{Tl} = (\partial\gamma/\partial E)_{T,P,\mu_{TlCl},\mu_{KCl}} \tag{53}$$

where the subscript to E is dropped because the nature of the reference electrode (left-hand electrode) is unimportant when the composition of the solution is kept constant.

The charge in (53) is the charge measured experimentally by chronocoulometry for example. The corresponding electrode capacity is

$$C_{Tl} = (\partial Q_{Tl}/\partial E)_{T,P,\mu_{TlCl},\mu_{KCl}} \tag{54}$$

However, Q_{Tl} is not the physical charge on the interphase because the location of this charge is not known.

The physical charge on the interphase may be estimated using the argument developed by Frumkin [10]:

If the last term on the right-hand side of (47) is replaced by its equivalent $Q_{Tl}dE_{Tl}$ and μ_{KCl} is kept constant as well as T,P, the cross differential relation

$$(\partial E_{Tl}/\partial\mu_{TlCl})_{Q_{Tl}} = -(\partial\Gamma_{TlCl}/\partial Q_{Tl})_{\mu_{TlCl}} \tag{55}$$

may be written. This is transformed directly into

$$(\partial E_{Tl}/\partial\mu_{TlCl})_{Q_{Tl}} = -(\partial\Gamma_{TlCl}/\partial E_{Tl})_{\mu_{TlCl}}/C_{Tl} \tag{56}$$

with the aid of (54).

Frumkin wrote the physical charge σ on this type of interphase as

$$\sigma = F(\Gamma_{Tl+} - A_{Tl+}) \tag{57}$$

where A_{Tl+} is the surface excess of Tl^+ actually present as Tl^+ ions in the interface. If the concentration of KCl is much greater than that of TlCl in the solution, then the diffuse part of the electrode double layer will contain a negligible amount of Tl^+. Specifically adsorbed Tl^+ make up the monolayer of Tl and are no longer ions. Hence A_{Tl+} will be zero under these conditions, since $\Gamma_{Tl+} = \Gamma_{TlCl}$, it follows from (57) that

$$\sigma = F\Gamma_{TlCl} \tag{58}$$

and the electrostatic capacity

$$C = (\partial\sigma/\partial E)_{T,P,\mu_{TlCl},\mu_{KCl}} = F(\partial\Gamma_{TlCl}/\partial E)_{T,P,\mu_{TlCl},\mu_{KCl}} \tag{59}$$

and (56) may be written

$$(\partial E_{Tl}/\partial\mu_{TlCl})_{Q_{Tl},\mu_{KCl}} = C/FC_{Tl} \tag{60}$$

or

$$C = (FC_{Tl}/RT)\,(\partial E_{Tl}/\partial \ln a_{TlCl})_{Q_{Tl},\mu_{KCl}} \tag{61}$$

This provides a route for the measurement of the electrostatic capacity of this interphase, provided the non-thermodynamic assumptions are valid. This route was exploited extensively by Frumkin's group for the adsorption of hydrogen on polycrystalline platinum metals [10] but it has rarely, if ever, been used for sub-monolayer metal deposition. It requires measurement of the charge Q_{Tl}, not just the change of Q_{Tl} with E. Such experiments have been done for Pb on Au [11] and show that E_{Pb} is independent of the activity of the lead salt under these conditions. Hence, in this system $C \ll C_{Pb}$. This means that Q_{Pb} measures the amount of Pb adsorbed to a good approximation. This assumption is frequently made.

The existence in this system of two types of charge: Q_{Tl} the total charge and σ the electrostatic charge (related to C_{Tl} and C respectively) means of course that there are two potentials of zero charge.

6. Adsorption Isotherms

The relation between amount of a substance present at the interphase and its chemical potential at equilibrium is often represented by a relationship which at constant temperature is called an adsorption isotherm

$$\Gamma = f(\mu) \tag{62}$$

When the equilibrium is between a vapour and a surface, there is a simplification in that the surface concentration and surface excess may be taken as the same. This is not so in condensed phases except as an approximation and there is a further problem in that the adsorption process is always a replacement of one species by another. This may be represented by a sort of chemical equilibrium [12]

$$A_{soln} + rS_{ads} \rightleftharpoons A_{ads} + rS_{soln} \tag{63}$$

where A is the adsorbing species and S is taken as the solvent. The subscript soln means in solution (bulk phase) and ads means in the interphase. The factor r allows for the fact that A and S occupy a different volume, i.e., A replaces r molecules of S.

The corresponding equilibrium constant may be written

$$K = a_A^{ads} \left(a_S^{soln}\right)^r / a_A^{soln} \left(a_S^{ads}\right)^r \tag{64}$$

where the a's represent activities of the species in the subscript in the location of the superscript. In electrified interphases there is a further complication in that K is dependent on the electrical state of the interphase and the interphase must remain electrically neutral.

The further development of (64) is really a branch of statistical, rather than classical, thermodynamics because it necessarily involves a model, even if crude, of the interphase. This may involve the assumption that A and S are incompressible, so that

$$r[A] + r[S] = N \tag{65}$$

where, in effect N sites in the interphase are defined in terms of the volume occupied by a solvent molecule, and the square brackets indicate surface concentration. If a fractional occupation θ is defined as

$$\theta = r[A]/N \tag{66}$$

and activities in the surface are taken as equal to concentrations (64) may be written

$$\theta/(1-\theta)^r = [rK/(a_S^{soln})^r] \, a_A^{soln} = \beta \, a_A^{soln} \tag{67}$$

since a_S^{soln} may often be taken as constant. β is called the adsorption coefficient. Equation (67) is similar to the Flory-Huggins expression for the activity of polymer solutions and so is often called the Flory-Huggins isotherm, though this form was derived by others earlier.

The calculation of θ requires a knowledge of the surface concentration. When the solution is dilute and the adsorption of A strong the difference between surface excess and surface concentration is small, but otherwise a further model assumption is necessary, e.g. that the adsorbed layer has a certain thickness; this may be a monolayer. Such an assumption was introduced for air/solution interphases [13] and used often for other types.

Adsorption isotherms of the type of (67) have been tested for electrified interfaces by assuming that K is essentially a function of the potential difference across the interphase or of its electrostatic charge. No universal behaviour has been established and the criteria for a choice between these two limits remain uncertain. Frequently (67) is unable to account for the observed behaviour using either assumption and the equation is modified in a way which amounts to allowing for the non-ideality of the interphase, i.e.

$$a_A^{ads} = [A].\gamma_A \tag{68}$$

$$a_S^{ads} = [S]\ \gamma_S \tag{69}$$

For example if it is assumed that

$$\gamma_A/\gamma_S = \exp a\theta \tag{70}$$

and $r = 1$ Frumkin's isotherm is obtained. This has been used widely and fits a great number of data reasonably well. The parameter α expresses the interaction between adsorbates in some way, but the isotherm is essentially empirical. Attempts to treat this problem on the basis of better analysis of a well-defined model usually lead to an isotherm with more parameters. With data of limited accuracy it is difficult if not impossible to obtain a meaningful fit.

Data for the adsorption of 2-butanol of very high accuracy were obtained by Mohilner et al. [14]. They analysed their results by assuming an interfacial thickness and then treating the interphase as a binary solution expressing log γ_A and log γ_S as power series in the interfacial composition. This provides a flexible fitting function but there has, as yet, been no physical interpretation of the coefficients of this power series.

It has to be concluded that the main use of isotherms is in fitting experimental data and that, as yet, little information about the true nature of interactions in the interphase can be obtained by this route. It is here that the more molecular approach of recent years can provide direct information.

Conclusion

The thermodynamic study of electrified interphases can provide useful information about the charge in the interphase. This is not available from the more recent methods for the study of interphases. However, the interpretation of the quantity obtained needs some care and this paper is an attempt to indicate the way in which such problems may be tackled.

References

1.   J.W. Gibbs, Trans Connecticut Academy, 3 (1878) 108, 343.

2.   A.N. Frumkin, Electrocapillary Phenomena and Electrode Potentials, Odessa 1919.

3.   A.N. Frumkin, Phil. Mag. 40 (1920) 363.

4.   R. Parsons in Comprehensive Treatise of Electrochemistry, ed. J.O'M. Bockris, B.E. Conway and E. Yeager, Vol 1, Plenum Press, New York, p 1.

5.   R. Defay, I Prigogine, A. Bellemans, and D.H. Everett, Surface Tension and Adsorption, Longmans, London (1966), Chap. 17.

6.   R.G. Linford, Chem. Soc. Rev. $\underline{1}$ (1972) 445, J. Electroanal. Chem., $\underline{43}$ (1973) 155.

7.   P.R. Couchman and C.R. Davidson, J. Electroanal. Chem. $\underline{85}$ (1977) 407.

8.   D.M. Mohilner and T.R. Beck, J. Phys. Chem., $\underline{83}$ (1979) 1160.

9.   R.G. Linford, Chem. Rev. $\underline{78}$ (1978) 81.

10.   A.N. Frumkin, J. Electroanal. Chem., $\underline{12}$ (1966) 504.

11.   A. Hamelin and J. Lipkowski, J. Electroanal. Chem., $\underline{171}$ (1984) 317.

12.   D.H. Everett, Trans. Faraday Soc., $\underline{60}$ (1964) 1803; $\underline{61}$ (1965) 2478.

13.   E.A. Guggenheim and N.K. Adam, Proc. Roy. Soc. A139 (1933) 218.

14.   D.M. Mohilner, H. Nakodomari and P.R. Mohilner, J. Physical Chem., $\underline{80}$ (1976) 1761; $\underline{81}$ (1977) 244.

ELECTRODE POTENTIALS AND ENERGY SCALES

S. TRASATTI
Department of Physical Chemistry
and Electrochemistry
University of Milan
20133 Milan, Italy

ABSTRACT. Only relative potentials can be thermodynamically measured in electrochemistry. Their numerical values do not provide any insight into the "absolute" energetics of the species involved, and into the structural details of the interface. The physical meaning of the measured electrode potential is disclosed by scrutinizing the physical implications of the various steps of the experimental approach for the measurement. It is shown that various forms of *single electrode potential* can be explicited. It is also demonstrated that only one of these single electrode potentials is the analogue of the electron work function for an isolated phase: hence, it enables to realize the alignment of electron energy levels with ion energy levels (redox couples). In the case of the SHE, such a potential has been denoted with $E^0(H^+/H_2)(abs)$. Different experimental values are available for this potential ranging from *ca* 4.4 to *ca* 4.8 V. The details of various experimental approaches are scrutinized and the major experimental shortcomings pointed out. It is shown that the value of 4.8 V leads to some conceptual difficulties as it is "deconvoluted" into the several interfacial components.

1. INTRODUCTION

Interfaces are tridimensional regions at the boundary between two dissimilar phases, whose properties differ from those of the adjoining phases [1]. Electrified interfaces are characterized by the presence of an electric potential drop across the boundary, affecting both the transfer of charged particles and adsorption phenomena [2]. A system consisting of an electronic conductor (metal or semiconductor) in contact with an ionic conductor (electrolyte) is what is known as an *electrode*.

The electrical state of an electrode is usually defined in term of its *potential* [3]. In an isolated electrode system, there exists only one interface, that between the electronic conductor (henceforth denoted with M) and the electrolyte (henceforth denoted with S). Consideration of an electrode out of the context of the experimental measurement of

R. Guidelli (ed.), Electrified Interfaces in Physics, Chemistry and Biology, 229–244.
© 1992 *Kluwer Academic Publishers.*

its potential, leads intuitively to identify the *electrode potential* with the electric potential drop across the interface, $\phi^M - \phi^S \equiv \Delta\phi(M/S)$, also called *Galvani* (or *inner* electric) potential difference [2]. It is the purpose of this paper to show that *this is in fact not the case*. The treatment will illustrate (i) the physical meaning of the *measured* electrode potential, (ii) its relation with the energetic parameters of the redox system in solution eventually governing it, (iii) the relation with the absolute energy of species in a vacuum, and (iv) the experimental approaches to measure the various components.

In consideration of the varied backgrounds of the participants in this conference, the treatment will be kept at an introductory level, although the conclusions will reflect the state-of-the-art in the field. Reference to the author's work will be mainly made; quotation of other papers can be found therein.

2. ELECTRODE POTENTIALS

The conceptual discussion given below is applied to the specific case of metal/electrolyte interfaces, but it is applicable to any kind of interfaces including electrolyte/gas and electrolyte/electrolyte interfaces. The determination of $\Delta\phi$ across interfaces entails common conceptual aspects which need not be discussed for each specific case.

2.1. The *measured* electrode potential

$\Delta\phi$ values can be measured experimentally within a homogeneous phase, *eg* a solid resistor or an electrolyte solution. $\Delta\phi$ values across an interface are not amenable to direct experimental determination [4]. This is because the two metallic probes from the measuring instrument (both made of the same electronically conducting material, R) have to put in contact with the bulk of the two phases to pick up the electric signal there. This will unavoidably create two new interfaces: a R/M interface and a S/R interface. The configuration of the system under the conditions of measurement becomes the following (Fig. 1):

$$R'/M/S/R ,\qquad\qquad (1)$$

where R' differs from R for the *electrical* state only. It thus happens that an attempt to measure a single $\Delta\phi$, *ie* $\phi^M-\phi^S$, results in the measurement of the sum of three $\Delta\phi$ in series. Various systems can be envisaged to try to avoid these shortcomings but the conclusion is that the situation described above is a quite general one ensuing from the very way electric potentials are measured.

From the experimental arrangement given in (1) it follows that the quantity we obtain as the electrode potential of M is not $\Delta\phi$. In fact, the total electric potential difference is:

$$\Delta\phi(R'/R) = \Delta\phi(R'/M) + \Delta\phi(M/S) + \Delta\phi(S/R) .\qquad\qquad (2)$$

While the M/S contact represents the electrode interface, the newly

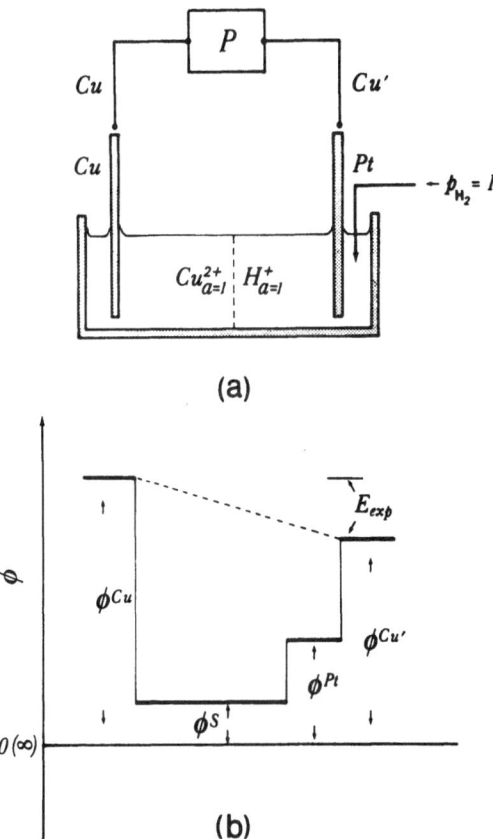

(a)

(b)

Figure 1. (a) An example of experimental arrangement to measure electrode potentials. P is the measuring instrument. (b) Graphical illustration of eqn. (2) in the text. The measured potential difference is the sum of three Δφ arising at the three interfaces.

created R/M interface does not pose any problems. It is a metal/metal contact quite easily realizable without any special artifice. This is not the case of the R/S contact which involves the formation of an additional electrochemical interface. The simple contact realized between a metal wire and an electrolyte solution would be irreproducible, unstable and would depend on the nature of R. It is thus expedient to replace the simple R/S contact with a specific electrochemical system which is known as the *reference electrode*.

A system including a reference electrode is customarily written as:

$$M/S/Ref ,\tag{3}$$

and the measured potential difference (ΔE) is customarily taken to measure the electrode potential of M *relative* to Ref:

$$\Delta E = E^M - E^R , \tag{4}$$

where superscript R replaces Ref for simplicity, and reminds us that the configuration of system (3) is in fact that given in (2). Thus, comparison of eqn. (4) with eqn. (2) readily shows not only that the so-called *"electrode potential"* is not $\Delta\phi$, but also that ΔE is not a *relative* measure of $\Delta\phi$:

$$\Delta E = E^M - E^R = \Delta\phi(R'/M) + \Delta\phi(M/S) + \Delta\phi(S/R) , \tag{5}$$

whence:

$$E^M \neq \Delta\phi(M/S) . \tag{6}$$

2.2. The *absolute* electrode potential

The term *"absolute potential"*, often used in the literature, issues precisely from the argument that, if only *relative* electrode potentials can be measured, it would be interesting to know the value of the electrode potential of a *single* electrode, *viz* not measured with respect to another electrode. This aspect has been exhaustively discussed in the literature; the conclusion has been that for a single electrode several options satisfying the concept of *"absoluteness"* are available [5,6]. However, only one of these is capable of bringing electrochemical quantities into the common frame of physical quantities as measured (in a non-relative way) in a vacuum [7].

The apparent inconsistency implicit in eqn. (6), *ie* two terms in E on the l.h.s. *vs* three terms in $\Delta\phi$ on the r.h.s., can be easily removed if it is considered that electronic equilibrium exists at the metal/metal contact. Under similar circumstances, the electronic energy (expressed by the electrochemical potential of electrons) must be the same in both phases [2]:

$$\tilde{\mu}_e^M = \tilde{\mu}_e^R \tag{7}$$

from which:

$$(\phi^R - \phi^M) = (\mu_e^R - \mu_e^M)/F . \tag{8}$$

Substitution of eqn. (8) into eqn. (5) gives:

$$E^M - E^R = [\Delta\phi(M/S) - \mu_e^M/F] - [\Delta\phi(R/S) - \mu_e^R/F] . \tag{9}$$

Each of the two terms in brackets on the r.h.s. contains parameters relative to a single interface only. They thus define single electrode potentials or, more commonly, *absolute* electrode potentials.

Equation (9) shows, beyond any doubt, that the measured electrode potential difference (ΔE) is not a difference of electric potential drops, *viz* $\Delta(\Delta\phi)$, but a difference, measured in Volt, of electronic energy between the metals constituting the two electrodes. This outcome cannot be regarded as unexpected. Electrons move under the action of gradients of electronic energy. There can well exist a gradient of electrical energy between two dissimilar phases at constant electronic energy. Under similar circumstances no potential difference is measurable and no electron flux (electric current) is observed. In other words, charge transfer is not governed by $\Delta\phi$ alone.

A practical example. A reversible hydrogen electrode at a given pH has a fixed E value as measured with respect to a given reference electrode. However, eqn. (9) shows that if two different metals are used to make the hydrogen electrode, since μ_e normally differs from metal to metal, different $\Delta\phi(M/S)$ exist at the two interfaces despite the same value of the measured electrode potential. In other words, the same value of E for two different electrodes entails in principle different values of $\Delta\phi$ at the two interfaces. Therefore, any conceptual description of the physical picture of electrochemical interfaces cannot be based on intuitive arguments.

2.3. Reference levels for "absolute" electrode potentials

Since ΔE is a relative measure of electronic energy, the expression defining its components already entails the identification of the reference level with respect to which electronic energies are measured. In other words, the choice of the zero energy level for electrons is not free, but it issues from the way the single electrode potentials of the two electrodes are separated [5]. Thus, eqn. (9) implies that the energy of electrons at the Fermi level of the metal M is measured by the work to extract electrons from the bulk of M ($-\mu_e M$), plus the work to cross the electrode interface, *viz* $F\Delta\phi(M/S)$. It follows that the total work implies (physically) taking an electron from the bulk of M and placing it in the bulk of S, *where its energy is taken as zero*. It is not taken to the vacuum above the solution, otherwise the surface of the solution should be crossed, whereas in fact no parameters of that region appear in eqn. (9).

From the above discussion it is clear that eqn. (9), while defining single electrode potentials, does not provide any physical values which can be compared with electronic energies as measured in a vacuum. It is useful to recall here that the electronic energy of a metal, as measured experimentally in a vacuum, is given by its work function, Φ [3]:

$$\Phi M = -\mu_e M + F\chi M ,\qquad\qquad(10)$$

where χM is the surface potential, *ie* the electric potential drop associated with the surface dipolar layer. Φ is commonly defined as the work to take an electron from the interior of M to the exterior, the measurement usually corresponding to the condition of zero surface charge (uncharged metal). If a charge is located at the surface (and χM is assumed to be charge density independent [2]), the work function is still definable as the work to take an electron from the metal to a point outside M *close* to its surface. The *closeness* is not a free choice but a consequence of the measurement. The point is located at a distance where image forces are no longer operative while the coulombic forces

234

related to the surface charge are still integral [2]. As the Volta
potential difference (cpd) is measured between two surfaces, the
measured $\Delta\psi$ is precisely that between two points *close* to the two
surfaces. Hence, the definition of the point where short range and long
range electrical forces can be split is not a conceptual option but an
experimental consequence.

2.4. Electrode potential and electronic energy

As shown above, eqn. (9) issues from eqn. (2) where the measured ΔE is
equated to the total $\Delta\phi$. However, since ΔE is in fact a relative measure
of electronic energy, it is clear that $\Delta\phi$ is not the most appropriate
quantity to start with, since it includes only the electrical part of
the total electronic energy. It follows that eqn. (9) does not contain
all relevant terms; some of them might be the same for the two
electrodes and cancel out as eqn.(2) is *"deconvoluted"*. It is thus
necessary to "deconvolute" an equation where all terms have been
appropriately accounted for.

It is clear from eqn. (9) that ΔE measures a difference in
electronic energy. However, it is also evident that the two terms in
square brackets do not contain all the actual contributions to the
electronic energy. The total electronic energy in a phase is measured by
the electrochemical potential, $\bar{\mu}_e$ [2]. Therefore, it must be:

$$\Delta E = E^M - E^R = (\bar{\mu}_e^R - \bar{\mu}_e^M)/F \; ; \qquad (11)$$

$\bar{\mu}_e^M$ does not depend on the nature of the other interface. In principle,
this quantity possesses the requisites of a single electrode potential.
However, $\bar{\mu}_e^M$ does not contain the details of the M/S interface and it is
thus very inconvenient as an "absolute" potential. This is because eqn.
(11) entails a work to transfer an electron from M to R through an
external circuit (or, better still, through a point at an infinite
distance from the phases where $\bar{\mu}_e$ is taken as zero). Moreover, the
absolute value of $\bar{\mu}_e^M$ depends on the surface charge density which is a
function of the size of the metal [7].
In fact:

$$\bar{\mu}_e^M = -(\phi^M + F\psi^M) \; , \qquad (12)$$

where $\psi^M = 0$ as the surface charge vanishes.

In order to indroduce in the expression of E all relevant terms, it
is necessary to account for all possible physical steps in the transfer
of a test electron from M to R *through the solution*. Therefore, eqn.
(11) must be replaced by:

$$\Delta E = E^M - E^R = [(\bar{\mu}_e^R - \bar{\mu}_e^S) - (\bar{\mu}_e^M - \bar{\mu}_e^S)]/F \; , \qquad (13)$$

where $\bar{\mu}_e^S$ is the energy of electrons in the liquid phase [3]. Since
electrons are used as test charges, eqn. (13) does not imply any real
experiments. Therefore, practical problems related to the actual state
of an electron in a solvent are irrelevant. Equation (13) implies the

transfer of an electron from M to R *across the two interfacial regions.*
In this way, it is possible to account for terms that do not appear in
eqn. (9) simply because they are the same for the two interfaces and
cancel out.

From eqn. (13), taking into account the expression of $\bar{\mu}_e$, we
obtain:

$$\Delta E = E^M - E^R = [\phi^M/F + \Delta\psi(M/S)] - [\phi^R/F + \Delta\psi(R/S)] . \qquad (14)$$

$\Delta\psi(M/S)$ is the Volta potential difference between the metal and the
solution of the given electrode. Equation (14) shows that the two terms
in square brackets are single electrode potentials measuring the work to
transfer a test electron from the bulk of the metal (electronic Fermi
level) to a point *close* to the surface of the solution.

2.5. "Absolute" electrode potential and surface science

Of the possible single electrode potentials which can be thermodynamically defined depending on the conceptual route followed to transfer a
test electron from M to R, only the one defined by eqn. (14) has a
pratical impact. Let us consider a metal phase in the form of a sphere
(Fig. 2). The electronic energy is measured by the work to transfer a
test electron from the interior of M to the exterior of the phase.
Absolute values of any other quantities are defined in the same way.
Thus, the energy of a metal ion in the metal lattice is measured by the
work (taken with the negative sign) involved in trasferring the metal
ion, used as a test particle, from the interior of M to the exterior.
Accordingly, the absolute solvation energy of an ion is measured by the
work (taken with the negative sign) to transfer it from the solution to
the gas phase above it. This implies that if the energy of various
species has to be measured on a common energy scale, it is

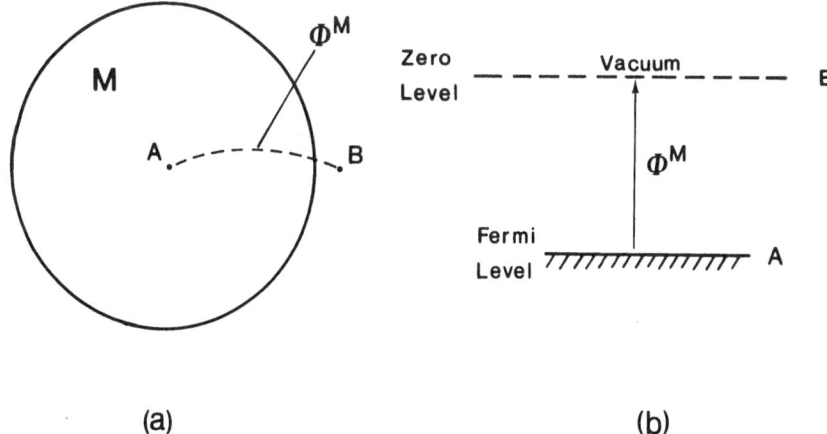

(a)   (b)

Figure 2. *(a)* Sketch to illustrate the definition of electron work
function. *(b)* The corresponding energy diagram.

necessary in electrochemistry to "deconvolute" *measured* potential differences (which give only relative values for the various quantities) into *single* electrode potentials on the basis of a conceptual approach providing the same grounds for the meaning of the energies obtained.

If the expression of the single electrode potential issuing from eqn. (14) is inspected:

$$E^M(abs) = \Phi^M + \Delta\psi(M/S) ,$$ (15)

(if Φ^M is measured in eV, it is not necessary, for the sake of simplification, to divide it by F since the value in V is numerically the same), it can be readily seen that eqn. (15) expresses the work involved in the extraction of an electron from M (Φ^M) and in its transfer to a point close to the solution, $\Delta\psi(M/S)$ - Fig. 3(a). If an electrode is represented as a metal sphere surrounded by a liquid layer of such a thickness as to possess *the properties of a bulk phase* - Fig. 3(b) - E^M measures the work to transfer a test electron from the interior of M to the exterior of the composite phase, *ie* of S. Thus, the conceptual route coincides with that used to define Φ^M for an isolated metal phase. This proves that E^M, as defined by eqn. (15), is *the analogue, for an electrode, of the work function of a metal*. It can therefore align the electrochemical energy scale (E) with the physical energy scale (Φ).

It is worth stressing once again that all the other possible single electrode potentials also give "absolute" values of electronic energy. But the reference level does not match that which issues from the way electronic energies are measured in physics. *Non-interrelated* energy scales would then be obtained. The concept of "absolute" electrode

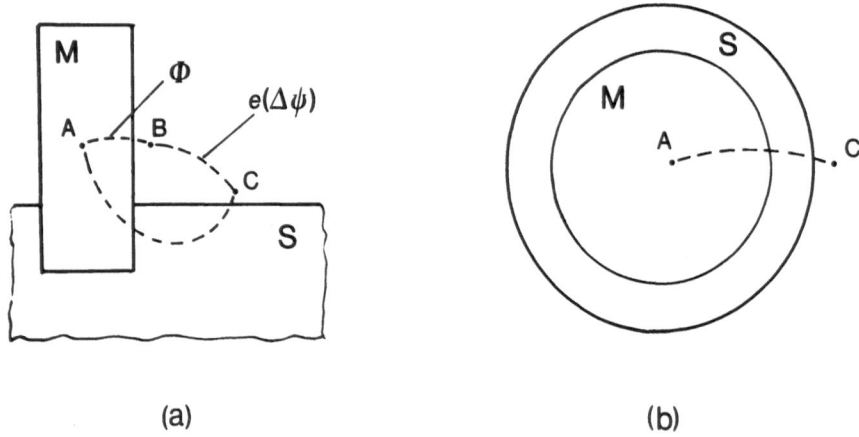

(a)   (b)

Figure 3. (a) Sketch to illustrate the physical meaning of eqn. (15). B and C are points close to the surface of M and S, respectively. (b) Physical configuration of the electrode system "seen" by a test electron transferred from A to C across the M/S interface. Note the conceptual parallelism with Fig. 2(a).

potential has long appealed to the interest of electrochemists because of its physical content. However, it is appropriate to restate that electrochemistry "works" even without the "absolute" potential. It is as the structure of the interfacial region is investigated, and as electronic energies have to be compared, that "absolute" values are necessary and the concept of "absolute" electrode potentials comes in. It is therefore convenient to find an expression relating the electrode potential with the properties of the species in solution with which the electrode is in equilibrium (in case an equilibrium exists), and which are thus responsible for the onset of the observed electrode potential.

2.6. Electrode potential and interfacial equilibrium

Equation (15) relates E^M to the electronic parameters. However, no discussion has thus far involved the *origin of the electrode potential*. If the origin is related to an interfacial equilibrium, the parameters of the species taking part in the equilibrium must be necessarily included, though not explicity, in eqn. (15). Let us assume that the interfacial equilibrium involves the metal ions, so that:

$$M^+(S) \leftrightarrow M^+(M) . \tag{16}$$

Equation (16) represents the actual equilibrium reaction consisting in the exchange of metal ions between the two phases. We can also write:

$$M^+(S) + e(M) \leftrightarrow M^+(M) + e(M) , \tag{17}$$

which is perfectly equivalent to eqn. (16). However, $M^+(M) + e(M)$ is nothing but a metal atom in the metal. Therefore:

$$M^+(S) + e(M) \leftrightarrow M(M) , \tag{18}$$

which is the interfacial equilibrium as usually written in textbooks.
　　When writing down interfacial equilibria, eqn. (16) is the relevant reaction. Therefore:

$$\tilde{\mu}_{M^+}{}^M = \tilde{\mu}_{M^+}{}^S \tag{19}$$

from which:

$$(\phi^M - \phi^S) = (\mu_{M^+}{}^S - \mu_{M^+}{}^M)/F . \tag{20}$$

Equation (20) turns out to be the usual Nernst relationship since:

$$\Delta\phi(M/S) \propto \ln a_{M^+}{}^S . \tag{21}$$

However, while $\mu_{M^+}{}^S$ defines the chemical solvation energy of M^+ in S and needs not be resolved, $\mu_{M^+}{}^M$ must be expressed in terms of more common quantities. Actually, $-\mu_{M^+}{}^M$ measures the chemical work to exctract M^+ from the metal, and this can be experimentally resolved into the work to form a metal ion in the gas phase (atomization + ionization), and the chemical work to bring the electron back into the metal (μ_eM) [4,5]. Thus:

$$-\mu_{M^+}M = \Delta_a GM + \Delta_i GM + \mu_e M \ . \tag{22}$$

Substituting eqn.(22) into eqn.(20) gives:

$$(\phi^M - \phi^S) = \Delta_a GM + \Delta_i GM + \mu_e M \ . \tag{23}$$

Since ϕ can be split into a surface and an electrostatic part [2]:

$$\phi^M - \phi^S = \chi^M + \psi^M - \chi^S - \psi^S \ , \tag{24}$$

it follows from eqn. (15):

$$\phi^M + \Delta\psi(M/S) = E^M(abs) = (\Delta_a GM + \Delta_i GM + \alpha_{M^+}S)/F \ . \tag{25}$$

Equation (25) relates the physical parameters (l.h.s.) of an electrode with the thermodynamic parameters (r.h.s.) of the redox couple involved in the electrode equilibrium [6]. If no interfacial equilibrium is operative, the r.h.s. has no relevance for the given system, while the l.h.s. maintains its validity which is independent of the origin of the electrode potential.

$\alpha_{M^+}S$ is the *real* solvation energy of M^+; it measures the work involved in bringing M^+ from the gas phase into the solution across the actual surface of the liquid phase [2]. It is worth stressing that $\alpha_{M^+}S$ is an *absolute* energy, while only *relative* solvation energies can be obtained from measured ΔE values. This provides a common example of pratical impact of the concept of "absolute" potential. Values of α_{M^+} for most common ions in aqueous solution are tabulated in [3].

If the metal electrode is in equilibrium with a redox couple in solution, the situation *resembles* an electronic equilibrium between two metallic phases. For this reason the "absolute" electrode potential has been pointed out to be a measure of the *Fermi level in solution* [8]. This terminology has been questioned on the grounds that Fermi statistics does not apply to solution species [9]. In fact, the concept of Fermi level is applied to the solution for the sole sake of conceptual exemplification.

2.7. Measurement of E(abs)

Equation (15) shows that $E^M(abs)$ is amenable to experimental measurement since both, Φ^M and $\Delta\psi$, can be measured. Two approaches are possible. In one case, $\Delta\psi$ is measured for the M/S interface in an experiment separately from the measurement of Φ^M for M in a vacuum. In the other case, $(\Phi^M + \Delta\psi)$ is measured in one and the same experiment using the configuration described above, *viz* with a metal covered by a solution layer. Both approches can be questioned.

In the first approach, $\Delta\psi$ can be measured using a jet electrode, if Hg is used [10,11], or using a vibrating condenser, if a solid electrode is employed [12]. It is customary to adopt for Φ the value measured for

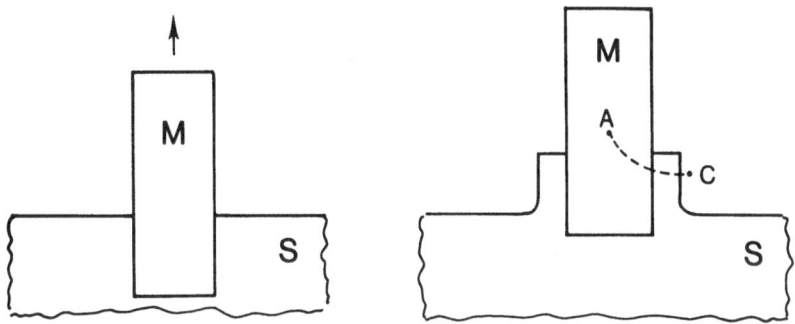

Figure 4. Sketch to illustrate the concept of *emersed* electrode. The electron energy difference between A and C is *in principle* the same as in Fig. 3(a).

the clean metal surface in a vacuum. A possible shortcoming is that the state of the metal surface is in fact not the same; in particular, the surface of M during the measurement of Δψ can be contamined by the solvent vapour, or by gas molecules (even if inert gas is used) [13]. Another unfavourable factor may be the contamination of the surface of the solution with surface active impurities if quiescent solutions are used during the measurement [7]. Thus, either ΦM and Δψ(M/S) might not refer to the appropriate surface conditions, and the resulting EM(abs) value could be erratic.

In the other approach, the realization of the appropriate solution layer is the main problem. It has been claimed [14] that the interface comes out intact from the solution as the electrode is *"emersed"* into the gas phase (Fig. 4). However, while the ionic composition is probably maintained, most of the the solvent seems to be lost during the emersion [14]. In other words, the problem is to ascertain if the layer of solution which undoubtedly exists on the emersed metal surface, possesses the features of a bulk phase. Otherwise, eqn. (15) could not be applied. The shortcoming of the presence of impurities on the surface of the solution possibly dragged off together with the emersed surface, exists also in this case.

2.7.1. *Experimental values of E⁰(H⁺/H₂)(abs).* Table 1 shows that two ranges of experimental values are available for the SHE: one (determined with higher accuracy) of 4.44 V, and the other close to 4.8 V. It is intriguing that the first value has been obtained with a Hg jet electrode in two different laboratories at about 30 years distance with a reproducibility of better than 3 mV [10,11]. All of the uncertainty comes from the value of the work function of Hg. On the other hand, the value close to 4.8 V has been obtained in three different laboratories [12,13,15] using two quite different approaches, and is apparently supported by indirect estimates of electronic energy levels [16].

The consistency of results around 4.8 V has led to point out [12,13] that the value of 4.44 V is probably due to the selection of a Φ value for Hg not reflecting the actual state of a Hg jet. Indeed, the Hg

Table 1. Experimental values of $E^0(H^+/H_2)$(abs)

Value / V	Experimental approach	Ref.
4.44±0.02	$\Delta\psi$ with Hg jet; flowing solution; Φ of Hg from literature	[10,11]
4.73±0.05	$\Delta\psi$ with solid electrodes; quiescent solution; Φ measured photoelectrically	[12]
4.7 (ca)	$\Delta\psi$ with emersed electrode; Φ from cpd of polycrystalline Au and In_2O_3	[14]
(4.7)	Indirectly estimated from surface states at Ag(111)/electrolyte interface	[17]
4.456±0.025	$\Delta\psi$ of SCE using static Hg; Φ of Hg from literature	[16]
4.85 (ca)	Φ of emersed electrode from the onset of He I spectrum; polycrystalline Au	[15]

stream is not a vacuum, but in an inert gas atmosphere presumably containing residual humidity. After some authors [12], the correct value would be 4.8 V in that the actual work function of the Hg jet would be ca 4.8 eV and not 4.5 eV as for "clean" Hg. The hypotesis is that the increment of 0.3 eV would be due to some oxidation of the Hg surface.

While the $\Delta\psi$ measured with a Hg pool is in fact different by about 0.5 V [12], it seems hard to support the above hypothesis on the basis of work function measurements for Hg in the presence of residual gases. Adsorption of water reduces the work function and this is also the case with inert gases. There remains the possibility of surface oxidation by residual oxygen, but the values of $\Delta\psi$ measured with the Hg stream have been shown [11] to be stable even in the presence of O_2 impurities, provided the gas flows rapidly as it is the case during the experiments. On the other hand, surface potential measurements at the free surface of purified water have shown [18] that the value for a flowing surface differs by about 0.3 V from that for a quiescent surface, as a result of adsorption of surface active residual impurities in the solution (probably also coming from the gas phase). Since emersed electrodes drag off precisely the surface layer of the solution as they come out of the liquid phase, the liquid layer attached to emersed solid surfaces might also be contaminated. There remains the estimated value based on binding energies for image-potential-induced surface states [17], which is however difficult to assess both quantitatively and qualitatively.

2.8. E(abs) and structure of the interface

Since the relative reliability of the two groups of values of $E^0(H^+/H_2)$(abs) is difficult to assess on the sole basis of the experimental arrangement, it is expedient to ascertain the impact of the two values on the picture of the structure of the metal/solution interface. In order to do that, reference to a system for which independent estimates of interfacial parameters exist, is a requirement. A possible system appears to be Hg at the potential of zero charge, for which a very precise value in the absence of specific adsorption is known, *ie* $E_{\sigma=0} = -0.192 \pm 0.001$ V *vs* SHE [19]. Since:

$$E_{\sigma=0}{}^{Hg}(SHE) = E_{\sigma=0}{}^{Hg}(abs) - E^0(H^+/H_2)(abs) , \qquad (26)$$

two values are obtained for $E_{\sigma=0}{}^{Hg}(abs)$, *ie* 4.25 V and 4.61 V, on the basis of the two values of $E^0(H^+/H_2)(abs)$, respectively. According to eqn. (15), these values lead to $\Delta\psi(Hg/H_2O) = -0.25$ V and $\Delta\psi = 0.11$ V respectively, for the Volta potential difference at the Hg/H₂O contact at the potential of zero charge of Hg. The former is the value actually measured with a Hg jet electrode [10,11], the latter is the value which *would be measured*, were (according to some views) the Hg surface of the jet really clean as in contact with a vacuum.

The two figures *per se* do not offer any clue to the resolution of the problem. It is first necessary to disclose the relationship between $\Delta\psi(M/S)$ and the electrical parameters of the M/S interface. Since, as shown by Fig. 3(a), the electrical work to bring an electron from M to S must be the same irrespective of the route followed, it follows that under the conditions of zero free charge at the M/S boundary:

$$(\phi^M - \phi^S)_{\sigma=0} = \chi^M + \psi^M - \psi^S - \chi^S = g^M(dip)_0 - g^S(dip)_0 . \qquad (27)$$

Electrons can be brought from M to S either across the metal/vacuum and the vacuum/solution interfaces where χ^M and χ^S are the respective surface potentials, or by going straight away across the M/S boundary, where no free charges are present ($\sigma = 0$) and the surface potentials are normally different from those at the boundary with a vacuum on account of the perturbations brought about by the physical contact between the two phases. Thus [20]:

$$g^M(dip)_0 = \chi^M + \delta\chi_0{}^M \qquad (28a)$$

$$g^S(dip)_0 = \chi^S + \delta\chi_0{}^S , \qquad (28b)$$

where $\delta\chi$ are perturbation terms, and subscript "0" indicates the charge density dependence of these parameters.

Comparison of eqn.(27) with eqns.(28) gives:

$$(\psi^M - \psi^S)_0 = \delta\chi_0{}^M + \delta\chi_0{}^S . \qquad (29)$$

Eqn.(29) means that a Volta potential difference is measured at the potential of zero charge as long as the surfaces of the two phases are structurally perturbed by the mutual contact. Note that this straightforward relation between the Volta potential difference and the modifications caused in the surface regions holds only at the potential of zero charge; at any other potential free surface charges are present at the M/S interface thus bringing in an additional contribution to the interfacial potential drop which cannot be disentangled from the structural term.

For Hg at the potential of zero charge we thus have:

$$\delta\chi_0{}^{Hg} - \delta\chi_0{}^{H2O} = -0.25 \text{ V or } 0.11 \text{ V} \tag{30}$$

depending on the assumed situation. It is now necessary to estimate one of the terms on the l.h.s. There is evidence [21] that $|\delta\chi^{H2O}| \ll |\delta\chi^{Hg}|$, which indicates that the perturbation of the electron distribution at the surface of Hg is much more effective than the change in water orientation in the surface layer. There is also evidence that χ^{H2O} is positive (ie the preferential orientation of water is with the oxygen atom towards the gas phase) and that $\delta\chi_0{}^{H2O} < 0$, ie $g^{H2O}(dip)_0 < \chi^{H2O}$. Since the estimates of χ^{H2O} range 0.13V to 0.03V, and those of $g^{H2O}(dip)_0$ at Hg range 0.08V to negligible values, $\delta\chi_0{}^{H2O}$ at Hg may be between -0.05V and -0.03V (note that since g(dip) and χ are interrelated, high estimates of χ lead to high estimates of g(dip) and vice versa).

If the estimated value of $\delta\chi_0{}^{H2O}$ is introduced into eqn. (30), the two values of $\Delta\psi$ give the following estimates for $\delta\chi^{Hg}$: -0.28V to -0.30V for the former, and 0.06 V to 0.08 V for the latter. A negative value of $\delta\chi^{Hg}$ means that the electron tail at the surface of the metal contracts as water molecules are approaching, as a consequence of electron-electron interactions. On the other hand, a positive value of $\delta\chi^{Hg}$ implies that electrons spill over even more as water molecules approch the metal surface.

Physical concepts, before than calculations, suggest that the correct physical picture should be the former since the electron clouds of water molecules will repel the surface electrons [22]. Thus, the value of 4.8V for $E^0(H^+/H_2)(abs)$ leads to some conceptual difficulties concerning the energetic and structural situation at a metal/water interface. The difficulty can be alleviated by making a different estimate for $\delta\chi_0{}^{H2O}$. In particular, in order to achieve a value $\delta\chi^{Hg} \simeq 0$ V, which is also unlikely, the value of $\delta\chi_0{}^{H2O}$ should be around -0.11 V if $\Delta\psi = 0.11$ V. For the time being, this value seems unlikely since it entails too a high difference between χ^{H2O} and $g^{H2O}(dip)_0$. Even higher values of $\delta\chi_0{}^{H2O}$ can be expected if a negative value of $g^{H2O}(dip)_0$ is assumed, ie water molecules would be preferentially oriented with the hydrogen atoms towards the Hg surface. However, presently no one piece of evidence suggests that such an orientation is realistic.

If a lower value of $E^0(H^+/H_2)(abs)$ is postulated, eg 4.7 V, which seems to be the lower limit of the second group of values, the Volta potential difference at the Hg/H_2O interface would be around zero, which would imply a possible small negative value of $\delta\chi^{Hg}$ of -0.05V. While the sign is that expected, the magnitude still seems small, considering that noble gases on Hg produce $\Delta\Phi$ values of the order of 0.25 eV [23].

3. CONCLUSION

The configuration of an isolated electrode as an analogue of an isolated metal phase, consists of a metal phase surrounded by a liquid layer thick enough as to possess bulk properties. Accordingly, the analogue of the electron work function (Φ^M) of an isolated metal phase is, for an isolated electrode, the work function of the system sketched in Fig. 3, *viz* the work to take an electron from the Fermi level of the metal to a point just outside the liquid surface. Such a work is quantitatively expressed by $\Phi^M + \Delta\psi$(M/S) where $\Delta\psi$(M/S) is the metal/solution conctact (or Volta) potential difference.

A configuration for an electrode like that described above is *in principle* realized with emersed electrodes. However, it is uncertain whether the condition of bulk properties for the emersed liquid layer is achieved. Alternatively, $\Delta\psi$ can be measured separately from Φ^M. However, it is in question whether the same surface conditions have been realized for M in the experiments reported in literature.

The electron work function of an electrode (as above envisaged) possesses the requisites for a single electrode potential. While other forms of single electrode potential can be disclosed, the above is the only one which permits to align electronic energy levels in the solid with those for solution species. Accordingly, absolute values for energy states of ions in a phase can be derived.

Experimental determinations of E^0(H^+/H_2)(abs) have provided two different extreme values: 4.44 V and *ca* 4.8 V. These two values entail different interfacial parameters. If the case of the Hg/water interface is discussed as a study case, it has been shown that the higher value leads to some conceptual difficulties concerning the interactions between Hg and water at the interface. Should the correct experimental approach be that leading to the higher value for E^0(H^+/H_2)(abs), it would be necessary to reconsider some aspects which are presently regarded as well established, in particular, the surface orientation of water at Hg, and the extent of polarization of the electron tail at the surface of Hg upon contact with water.

Acknowledgements. Financial support from the Ministry of Education (40% Funds) and the C.N.R. (Rome) is gratefully acknowledged.

4. REFERENCES

1. Trasatti, S. and Parsons, R. (1986) 'Interphases in systems of conducting phases', *Pure Appl. Chem.* **58**, 437-454.
2. Parsons, R. (1954) 'Equilibrium properties of electrified interfaces', in J. O'M. Bockris (ed.), *Modern Aspects of Electrochemistry*, Butterworths, London, pp. 103-179.
3. Trasatti, S. (1980) 'The electrode potential', in J. O'M. Bockris, B. E. Conway and E. Yeager (eds.), *Comprehensive Treatise of Electrochemistry*, Vol. 1, Plenum Press, New York, pp. 45-81.
4. Trasatti, S. (1974) 'The concept of absolute electrode potential. An attempt at a calculation', *J. Electroanal. Chem.* **52**, 313-329.
5. Trasatti, S. (1982) 'The concept and physical meaning of absolute electrode potential. A reassessment', *J. Electroanal. Chem.* **139**, 1-13.
6. Trasatti, S. (1990) 'The absolute electrode potential. The end of

the story', *Electrochim. Acta* **35**, 269-271.

7. Trasatti, S. (1986) 'The absolute electrode potential. An explanatory note', *Pure Appl. Chem.* **58**, 955-966.

8. Gerischer, H. and Ekard, W. (1983) 'Fermi levels in electrolytes and the absolute scale of redox potentials', *Appl. Phys. Lett.* **43**, 393-395.

9. Bockris, J.O'M. and Khan, S. U. M. (1983) 'Fermi levels in solution', *Appl. Phys. Lett.* **42**, 124-125.

10. Randles, J. E. B. (1956) 'The real hydration energies of ions', *Trans. Faraday Soc.* **52**, 1573-1581.

11. Farrell, J. R. and McTigue, P. (1982) 'Precise compensating potential difference measurement with a voltaic cell. The surface potential of water', *J. Electroanal. Chem.* **139**, 37-56.

12. Gomer, R. and Tryson, G. (1977) 'An experimental determination of absolute half-cell emf's and single ion free energies of solvation', *J. Chem. Phys.* **66**, 4413-4424.

13. Hansen, W. N. and Hansen, G. J. (1988) 'Implications of double layer emersion', in M. P. Soriaga (ed.), *Electrochemical Surface Science* (ACS Symposium Series 378), American Chemical Society, Washington, DC, pp. 166-173.

14. Kolb, D. M. and Hansen, W. N. (1979) 'Electroreflectance spectra of emersed metal electrodes', *Surf. Sci.* **79**, 205-211.

15. Kötz, E. R., Neff, H. and Müller, K. (1986) 'A UPS, XPS and work function study of emersed silver, platinum and gold electrodes', *J. Electroanal. Chem.* **215**, 331-344.

16. Hansen, W. N. and Hansen, G. J. (1987) 'Absolute half-cell potential: A simple direct measurement', *Phys. Rev. A* **36**, 1396-1402.

17. Schneider, J., Franke, C. and Kolb, D. M. (1988) 'Image-potential-induced surface states at the Ag(111)-electrolyte interface', *Surf. Sci.* **198**, 277-284.

18. Koczorowski, Z., unpublished results. Personal communication.

19. Grahame, D. C., Coffin, E. M., Cummings, J. I. and Poth, M. A. (1952) 'The potential of the electrocapillary maximum of mercury. II', *J. Am. Chem. Soc.* **74**, 1207-1211.

20. Trasatti, S. (1986) 'Components of the electrode potential. Concepts and problems', in A. F. Silva (ed.), *Trends in Interfacial Electrochemistry*, D. Reidel, Dordrecht, pp. 1-24.

21. Trasatti, S. (1979) 'Solvent adsorption and double-layer potential drop at electrodes', in B. E. Conway and J. O'M. Bockris (eds.), *Modern Aspects of Electrochemistry*, Vol. 13, Plenum Press, New York, pp. 81-206.

22. Trasatti, S. (1986) 'Potentials of zero charge. What they suggest about the structure of the interfacial region', in A. F. Silva (ed.), *Trends in Interfacial Electrochemistry*, D. Reidel, Dordrecht, pp. 25-48.

23. Ford, R. R. and Pritchard, J. (1971) 'Work functions of Au and Ag films. Surface potentials of Hg and Xe', *Trans. Faraday Soc.* **67**, 216-221.

PHENOMENOLOGICAL APPROACH TO METAL/ELECTROLYTE INTERFACES

S. TRASATTI
*Department of Physical Chemistry
and Electrochemistry
University of Milan
20133 Milan, Italy*

ABSTRACT. Recent experimental results in the field of double layer are analyzed in the light of a number of correlations developed by the present author over the past twenty years. In particular, the behaviour of *sd*-metals (Au,Ag,Cu) is discussed in the context of the group of *sp*-metals. The potential of zero charge *vs* work function plot shows that *sd*-metals cannot be gathered in a single group. The parameter ΔX, derived from the plot, and measuring the perturbation of the structure of the surface regions as the two phase come in contact, is used to correlate other experimental data. Thus ΔX correlates with the affinity of metals for oxygen, with the temperature coefficient of the potential of zero charge, with the differential capacitance, and with the energy of adsorption of neutral species from the solution. From the analysis of these plots insight into the contribution of the surface of the metal vs the contribution of the interfacial solvent can be gained. Inspection of data for metal single crystal faces brings into evidence regularities in the variation of the behaviour with the atomic density of the crystal plane. However, opposite trends can be observed with differently prepared surfaces. Scrutiny of different approaches emphasizes the crucial stage of the selection of the physical properties to be corre-lated with the electrochemical properties. The peculiar behaviour of Au in DMSO is also discussed. The need for further experimental work is stressed and directions of research are suggested.

1. INTRODUCTION

The distribution of charged and neutral species at the electrode/solution interface is governed, toghether with the electric field, by a number of factors which can be expressed in terms of pairwise chemical interactions: particle-particle, particle-solvent, solvent-solvent, particle-metal and metal-solvent interactions [1]. Thus, specific adsorption is customarily described as arising from a competition between particle-solvent and particle-metal interactions [2,3]. Similarly, surface condensation of adsorbates can be understood in terms of competition between particle-particle and particle-solvent

R. Guidelli (ed.), Electrified Interfaces in Physics, Chemistry and Biology, 245–273.
© 1992 *Kluwer Academic Publishers.*

interactions [4-6].

Although any similar description is only a first approximation (for instance, particle-solvent interactions are expected to differ in the bulk of the solution and at the interface, so that these interactions are anisotropic in the interfacial region) [1,7], the representation of the energetic situation in terms of separate single contributions is a great conceptual aid especially in double layer modeling. In this respect, this paper is devoted *(i)* to the discussion of the implications of the correct description of the experimental parameters in terms of the various factors mentioned above, and *(ii)* to a survey of recent experimental results in this area with the aim to emphasize the insight which can be gained into the structure of the metal/solution interface.

2. PHENOMENOLOGICAL APPROACH

The various factors which are responsible for the structure of the electrode/solution interface are strongly interrelated and cannot be determined separately. However, the additional variable which character- izes electrochemical interfaces, *ie* the electric field, can be controlled separately. Therefore, the discussion in this paper will be restricted to the conditions of *"zero surface charge density"*, *viz* with the electrode at the potential of zero charge ($E_{\sigma=0}$). Any effects of the electric field related to the presence of free charges on either sides of the interface are thus cancelled, and only short-range chemical or dipolar contributions remain operative.

Electrochemical interfaces can be studied either *in situ* or *ex situ*. While it is obvious that the actual situation can be observed *only in situ*, there have been attempts to use *ex situ* investigations to reduce the number of variables, thus trying to disentangle the various factors which operate at the interface [8]. In other words, the electro- chemical interface can be possibly simplified either by removing one of the bulk phases so as to be able to look directly at the electrode surface (*eg* UHV studies, emersed electrodes [9]), or by synthesizing the liquid phase on the metal surface particle by particle (*eg* adsorption of species from the gas phase [10]). While *ex situ* investigations always produce more or less approximate situations, they are of great help to compare quantities or model predictions based on *in situ* thermodynamic approaches. On the other hand, purely electrochemical techniques lack molecular specificity, which is the main advantage of spectroscopic techniques.

Two different groups of *in situ* approaches are to be considered: those based on conventional AC and DC techniques (electrochemical techniques) [11] and those based on vibrational spectroscopies and optical probing of the electrode/solution interface [12]. This paper is devoted to the analysis of the response of traditional electrochemical approches to the study of electrochemical interfaces, *eg* the measurement of the potential of zero charge and its variation with adsorption or with temperature, and the measurement of the differential capacitance.

Since an electrode/solution interface behaves *in principle* as an electric condenser, the structure and the properties of this capacitor can be deduced from the electric potential drop across it and its dependence on charge density, *ie* the differential capacitance, as a function of temperature, metal, solvent and solute nature. This paper will consider in particular the following parameters: *(1)* the dependence the potential of zero charge and its temperature coefficient on the nature and the structure of the metal surface; *(2)* the capacitance at the potential of zero charge; *(3)* the dependence of the adsorption energy of solutes and solvents on the nature and the structure of the metal surface.

These aspects have been discussed several times in the literature and the basic principles are well established [1]. In this paper, emphasis will be placed especially on some recent developments, particularly with single crystal faces of Ag and Au.

2.1. The Potential of Zero Charge

According to the previous lecture [13], the electrode potential measures the value of the energy of the electrons at the Fermi level of the given metal electrode relative to the metal of the reference electrode. Taken separately, the "absolute" value of the electrode potential measures the work to transfer an electron from the metal surrounded by a layer of solution to the exterior of this system, *ie* to a point in a vacuum outside the solution. This work entails crossing the metal/solution interface and the solution/air interface. Since, the latter does not matter for the discussion of the electrochemical interface, the relevant term can be incorporated into a costant including all the contributions from the reference electrode. Therefore:

$$E^M_{\sigma=0} = -\mu_e^M/F + (\phi^M - \phi^S)_0 + \text{const} , \qquad (1)$$

where μ_e^M is the electronic energy in the bulk of M, and subscript "zero" indicates the particular charging condition. Since the electric potential drop at $\sigma = 0$ (σ is the surface charge density of the metal) consists only of dipolar terms:

$$(\phi^M - \phi^S)_0 = g^M(\text{dip})_0 - g^S(\text{dip})_0 , \qquad (2)$$

eqn. (1) can also be written as:

$$E^M_{\sigma=0} = \phi^{M'} - g^S(\text{dip})_0 + \text{const} , \qquad (3)$$

where:

$$\phi^{M'} = \phi^M + \delta\chi^M . \qquad (4)$$

ϕ' (expressed in eV in all the above and successive equations so that the conversion factor F can be omitted) is the electron work function of M in a vacuum with the same surface electron distribution as in the presence of the solution. $g^S(\text{dip})_0$ is the contribution to the interfacial potential drop due to any preferentially oriented (or chemically polarized) solvent molecules. Eqn. (3) contains the parameters which are induced by the interaction between the metal and the solvent.

Since both Φ' and $g^S(dip)_0$ cannot be determined separately, the numerical value of $E_{\sigma=0}$ for a single metal does not help very much to figure out the interfacial structure. It is thus expedient to have recourse to correlations, from which relative values of the parameters can be obtained. To this aim eqn. (3), with eqn. (4), is better resolved into the following equation [14]:

$$EM_{\sigma=0} = \Phi^M + [\delta\chi^M - g^S(dip)_0] + const .\tag{5}$$

Φ^M is an experimentally accessible quantity which can be measured with great accuracy (± 0.02 eV). The term in square brackets can be seen as a quantity measuring the extent of modification occurring in the surface regions of the phases as they are brought into contact. Eqn. (5) can be further simplified [15]:

$$EM_{\sigma=0} = \Phi^M + X + const .\tag{6}$$

Equation (6) shows that X cannot be determined directly from the experimental values of $EM_{\sigma=0}$ and Φ^M since the value of the constant is also unknown. However, relative values of X can be obtained by comparing, for instace, $E_{\sigma=0}$ and Φ for two metals. Thus:

$$\Delta^1{}_2E_{\sigma=0} = \Delta^1{}_2\Phi + \Delta^1{}_2X .\tag{7}$$

However, the accuracy for only two experimental data points is usually insufficient to bring the actual trend into evidence since for metals the above quantities do not change very much. A better approach is therefore a graphical correlation of $E_{\sigma=0}$ *vs* Φ^M.

Plots of $E_{\sigma=0}$ *vs* Φ have been scrutinized several times with the aim to disclose a possible difference between the structure of the metal/vacuum and that of the metal/solution interface [1,14-17]. From eqn. (7) it is clear that $\Delta E_{\sigma=0} \approx \Delta\Phi$ would mean that the main factor behind the change in $E_{\sigma=0}$ from metal to metal is the total electronic energy while interfacial parameters are less important. However, a problem with $E_{\sigma=0}$ *vs* Φ plots is that the two quantities usually are measured with different samples. Since the value of Φ is very sensitive to the metal surface structure, the choice of Φ to be used in the plot becomes a crucial step in the approach, which can heavily condition the picture emerging from the correlation.

The first plot compiled by the present author in 1971 [14] clearly showed that $E_{\sigma=0}$ and Φ are linearly related within well identified groups of metals. Since then, the few new data produced experimentally have enabled the general picture to be improved while the main conclusions have remained unchanged over the past twenty years [15].

The solid straight line A of unit slope passing through the point of Hg in Fig. 1 is a reference line. Hg is taken as a reference metal in view of the accuracy of its experimental data. According to eqn. (7), a metal with the same value of X as Hg would be located on the straight line A. If the point for a metal falls away from line A, this would

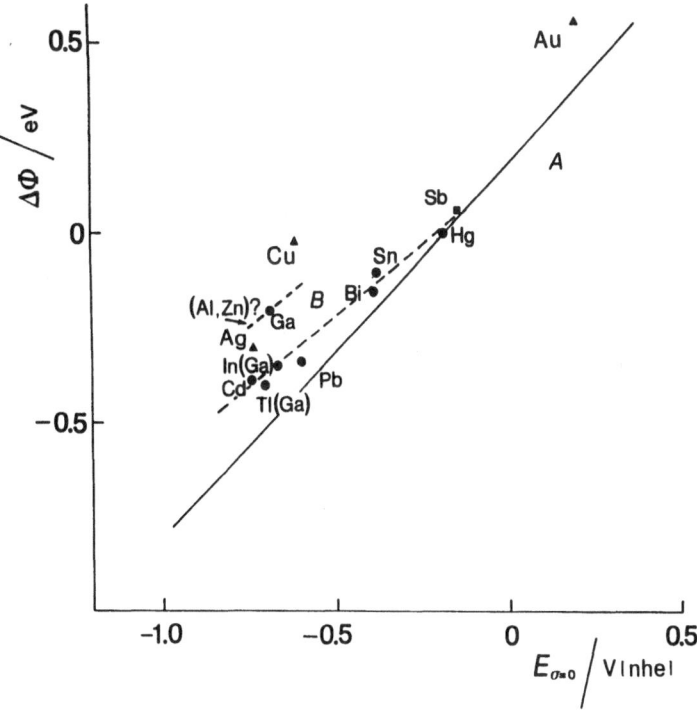

Figure 1. Work function-potential of zero charge plot for polycrystalline *sp*-metals and the (110) face of *sd*-metals in aqueous solution. Line
A is of unit slope. Line B is drawn through the points of *sp*-metals. Φ
values referred to that of Hg.

imply that $X^M \neq X^{Hg}$, *ie* the metal/water contact produces at the interface electrical effects *quantitatively* different from those at Hg.

For the sake of simplification, *d*-metals are not shown in Fig. 1.
They have been discussed elsewhere [14,18]. *sp*-Metals can be seen to
fall in a group within which two sub-groups can be possibly identified.
Although the two sub-groups are rather near and a broad co-variation of
the two variables might justify the actual situation, the metals in the
two sub-groups are recognized to belong to different Periods of the
Periodic Table. Therefore, the observed distribution may have a rationale behind it.

Fig. 1 shows that, while the main factor responsible for the value
of $E_{\sigma=0}$ is in fact Φ^M, the observed deviations observed from line A
suggest that the term X also varies with the metal in a way systematically depending on Φ^M. It is intriguing that all points lie on the left
of the reference line. This means that X for all other metals
contributes a more negative term than for Hg. In other words, the actual
value of $E_{\sigma=0}$ at a given Φ is always more negative than for Hg-like
metals. This can be due to a more negative contribution of $g^S(dip)_0$ – *cf*
eqn. (5) –, *ie* to a more specific orientation of solvent molecules, but
it can also be related to a more negative value of $\delta\chi_0^M$ [19,20]. It is
in principle difficult to assess which factor is responsible for the

observed behavior: whether the polarizability of the electron cloud at the metal surface, or the polarizability of the solvent molecules in contact with the metal. The two terms sum up to define X which can be regarded as the electrical response of the surface to the metal-water interaction. X can also be taken as measuring the *degree of coupling* of solvent molecules with the metal surface.

Why negative values of X are usually observed can be understood by considering that surface electrons tend to retracts as water molecules approach the metal thus producing a lower χ^M, *viz* a negative $\delta\chi^M$. Moreover, water molecules are normally attracted by metal surfaces through their oxygen atom thus contributing a negative dipolar layer, *viz* with the negative side facing the metal. The obtained plot also suggests that Hg is, among the metals investigated, the one with the smallest negative value of X. It can be stated that metal-water interactions produce small effects on Hg, or that Hg is a *"hydrophobic"* metal.

The concept of *"hydrophilicity"* [15,21] has often been discussed in relation to $g^S(dip)_0$. A higher value of $g^S(dip)_0$ can be the result of stronger metal-water interaction, *viz* of greater *hydrophilicity*. However, the concept of hydrophilicity is less straightforward if $\delta\chi^M$ is considered since this quantity does not involve concepts of chemical interaction with bond formation. Nevertheless, $\delta\chi^M$ can also be regarded as measuring the difference between the presence and the absence of water: it is therefore a result of water-metal interaction. The difference between $\delta\chi^M$ and $g^S(dip)_0$ is that, while a high value of the latter implies the idea of bond strength, a large value of $\delta\chi^M$ not necessarily involves the concept of strong metal-water chemical bond.

Fig. 1 includes the points for Ag,Au and Cu. Since the Φ of these metals is largely anisotropic, (*ie* it depends on the crystallographic orientation), $E_{\sigma=0}$ depends on the crystal face. While this particular aspect will be discussed in the next section, the points for the (110) face of these metals (whose behaviour is close to that of the polycrystalline surface) are reported in the plot with the aim to discuss their position in the context of the *sp*-metals. Apparently, the points for these three metals are scattered since they do not fall in the groups formed by the *sp*-metals although they belong to the same Periods. The data in Fig.1 anticipate a different behaviour of *sd*-metals (Cu, Ag, Au) with respect to all other *sp*-metals. Since transition metals form another separate group, it seems evident that the electronic structure of the external bands is the main factor responsible for metal-water interactions.

2.2. The Metal-Water Interaction Term, X

The horizontal distance, in Fig. 1, of each point from the solid straight line measures, according to eqn. (7), the relative value of X for the given metal with respect to Hg taken as a reference. Absolute values of X for Hg have been estimated [14,18], therefore absolute values of X for all other metals can be derived. However, such a step does not add anything to the physical meaning of successive analyses, while it does introduce questionable model assumptions in a phenome-

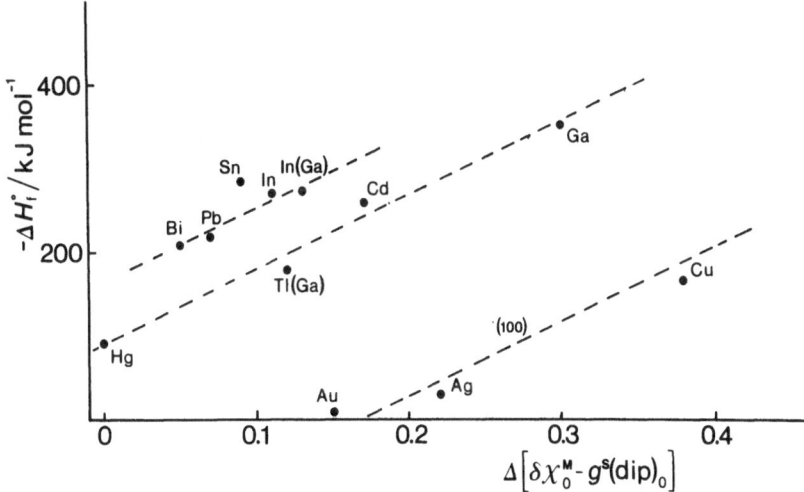

Figure 2. Plot of ΔX (from Fig. 1) *vs* the enthalpy of formation of the oxide MO.

nological approach. The values of ΔX derived from Fig.1 cannot be questioned (since they are based on a thermodynamic equation) except that (eventually) for the criteria of choice of the values of Φ.

The existing evidence suggests that X for Hg is very likely negative [14]. Fig.1 points out that X is even more negative for all other metals. A model assumption is now needed to try to find out some rationale behind the trend in the X values. Since X measures the impact of metal-water interactions on the value of $E_{\sigma=0}$, X should be proportional to the thermodynamic affinity of metals for water. There exist no tabulated data for a hypotetical $M-OH_2$ compound, but since interaction is expected to take place through the oxygen atom of water molecules, the present author has long ago [22] suggested that the needed parameter can be $\Delta_f H^0$, the entalpy of formation of the MO oxide. This parameter has been used also to make thermodynamic predictions regarding the type of water adsorption on metals from the gas phase, and it has been shown to work in that case too [23].

Fig. 2 shows a plot of ΔX against $\Delta_f H^0$. As expected, the broad trend is that ΔX increases with the negative value of $\Delta_f H^0$. It is even more interesting that metals can be gathered into different groups with *sp*-metals in two distinct groups, and *sd*-metals positively separated. A possible objection may be that the separation arises because $\Delta_f H^0$ refers to the formation of bulk compounds and not to adsorbate monolayers only. $\Delta_f H^0$ for bulk compounds would include metal-metal bond breaking which is not expected to occur on adsorption. However, this is only partly true: as chemisorption takes place, surface electrons will be concentrated (or diluted) at the surface sites where the adsorbate is placed, with delocalized effects on nieghbouring sites. The effect of chemisorption on surface conductivity is a pratical example, the other being the difference between M-H bond strength in solution and in the gas phase [24]. Therefore, the $\Delta_f H^0$ values are very likely to be consistent with

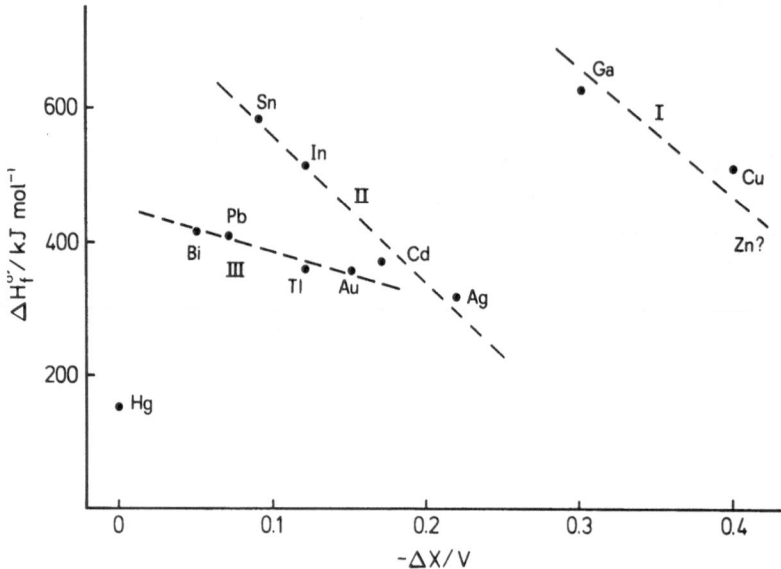

Figure 3. Plot of ΔX (from Fig. 1) vs the enthalpy of formation of the oxide MO, corrected for the work to break metal-metal bonds. I, II, III indicate the Periods of the Periodic Table.

the original concepts. However, an attempt to correct the $\Delta_f H^0$ values for the metal-metal surface bond by subtracting the metal sublimation heat, produces an intriguing arrangement of the metals, as shown in Fig. 3. The metals can still be gathered in three main groups but, quite interestingly, according to the Periods of the Mendeleev Table.

It is very intriguing that the slopes of the straight lines tentatively drawn in Fig. 3 are now of opposite sign with respect to Fig. 2. Within a given group the value of ΔX increases as the affinity for water (actually, $\Delta H^0{}'$) decreases. Au and Ag are known to be among the very few metals which adsorb water associatively from the gas phase [23]. Nevertheless, they show a large value of ΔX. A possible explanation is that, in the case of Ag and Au, the polarizability of the surface electrons, as mesured by $\delta\chi^M$, is more important than water orientation, as measured by $g^S(dip)_0$. On the other hand, Fig. 3 places Au in a position indicating a slightly higher value of $\Delta_f H^0{}'$ than for Ag despite a lower ΔX value. This could indicate that the "hydrophilicity" of Au is slightly higher than that of Ag. However, the difference is within the accuracy of the calculation of $\Delta_f H^0{}'$.

The quite isolated position of Hg in Fig. 3, is noteworthy. This may be primarily related to the fact that the liquid metal is the reference state for thermodynamic parameters at 25°C. In other words, the point of Hg simply cannot fit the plot of Fig.3.

2.3. Non-aqueous Solvents

Plots like the one in Fig.1 can be obtained for non-aqueous solvents too [25]. Recent results have enriched the picture for DMSO [26,27]. The

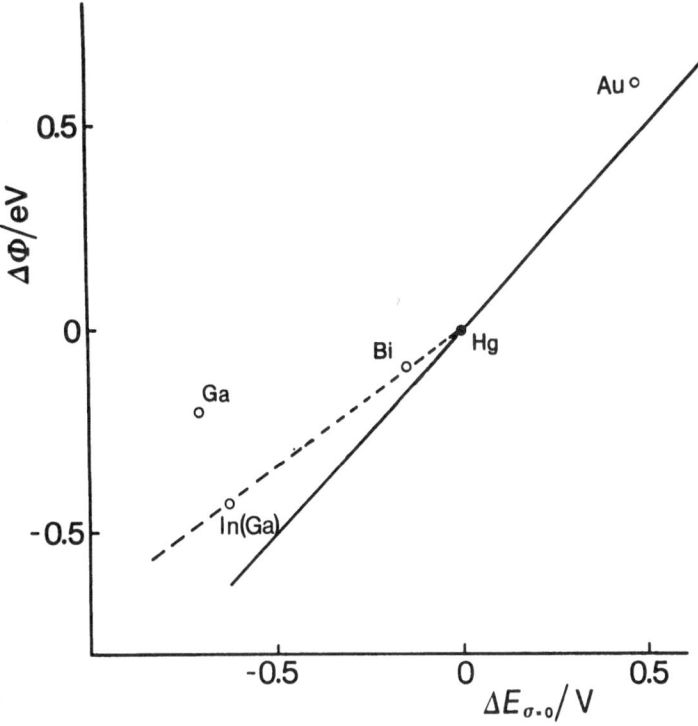

Figure 4. Work function-potential of zero charge plot for polycrystalline *sp*-metals and Au in DMSO solution. Line A is of unit slope.

plot for this solvent is shown in Fig. 4. Although the available data
points are only few, the evidence is for a behaviour resembling that in
aqueous solutions for all sp-metals. In particular, they exhibit a more
negative value of X with respect to Hg. Since DMSO can interact with the
metal surface through the oxygen atom, the picture emerging for *sp*-
metals is not surprising. The only *sd*-metal in this plot is Au, whose
position appears to be consistent with that in aqueous solutions (Fig.
1). There exist no gas phase adsorption data with this solvent.

2.4. Contact (Volta) Potential Difference

In the previous lecture [18] it has been shown that the cpd between a
metal and a solution at $E_{\sigma=0}$ measures the modifications occuring in the
surfaces of the two phases upon contact. In more practical terms, $\Delta\psi_0$
measures precisely the change in the electron work function of M as it
is covered with a macroscopic film of solution at $\sigma = 0$. $\Delta\psi_0$ includes a
contribution from metal electrons and a contribution from solvent
dipoles. It is thus difficult to compare the behaviour of different
metals, as Figs. 2 and 3 have shown. However, for a given metal in
contact with different solvents, $\delta\chi^M$ may be less important than $g^S(dip)_0$
[25]. Thus, $\Delta\psi_0$ could be probably compared as a function of the nature
of the various solvents.

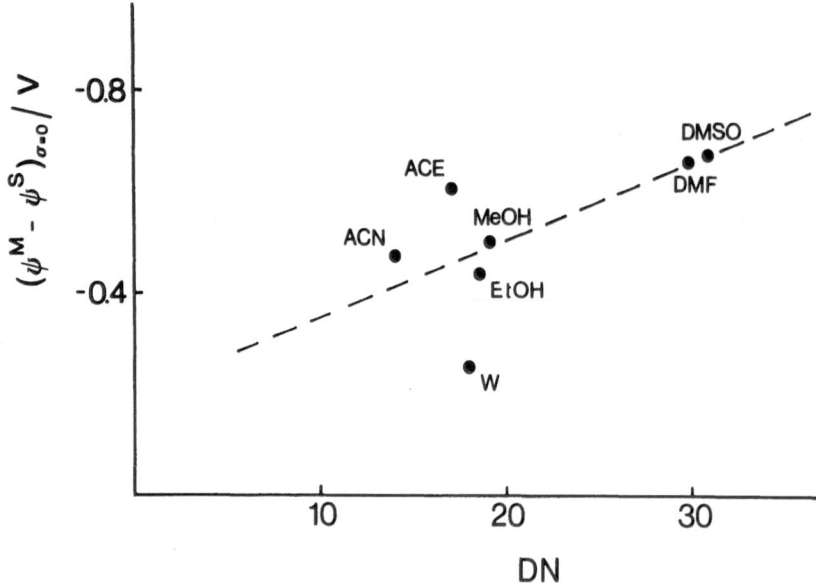

Figure 5. Contact (Volta) potential difference between Hg at the potential of zero charge and the indicated solvents against the Donor Number of the solvents.

The tendency of a solvent to form a bond may be measured to a first approximation by its donor number (DN). Fig. 5 shows that $\Delta\psi_0$ for Hg is broadly related to DN in a linear manner: the higher DN the higher $\Delta\psi_0$ as a result of stronger metal-solvent interactions [25]. This plot has been improved by Jaworski [28] who has pointed out that, like for the electronegativity of a metal, the donor-acceptor properties of solvents are more appropriate than just the donicity. Jaworski has thus been able to produce a plot where especially the point for water, scattered in Fig. 5, falls nearer to the common line. Thus, Fig. 5 proves that $\Delta\psi_0$ is related to the strength of the interaction between the metal surface and the solvent molecules.

2.5. Effect of the Crystallographic Orientation

While no detectable effects are produced by different crystallographic orientations for Pb, Sn and Cd which are low-melting metals, anisotropy is large for Ag, Au and Cu. These metals must be discussed separately.

According to eqn.(6), since Φ is an anisotropic quantity, *ie* it depends on the crystallographic orientation, also all other terms are expected to be anisotropic, in particular $E_{\sigma=0}$:

$$E_{\sigma=0}(hkl) = \Phi(hkl) + X(hkl) + const . \qquad (8)$$

Therefore, depending on the value of X, $E_{\sigma=0}$ is expected to vary with the crystallographic orientation in a way paralleling that of Φ. This has been verified exhaustively [29,30]. In particular, Fig. 6 shows that

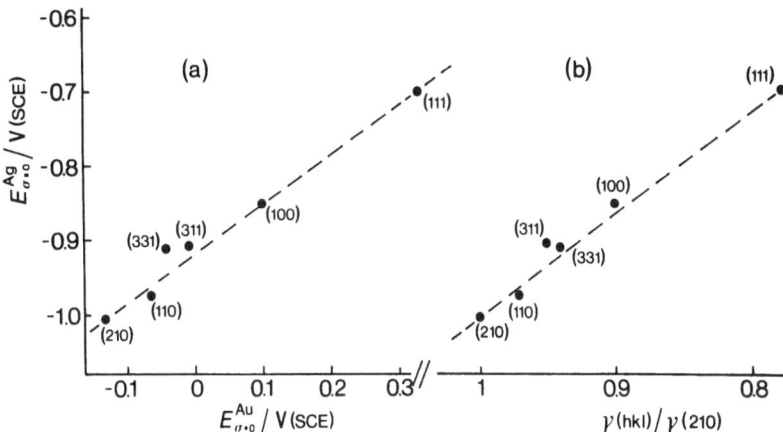

Figure 6. Plot of the potential of zero charge of Ag single face electrodes in aqueous solution *vs* E$_{\sigma=0}$ for the same faces of Au *(a)*, and the *calculated* relative surface tension of FCC metals *(b)*.

E$_{\sigma=0}$ and the *calculated* surface tension show quite similar patterns as the crystallographic orientation is varied [31].

As for the case of *sp*-metals, a more quantitative approach to X for single crystal faces involves a relative comparison of E$_{\sigma=0}$ and Φ. The choice of Φ is even more critical, since the variation of Φ from face to face is rather small and the perfection of the surface structure is too a crucial condition to be experimentally achieved or reproduced. A small difference in Φ may lead to opposite conclusions, as shown later on.

Figure 7 shows a plot of E$_{\sigma=0}$ *vs* Φ for the three main faces of Ag, Au and Cu [15,32]. It is seen that the two quantities varies in parallel, *ie* the main factor producing the variation of E$_{\sigma=0}$ with the crystal face is the variation of the electron work function. If a straight line of unit slope passing through Hg is drawn, the position of E$_{\sigma=0}$ for Au, Ag and Cu shows systematic deviations in the same direction for all three metals. In particular, the slope of the relationships is probably very similar to that of the lines gathering *sp*-metals in Fig. 1 (line B).

It is interesting to argue about the way X varies from face to face. In this respect, the choice of Φ becomes the outstanding factor. In Fig. 7, the horizontal distance of each data point (or the point

Table 1. **Interfacial parameter for main crystal faces**

	\-ΔX/V		
	(111)	(100)	(110)
Cu	0.32	0.19	0.12
Ag	0.35	0.22	0.14
Au	0.42	0.24	0.18

256

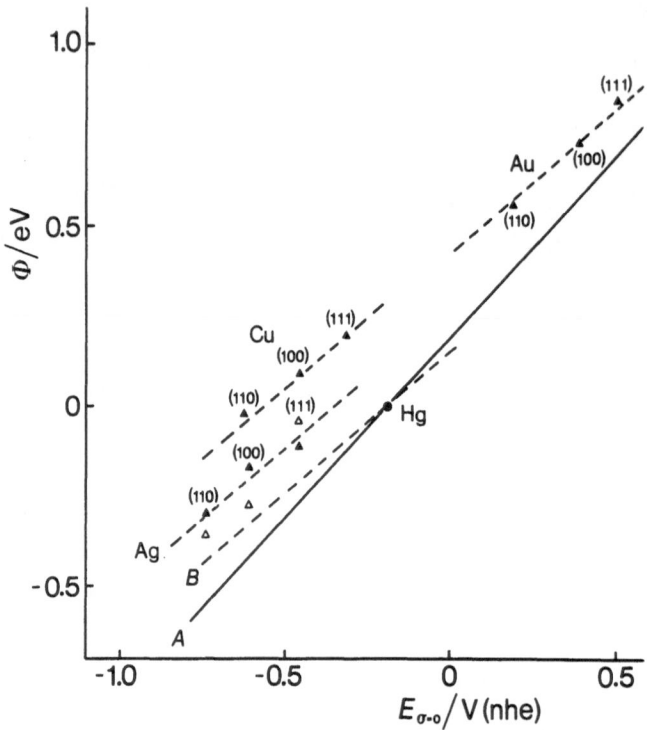

Figure 7. Work function-potential of zero charge plot for single crystal faces of Cu, Ag and Au in aqueous solution. Lines A and B as in Fig. 1.

interpolated on the dashed lines) from line A for Hg gives the relative value of X. These have been summarized in Table 1. It can be seen that ΔX increases in the order $(111) < (100) < (110)$ for all three metals.

This conclusion has been questioned by Valette [33] on the basis of different choices of Φ, mostly based on a single set of Φ values, and somewhat "massaged". Figure 8 shows the specific case of Ag. The solid line is of unit slope. Line 1 has the same slope as that in Fig.7 and is based on the Φ values selected by the present author [34]. It is clear that the slope of line 1 is <0, so that $\Delta X(110) > \Delta X(100) > \Delta X(111)$. Lines 2 to 4 are drawn using the Φ values recommended by Valette in three papers, in 1982 [35], 1984 [33] and 1987 [36], respectively. It is clear that in all three cases the slope is >1, *viz* $\Delta X(110) < \Delta X(100) < \Delta X(111)$. Valette's 1987 plot is based on a single set of Φ values for Ag. A recent experimental determination of Φ for the main faces of Ag [37,38], although producing systematically higher Φ values by a costant term (*ie* the variation from face to face is not affected) - *cf.* [39] for the reasons - confirms that the most probable slope is <1 (curve 5 in Fig. 8), as already observed with the Φ values selected earlier by the present author (curve 1 in Fig. 8). It can thus be concluded that the major evidence is in favour of $\Delta X(110) > \Delta X(100) > \Delta X(111)$ for Ag single crystal face electrodes.

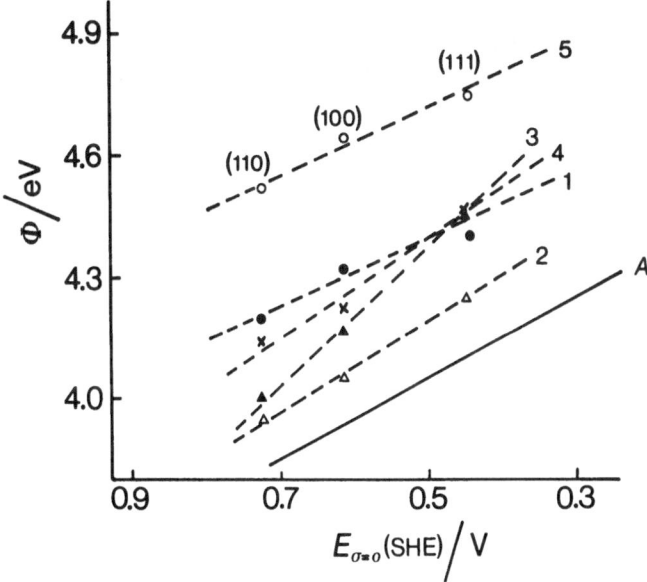

Figure 8. Work function-potential of zero charge plot for single crystal faces of Ag in aqueous solution. Line A as in Fig. 1. (1) Work functions from [34]; (2) From [35]; (3) From [33]; (4) From [36]; (5) This work. See text for details.

2.6. Electrochemical *vs* Gas Phase Data

It is interesting to compare the electrochemical results with gas phase *ex situ* experiments of "molecular" synthesis of an electrochemical interface. Work function variations upon water adsorption are available for a number of transition metals [40] but unfortunately no reliable data exist for *sp*-metals [23,41]. However, recent results for Ag (110) and Cu (110) (*sd*-metals) [42] can be discussed, although the issue of the sequence of the faces cannot be resolved with these data. In order to compare $E_{\sigma=0}$ data with $\Delta\Phi$ data, it is necessary to calculate first the "*absolute*" value of $E_{\sigma=0}$ - denoted with $E_{\sigma=0}(abs)$ - and then to subtract the experimental Φ. Using the data in Fig. 7, Table 2 has been compiled for the main faces of Ag. It is seen that negative $\Delta\Phi$ values are predicted upon adsorption of water from the gas phase. Data reported in the literature show that this is indeed the case. Data are available

Table 2. Calculated cpd for **Ag** main crystal faces

	$E_{\sigma=0}(SHE)/V$	Φ/eV	$E_{\sigma=0}(abs)/V$	$\Delta\psi/V$
(111)	−0.46	4.42	4.00	−0.42
(100)	−0.62	4.28	3.84	−0.45
(110)	−0.73	4.19	3.72	−0.47

only for the (110) face of Ag [42], so the sequence of the various faces cannot be discussed. However, it ensues from the preceding discussion that the main contribution to $\Delta\Phi$ is probably $\delta\chi^M$ rather than $g^S(dip)_0$.

It is remarkable that the experimental $\Delta\Phi$ is higher than the $(E_{\sigma=0}(abs) - \Phi)$ value. However, if the data for the (110) face of Ag and Cu are compared, Table 3 shows that the experimental $\Delta\Phi$ is ca. 0.2 eV higher for Cu, as it is the case for $(E_{\sigma=0}(abs) - \Phi)$. Thus, the same difference of $\Delta\Phi$ is observed between Ag and Cu, although from quantitatively different original values. Several reasons can be invoked; none can be definitely proven for the time being. The main point may be that the values of $\Delta\Phi$ as derived from $E_{\sigma=0}(abs)$ entail that the liquid layer on the metal has the features of a bulk phase [13]. It is hard to say if this is the case for the water adsorbed from the gas phase at very low temperature (<130 K). If not, the term missing would be metal independent as observed.

Although no complete $\Delta\Phi$ data are available to check the crystal face sequence, TDS measurements have shown [23] that while for Ag T_{des} ≈ 150K for the three main faces, ie the metal/solvent interaction is weaker than the solvent-solvent interaction (which means that the surface orientation of the solvent is governed by the liquid phase rather than by the heterogeneous interactions. However, electrical effects are much more sensitive than chemical effects), T_{des} ≈ 150 K also for the (111) and the (100) face of Cu, but the value goes up to 175 K for the (110) face of Cu. These data suggest that the metal-water interaction increases in the series (111) < (100) < (110) and that only on Cu (110) the strength of interaction exceeds the water-water interaction in frozen water so as to be detected by TDS. Thus, TDS corroborates the ΔX values derived from Fig. 7.

Table 3. $\Delta\Phi/eV$ due to water adsorption

	(a)	(b)
Ag(110)	-0.42	-0.70
Cu(110)	-0.65	-0.95

(a) Calculated from the potential of zero charge.

(b) Measured experimentally [42].

2.7. Single Crystal Faces in Non-aqueous Solvents

There exists only one set of data for single crystal faces of *sd*-metals in DMSO [27]. $E_{\sigma=0}$ values for Au have been found to vary the other way round as a function of the crystallographic orientation, *viz* (111) < (100) < (110). Thus, if plotted *vs* the crystallographic orientation (Fig. 9), these data show that the parallelism between $E_{\sigma=0}$ and γ on one side, and $E_{\sigma=0}$ and Φ on the other side fails to show up, so that the pattern of Fig. 6 may not be a general case.

It is intriguing that the pattern for Au in DMSO in Fig. 9 is exactly the negative image of the pattern in water. Thus, the crystal

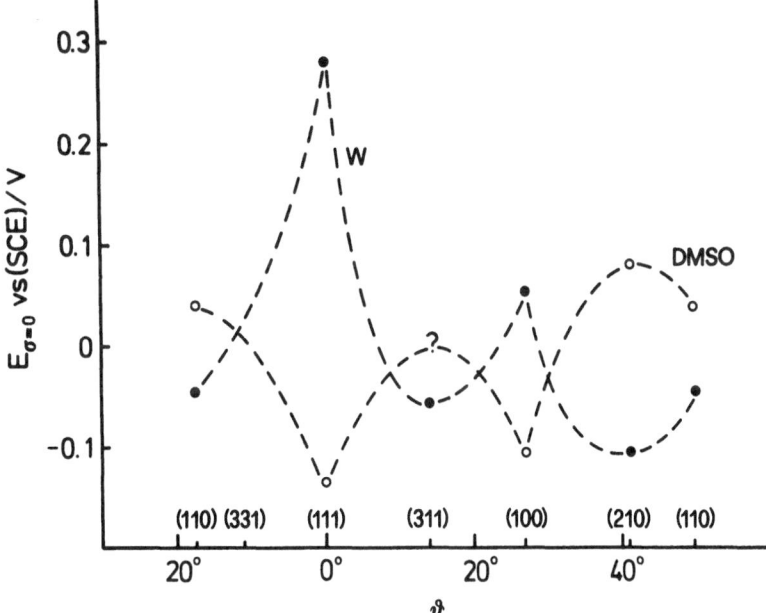

Figure 9. Variation of the potential of zero charge of Au single crystal faces with the crystallographic orientation in water (●) and in DMSO (O). (?) No data yet available.

face structure still arranges the faces in a regular sequential order, but now the main quantitative contribution must come from the solvent, $g^S(dip)_0$. The regularity in the crystal face sequence can be appreciated better in Fig.10 where $E_{\sigma=0}$ in water has been plotted *vs* $E_{\sigma=0}$ in DMSO. Despite the few points and some scatter, a linear correlation can be recognized. The slope is ca. 0.5, *ie* the effect of DMSO dipoles over-compensates by 50% the variation in work function. This also means that the effect varies in the sequence (111) > (100) > (110). Why it is so is not yet clear and needs further experimental verification, especially in view of the "regular" position of Au in Fig. 4. Some apparent inconsis-tency of the values of $E_{\sigma=0}$ for the polycrystalline Au surface can in fact be noticed between the data in two different papers [26,27]. DMSO adsorbs from aqueous solutions on polyscrystalline Au more strongly than on Hg [43]. However, the details of the mechanism of adsorption have not been clarified.

2.8. Temperature Coefficient of $E_{\sigma=0}$

Further insight into the factors responsible for the interfacial structure can be gained from the analysis of the temperature coefficient of $E_{\sigma=0}$. From eqn.(3) it follows that

$$dE_{\sigma=0}/dT = d\Phi'/dT - dg^S(dip)_0/dT . \qquad (9)$$

Equation (9) shows that the experimental temperature coefficient of $E_{\sigma=0}$

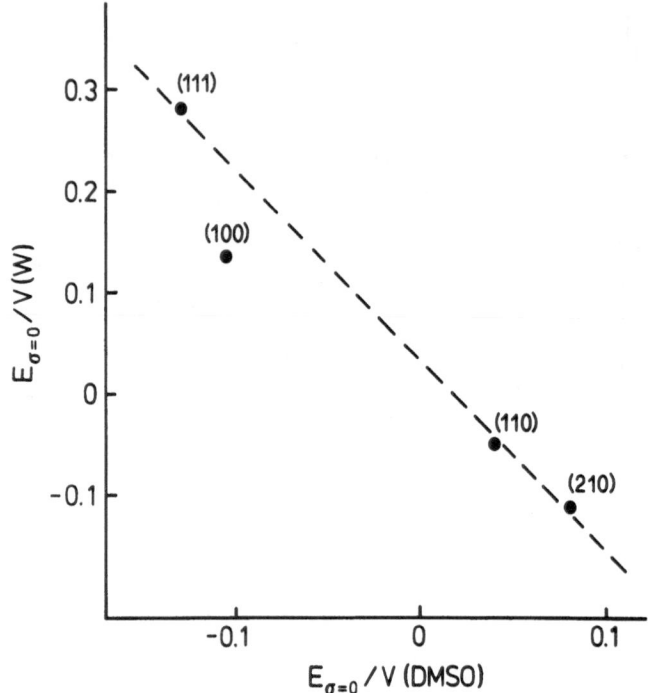

Figure 10. Plot of the potential of zero charge in aqueous solution *vs* the potential of zero charge in DMSO solution for Au single crystal faces. (-----) Line of unit slope.

contains two contributions, one from the metal surface electrons and the other one from the interfacial solution [44]. Values of $d\Phi'/dT$ can be calculated, but it is clear from eqn. (4) that this coefficient does not coincide with the experimentally measurable $d\Phi/dT$. Values of the latter term are available for a number of metals and they appear to be usually negative and comparable with $dE_{\sigma=0}/dT$ [45].

It may be interesting to investigate whether $dE_{\sigma=0}/dT$ is determined by the properties of the interfacial solvent or by those of the metal surface. If the metal contribution is neglected, a positive value of $dE_{\sigma=0}/dT$ is normally predicted on the basis of a simple model of dipoles oriented with the negative end towards the metal and disorientated by the increasing temperature. Thus, $dE_{\sigma=0}/dT$ would parallel $g^S(dip)_0$, in the sense that a more positive value would imply a higher metal-solvent interaction (*ie* a higher value of $g^S(dip)_0$).

The above conceptual approach is however only a rough approximation. The real situation is much more complex. For the sake of simplification, various solvents can be used with the same metal so as to keep the contribution from $d\Phi/dT$ *in principle* constant. This is possible with Hg, for which the estimate of $g^S(dip)_0$ has also been accomplished [25]. Fig. 11 shows that indeed $dE_{\sigma=0}/dT$ increases with $g^S(dip)_0$. However, negative values have also been observed [46] with solvents for which the $g^S(dip)_0$ value has been estimated to be negative, though very small.

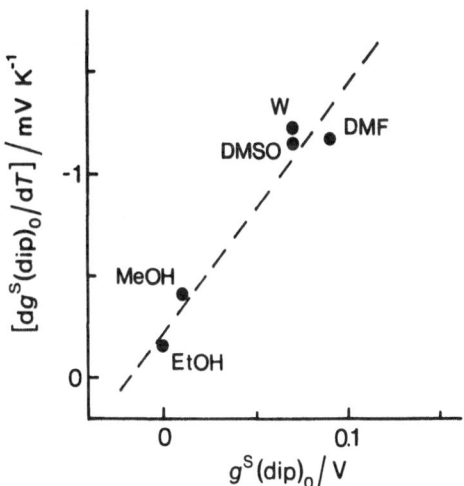

Figure 11. Temperature coefficient of the potential of zero charge of Hg in the indicated solvents as a function of the *estimated* interfacial potential of the solvents.

Thus, negative temperature coefficients can be observed even with negative values of $g^S(dip)_0$.

No complete analysis is possible for the whole group of *sp*-metals in water since only two data exist: 0.57 mV K^{-1} for Hg [47] and 0.15 mV K^{-1} for Cd [48]. From these one could draw the conclusion that more hydrophilic metals (Cd) exhibit lower temperature coefficients of $E_{\sigma=0}$, but this conclusion is also rough since we do not know if the two metals fall in a homogeneous group. Another conclusion could be that $dE_{\sigma=0}/dT$

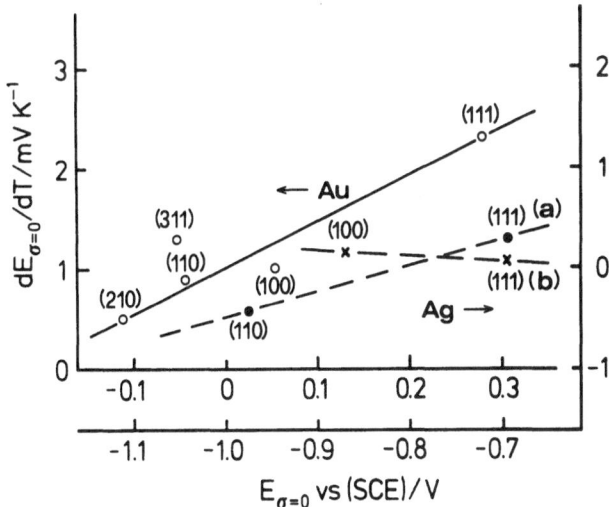

Figure 12. Temperature coefficient of the potential of zero charge as a function of the potential of zero charge. (O) Au [49]; (●) Ag [52,53]; (x) Ag [54].

parallels Φ, in the sense that the higher Φ the higher the measured temperature coefficient of $E_{\sigma=0}$. In other words, the Φ term prevails over the $g^S(dip)_0$ term.

A set of $dE_{\sigma=0}/dT$ has recently been measured for Au single crystal faces [49]. They have been found to parallel the potentials of zero charge. This is shown in Fig. 12 where $dE_{\sigma=0}/dT$ is plotted against $E_{\sigma=0}$. With the exception of the (311) face whose temperature coefficient is scattered, the other points appear to fall on a straight line. Thus, at first sight, the electron work function is probably the main factor governing not only the sequence of $E_{\sigma=0}$ but also its temperature coefficient. On the other hand, we have seen in Fig. 7 that ΔX is larger for the (110) than for the (111) face. Thus, the correlation observed in Fig. 11 with Hg is not valid here.

The effect of dΦ/dT is difficult to be accounted for quantitatively. Calculations for Ag and Cu single crystal faces [50], and experiments [51], show that dΦ/dT is negative with (110) > (100) > (111) (absolute value). It is rasonable to argue that the value of dΦ'/dT might overcompensate the value of $dg^S(dip)_0/dT$ and be responsible for the observed sequence. Fig. 12 shows also some data for Ag single crystal faces. The temperature coefficient is positive for Ag (111) [52] and negative for Ag (110) [53]. In view of the larger values of dΦ/dT calculated for Ag than for Cu, and of the weaker interaction of Ag with water, the observation of a negative temperature coefficient for the (110) face is not unreasonable.

If plotted *vs* the crystallographic orientation, $dE_{\sigma=0}/dT$ would exhibit the same pattern as $E_{\sigma=0}$. Since a more positive value of $E_{\sigma=0}$ does not imply *per se* a higher hydrophilicity of the surface, thus a more positive value of $dE_{\sigma=0}/dT$ does not necessarily entail a stronger metal-water interaction.

Popov *et al.* [54] have measured the temperature coefficient of $E_{\sigma=0}$ for Ag (100) and (111). They have found positive values but (100) > (111), a reverse sequence. The two sets of data for Ag in Fig. 12 refer to differently prepared single crystal surfaces. Other results discussed later on will show that the sequence of Ag faces obtained from the two types of crystal surfaces is always opposite. The reason for that has still to be clarified.

2.9. Interfacial Permittivity

A metal/solution interface behaves as an electric condenser; its capacitance (C) constitutes an additional experimental parameter to gain insight into the structure and the properties of the interface. Since, by definition, $C = d\sigma/dE$, the expression of the electrode potential must be completed to account for the presence of free charges. Thus, eqn. (3) becomes [55]:

$$E_\sigma = \Phi' - g^S(dip)_\sigma + g(ion)_\sigma + const ,\qquad (10)$$

where $g(ion)_\sigma$ is the potential drop associated with the presence of free charges at constant solvent orientation. Differentiating with respect to charge density:

$$dE/d\sigma = 1/C = d(\chi^M + \delta\chi^M)/d\sigma - dg^S(dip)/d\sigma + dg(ion)/d\sigma \; . \quad (11)$$

$$\quad\quad\quad\quad\quad\quad\quad I \quad\quad\quad\quad\quad\quad\quad\quad II \quad\quad\quad\quad\quad III$$

Equation (11) shows that the interfacial permittivity includes contributions from the electron gas of the metal surface (I), the reorientation polarizability of solvent molecules (which is expected to be influenced by the metal-water interaction) (II), and the electronic and atomic (molecular) polarizability of the same molecules (III). Therefore, the measured capacitance can be resolved into three capacitors in series which cannot be separated either physically or spatially:

$$1/C = 1/C_M + 1/C_{or} + 1/C_{mol} \; . \quad\quad\quad\quad\quad\quad\quad\quad\quad\quad (12)$$

Although the existence of C_M was perceived long ago [56], this term has often been conceptually neglected so that all the variation of the measured C has often been attributed to $1/C_{or}$, *viz* to the term governed by the metal-water interaction. Thus, C at $E_{\sigma=0}$ has been regarded as a parameter proportional to the hydrophilicity [1,57,58]. Fig. 13 shows in fact that C_0 (C at $E_{\sigma=0}$) increases monotonically with the increase of the parameter $-\Delta_f H^0$, the metal-water chemical interaction term. In such a plot the points for Ag and Au (here the (110) face in the absence of specific adsorption) are seen to fall very far from the common line.

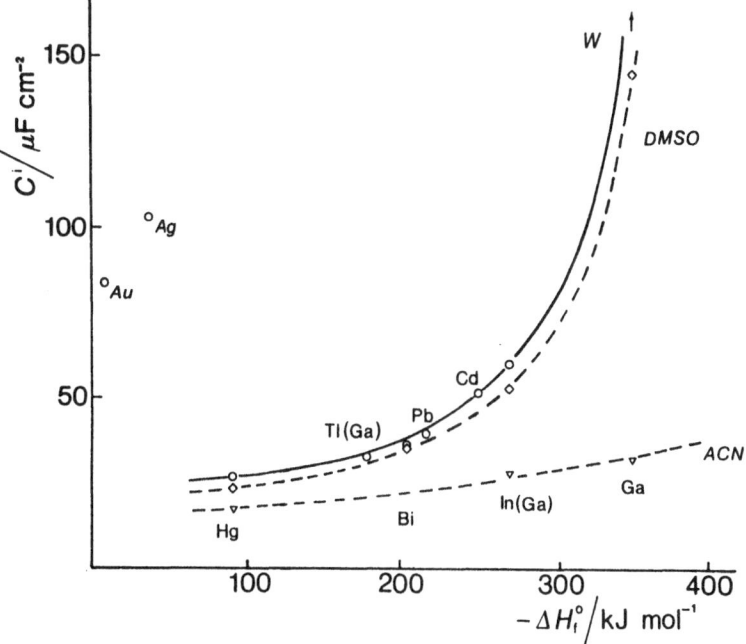

Figure 13. Inner layer capacitance at the potential of zero charge for polycrystalline *sp*-metals and the (110) face of *sd*-metals in water (o) DMSO (◇) and ACN (▽), respectively, against the enthalpy of formation of the oxide MO.

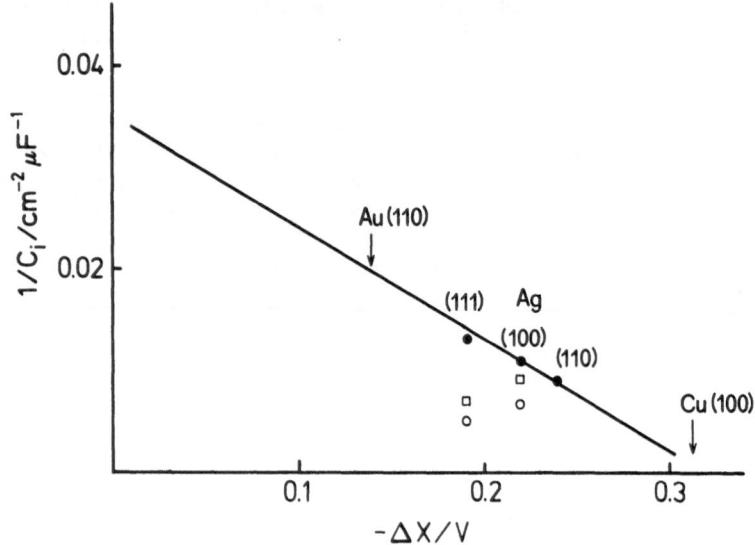

Figure 14. (———) Line resulting from a plot of the reciprocal of the inner layer capacitance from Fig. 13 *vs* ΔX from Figs. 1 and 7. (●) From [59]; (O) From [60]; (□) From [61]. The approximate positions for Au (110) and Cu (100) are also indicated.

This suggests that most of the interfacial permittivity is probably due to the surface electrons, consistently with the data in Fig. 2. Thus, it may be more appropriate to plot C as a function of ΔX.

The present author has found [55] that a strict linear correlation exists between $1/C$ and ΔX in the case of *sp*-metals. The experimental capacitance for the three main faces of Ag fit very nicely such a straight line. This is shown in Fig. 14. Thus, the values of ΔX derived from Fig.7 are consistent with the values of C_0 determined by Valette [59], whose sequence is (111) < (100) < (110), *ie* the measured C_0 parallels ΔX. In this respect, it is obvious that the same values of C_0 (measured by Valette) cannot fit the same line using the values of ΔX as estimated by Valette [36]. The same is the case for the values of C_0 as reported by Vitanov *et al.* [60]. They observed the opposite sequence, *viz* $C_0(111) > C_0(100)$ [61]. Since the measured $E_{\sigma=0}$ are the same as those of other authors, the ΔX values are also the same. Fig. 14 shows that the C_0 values of Vitanov *et al.* [60,61], besides not fitting the common curve, are even too high. This does not entail that they are unreliable. It means that differently prepared surfaces of Ag single crystals behave not only quantitatively but also qualitatively differently. This point needs to be further explored experimentally.

2.10. Adsorption of Neutral Compounds

The parameters discussed above are electrical in nature and represent the electrical response of the interface. They contain information about the orientation of dipoles and their polarizability but not directly

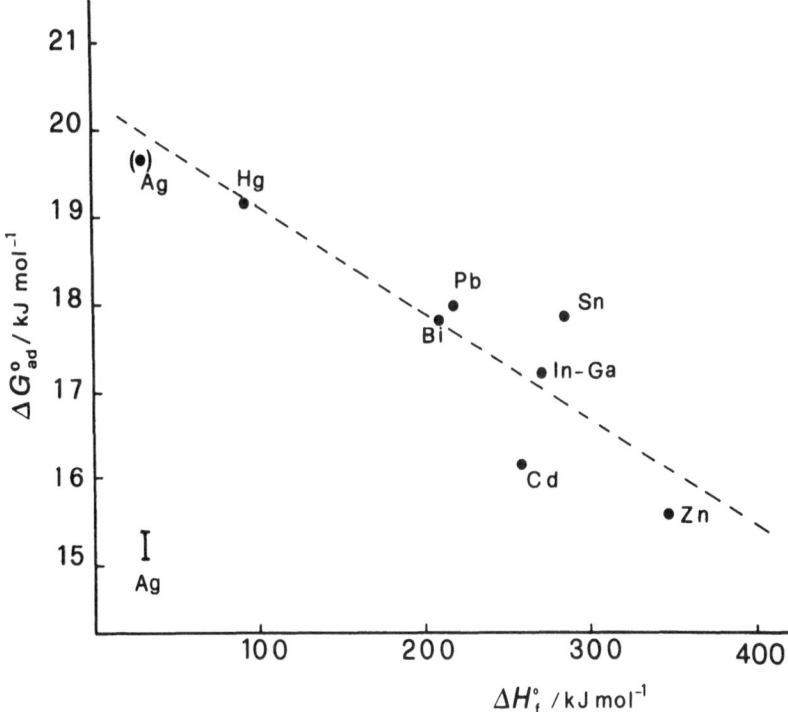

Figure 15. Gibbs energy of adsorption of *n*-amyl alcohol from aqueous solution as a function of the enthalpy of formation of the oxide MO. The two data points for Ag(100) are from [24] (I) and from [62], respectively.

about particle-particle interaction strengths. Some insight into the metal-water interaction strength can *in principle* be gained by investigating the adsorption of neutral compounds whose electrical and chemical interaction with metal surfaces can be assumed to be negligible [24].

Adsorption at an interface is regarded as a solvent replacement reaction. Therefore, the energy of adsorption of a compound from the solution contains also a contribution related to the desorption of solvent molecules. This approach has often been used in the literature to discuss the hydrophilicity series. However, the interpretation of the data is not straightforward in this case as well.

Fig. 15 shows a plot of $\Delta_a dG^0$ for *n*-pentanol adsorption from aqueous solutions onto various metals as a function of $\Delta_f H^0$ of the oxide MO. The plot shows a broad trend which suggests that the adsorption of *n*-pentanol decreases as $\Delta_f H^0$ becomes more negative, as a consequence of the increased difficulty for the adsorbate to displace solvent molecules. *n*-Pentanol adsorbs with the hydrocarbon chain facing the metal so that the interaction with the solid surface is expected to be minor and almost metal indipendent. It has been suggested [24] that the plot can represent the hydrophilicity scale increasing from Ag to Zn.

Recent results have shown that the position of Ag in Fig. 15 is probably false. The high $\Delta_a dG^0$ value measured earlier has been attributed to the presence of sulfate ions specifically adsorbed on Ag. If fluoride is used, a much lower value of $\Delta_a dG^0$ is observed [62]. The

266

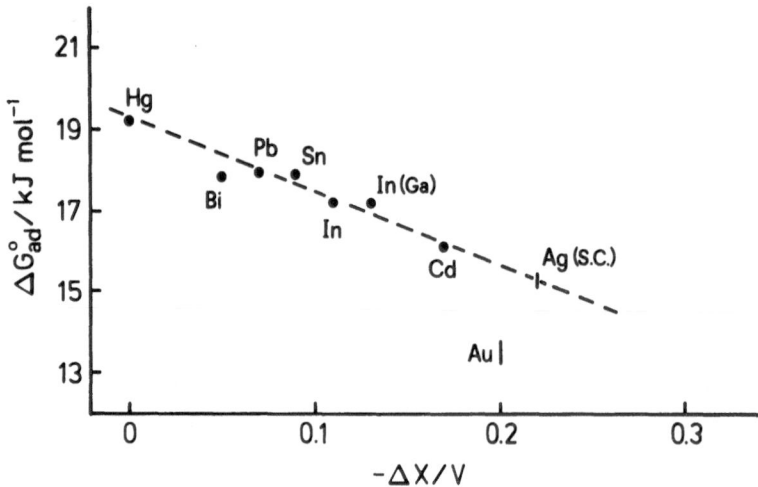

Figure 16. Gibbs energy of adsorption of *n*-amyl alcohol from aqueous solution as a function of ΔX from Fig. 1. The data for Au are for the polycrystalline surface [63,64]. Ag (100) from [62].

same is the case with Au: adsorption of *n*-pentanol on the polycrystalline surface [63,64] is observed with a lower $\Delta_{ad}G^0$ than for Ag. Another observation in connection with Fig. 15 is that the points are rather scattered, although it is to be considered that the experimental accuracy in these parameters cannot be very high.

Since metal-water interactions are measured *globally* by ΔX, $\Delta_{ad}G^0$ has been plotted vs ΔX in Fig. 16. It is intriguing that the plot turns out less scattered, and that the point for Ag now falls on the common line. The point of Au remains distant, but the estimated value of ΔX is less reliable since the electrode used was polycrystalline.

If the correlation in Fig. 16 possesses a general significance, in the case of single crystal faces $\Delta_{ad}G^0$ should increase with the value of ΔX; for FCC metals adsorption should be higher in the sequence (111) > (100) > (110) [32]. Experimental data are scanty. Since ΔX values are not readily available for all faces, $E_{\sigma=0}$ can be used as the correlating variable.

Fig. 17 shows that in the case of adsorption of diethylether on Au [65] and of cyclohexanol on Bi [66], $\Delta_{ad}G^0$ increases as the $E_{\sigma=0}$ becomes more positive, *ie* the trend is the expected one. Results for Pb, Zn and others for Bi exhibit the same trend [32]. Qualitatively, adsorption of ethylether on Ag appears to be stronger on the (111) than on the (110) face [67]. However, opposite trends have also been observed. *t*-Amyl alcohol has been reported [68,69] to adsorb on Au in the sequence (110) > (100). Adsorption has been observed to produce a potential shift towards more negative values on Au and towards more positive values on

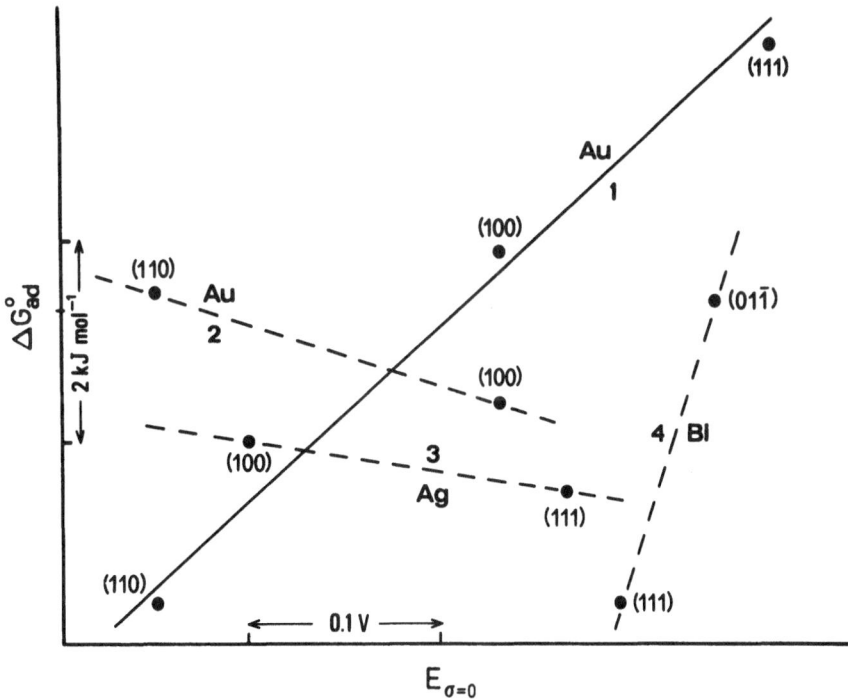

Figure 17. Variation of the Gibbs energy of adsorption of neutral compounds from aqueous solution with the crystallographic orientation. (1) From [65]; (2) From [68,69]; (3) From [62]; (4) From [66].

Hg [70]. Therefore, the interfacial orientation is different and apparently involves some specific interaction of the alcoholic group with the Au surface. Similarly, pyridine adsorbs slightly more on polycrystalline Au [71] than on the (100) face [72], but the interaction with the π-electrons of the ring may be involved in this case. A qualitative comparison suggests that adsorption of pyridine on Ag [73] is higher than on Au. Finally, Popov *et al.* [62] have reported that in KF solutions the surface activity of *n*-amyl alcohol is higher on the (111) than on the (100) faces (*cf.* Fig. 17). The authors conclude that the (111) face is more hydrophilic than the (100). However, the entire picture emerging from the results with Vitanov and Popov's electrodes is opposite to that with Hamelin and Valette's electrodes. More experimental investigations are therefore needed to discriminate among the various factors.

3. CONCLUSIONS

The correlative analysis carried out in this paper shows that the interfacial behaviour of *sp-* and *sd*-metals follows regularities that can be rationalized in terms of metal-water interaction. The phenomenological parameter ΔX, derived by comparing the potential of zero charge at the metal/solvent interface with the electron work function at the metal/vacuum interface, can arrange a number of experimental quantities for several metals in regular sequences showing internal consistency. ΔX, whose physical meaning is the deviation of the interfacial structure for the two phases in contact from the situation typical of the two separate phases, is found to increase as Φ decreases (and $E_{\sigma=0}$ becomes more negative). This is also the case for the various single crystal faces of a given metal, for which ΔX appears to increase as the atomic packing of the surface decreases.

The approach based on the $E_{\sigma=0}$ *vs* Φ plot leads to further predictions. The inner layer capacitance of *sp*-metals at $\sigma = 0$ is found to increase as ΔX increases, which predicts for f.c.c. metals that the capacitance should increase in the order (111) < (100) < (110). This is found experimentally [59]. However, the opposite behaviour observed with crystal faces grown in Teflon capillaries [60] suggests that great attention is to be paid to the preparation of the surface. More experiments are needed to assess this aspect.

While $E_{\sigma=0}$ varies with the crystallographic orientation showing the same pattern as the surface tension and the electron work function, the temperature coefficient of $E_{\sigma=0}$ is observed to follow the same behaviour. This indicates that the surface structure of the metal rather than the interfacial properties of the solvent is responsible for the crystal face sequence. The possibility that the temperature coefficient of Φ outweighs that of $g^S(dip)_0$ is also pointed out, and probably verified with Hg in some non-aqueous solvents [25].

Finally, while ΔX is an electrical parameter not directly measuring the metal-water interaction strength, insight into energetics can be gained by using neutral adsorbates as probes of the state of water at the interface. A linear inverse correlation between the energy of adsorption of neutral compounds and ΔX suggests that water can be displaced from the surface of metals with low values of ΔX more easily. Evidence has been given that this is probably the case of metal single crystal faces. However, the paucity of experimental data calls for further detailed investigations.

Au shows very intriguing results. In DMSO the sequence of $E_{\sigma=0}$ with the crystallographic orientation is utterly (and systematically) reversed, a unique case in the present literature. Moreover, the metal exhibits adsorption parameters for pyridine and alcohols that upsets the expected behaviour as well. This draws particular attention to this metal which seems to be at the border between two limiting behaviours.

Despite the extensive use of *in situ* and *ex situ* physical techniques, classical thermodynamic approaches still provide the major wealth of experimental data. The slow-down of the research with *sp*-metals after Prof. Frumkin's passing, has been compensated by investigations with *sd*-metals, particularly Au and Ag. But in this very field the available results raise questions calling for further, more aimed, experimental work. This includes adsorption of neutral compounds, capacitance in non-aqueous solvents, and temperature coefficients of interfacial properties. The time of surface thermodynamics has not yet expired for electrochemical interfaces.

Acknowledgements. Financial support to this work from the Ministry for University and Scientific Research(40% funds) and C.N.R. (Rome) is gratefully acknowledged.

4. REFERENCES

1. Trasatti, S. (1979) 'Solvent adsorption and double-layer potential drop at electrodes', in B. E. Conway and J. O'M. Bockris (eds.), *Modern Aspects of Electrochemistry*, Vol. 13, Plenum Press, pp. 81-206.
2. Trasatti, S. (1975) 'A discussion of the energies involved in specific adsorption of ions', *J. Electroanal. Chem.* **65**, 815-829.
3. Levi, M. D., Shlepakov, A. V., Damaskin, B. B. and Bagotskaya, I. A. (1982) 'Metal-solvent interaction and ion adsorption. Influence of solvent and metal nature', *J. Electroanal. Chem.* **138**, 1-27.
4. Amadelli, R., Daghetti, A., Vergano, L., De Battisti, A. and Trasatti S. (1979), 'Adsorption of butyronitrile at the Hg/aqueous solution interface. Evidence for field dependent solvent-adsorbate surface interaction', *J. Electroanal. Chem.* **100**, 379-393.
5. Nikitas, P. (1986) 'A two-state solvent model for rigid rod adsorption at charge interfaces', *J. Electroanal. Chem.* **197**, 29-47.
6. Sangaranarayanan, M. V. and Rangarajan, S. K. (1984) 'Adsorption isotherms: microscopic modelling', *J. Electrocanal. Chem.* **176**, 119-137.
7. Trasatti, S. (1981) 'Effect of the nature of the metal on the dielectric properties of polar liquids at the interface with electrodes. A phenomenological approach', *J. Electroanal. Chem.* **123**, 121-139.
8. M. P. Soriaga (ed.) (1988) *Electrochemical Surface Science (ACS Symposium Series 378)*, American Chemical Society, Washington, DC.
9. Kötz, E. R., Neft, H. and Müller, K. (1986) 'A UPS, XPS and work function study of emersed silver, platinum and gold electrodes', *J. Electroanal. Chem.* **215**, 331-344.
10. Sass, J. K., Schott, J. and Lacky, D. (1990) 'Prospects for the direct determination of the dielectric constant of water in the double layer using model adsorption experiments in UHV: co-adsorption of Cs and H_2O on Cu(110)', *J. Electroanal. Chem.* **283**, 441-448.
11. Southampton Electrochemistry Group (1985) *Instrumental Methods in Electrochemistry*, Hellis Horwood, Chichester.
12. R. J. Gale (ed.) (1988) *Spectroelectrochemistry. Theory and Practice*, Plenum Press, New York.
13. Trasatti, S. (1991) 'Electrode potentials and energy scales', this book.
14. Trasatti, S. (1971) 'Work function, electronegativity, and electrochemical behaviour of metals. II. Potentials of zero charge and 'electrochemical' work function', *J. Electroanal. Chem.* **33**, 351-378.
15. Trasatti, S. (1987) 'Progress in the understanding of the structure of the metal electrode/solution interface. Evolution of the concept of hydrophilicity', *Croat. Chem. Acta* **60**, 351-378.

16. Parsons, R. (1954) 'Equilibrium properties of electrified inter-
    faces' in J. O'M Bockris (ed.), *Modern Aspects of Electrochemistry*,
    Butterworth, London, pp. 103-179.
17. Frumkin, A. N. (1979) *Potentials of Zero Charge*, Nauka, Moscow.
18. Trasatti, S. (1986) 'Potential of zero charge. What they suggest
    about the structure of the interfacial region', in A. F. Silva
    (ed.), *Trends in Interfacial Electrochemistry*, D. Reidel,
    Dordrecht, pp. 25-48.
19. Badiali, J. P., Rosinberg, M. L. and Goodisman, J. (1981) 'Effect
    of solvent on properties of the liquid metal surface', *J. Electro-
    anal. Chem.* **130**, 31-45.
20. Schmickler, W. (1983) 'The potential of zero charge of jellium',
    Chem. Phys. Lett. **99**, 135-139.
21. Frumkin, A., Damaskin, B., Grigoryev, N. and Bagotskaya, I. (1974)
    'Potentials of zero charge, interaction of metals with water and
    adsorption of organic substances. I. Potentials of zero charge and
    hydrophilicity of metals', *Electrochim. Acta* **19**, 69-74.
22. Trasatti, S. (1974) 'Interaction of water with metal surfaces. A
    theory on the role of the solid surface', *J. Electroanal. Chem.* **54**,
    437-441.
23. Thiel, P. A. and Madey, T. E. (1987) 'The interaction of water with
    solid surfaces: fundamental aspects', *Surf. Sci. Rep.* **7**, 211-385.
24. Trasatti, S. (1983) 'Physical, chemical and structural aspetcs of
    the electrode/solution interface', *Electrochim. Acta* **28**, 1083-1093.
25. Trasatti, S. (1987) 'Interfacial behaviour of non-aqueous
    solvents', *Electrochim. Acta* **32**, 843-850.
26. Jarzabek, G. and Borkowska, Z. (1987) 'Electrochemical study of the
    polycrystalline gold electrode in dimethylsulphoxide', *J. Electro-
    anal. Chem.* **226**, 295-303.
27. Borkowska, Z. and Hamelin, A. (1988) 'The influence of the crystal-
    lographic orientation on the double layer parameters of the Au/
    dimethylsulphoxide interface', *J. Electroanal. Chem.* **241**, 373-377.
28. Jaworski, J. S. (1989) 'Solvent effects on the Volta potential
    difference at the uncharged metal-solvent interface', *Electrochim.
    Acta* **34**, 485-487.
29. Lecoeur, J. Andro, J. and Parsons, R. (1982) 'The behaviour of
    water at stepped surfaces of single crystal gold electrodes', *Surf.
    Sci.* **144**, 320-330.
30. Hamelin, A. Stoicoviciu, L., Doubova, L. and Trasatti, S. (1988)
    'Influence of the crystallographic orientation of the surface on
    the potential of zero charge of silver electrodes', *Surf. Sci.* **201**,
    L498-L506.
31. Bacchetta, M., Trasatti, S., Doubova, L. and Hamelin, A. (1986)
    'The dependence of the potential of zero charge of silver
    electrodes on the crystallographic orientation of the surface', *J.
    Electroanal. Chem.* **200**, 389-396.
32. Trasatti, S. (1985) 'Crystal face specificity of double-layer
    structure and electrocatalysis', *Mat. Chem. Phys.* **12**, 507-527.
33. Valette, G. (1984) 'Surface potential variations at the water/
    single-crystal of silver interface', *J. Electroanal. Chem.* **178**,
    179-183.

34. Trasatti, S. (1984) 'Prediction of double layer parameters. The case of silver', *J. Electroanal. Chem.* **172**, 27-48.

35. Valette, G. (1982) 'Hydrophilicity of metal surfaces. Silver, gold and copper electrodes', *J. Electroanal. Chem.* **139**, 285-301.

36. Valette, G. (1987) 'Silver-water interactions. Part I. Model of the inner layer at the metal/water interface', *J. Electroanal. Chem.* **230**, 189-204.

37. Wandelt, K. (1989), *Croat. Chem. Acta*, in the press (personal communication).

38. Wandelt, K. (1987) 'The local work function of thin metal films: definition and measurement', in P. Wißmann (ed.), *Thin Metal Films and Gas Chemisorption*, Elsevier, Amsterdam, pp. 280-368.

39. Eickmans, J., Otto, A. and Goldmann, A. (1986) 'On the annealing of SERS active Ag films: UPS and TDS of adsorbed Xe', *Surf. Sci.* **171**, 415-441.

40. Heras, J. M. and Viscido, L. (1980) 'Work function changes upon water contamination of metal surfaces', *Appl. Surf. Sci.* **4**, 238-241.

41. Heras, J. M. and Viscido, L. (1988) 'The behaviour of water on metal surfaces', *Catal. Rev.-Sci. Eng.* **30**, 281-338.

42. Stuve, E. M. Bange, K. and Sass, J. K. (1986) 'Surface science model studies of the electrochemical interface' in A. F. Silva (ed.), *Trends in Interfacial Electrochemistry*, D. Reidel, Dordrecht, pp. 255-280.

43. Jarzabek, G. and Borkowska, Z. (1988) 'Adsorption of DMSO on the polycrystalline Au electrode from aqueous solutions', *J. Electroanal. Chem.* **248**, 399-410.

44. Trasatti, S. (1977) 'The temperature coefficient of the water dipole contribution to the electrode potential', *J. Electroanal. Chem.* **82**, 391-402.

45. Trasatti, S. (1978) 'Temperature coefficient of Hg work function from electrochemical data', *Appl. Surf. Sci.* **1**, 341-346.

46. Borkowska, Z., Fawcett, W. R. and Anantawan, S. (1980) 'Temperature dependence of the surface potential at the Hg/non-aqueous solvent interface', *J. Phys. Chem.* **84**, 2769-2774.

47. Randles, J. E. B. and Whiteley , K. S. (1956) 'The temperature dependence of the electrocapillary maximum of Hg', *Trans. Faraday Soc.* **52**, 1509-1512.

48. Rybalka, K. V. and Panin, V. A. (1973) 'Influence of temperature on the properties of the electrical double layer on Cd', *Elektrokhimiya* **9**, 172-176.

49. Silva, F., Sottomayor, M. J. and Hamelin, A. (1990) 'The temperature coefficient of the potential of zero charge of the Au single-crystal electrode/aqueous solution interface. Possible relevance to Au-water interactions', *J. Electroanal. Chem.* **294**, 239-251.

50. Kiejna, A. (1986) 'On the temperature dependence of the work function', *Surf. Sci.* **178**, 349-358.

51. Hölzl, J. and Schulte, F. K. (1979) 'Work function of metals', *Springer Tracts Mod. Phys.* **85**, 1-150.

52. Hamelin, A. Doubova, L., Stoicoviciu, L. and Trasatti, S. (1988) 'The temperature dependence of the double layer parameters of the

(111) face of silver', *J. Electroanal. Chem.* **218**, 133-145.

53. Bacchetta, M. Francesconi, A., Trasatti, S., Doubova, L. and Hamelin, A. (1987) 'The temperature coefficient of the potential of zero charge of the (110) face of silver', *J. Electroanal. Chem.* **218**, 355-360.

54. Popov, A., Velev, O., Vitanov, T. and Tonchev, D (1988) 'Temperature dependence of some electric double-layer parameters of Ag single-crystal electrodes', *J. Electroanal. Chem.* **257**, 95-100.

55. Trasatti, S. (1978) 'Inner layer capacity in the absence of metal water interaction', *J. Electroanal. Chem.* **91**, 293-298.

56. Rice, O. K. (1928) 'Application of the Fermi statistics to the distribution of electrons under fields in metals and the theory of electrocapillarity', *Phys. Rev.* **31**, 1051-1059.

57. Damaskin, B. B. and Frumkin, A. N. (1974) 'Potentials of zero charge, interaction of metals with water and adsorption of organic substances. III. The role of the water dipoles in the structure of the dense part of the electrical double layer', *Electrochim. Acta* **19**, 173-176.

58. Bagotskaya, I. A. and Schleparov, A. V. (1980) 'Hydrophilicity series of weakly hydrogen-adsorbing metals, and the positions of the Au and Ag electrodes in this series', *Elektrokhimiya* **16**, 565-569.

59. Valette, G. (1987) 'Inner layer capacity at the pzc for perfect (111), (100) and (110) face of Ag. Surface area and capacitance contributions of superficial defects for real electrodes', *J. Electroanal. Chem.* **224**, 285-294.

60. Vitanov, T., Popov, A. and Sevastyanov, E. S. (1982) 'Electrical double layer on (111) and (100) faces of silver single crystal in solutions containing ClO_4^- and F^{-1}', *J. Electroanal. Chem.* **142**, 289-297.

61. Naneva, R., Bostanov, V., Popov, A. and Vitanov, T. (1989) 'Zero charge potential and structure of the electric double layer of the (0001) face of a Cd single crystal in NaF solution' *J. Electroanal. Chem.* **274**, 179-183.

62 Popov, A., Velev, O. and Vitanov, T. (1988) 'Adsorption of *n*-amyl alcohol on the (111) and (100) faces of Ag', *J. Electroanal. Chem.* **256**, 405-410.

63. Holze, R. and Beltowska-Brzezinska, M. (1985) 'On the adsorption of aliphatic alcohols on Au', *Electrochim. Acta* **30**, 937-939.

64. Tucceri, R. I. and Posadas, D. (1987) 'Capacitance and surface conductance study of the adsorption of pentan-1-ol on Au', *Electrochim. Acta* **32**, 27-31.

65. Lipkowski, J., Nguyen Van Huong, C., Hinnen, C. Parsons, R. and Chevalet, J. (1983) 'Adsorption of diethylether on single crystal Au electrodes. Calculation of adsorption parameters', *J. Electroanal. Chem.* **143**, 375-396.

66. Lust, E. J. and Palm, U. V. (1985) 'Adsorption of cyclohexanol on singular faces (001), ($1\overline{0}1$), and ($01\overline{1}$) of Bi single crystals', *Elektrokhimiya* **21**, 1381-1384.

67. Hinnen, C., Nguyen Van Huong, C. and Dalbera, J.-P. (1982) 'Etude comparative de l'adsorption du diethyl-ether sur les electrodes

monocristallines d'Au at d'Ag d'orientation (111), (100) et (110)', *J. Chim. Phys.* **79**, 37-43.

68. Richer, J., Stolberg, L. and Lipkowski, J. (1986) 'Quantitative investigations of adsorption of *t*-amyl alcohol at the Au (110)-aqueous solution interface', *Langmuir* **2**, 630-638.

69. Ricer, J. and Lipkowski, J. (1988) 'Quantitative investigations of the adsorption of *t*-amyl alcohol at the Au(100)/aqueous solution interface', *J. Electroanal. Chem.* **251**, 217-234.

70. Damaskin, B. B. and Grigoriev, N. B. (1962) 'The effect of potential on the attractive force between adsorbed organic molecules', *Dokl. Akad. Nauk SSSR* **147**, 135-137.

71. Stolberg, L., Richer, J. Lipkowski, J. and Irish, D. E. (1986) 'Adsorption of pyridine at the polycrystalline Au-solution interface', *J. Electroanal. Chem.* **207**, 213-234.

72. Stolberg, L., Lipkowski, J. and Irish, D. E. (1987) 'Adsorption of pyridine at the Au(100)-solution interface', *J. Electroanal. Chem.* **238**, 333-353.

73. Hamelin, A., Morin, S., Richer, J. and Lipkowski, J. (1989) 'Adsorption of pyridine on the (110) face of Ag', *J. Electroanal. Chem.* **272**, 241-252.

THE APPLICATION OF SCANNING TUNNELING MICROSCOPY TO ELECTROCHEMISTRY

D. M. KOLB * , R. J. NICHOLS
Fritz-Haber-Institut der Max-Planck-Gesellschaft
Faradayweg 4-6
D-1000 Berlin 33
Federal Republic of Germany

AND

R. J. BEHM
Institut für Kristallographie und Mineralogie,
Universität München
Theresienstr. 41
D-8000 München 2
Federal Republic of Germany

ABSTRACT. Scanning tunneling microscopy is becoming an increasingly popular technique for examining electrodes in-situ in the electrolyte solution. This chapter outlines the basic principles of STM and the special experimental considerations for operation in electrolyte. Four case studies, performed at our laboratories, are discussed, with the aim of providing an insight into the valuable information which can be obtained by using this technique.

1. Introduction

There has been hardly any technique in surface science in the last 10 or 20 years which has given rise to so much interest and has received so much attention as Scanning Tunneling Microscopy (STM). The presentation by Binnig and Rohrer of their first STM results at conferences during 1982 to 1983 led to an audience response ranging from frank disbelief to great enthusiasm. At these conferences they showed real-space images of surfaces which were obtained at an unprecedented, and in some ways unbelievably good resolution. Such an image is shown in figure 1, which is of a reconstructed Au(100) surface in UHV [1]. Atomic steps are clearly seen. Even a slight ripple of about 0.3 Å height and 12 Å breadth is apparent and this arises from the surface reconstruction, which causes the top layer to be buckled.

---

* *New permanent address: Dept. of Electrochemistry, University of Ulm,*
D-7900 Ulm, Federal Republic of Germany.

R. Guidelli (ed.), Electrified Interfaces in Physics, Chemistry and Biology, 275–292.
© 1992 *Kluwer Academic Publishers.*

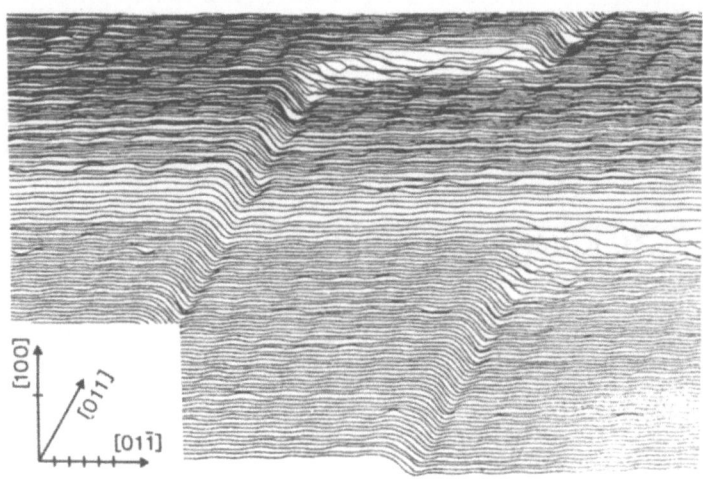

Figure 1. STM image of a clean (1x5) reconstructed Au(100) surface in vacuum, showing monoatomic steps. 5 Å per division [1].

The lateral resolution of the STM is much less than that normal to the surface. Nevertheless, this has been improved over the years to an extent that now even single atoms, on semiconductor or metal surfaces, can be imaged routinely in UHV. This is demonstrated in figure 2 for Al(111), where the individual atoms in the surface layer are clearly resolved [2].

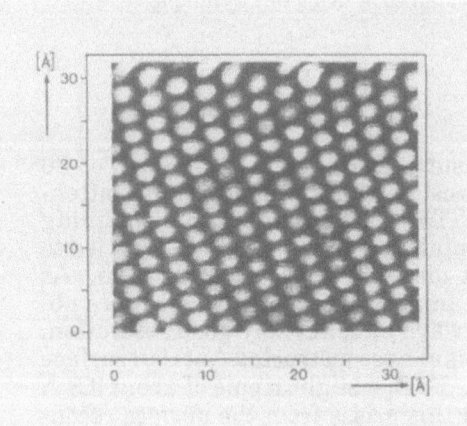

Figure 2. Grey scale representation of an STM image of the clean Al(111) surface in vacuum (34 Å x 34 Å, corrugation amplitude 0.3 Å, $V_t = -50\,\mathrm{mV}$, $I_t = 6\,\mathrm{nA}$) [2].

At a very early stage it was realized that this technique should also work in electrolyte [3], and hence the STM quickly arose the interest of electrochemists. Today it has been well established that the STM works in an electrochemical environment even with atomic-scale resolution. Consequently, the number of electrochemists who are using the technique, is increasing rapidly. Cataldi et al. have recently reviewed progress in the field of in-situ STM [4].

In this chapter we will first briefly recall the principle of STM and then describe the conditions necessary for an in-situ operation. Finally we will describe some studies performed by our groups in Berlin and Munich which demonstrate the potential of the technique.

2. Principles

The principle of STM is sketched in figure 3. When a metal tip is brought into close proximity of the surface under study, electrons will tunnel through the energy barrier, which is constituted by the gap. The tunnel current I_t will depend on the tunnel voltage V_t, on the barrier height ϕ_t (which is not identical with, but is related to the local work function ϕ - typical values for ϕ_t of metals in UHV are 3-4 eV), and on the separation s. For two planar electrodes the relationship between the tunnel current and these parameters is described by the Fowler-Nordheim equation:

$$I_t \simeq V_t \exp\left(-A\sqrt{\phi_t} \cdot s\right) \quad \text{with } A = 1.028 \text{ Å}^{-1} \text{ eV}^{-1/2}$$

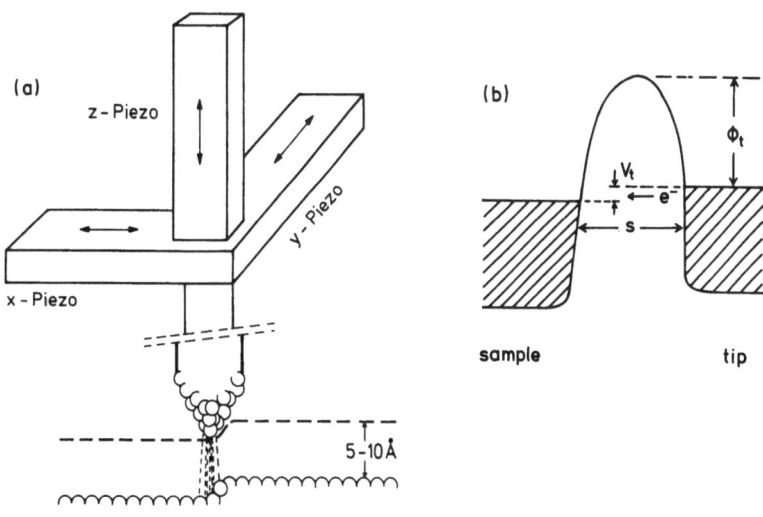

Figure 3. (a) Schematic diagram showing the principle of STM; (b) energy diagram for sample and tip in tunneling position. V_t: tunnel voltage; ϕ_t: tunnel barrier; s: tunnel gap.

Note that this equation is valid only for the low field case $V_t \ll 1$ V, while for larger voltages (several V) the so-called field-emission case applies. An important point to note is the exponential dependence of the tunnel current on the distance s, which forms the basis for the high resolution of the STM.

An image of the surface topography can be obtained if we scan the tip across the surface and operate the STM in either constant current or constant height mode, as illustrated in figure 4. In the first case the tunnel current is

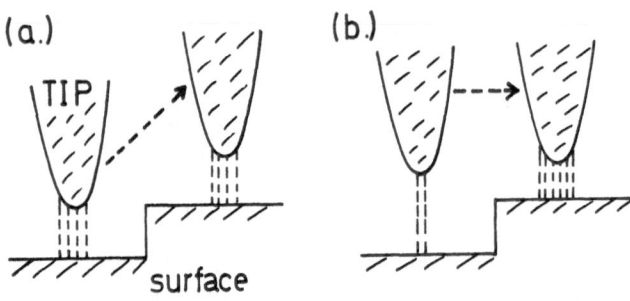

Figure 4. A schematic representation of the two different modes which can be used to produce STM images of the surface. (a) Constant current mode (z-feedback on); (b) constant height mode.

maintained at a constant value via a feed-back circuit which controls the vertical position of the tip by applying a voltage to the z-piezo element. Consequently the z-position of the tip changes as it scans across an atomic step and an image of the surface can be directly obtained from the z-piezo voltage as a function of the x, y-position of the tip. In contrast the constant height mode involves holding the z-position of the tip constant, with the surface topography being directly reflected in the change of the tunnel current. In most cases the first mode is used, where one follows lines of constant tunnel current. In both cases, however, the STM image represents electron density profiles, rather than actual geometric structure; but usually these are closely related to each other.

Figure 5 shows a schematic diagram of a so-called "pocket-size" STM, which we used throughout our work [5]. This microscope is constructed around a piezoelectric tripod assembly with the tip mounted on the z-piezo. The sample rests on a table, which can be moved up and down by a lever which is controlled by a micrometer screw (as depicted). This enables a course adjustment of the tip-to-sample distance. The fine approach is then facilitated by a gradual expansion of the z-piezoelectric crystal. This is controlled by a high voltage amplifier, with a ± 150 V bias voltage producing a displacement in the range of ± 1.5 μm. If the tip is still too far from the surface, even after full expansion of the piezo, then the piezo is contracted and the micrometer screw is moved by 1/8 of a turn (1 turn = 12 μm). This cycle, consisting of an approach using a

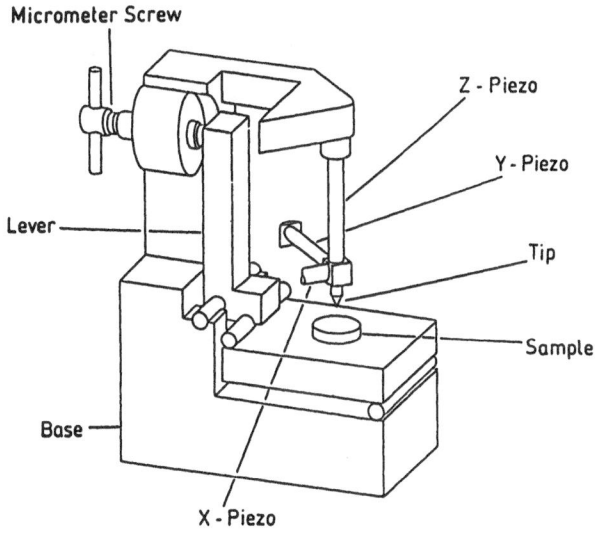

Figure 5. A schematic diagram of the type of STM which we have used throughout our work [5].

Figure 6. Detail of the Scanning Tunneling Microscope showing an Apiezon-wax coated tungsten tip and the electrochemical cell with a gold sample [6].

combination of piezoelectric and mechanical methods, is repeated until the tip comes to within the required distance from the surface, i.e. when the tunneling current reaches its selected value. At this point the feedback circuit comes into operation and maintains the tunneling current at a constant value. The microscope rests on a stack of copper plates, separated by Viton spacers which dampen out vibrations above 100 Hz very efficiently. The microscopy body has a resonance frequency of 5 kHz and hence it is efficiently decoupled from vibrations from the surroundings.

Figure 6 shows a close-up of the tip and electrochemical cell, a macor vessel with a gold single-crystal disc mounted onto the base. The rim of the macor potlet is gold plated and serves as a counter electrode. Various reference electrodes have been used including a hydrogen-loaded palladium wire or a glass capillary connected to an SCE reference electrode. For metal ion-containing electrolytes (e.g. Cu^{2+} ions in the study of copper deposition) the most suitable reference electrode system is simply a matching metal wire.

Since the tip inevitably acts as a fourth electrode, there will be an electrochemical current flowing at the tip-electrolyte interface, which is superimposed onto the tunnel current and cannot be distinguished from the latter by the feedback control. Therefore we must control the tip potential independently and we choose the potential such that the faradaic current at the tip is minimized. In addition a large portion of the tip is isolated to reduce the area which is in contact with the electrolyte. In our case the tip is covered with Apiezon (a thermoplastic wax), while others have used glass to seal the tip. This leaves

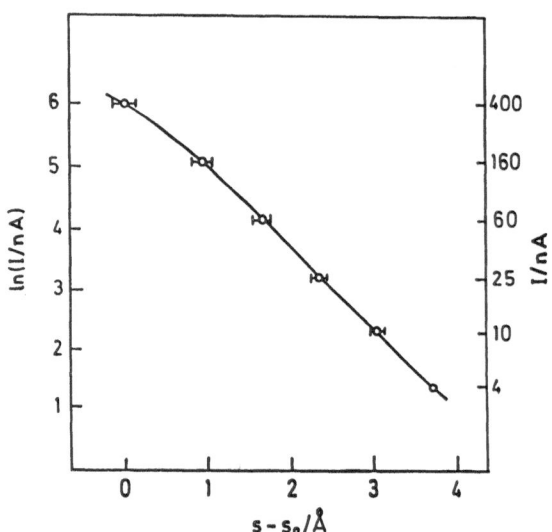

Figure 7. Current between tungsten tip and gold sample in 0.01 M Na_2SO_4 as a function of relative distance s-s_0. The exponential dependence of $I(s)$ proves that the measured current indeed results from a tunnel process. Voltage between tip and sample: 758 mV. Potential of the gold sample: +425 mV vs. Pd-H. A tunnel barrier of about 2.15 eV has been derived from the Fowler-Nordheim equation [6].

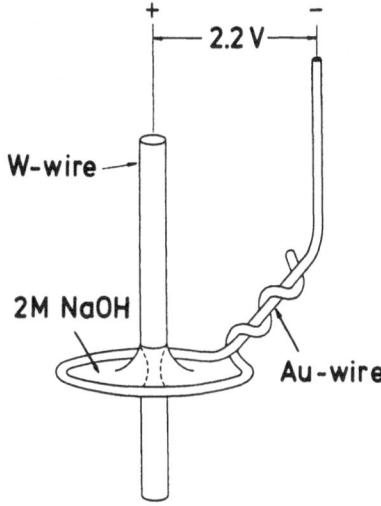

Figure 8. A pictorial representation showing the method used in producing tips by electrochemical etching of tungsten wire [7].

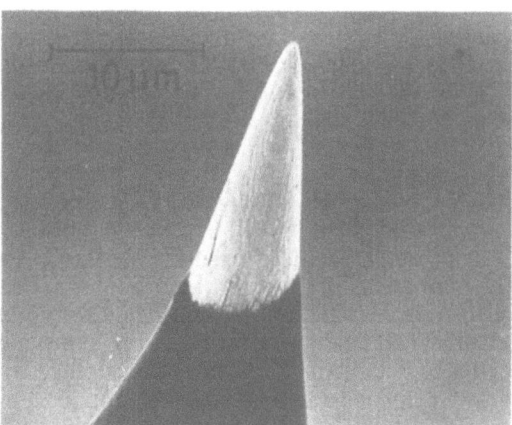

Figure 9. A scanning electron micrograph of a tungsten tip which has been coated with Apiezon wax. The tip radius is less than 2 μm.

typically an area of about 10^{-4} to 10^{-5} cm^2 (or even smaller) accessible to the electrolyte (see figure 9). With these precautions, control of the tip potential and reduction of the exposed area, the electrochemical current was generally below 100 pA, while tunneling currents were in the range of 1 - 250 nA. The graph in figure 7 shows the tip current in the tunnel position as a function of distance s under electrochemical conditions [6]. The logarithmic dependence of I on s confirms that the current indeed results from a tunnel process.

A wide variety of tip material has been used for in-situ STM measurements and these have included W, Ta, Au, Pt and Pt-Ir alloys. Most of our experiments have been conducted with tungsten tips, which are prepared by electrochemical etching, as depicted in figure 8 [7]. The tungsten tips were etched with a d.c. voltage of 2.2 V and a 2M NaOH electrolyte. Tips produced from 0.2 mm diameter wire took typically 10 min to etch. A scanning electron microscopy (SEM) image of such a tungsten tip, with a tip radius of less than 2 μm and already coated with Apiezon, is shown in figure 9. However, if the tip were geometrically smooth then even a tip radius of 100 Å would yield a 30 Å broad tunnel region. Since atomic resolution can be achieved then we must conclude that we have a microtip on top of the tip, with ideally only a single tip atom contributing to the tunneling process. Obviously, microtips for imaging atomic steps are produced more or less by accident. Maybe all that one needs is a rough tip surface, and the extreme distance dependence of the tunnel current causes just one microtip to determine the image. Special methods have been devised which enable high quality tips to be more consistently produced for UHV-STM studies. One method involves a kind of sputtering of gold clusters onto the tip [7]. In analogy, Cu^{2+} ions added to the electrolyte have greatly improved stability and lateral resolution of the tungsten tip in our experiments. Since the tip potential was always kept relatively negative, some Cu deposition onto the tip may have occurred. Nevertheless, tip stability can be a major problem in electrolytes, since such a microtip may be easily attacked by electrochemical reactions. For general purposes, the sharpness of atomic steps in the image can serve as a quick test for the lateral resolution.

3. Case Studies

3.1 AN STM STUDY OF FLAME-TREATED GOLD SURFACES

The flame-treatment is now routinely used for surface preparation by many electrochemists who deal with single crystal metal electrodes. This method involves heating the sample in a Bunsen flame and subsequently quenching it with triply distilled water. Further handling of the sample is done with a droplet of water adhering to it in order to protect the surface from contamination by the laboratory air. Figure 10 shows an STM image of a 1000 Å x 1000 Å area of a flame-treated Au(111) electrode in an aqueous electrolyte. Monoatomic high steps are clearly resolved and these are separated by atomically smooth (111) terraces, which often extend over many hundreds of Å. Such areas could be readily located and this testifies to the high quality of the flame-annealed gold surfaces.

The topography of flame-treated Au(100) electrodes is somewhat different. On this surface gold islands with a monoatomic height, are often observed, as is shown in figure 11. We believe that these islands result from the

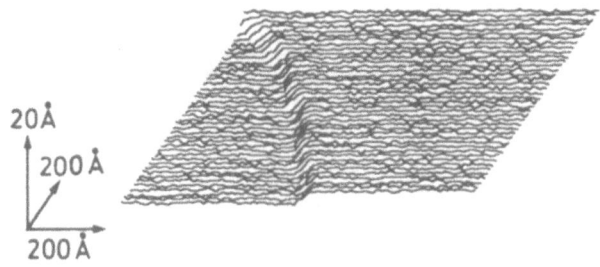

Figure 10. An STM image (3-D line scan representation) of a 1000 Å x 1000 Å area of a flame-treated Au(111) electrode at +400 mV (SCE) in 0.05 M H_2SO_4 + 5 mM HCl [8].

(5x20) → (1x1) structural transition [8]. The clean (flame-annealed) Au(100) possesses a (5x20) reconstructed form [9]. However, upon transfer of the electrode to the electrolyte and immersion without potentiostatic control the (5x20) surface reverts to the (1x1) structure [9]. The (5x20) surface has a hexagonal structure which is more densely packed than the square unit cell of the (1x1) surface. This expansion of the surface lattice, due to the (5x20) → (1x1) structural transition, leads to a 20 % excess of gold atoms on the surface, and these atoms presumably coalesce by surface diffusion to form the observed islands. Such a process is analogous to the (5x20) → (1x1) restructuring of Pt(100), occurring upon CO adsorption from the gas phase, which has been observed by STM in UHV studies [10].

Very recently it has been possible to obtain even lateral atomic resolution for electrode surfaces in-situ [11,12]. An example is given in figure 12 where the individual atoms of a (1x1) Au(100) surface are resolved, which have a nearest-neighbor distance of 2.9 Å. Such images could only be obtained with high tunneling currents, which corresponds to a very close proximity of electrode surface and tip. It should be noted that the observed atomic corrugation is rather high and may to a certain extent result from tip-surface interactions

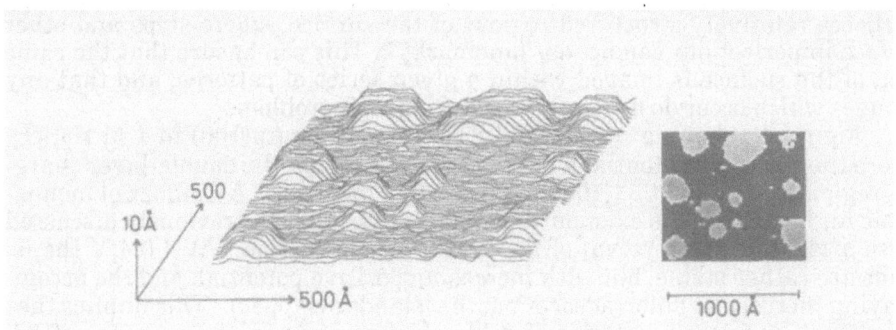

Figure 11. 3-D line scan (left) and grey scale (right) representations of a Au(100) surface in 1 M H_2SO_4 at +500 mV (SCE).

2 nm

Figure 12. Atomic-resolution STM image of the clean, unreconstructed Au(100) surface in 0.05 M H_2SO_4 + 5 mM $CuSO_4$ (atomic corrugation 0.5 Å, U_{tip} = -60 mV vs. SCE, I_t = 250 nA, U_{sample} = 0.3 V vs. SCE) [12].

rather than reflecting the corrugation of the electron charge density distribution in front of the surface [13].

3.2 ELECTROCHEMICALLY INDUCED CHANGES IN ELECTRODE SURFACE STRUCTURE

This section deals with electrochemically induced changes in the topography of gold surfaces. To be more precise, it describes changes which are caused by an anion-induced mobility of the surface atoms. For such STM studies it is useful to choose relatively structured regions of the surface, where steps and other surface imperfections can act as "landmarks". This can ensure that the same area of the surface is imaged within a given series of patterns, and that any changes which occur do not simply arise from drift problems.

Figure 13 shows a sequence of STM images for Au(100) in 1 M H_2SO_4, where the electrode potential has been changed within the double-layer charging region from +400 to +1000 mV (vs. SCE) and back. A number of monoatomic high gold islands exist on the upper terrace and as previously discussed these arise from the removal of the (5x20) reconstruction. At +0.4 V the islands are rather stable, but with increasing positive potential, and the accompanying increase in anion adsorption, the islands disappear. This implies that there is an increased mobility of gold atoms at steps in the presence of adsorbed anions.

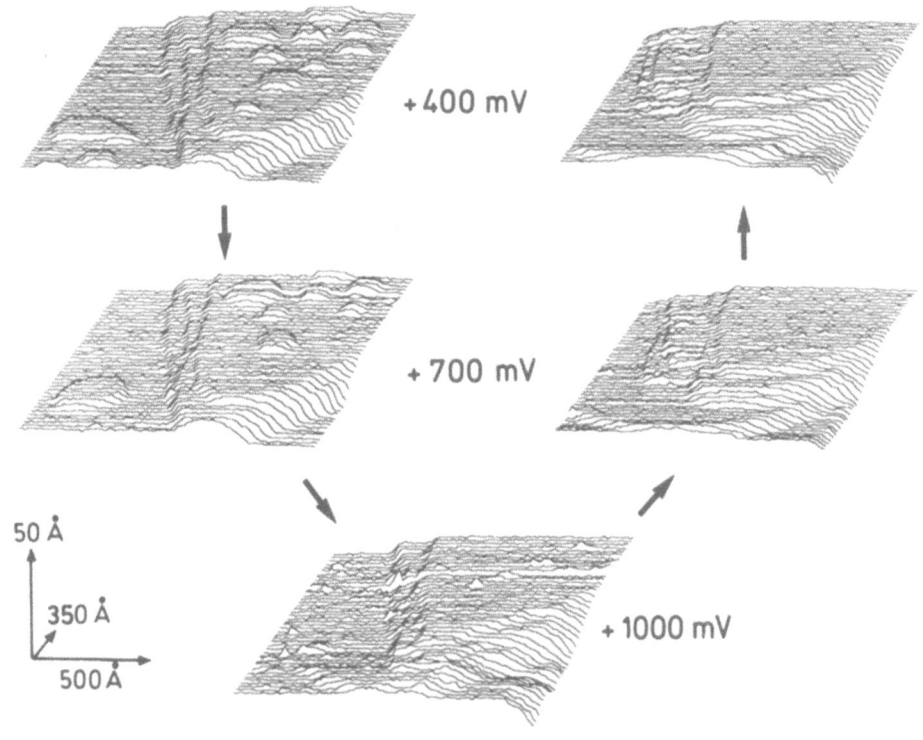

Figure 13. A series of STM images of a Au(100) electrode in 1 M H₂SO₄, showing changes in surface topography which occur within the double-layer charging region [8].

A similar situation for Cl⁻ adsorption on Au(100) in 50 mM H$_2$SO$_4$ + 5 mM HCl is shown in figure 14. While the islands are relatively stable at +500 mV, they are readily dissolved at +700 mV. Although the Cl⁻ coverage is only slightly higher at +700 mV, the Cl⁻ influence seems to be markedly increased. This effect could be correlated with a pronounced spike at about +670 mV in the capacity/potential curve [14], which could indicate a change in the chloride adsorption (e.g. a structural transition within the Cl⁻ adlayer).

The next sequence (figure 15) shows a series of images recorded during a potential scan into the region of oxide formation for Au(100) and back to the double layer region [in 50 mM HClO$_4$ + 1 mM Cu(ClO$_4$)$_2$]. At these potentials Cu is not deposited onto the gold surface, but we noted a significant improvement of the tip stability when Cu^{2+} ions were added to the solution. As in the previous example a dissolution of the small islands occurs within the double-layer region. At potentials of oxide formation the surface takes up a much rougher morphology (a closer look in this region reveals little hills of variable height up to about 10 Å). STM images taken during the growth of the oxide indicate a three-dimensional island growth mechanism [8]. The sequence in fig-

286

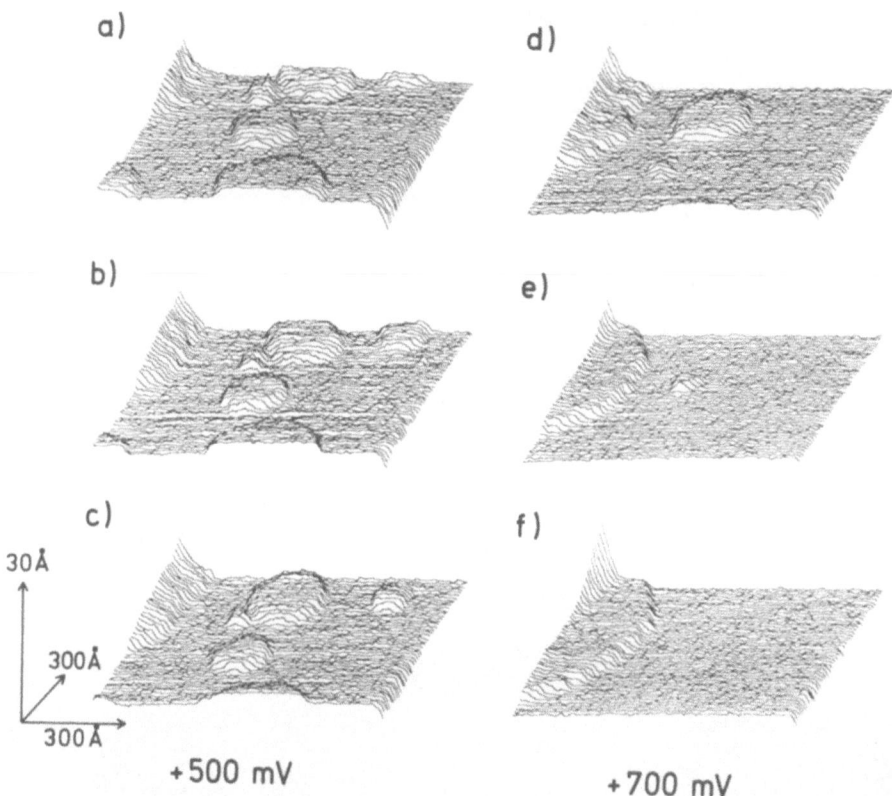

a)

d)

b)

e)

c)

f)

30Å

300Å

300Å

+500 mV

+700 mV

Figure 14. STM images of Au(100) in 0.05 M H$_2$SO$_4$ + 5 mM HCl as a function of time, at two different electrode potentials. Images were taken consecutively with a time lapse between successive images of ca. 100 s [8].

ure 15 is completed by returning to the double-layer region. These images indicate, quite surprisingly, that a substantial smoothening of the surface has occurred during this oxidation-reduction cycle. Obviously surface roughening by oxidation-reduction cycles (ORCs) requires a more extensive surface oxidation.

One conclusion from all these examples certainly is that the electrode surface is rather mobile at potentials positive of the pzc. This long-term change in surface topography may account for the slow change in the double-layer capacity, which has been noted by several authors in the past [15,16].

3.3 THE UNDERPOTENTIAL DEPOSITION OF Cu ON Au

When a metal is electrodeposited from solution onto a foreign metal substrate (e.g. Cu on Au instead of Cu on Cu), then the first monolayer is often deposited at potentials positive of the so-called Nernst potential for bulk deposition. This is simply a consequence of a stronger degree of bonding between Cu and Au than between Cu and Cu. Current-potential curves for the deposition of the

287

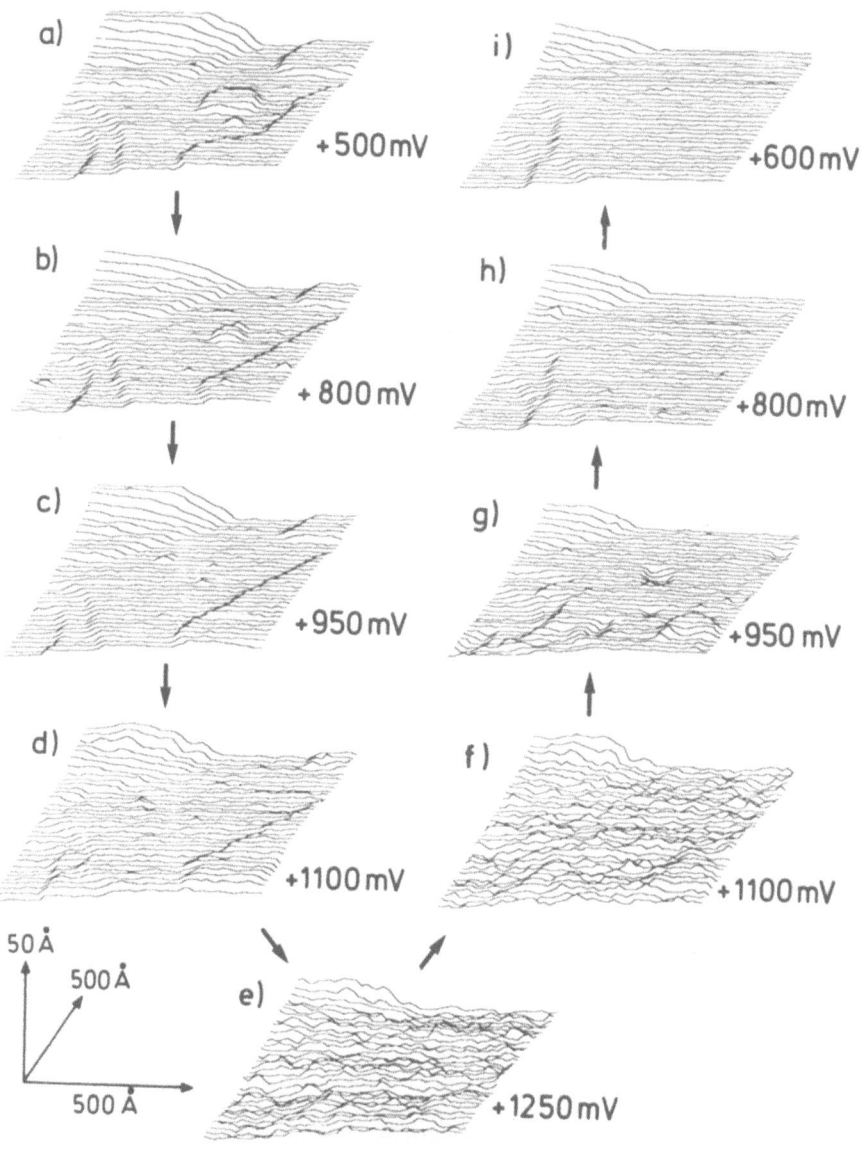

Figure 15. STM images of Au(100) in 0.05 M HClO$_4$ + 1 mM Cu(ClO$_4$)$_2$, as a function of electrode potential, showing changes induced by oxide formation [8].

first monolayer of Cu on Au(111) and on Au(100) are shown in figure 16 [17].

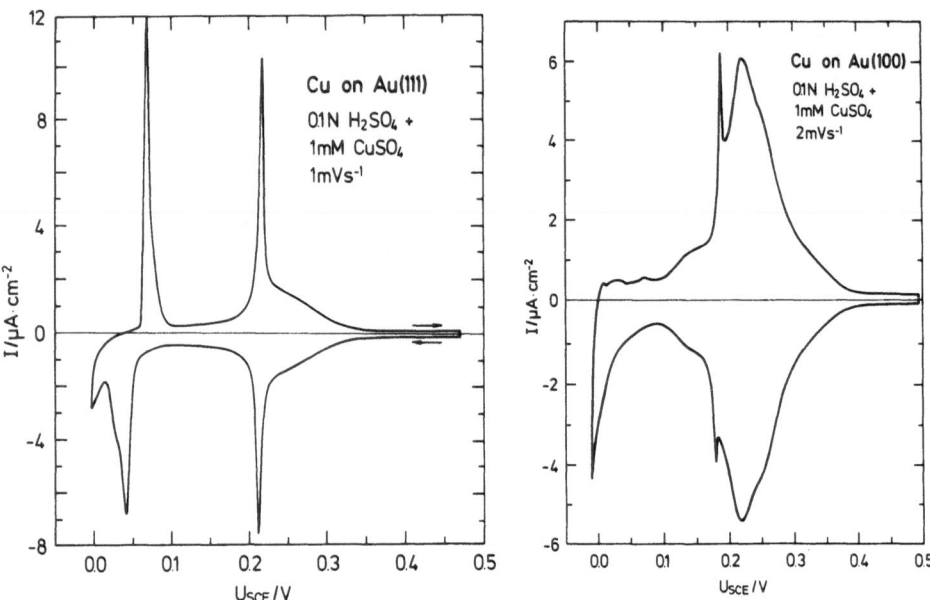

Figure 16. Current-potential curves for the underpotential deposition of Cu on Au(111) and Au(100) in 0.05 M H_2SO_4 + 1 mM $CuSO_4$ [17].

As can be inferred from the voltammogram, the deposition of Cu on Au(111) occurs in two energetically well-separated adsorption steps, which is an indication of ordered adsorption and the existence of a structural transformation at medium coverages. This has indeed been verified by ex-situ LEED experiments [18]. It has been deduced from these data that at very low coverages there is a random adsorption of copper atoms. At medium coverages the copper overlayer forms a ($\sqrt{3}x\sqrt{3}$) R 30° (honeycomb) structure, while a full monolayer acquires a (1x1) pseudomorphic phase. Additional interesting data has been obtained by Auger spectroscopy, which indicates a substantial coadsorption of anions [19]. This consolidates the conclusion that the coadsorbed anions play a critical role in determining the structure of the Cu overlayer.

Figure 17a shows a quite remarkable STM image of a Au(111) electrode covered with roughly half of a monolayer of Cu, deposited from a sulphuric acid electrolyte. Not only are the individual Cu adatoms on Au(111) seen in-situ, but also the ($\sqrt{3}x\sqrt{3}$)R 30° structure is clearly resolved [12]. So this is in good agreement with the ex-situ LEED observations.

However, there is an interesting complication: if the potential is held for several minutes at slightly more negative values (i.e. at slightly higher coverages), then the ($\sqrt{3}x\sqrt{3}$) R 30° structure (phase I, figure 17a) is gradually converted into a second more densely-packed structure, which turns out to be a (5x5) superstructure (phase II, figure 17b). Such a structure had not been de-

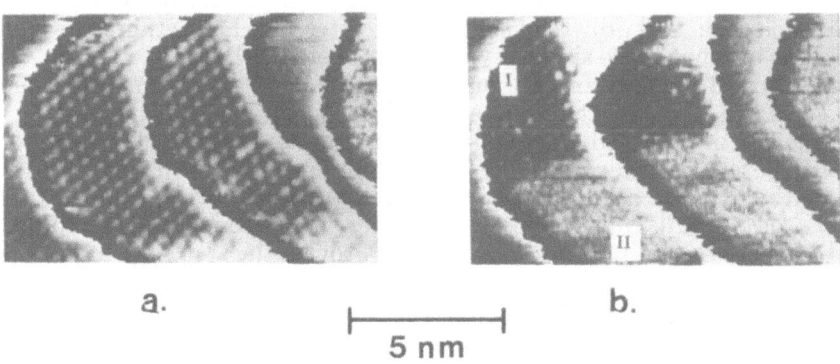

a.

b.

|———————————|
5 nm

Figure 17. STM images of a Cu-covered Au(111) surface in 0.05 M H_2SO_4 + 5 mM $CuSO_4$, taken on the same area at slightly different potentials. (a) U_{SCE} = 0.215 V; (b) U_{SCE} = 0.20 V. Two Cu-adlayer structures - a strongly corrugated $(\sqrt{3} \times \sqrt{3})$ R 30°, phase I, and a weakly corrugated (5x5), phase II - coexist on the surface (U_{tip} = -170 mV, I_t = 2.5 nA, phases I and II as indicated). [12].

tected in the previous LEED experiments, but the slow structural change with time pointed towards an impurity problem, with trace amounts of Cl⁻ being the most likely culprit. Such a hypothesis was confirmed by recent ex-situ LEED studies for Cu on Au(111) in sulphuric acid with small amounts of Cl⁻ added (10^{-4} M). A LEED pattern was observed, which was consistent with a (5x5) structure [20].

So far three different structures have been reported for Cu on Au(111) at medium coverages: a (2.2x2.2) in perchlorate solutions [19], a $(\sqrt{3}x\sqrt{3})$ R 30° in sulphate [12,18] and a (5x5) in chloride containing electrolytes [12,20]. These results contrast strikingly with copper deposition on Au(111) under vacuum conditions, where Cu adlayers are seen to grow in islands of a pseudo-morphic (1x1) structure [18]. This clearly indicates that coadsorbed anions play an important structure-determining role in Cu UPD. Similar results have been obtained for Cu on Au(100) in sulphuric acid, where a sequence of quasi-hexagonal structures have been attributed to repulsive interactions between neighboring copper atoms, which are induced by coadsorbed anions [12].

The agreement between the in-situ STM and the ex-situ LEED measurements is most pleasing. This provides a cross-correlation between the two techniques and alleviates any doubts about both the STM (no influence of the tip on the structure) and the ex-situ LEED experiments (no structural change due to the missing bulk electrolyte).

3.4 BULK METAL DEPOSITION

The deposition of metals is a field of great technological importance. Traditionally a great deal of information has been obtained by the use of electrochemical as well as optical techniques for examining the growth of metal deposits [21]. However, STM now provides a direct method for imaging the very initial stages of the bulk metal deposition, with the possibility for observation of the nucleation and growth of small crystallites. We have studied in-situ the deposition of copper onto single crystal gold substrates [22].

Figure 18 shows a series of images taken before, during and after deposition of bulk copper (the reference electrode is Cu/Cu^{2+}). It is interesting that copper deposition initially occurs around the step in the top right hand corner of the picture. However, in a later image the formation of crystallites is also seen on the terraced region. The surface is returned to its original state upon removal of the bulk copper deposit. Another sequence is shown in figure 19. Small copper crystallites are clearly seen to grow at the step edge in preference to the terraced region. Although such images were recorded at high overpotentials the growth rate of the Cu deposits was slow enough to observe the initial stages inspite of the relatively long time needed to record one image (ca 1 min). Several factors could contribute towards this slow deposition rate. These include tip shielding, a substantial kinetic hinderance in perchlorate solutions [23] and a high nucleation overpotential on these atomically "smooth" areas of the surface. Indeed a higher deposition rate was observed on rougher areas, once again indicating a preferential deposition at sites with a higher surface free energy.

In conclusion we are now able to observe the very initial stages of bulk copper deposition from electrolytes containing a rather low concentration of Cu^{2+} ions. The growth mechanism is certainly *not* a layer-by-layer two-dimensional growth. Deposition occurs preferentially at: rough surface features, step edges and dislocations. It is also interesting to note that at the ear-

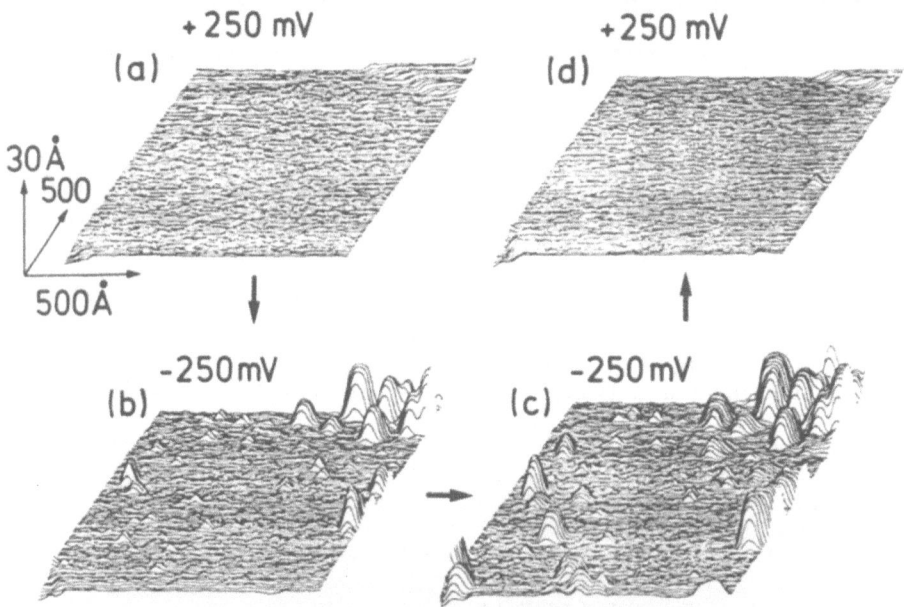

Figure 18. Images taken before, during and after bulk deposition of Cu on a Au(111) electrode in 0.1 M $HClO_4$ + 5 x 10^{-5} M $Cu(ClO_4)_2$. Potentials vs. Cu/Cu^{2+} [22].

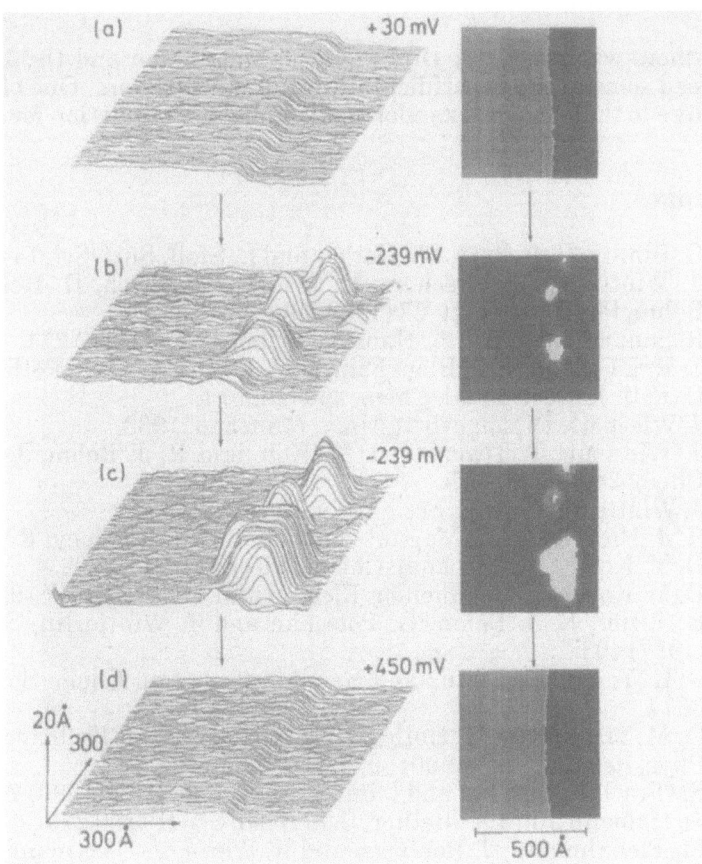

Figure 19. Images taken before, during and after bulk deposition of Cu on a Au(111) electrode in 0.1 M HClO$_4$ + 5 x 10^{-5} M Cu(ClO$_4$)$_2$. Growth of small copper crystallites occurs preferentially at step edges. Note that the scale in the z direction is greatly expanded with respect to the scales in the x, y directions and consequently the crystallites are in reality rather flat and broad in shape. Potentials are given vs. Cu/Cu^{2+} [22].

ly stages of their growth the small metal crystallites seem not to have taken up a particularly well defined geometry.

Although it is still the beginning of STM entering electrochemistry, it seems safe to state that this technique will have a tremendous impact on interfacial electrochemistry and it will add favorably to the already impressive array of methods in the field.

Acknowledgements

The authors wish to thank Dipl-Phys. O. Magnussen and Dr. J. Hotlos who performed some of the experimental work presented here. One of us (R. J. N.) would like to thank the Alexander von Humboldt-Foundation for a grant.

References

[1] G. Binnig, H. Rohrer, Ch. Gerber and E. Stoll, Surf. Sci. **144** (1984) 321.
[2] J. Wintterlin, J. Wiechers, H. Brune, T. Gritsch, H. Höfer and R. J. Behm, Phys. Rev. Lett. **62** (1989) 59.
[3] R. Sonnenfeld and P. K. Hansma, Science, **232** (1986) 211.
[4] T. R. I. Cataldi, I. G. Blackham, G. A. D. Briggs, J. B. Pethica and H. A. O. Hill, J. Electroanal. Chem. **290** (1990) 1.
[5] J. Wiechers, Diplomarbeit, Univ. München, 1988.
[6] J. Wiechers, T. Twomey, D. M. Kolb and R. J. Behm, J. Electroanal. Chem. **248** (1988) 451.
[7] J. Wintterlin, Thesis, Freie Univ. Berlin, 1989.
[8] R. J. Nichols, O. M. Magnussen, J. Hotlos, T. Twomey, R. J. Behm and D. M. Kolb, J. Electroanal. Chem. **290** (1990) 21.
[9] D. M. Kolb and J. Schneider, Electrochim. Acta **31** (1986) 929.
[10] E. Ritter, R. J. Behm, G. Pötschke and J. Wintterlin, Surf. Sci. **181** (1987) 403.
[11] S.-L. Yau, C. M. Vitus and B. Schardt, J. Am. Chem. Soc. **112** (1990) 3677.
[12] O. M. Magnussen, J. Hotlos, R. J. Nichols, D. M. Kolb and R. J. Behm, Phys. Rev. Lett. **64** (1990) 2929.
[13] S. Ciraci, A. Baratoff and I. Batra, Phys. Rev. B **42** (1990) 7618.
[14] A. Hamelin and J. P. Bellier, C-R. Acad. Sci. (Paris) **279C** (1974) 371.
[15] M. Fleischmann, J. Robinson and R. Waser, J. Electroanal. Chem. **117** (1981) 257.
[16] R. Waser and K. G. Weil, J. Electroanal. Chem. **150** (1983) 89.
[17] D. M. Kolb, K. Al Jaaf-Golze and M. S. Zei, DECHEMA-Monographie, Vol. 102, (VCH, Weinheim, 1986), p. 53.
[18] Y. Nakai, M. S. Zei, D. M. Kolb and G. Lehmpfuhl, Ber. Bunsenges. Phys. Chem. **88** (1984) 340.
[19] M. S. Zei, G. Quiao, G. Lehmpfuhl and D. M. Kolb, Ber. Bunsenges. Phys. Chem. **91** (1987) 349.
[20] D. M. Kolb, R. Michaelis and M. S. Zei, to be published.
[21] E. B. Budevski, in: *Comprehensive Treatise of Electrochemistry; Part 7: Kinetics and Mechanisms of Electrode Processes*, eds. B. E. Conway et al., (Plenum, New York 1984).
[22] R. J. Nichols, D. M. Kolb and R. J. Behm, J. Electroanal. Chem. (1991)
[23] T. A. Twomey and D. M. Kolb, 38th ISE-Meeting (Maastricht, 1987), Extended Abstracts. Vol. I, p. 334.

SINGLE CRYSTAL ELECTRODES

ROGER PARSONS,
Department of Chemistry,
University of Southampton,
Southampton, SO9 5NH,
United Kingdom.

ABSTRACT. It has become clear that an understanding of the behaviour of solid metal/electrolyte interphases, and of many reactions occurring in them, can be achieved only if the structure and composition of the interphase is well-defined. Techniques for the production of such interphases will be described as well as methods for verifying that a satisfactory interphase has been achieved. Examples of the information about interfacial structure which may be deduced from electrochemical experiments will be drawn largely from work on platinum electrodes.

1. Why should experiments on single crystal electrodes be done?

As soon as electrochemical experiments on solid electrodes are considered, it becomes necessary to assess the extent to which the surface structure of the solid affects the processes occurring at the electrode/solution interface. It will be seen here that these effects can be quite dramatic. A striking example is given by studies on iridium electrodes. The linear sweep voltammogram (which will be explained below) of a polycrystalline iridium electrode [1] has a rather featureless shape (Figure 1) whereas

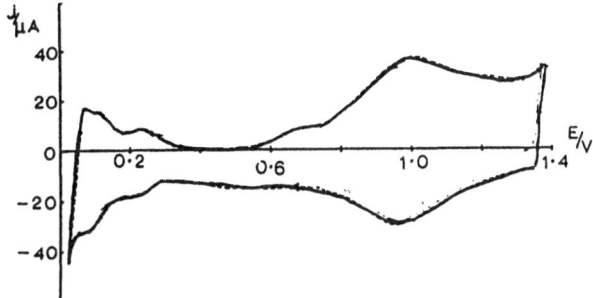

Figure 1. Polycrystalline Ir electrode in 0.5 M H_2SO_4. Sweep speed $v = 140$ mV s^{-1}. Potential with respect to a hydrogen electrode in the same solution (RHE). (From Ref. 1)

R. Guidelli (ed.), Electrified Interfaces in Physics, Chemistry and Biology, 293–308.
© 1992 Kluwer Academic Publishers.

294

the corresponding measurement on the low index planes of an iridium single crystal [2] shows very marked features (Figure 2). Without any detailed

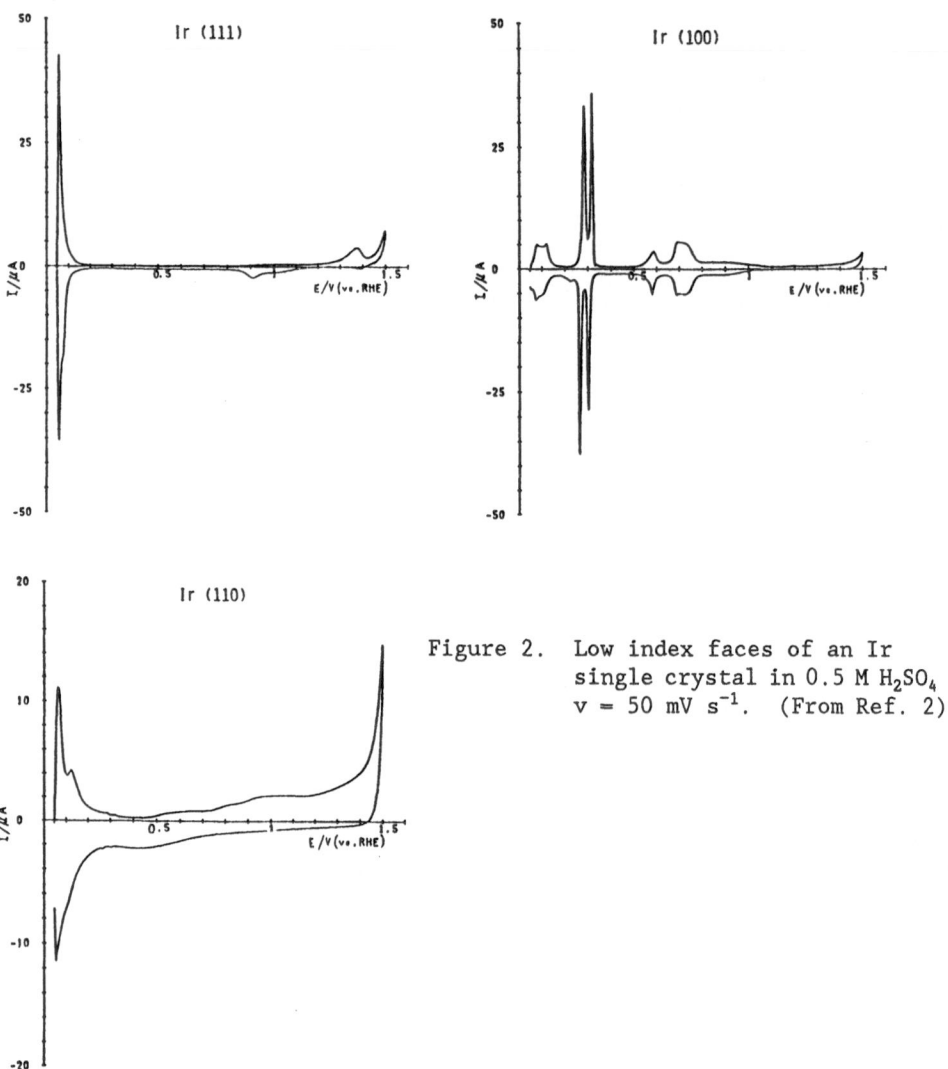

Figure 2. Low index faces of an Ir single crystal in 0.5 M H_2SO_4 $v = 50$ mV s^{-1}. (From Ref. 2)

analysis, it should be clear that the use of a polycrystalline electrode obscures the fine detail of the behaviour on the different types of surface structure and that there is no way in which this detail can be deduced from measurements on a polycrystalline surface.

This example uses behaviour which arises largely from the processes of adsorption of species like H and OH up to monolayer coverage and this type of process will be the principal subject of this chapter. In particular the behaviour of hydrogen adsorption on platinum will be used as an illustrative case history. However, other types of process are also sensitive to surface structure, for example, because of the fact that the potential of zero charge of a solid depends on its surface structure. Hence the electrical state of the interphase at a given applied potential will depend on the surface structure and this will affect adsorption which is primarily electrostatic and consequently the kinetics of even reactions which do not involve bonding to the electrode surface (see ref.3,4 for example). The adsorption of essentially neutral monolayers of H or OH and various metals, however, is more like a chemisorption process in a gas/solid system and the structure sensitivity arises from the variation of the bonding between the adsorbed species and the surface.

2. What can one learn from linear sweep voltammetry?

In the experiment of linear sweep voltammetry (LSV) the electrode potential is varied linearly with time and the current passing through the electrode is recorded. The direction of the sweep is usually reversed and the sweep may be repeated; the result is then called a cyclic voltammogram (CV).

Since the origin of this technique for the study of surface processes [5] it has been well known that the CV for a clean polycrystalline Pt electrode in 0.5 M H_2SO_4 has the form shown in Figure 3. Here the potential ranges from just above the reversible hydrogen

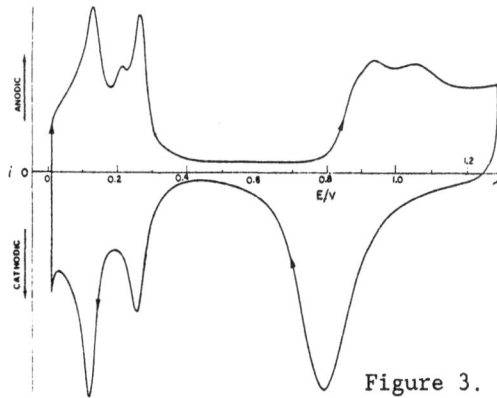

Figure 3.   Polycrystalline Pt in 0.5 M H_2SO_4
$v = 100$ mV s^{-1} T = 25°. Potential with respect to the RHE.   From Ref. 6)

potential (left-hand limit) to somewhat below the reversible oxygen potential. It is generally accepted that the region at the lower potential end (0.05 to 0.30 V) consisting of two principal peaks corresponds to the process of formation and removal of adsorbed hydrogen atoms, i.e.

$$H^+_{(aq)} + e^- \rightleftharpoons H_{(ads)} \tag{1}$$

where $H^+_{(aq)}$ is a hydrated proton and $H_{(ads)}$ is a chemisorbed hydrogen atom. Thus, on the negative sweep a platinum surface covered with water molecules and with some adsorbed HSO_4^- ions begins to adsorb H by process (1) at about 0.3 V and this process is completed by about 0.05 V. (Extension of the sweep to more negative potentials would result in formation of molecular hydrogen and a continuously increasing (negative) current. The total amount of adsorbed H can be determined by integrating the current in this region because the potential axis is also a time axis. The charge obtained can then be interpreted as an amount of H adsorbed if reaction (1) is assumed. The result, about 210 μC cm^{-2}, corresponds to a complete monolayer of H and this (among other evidence) supports this attribution of the current in this region.

On the positive sweep in this potential region process (1) is reversed and the fact that the voltammogram is almost exactly symmetrical about the potential axis shows that this adsorption reaction is sufficiently fast that it remains at equilibrium at each point of the voltammogram. This important result means that the potential (E) is directly related to the Gibbs energy of the adsorption reaction (ΔG) by the well-known equation

$$\Delta G = -nFE \tag{2}$$

where n is the number of electrons transferred in the reaction and F is Faraday's constant. Thus the position on the potential scale indicates the energy of adsorption; in particular the peak at 0.26 V corresponds to more strongly bound H while that at 0.12 V corresponds to more weakly bound hydrogen. The analogy can be made between these voltammograms and thermal desorption curves (TDS) in gas/solid systems. However, there is an important difference in the example discussed here because the energy of adsorption derived here is a well-defined thermodynamic quantity, while that from TDS depends on the assumption of a model for the irreversible process of thermal desorption.

In the region of more positive potentials the situation is different. On the positive sweep the current begins to flow at about 0.8 V which corresponds to the adsorption of an oxygen species, probably OH in the initial stages. If the potential is reversed at 1.2 V, the negative sweep shows quite different characteristics as seen in Figure 3. This irreversibility is due to a slow process, probably a place exchange between the adsorbed species and the underlying Pt atom [6].

This chapter will be concerned mainly with the region of hydrogen adsorption which has been found to be a sensitive indicator of surface structure as well as having great intrinsic interest.

3. Experimental problems in the use of single crystals.

The first experiments with Pt single crystals were carried out by Will in 1965 [7] with the object of understanding why there are two main regions of H adsorption on Pt. He was able to show that the peak of weakly bound hydrogen predominated on the (111) face while that of the strongly bound hydrogen predominated on the (100), thus supporting the idea that the different states had a structural origin. These results were obtained on crystals whose surface were subjected to an electrochemical pre-treatment consisting of cycling the potential many times over a range including a substantial amount of oxide layer formation. It is now known that such treatment perturbs the Pt surface to a great extent so that the state of the surface layer to which these voltammograms refer is not known.

Following this work, several groups repeated Will's experiment without a clear consensus of results. A major step forward was achieved by Hubbard and his colleagues [8] using an electrochemical cell integrated with an ultra-high vacuum (UHV) analysis apparatus. They were able to clean the Pt crystals in UHV and to verify their cleanliness with Auger spectroscopy (AES) and their surface structure with low energy electron diffraction (LEED). The crystal could then be transferred to the electrochemical cell for experiments and then returned to UHV for further characterization with the minimum of contamination. The results obtained with (111) and (100) orientations gave clear confirmation of Will's conclusions (Figure 4). However, it must be remarked that the transfer

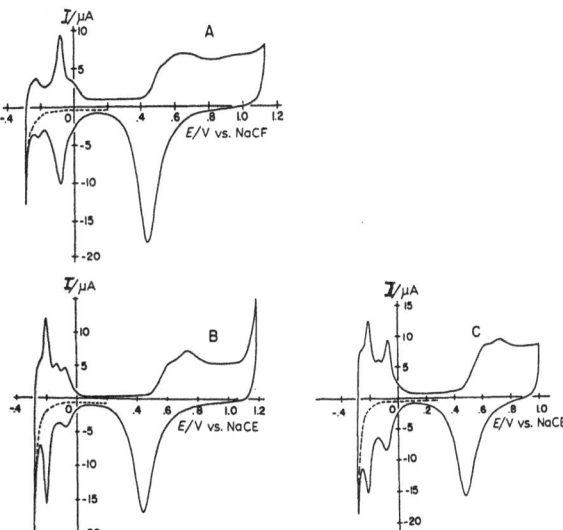

Figure 4.   Pt electrodes in IM H_2SO_4 v = 10 mV s^{-1} T = 23°
A: Pt (100) B: Pt(111) C: Polycrystalline Pt
Electrode area A: 0.622 B: 0.579 C: 0.800 cm^2
Potential with respect to a saturated NaCl calomel electrode;
to compare with RHE add 0.25 V.   (From Ref. 8)

was not totally free of contamination since before the electrochemical experiments, a small number of cleaning cycles was necessary to obtain CVs characteristic of a clean surface and on the reverse transfer, a brief argon ion bombardment was necessary to remove impurities.

The significance of this minimal cleaning became clear as a result of the development of a quite different and novel technique by Clavilier [9]. This uses small spherical single crystals about 2mm diameter. These are easily prepared by fusion of a Pt wire in a blowpipe. They have visible (111) facets which may be used for orientation using the reflection of a laser beam. About half the crystal is ground away and the polished flat surface forms the desired oriented face. Before each experiment the crystal is heated to redness in a blowpipe and quenched with pure water (or first cooled in hydrogen [10]). This enables it to be transferred to the electrochemical cell without contamination. It can also be brought into contact with the electrolyte at a pre-determined potential, set on the potentiostat. This may be chosen so that no transient current occurs on contact, i.e. no oxidation or reduction of the surface occurs. The hemispherical electrode makes a meniscus contact with the electrolyte so that only the face of interest is in contact with the electrolyte. The first sweep of the voltammogram can then give information about the initial state of the surface, e.g. the reduction of thermal oxide formed during pre-treatment. Apart from this, the remaining part of the first cycle is identical to subsequent cycle giving a strong indication that the surface is clean. Further support for the well-defined nature of these surfaces was obtained by examining the LEED pattern which was found to be exceptionally sharp on surfaces prepared in this way [11].

Use of this technique for the (111) surface gave an entirely new form of voltammogram (Figure 5) and this led to much controversy. Some

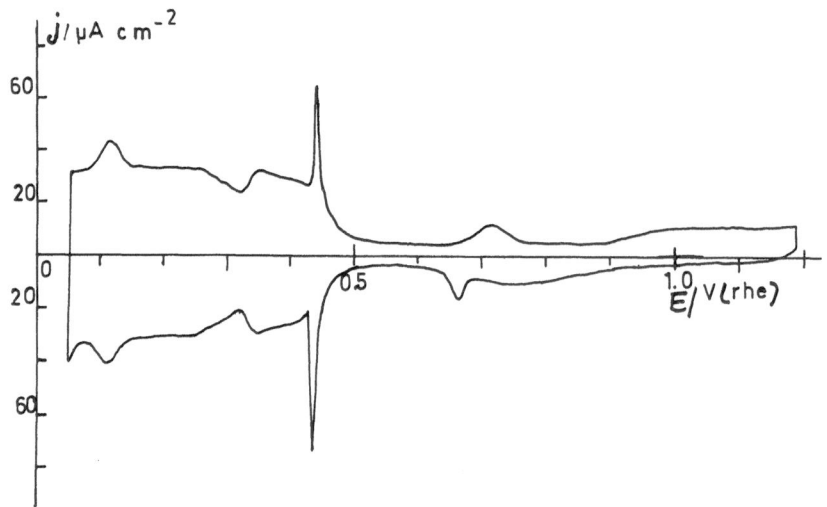

Figure 5. Pt(111) electrode in 0.5 M H$_2$SO$_4$ v = 50 mV s^{-1} (From Ref. 9)

of the key observations were:- that the LEED experiments indicated that surfaces prepared in this way showed a high degree of long-range order; that formation and removal of a monolayer of oxide resulted in the change in form to a voltammogram virtually identical to that found by Hubbard et al. for Pt(111) for which the LEED indicated a random stepped surface [11, 12]. It is evident that any experiment which is preceded by a decontamination cycle involving oxide formation will perturb the Pt surface such that the first stable CV to be seen resembles that for Pt(111) in Figure 4. This result emphasizes the mobility of Pt surfaces under these conditions. Somewhat similar, though less dramatic, effects are found for the other low-index planes.

4. Verification of Surface Structure

In view of the high mobility of surface atoms, even of a relatively high melting metal like Pt, it is of great importance to have techniques by which the surface structure can be verified in situ. The UV-visible electroreflectance (ER) spectrum can give useful information in some cases. In the case of Au electrodes, the binary symmetry of the (110) surface leads to an asymmetry of the ER spectrum but surfaces of higher symmetry are isotropic. More detailed information about surface reconstruction can be obtained from the detailed spectrum interpreted in terms of surface states, as shown by Kolb elsewhere in this volume. This technique is less useful for Pt because of the absence of appropriate surface states.

Of more general applicability is the technique of second harmonic generation (SHG) which depends on the production of light at a higher harmonic from atoms in an asymmetric environment, illuminated by a laser beam [13]. Recent work by Corn et al. [14] has demonstrated the six fold symmetry of a Pt(111) in situ as well as the change in this symmetry resulting from the chemisorption of species including H on this surface. Since this technique is not only surface specific, but has the capability of fast response [15], it has great promise for use in electrochemistry.

Also of wide applicability is the scanning tunnelling microscope (STM) which forms the subject of one of Kolb's chapters in this volume. This has provided direct evidence for the processes described above for Pt(111) [16] (Figure 6). Itaya et al. studied the (111) facets on the Pt spherical single crystal prepared as described above and observed the high degree of long range order in the initial state. After the crystal was cycled into the oxide region, the surface became highly disordered. It is worth mentioning that, in unpublished work, Clavilier obtained CVs for these facets which behaved in the same way as the cut (111) faces. Hence there is a direct correlation with the STM work.

In situ X-ray diffraction has also been shown to provide evidence for structural changes caused by adsorption of H [17] but not so far under the conditions discussed here.

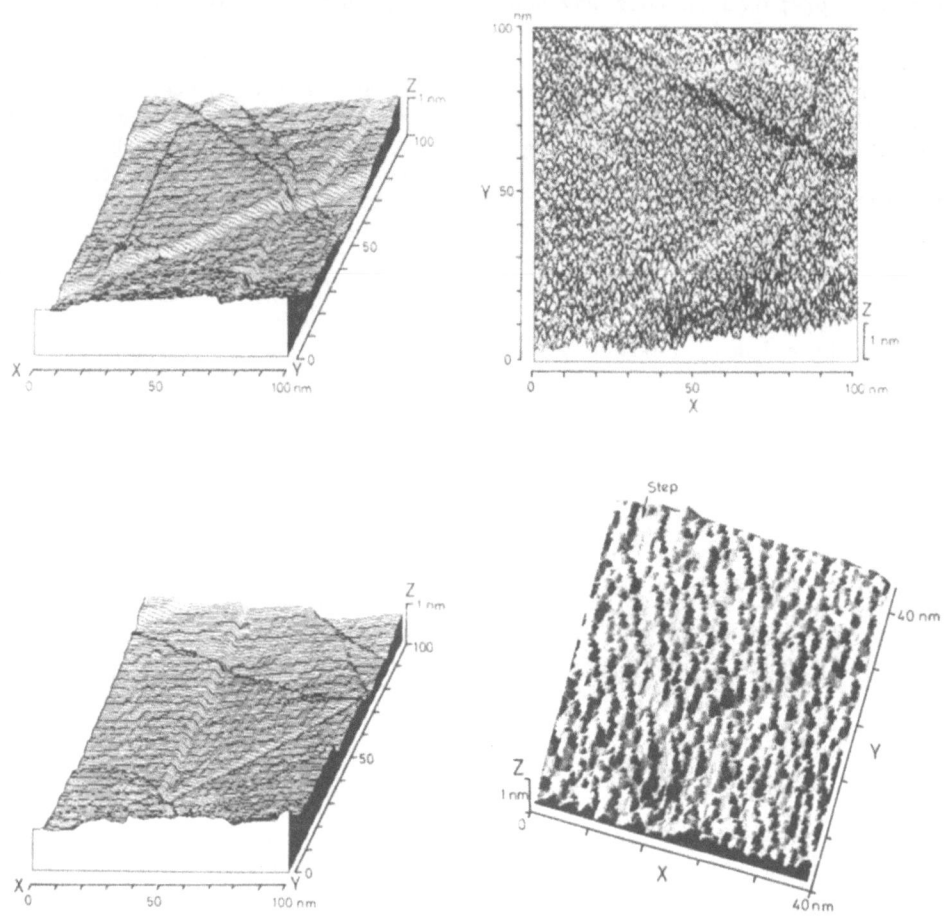

Figure 6. STM images of a Pt(111) facet obtained in 0.05 M H_2SO_4. The
electrode potentials of the Pt and tip electrodes were 0.95 and
0.9 V (RHE) respectively. The tunneling current was 2 nA. Scan
speed 200 nm s^{-1}.
a: b: after high temperature annealing but no electrochemical
oxidation.
c: after 5 cycles between 1.5 and 0.05 V (RME)
d: as c but with higher magnification.
(From Ref. 16)

5. Electrochemical Identification of Surface Structure

The sensitivity of the form of the CV in the region of H adsorption has
led to the proposal that this can be used to identify the nature of a

given surface. In the first place a clear answer can be given about the origin of the principal features of the voltammogram of polycrystalline Pt. This is most evident for the polyoriented surface of a single crystal sphere (Figure 7), which shows the two principal peaks characteristic of

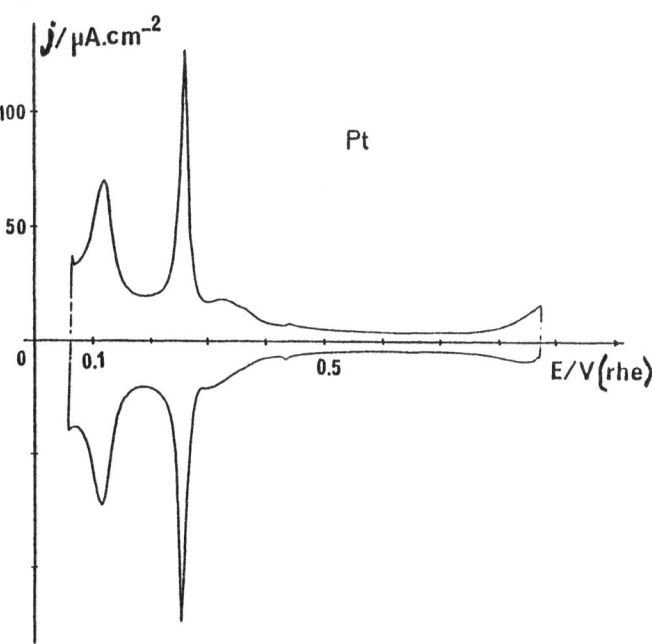

Figure 7.   Polyoriented Pt (spherical single crystal) in 0.5 M H_2SO_4
v = 50 mV s^{-1} (From S.G. Sun, Thesis, Paris 1986)

the usual polycrystalline electrode.   The peak at higher potentials corresponds to that on a (100) surface (Figure 4), which has a low degree of long-range order [18] and so may be identified with adsorption on a 4-fold Pt site.   The peak at lower potentials corresponds to that found for the (111) surface with random steps (Figure 4), but also to that for the (110) surface. Since steps on a (111) will have a (110) structure, it is reasonable to attribute this peak to a (110)-like site. Since the spherical single crystals have (111) facets which initially have a high degree of long-range order, the small peak visible at 0.43 V can be identified with the spike found on the (111) surface (Figure 5). Further details can be identified in a similar way.
      The Clavilier technique of producing surface of desired orientation allows the production of a series of surfaces of systematically varying structure.   This has now been done by several groups [19-23] with good agreement on the experimental results. Clavilier et al. [22, 23] have provided a convincing quantitative treatment of

series of stepped surfaces with systematic change of terrace width. One such series is shown in Figure 8 for the (110) zone. Hence the surface can

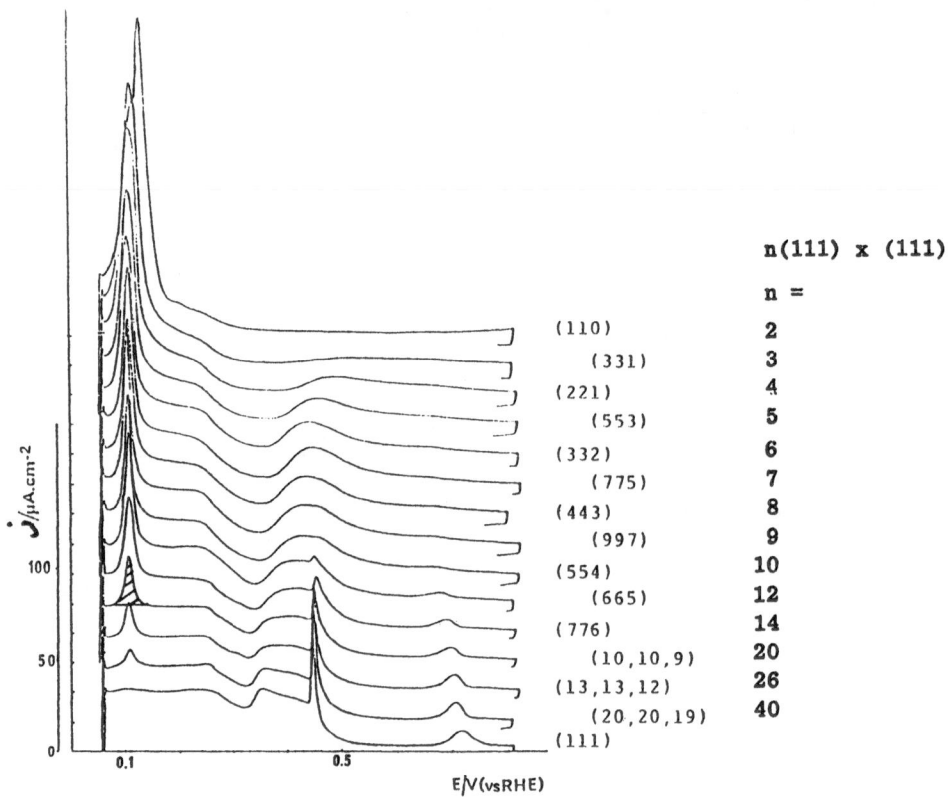

Figure 8. Voltammograms recorded on the positive sweep for Pt single crystals in the (ITO) zone in 0.5 M H_2SO_4 at 50 mV s^{-1}. The Miller index of the surface being studied is shown to the right of each curve as well as the step and terrace structure. The shaded region on the curve for Pt (10,10,9) indicates the charge attributed to the adsorption on the step. (From Ref.22)

be regarded as consisting of monatomic (111) steps separating terraces of (111) structure n atoms wide. This is denoted as an n(111) x (111) structure. One example of such a structure is shown in Figure 9. From Figure 8 it can be seen that the peak at 0.11 V, which is absent for Pt(111) where there are no steps (ideally), increases in area progressively as the step density increases. Hence, it is reasonable to attribute this hydrogen to adsorption on the steps. The remaining hydrogen may be attributed to the terraces.

From the structure of Figure 9, a unit cell may be defined which

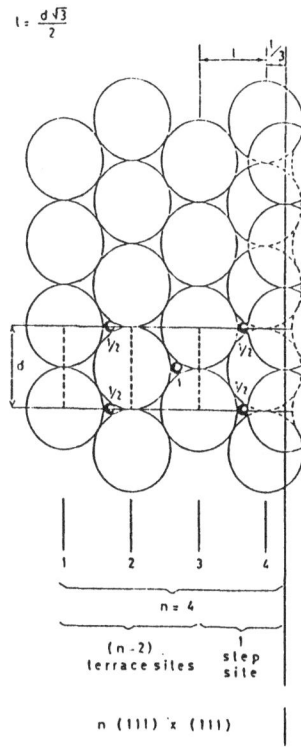

$l = \frac{d\sqrt{3}}{2}$

$n = 4$

$(n-2)$ terrace sites $\begin{array}{c}1\\ \text{step}\\ \text{site}\end{array}$

$n\ (111)\ x\ (111)$

Figure 9. Structure of a surface with (111) steps and (111) terraces:- n(111) x (111) where n = 4 in this diagram. The large circles represent Pt atoms while the small circles represent sites for H adsorption. (From Ref. 22).

is one atom wide and (n-2/3) atoms long. If d is the atomic diameter, the area of this unit cell is $(n-2/3)\ d^2\ \sqrt{3}/2 = S$. In one unit cell there is one step site (i.e. 2 halves) so that the charge per unit area due to step sites is e/S. Since the charge per unit area of a perfect (111) surface Q_{111} is $2e/d^2\ \sqrt{3}$, the step charge may be expressed as $Q_{111}/(n-2/3)$.

There are n-2 terrace sites per unit cell and the corresponding charge per unit area is (n-2)e/S or

$$\frac{n-2}{n-2/3}\ Q_{111} = Q_{111} - \frac{4\ Q_{111}}{3(n-2/3)} \tag{3}$$

Thus both the step charge and the terrace charge should vary linearly with 1/(n-2/3) as is confirmed by the experimental result shown in Figure 10. The open circles of plot a correspond to the total charge between 0.05 and 0.5V less the charge in the peak attributed to the steps. Thus this includes the double layer charge. For the (111) surface (left-hand intercept) this amounts to 292 $\mu C\ cm^{-2}$ while that calculated for a monolayer (one H to each Pt) of a perfect (111) surface is 241 $\mu C\ cm^{-2}$ which suggests a reasonable double layer contribution of 51 $\mu C\ cm^{-2}$. If this latter is taken as independent of crystal orientation, the filled circles are obtained. The slopes of both lines are in reasonable agreement with the predictions of the above model.

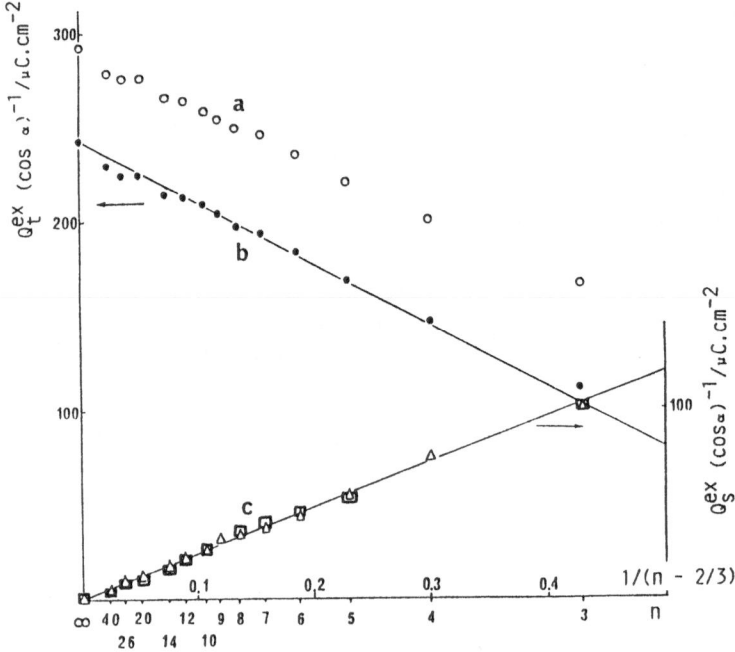

Figure 10. a: experimental terrace charge density measured in 0.5 M H₂SO₄ as a function of 1/(n-2/3). b: the same after correction for double layer charge. c: experimental step charge density measured in 0.5 M H₂SO₄ (Δ) and in 0.1 M HClO₄ (⊡). The cos α factor allows for the projection of the surface in a plane parallel to the terraces. (From Ref. 22)

This type of analysis, together with that of deconvoluting the experimental voltammograms [24], shows the way to the electrochemical characterization of surface. What remains to be done is the development of an understanding of the interactions involved and the energetics of these processes.

6. Can the unusual states be identified with adsorbed hydrogen?

The most surprising result from the study of Pt(111) by Clavilier's method was the appearance of an apparent extension of the hydrogen region up to about 0.5 V and the absence of any current due to oxide formation before 1.2 V. The former result shown in Figure 5 was originally interpreted by Clavilier et al. [9] as being due to the presence of hydrogen adsorbed at an unusually high energy. This was supported by various pieces of evidence. The presence of impurities could be ruled out by the identity of the initial voltammograms to subsequent CV's and the absence of contamination in LEED and Auger. The charge density, including the unusual region, was close to the 243 μC cm⁻² which corresponds to one H per Pt atom on an ideal (111) surface. The process involved was very fast; the voltammogram was unchanged up to 100 V s^{-1} and later work by pulse [25] and AC [26] has confirmed this.

The analysis of the stepped surfaces outlined above is self-consistent only if this unusual region is assumed to be hydrogen. Nevertheless, this interpretation has been criticized, first on the grounds that no such strongly bound hydrogen has ever been seen in gas phase experiments, and secondly on the grounds that there is clear experimental evidence for the involvement of anions in the electrochemical process. The first experiments by Clavilier [27] showed that the more strongly adsorbed the anion, the more the hydrogen region was pushed towards negative potentials, i.e. there appeared to be competition for sites between anions and hydrogen and only as the anions were desorbed by a negative shift of potential could the hydrogen adsorb. The results shown above are for sulphuric acid in which the anion, predominantly HSO_4^- is moderately strongly adsorbed. For $HClO_4$ [27], HF [28] and NaOH [29] where the anion is more weakly adsorbed, the unusual region appears to move to even higher energies, i.e. to the potential region 0.6 to 0.8 V (RHE) (Figure 11).

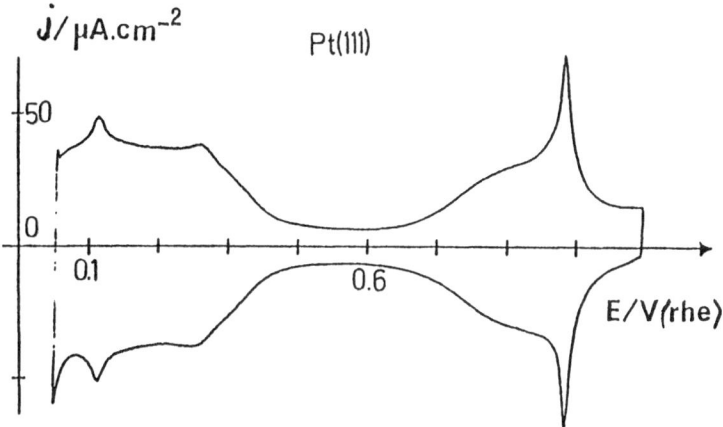

Figure 11. Pt(111) in 0.1 M $HClO_4$ v = 50 mV $^{-1}$. (From Ref. 27)

At the present time there appears to be general agreement about the experimental results and that the unusual results occur on a surface with a high degree of long range order as described above. However, the interpretation in terms of hydrogen adsorption has been challenged. The unusual region in weakly adsorbing electrolytes has been claimed to have a different origin from that in sulphuric acid. The former has been ascribed to an unusual state of OH adsorption [12] while the latter is ascribed to HSO_4^- adsorption [30]. The involvement of structured water in the interphase has also been proposed [28]. There is considerable difficulty in verifying any of these proposals by purely electrochemical techniques, partly because of the interdependence of the various solution species which may be adsorbed. At present too little is known about the adsorption of the various anions on single crystal surfaces. Much of the qualitative evidence comes from older radiotracer work on high area Pt [31]. However, a recent study of HSO_4^- adsorption on well ordered Pt(111) [32] leads to the conclusion that the whole of the unusual region cannot be accounted for by adsorption of sulphate species. A more likely suggestion is that this region is due to hydrogen adsorption under the control of simultaneous HSO_4^- adsorption with the latter providing a structural contribution, probably with the water. The sharp spike would then be caused by a phase change in this hydrated HSO_4^- structure. This is consistent with the infra-red study of Pt(111) in sulphuric acid solution by Bewick and Nichols, described elsewhere in this volume. Such an explanation is supported by a recent study of Tl adsorption on Pt(111) [33]. Here it was shown that the profile of the voltammogram for the later stages of Tl adsorption resembles closely that of H adsorption with a spike at almost the same potential (Figure 12), suggesting that this

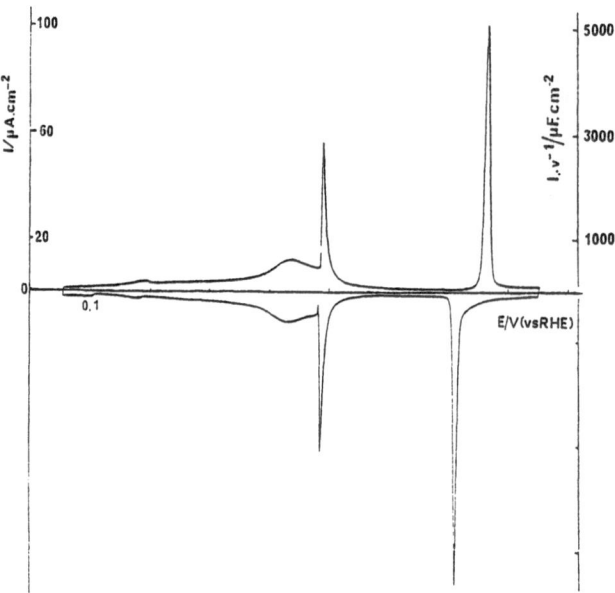

Figure 12. Pt(111) in 0.5 M H_2SO_4 containing 5×10^{-3} M Tl^+ $v = 20$ mV s^{-1}. (From Ref. 33)

process is affected by the double layer structure in a very similar way. It must be noted that it was verified that Tl adsorption caused no change in the surface structure by studying the voltammogram in the absence of Tl^+ after these experiments.

It is more difficult to elucidate the nature of the unusual peak in more weakly adsorbed electrolytes (Figure 11) since the distinction between OH and H adsorbed is not easy. Recently the use of N_2O as a specific reagent for adsorbed H has been proposed [34]. These experiments appear to demonstrate the common origin of the unusual region in sulphuric and perchloric acids. Similar conclusions were reached in measurements of the UV-visible electroreflectance spectrum of single crystal surfaces [35] as well as from a study of the second harmonic signal [14].

It is clear that the controversy over the nature of these interesting results will continue and that new methods, as well as studies on other metals, will be used in attempts to solve the problem. At present, it appears that the weight of the evidence is on the side of Clavilier's original explanation.

7. Conclusions

The brief and incomplete outline of recent work on single crystal Pt electrodes given here should indicate not only that use of such surfaces is essential for the full understanding of electrode processes involving adsorbed states, but that such studies lead to results of interest in surface science in general. They also lead to fundamental questions about the behaviour of surface species, which still need much work for their full explanation.

8. Acknowledgement

I would like to acknowledge, with much pleasure, the elegant experimental work and ingenious design of experiments by my friend and colleague Jean Clavilier. Without him, this fascinating area of electrochemistry might well have remained unexplored.

References

1.   A. Capon and R. Parsons, J. Electroanal. Chem., (1973) 44 239.
2.   S. Motoo and N. Furuya, J. Electroanal. Chem., (1984) 167 309.
3.   R. Parsons, in Aquatic Surface Chemistry. Ed. Werner Stumm, Wiley, New York, (1987) Chapter 2.
5.   F.G. Will and C.A. Knorr, Z. Elektrochem., (1959) 63 1008, (1960
6.   H. Angerstein-Kozlowska, B.E. Conway and W.B.A. Sharp, J. Electroanal. Chem., (1973) 43 9.
7.   F.G. Will, J. Electrochem. Soc., (1965) 112 451.
8.   A.T. Hubbard, R.M. Ishikawa and J. Katekary, J. Electroanal. Chem., (1978) 86 271.
9.   J. Clavilier, R. Faure, G. Guinet and R. Durand, J.

308

10.   S. Motoo and N. Furuya, J. Electroanal. Chem. (1984) <u>172</u> 339.

11.   D. Aberdam, C. Corotte, D. Dufayard, R. Durand, R. Faure and G. Guinet, Proc. 4th Int. Conf. Solid Surf., Cannes, 1980, p622.

12.   F.T. Wagner and P.N. Ross, J. Electroanal. Chem., (1983) <u>150</u> 141.

13.   G.L. Richmond, J.M. Robinson and V.L. Shannon, Progress in Surface Science (1988) <u>28</u> 1.

14.   M.L. Lynch, B.J. Barner and R.M. Corn, J. Electroanal. Chem., (1991) 300.

15.   R.M. Corn, M. Romagnoli, M.D. Levenson and M.R. Philpott, J. Chem. Physics, (1984) <u>81</u> 4127.

16.   K. Itaya, S. Suguwara, K. Sashikata and N. Furuya, J. Vac. Sci. Technol., (1990) <u>148</u> 515.

17.   M. Fleischmann and B.W. Mao, J. Electroanal. Chem. (1987) <u>229</u> 125.

18.   J. Clavilier and D. Armand, J. Electroanal. Chem., (1986) <u>199</u> 187.

19.   B. Love, K. Seto and J. Lipkowski, J. Electroanal. Chem., (1986) <u>199</u> 219.

20.   S. Motoo and N. Furuya, Ber. Bunsenges. Phys. Chem., (1987) <u>91</u> 457.

21.   N.M. Markovic, N.S. Marinkovic and R.R. Adzic, J. Electroanal. Chem., (1988) <u>241</u> 309.

22.   J. Clavilier, K. El Achi and A. Rodes, J. Electroanal. Chem., (1989) <u>272</u> 253.

23.   J. Clavilier, K. El Achi and A. Rodes, Chem. Phys. (1990), <u>141</u> 1.

24.   D. Armand and J. Clavilier, J. Electroanal. Chem. (1987) <u>225</u> 205; (1987) <u>233</u> 251.

25.   F.E. Woodard and C.N. Reilley, J. Electroanal. Chem. (1984) <u>167</u> 65.

26.   G. Bootle and R. Parsons, unpublished work.

27.   J. Clavilier, J. Electroanal. Chem., (1980) <u>107</u> 211.

28.   F.T. Wagner and P.N. Ross, J. Electroanal. Chem., (1988) <u>250</u> 301.

29.   C. Lamy, J. Leger and J. Clavilier, J. Electroanal. Chem., (1982) <u>135</u> 321.

30.   D.A. Scherson and D.M. Kolb, J. Electroanal. Che., (1984) <u>176</u> 353.

31.   N.A. Balashova and V.E. Kazarinov in Electroanalytical Chemistry, Ed. A.J. Bard, Dekker, New York (1969) Vol. 3, 135.

32.   E.K. Krauskopf, L.M. Rice and A. Wieckowski, J. Electroanal. Chem., (1988) <u>244</u> 347.

33.   J. Clavilier, J.P. Ganon and M. Petit, J. Electroanal. Chem., (1989) <u>265</u> 231.

34.   H. Ebert, R. Parsons, G. Ritzoulis and T. VanderNoot, J. Electroanal. Chem., (1989) <u>264</u> 181.

35.   F.V. Molina and R. Parsons, J. de Chim. Physique to be published.

MODELING OF METAL-WATER ELECTRIFIED INTERFACES

ROLANDO GUIDELLI and GIOVANNI ALOISI

Dipartimento di Chimica - Università di Firenze

Via G.Capponi,9 - 50121 FIRENZE - Italy

ABSTRACT. A number of models of metal-water electrified interfaces are described and their predictions are compared with the experimental charge dependence of the potential difference across the interface, the differential capacity and the entropy of formation of the interface, after correcting these thermodynamic quantities for the diffuse layer on the basis of the Gouy-Chapman theory. The contribution to the entropy of formation of the interface from the inhomogeneous electron gas is examined. After a brief description of monolayer models, particular emphasis is devoted to three-dimensional models of TIP4T water molecules near a charged wall, in which the statistical mechanical treatment is carried out by the Monte Carlo method as well as by a first-order approximation analogous to the quasi-chemical approximation; in the latter case the water molecules are arranged in a three dimensional lattice. The effect of electron spill over upon the modelistic predictions is briefly examined.

1. Some Experimental Features of Metal-Water Interfaces: Gouy-Chapman Corrected Thermodynamic Quantities

It is well known that electrified interphases between metals and dilute solutions of nonspecifically adsorbed 1-1 valent electrolytes exhibit thermodynamic properties which can be satisfactorily corrected for the so called diffuse layer on the basis of the Gouy-Chapman (GC) theory, yielding pseudothermodynamic quantities which are practically independent of the electrolyte concentration. Thus, if we subtract the potential difference φ_2 across the diffuse layer as calculated on the basis of the GC theory from the experimental potential difference φ^{M-S} across the whole interphase at a constant charge σ for different electrolyte concentrations, we obtain a potential difference φ^{M-2}, commonly called the inner-layer potential difference, which is independent of the electrolyte concentration:

$$\varphi^{M-S} - \varphi_2 = \varphi^{M-2}$$

On experimental grounds, this behaviour is usually determined from differential capacity measurements. If we differentiate both members of the previous equation with respect to the change density σ, we obtain:

R. Guidelli (ed.), Electrified Interfaces in Physics, Chemistry and Biology, 309–336.
© 1992 *Kluwer Academic Publishers.*

$$\frac{d\varphi^{M\text{-}S}}{d\sigma} - \frac{d\varphi_2}{d\sigma} = \frac{1}{C} - \frac{d\varphi_2}{d\sigma} = \frac{d\varphi^{M\text{-}2}}{d\sigma} = \frac{1}{C_i}$$

where the first term is, by definition, the reciprocal of the experimental differential capacity C, whereas the second one is directly calculated from the GC theory. The difference between these two terms turns out to be independent of the electrolyte concentration. The reciprocal of this quantity has the dimensions of a capacity and is called the *inner-layer differential capacity* C_i. For the interphase between mercury and aqueous solutions of the nonspecifically adsorbed NaF, C_i shows a rounded maximum at a positive charge of about +3 $\mu C/cm^2$ [1] (Fig. 1).

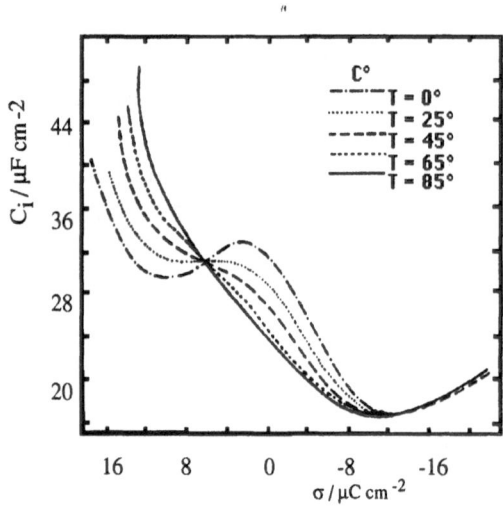

Figure 1. GC-corrected capacity at the Hg / NaF solution interphase vs. charge at different temperatures [1].

By analogous arguments, upon noting that the interfacial tension γ is the opposite of the integral of σ over $\varphi^{M\text{-}S}$, we may separate once again $\varphi^{M\text{-}S}$ into the two contributions $\varphi^{M\text{-}2}$ and φ_2. If we now subtract the quantity $-\int\sigma d\varphi_2$ as calculated on the basis of the GC theory from the experimental value of γ we obtain a concentration independent quantity which can likewise be called the *inner-layer interfacial tension*. Figure 2 shows such a quantity for the interphase between mercury and aqueous NaF [2]. Finally, we may observe that the *entropy ΔS of formation of the interphase*, namely the difference in entropy of the components when they are present at the interphase and when they are present in the bulk of the adjoining phases, is related to the temperature coefficient $(\partial\varphi^{M\text{-}S}/\partial T)_{\sigma,m}$ of the potential difference across the interphase via the pseudothermodynamic relationship $(\partial\Delta S/\partial\sigma)_{T,m} = -(\partial\varphi^{M\text{-}S}/\partial T)_{\sigma,m}$, where m denotes the molality of the

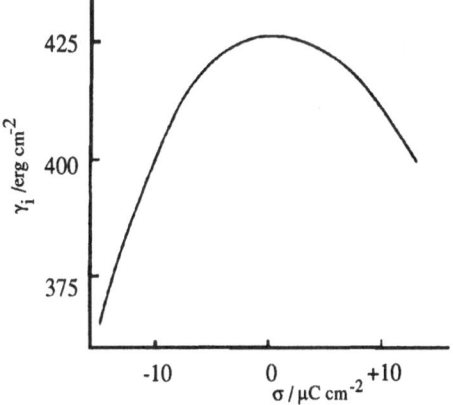

Figure 2. GC-corrected interfacial tension at the Hg / NaF solution interphase vs. charge [2].

electrolyte. If we separate once again $\varphi^{M\text{-}S}$ into the two contributions $\varphi^{M\text{-}2}$ and φ_2 and we subtract the value of $-(\partial\varphi_2/\partial T)_{\sigma,m}$ as calculated on the basis of the GC theory from the experimental value of $(\partial\Delta S/\partial T)_{T,m}$, we obtain a quantity which is approximately independent of the electrolyte concentration. This quantity is the derivative with respect to charge of an entropy which is commonly called the *entropy of formation of the inner layer*, ΔS_i. For the interface between mercury and aqueous NaF this quantity shows a rounded maximum of a negative charge of about -5 $\mu C\ cm^{-2}$ (Fig. 3) [3].

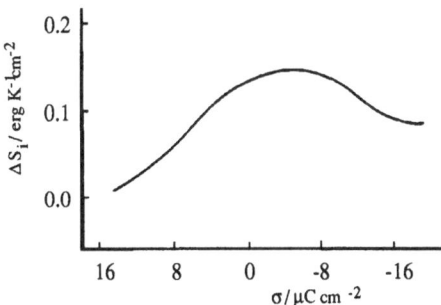

Figure 3. GC-corrected entropy surface excess at the Hg / NaF solution interphase vs. charge [3].

All these GC corrected quantities are usually ascribed to the existence of an ion-free layer adjacent to the electrode surface, called the *inner layer*. However, to explain the dependence of these quantities upon charge it is necessary to ascribe to the inner layer special dielectric properties. This is true not only for macroscopic models which represent the inner layer as a slab of dielectric continuum whose dielectric constant and thickness are considered to depend upon charge, but also for molecular models of the inner layer. The latter models, called *monolayer models*, represent the inner layer as a two-dimensional array of spherical water molecules [4]. The distortional polarization of these molecules is accounted for either by regarding them as immersed in a fictitious fluid with a dielectric constant $\varepsilon_\infty \approx 5$ or, which is substantially the same, by ascribing to these molecules a distortional polarizability about one order of magnitude greater than the value attributed to bulk water. These water molecules are allowed to rotate under the influence of the external electric field, $4\pi\sigma$, created by the charge density σ on the metal surface, and hence to polarize in such a way as to oppose the external field. In other words, the contribution to the potential difference φ^{M-S} across the interphase from the polarization of this monolayer of water molecules (namely the surface dipole potential due to water) becomes progressively more positive as σ shifts towards more negative values.

In recent years *nonprimitive, civilized models* of the interphase have been developed in which solvent molecules and ions are treated on the same footing as dipolar or charged hard spheres extending up to the bulk solution [5]. A major merit of these models is that of having explained the independence of the GC-corrected quantities from the electrolyte concentration without having to introduce an inner layer with special dielectric properties. In particular, the GC potential φ_2 is ascribed nothing but the mere significance which is involved in its derivation, namely the contribution to the potential difference φ^{M-S} across the interphase both from a dielectric continuum with the same dielectric constant as the bulk solvent and from point charges embedded in this medium. Incidentally, the representation of ions by point charges is satisfactorily justified on theoretical grounds, since several sophisticated statistical mechanical treatments of a model of charged spheres embedded in a dielectric continuum yield results in good agreement with those of the simple GC theory up to concentrations of a 1-1 valent electrolyte as high as 1 M.

We might therefore be tempted to conclude that what is left after the GC correction measures that partial contribution from the solvent which is not accounted for by a dielectric continuum, namely the contribution from solvent reorganization close to the metal surface. We must not forget, however, that the GC-corrected quantities also include a contribution from the inhomogeneus electron gas, which may be not entirely negligible, as frequently stressed in the last decade. The latter contribution should be eliminated from the GC-corrected quantities if we wish to obtain quantities exclusively ascribable to water reorganization, or else it should be incorporated into the model.

2. The Electron Gas and its Contribution to the Entropy of Formation of the Interface

It is worth noting that the valence electrons of the metal, due to their small mass, spill over the metal surface. The interaction of the electron gas with the lattice of metal ions is frequently simplified by smearing out the positive charge of the metal ions into a positive background charge, which is called *jellium*. The electron spill over creates a surface dipole, whose positive end is constantly directed towards the metal; in this respect it differs from the surface dipole due to water reorientation, whose positive end is turned towards the solution at positive charges and towards the metal at sufficiently negative charges. If we represent the electron density profile by a centrosymmetrical curve tending asymptotically to the bulk electron density in the direction towards the metal and to zero in the direction towards the solution, the centre of symmetry of this curve lies on the jellium edge when the charge density equals zero (solid curve in Fig. 4). A shift of the centre of symmetry towards the interior of the jellium will obviously create a positive charge density (dashed curve in Fig. 4); in this case the stronger electrostatic interaction between the jellium and the inhomogeneus electron gas will make the latter less deformable. Hence, the slope of the electron density profile will increase and the magnitude of the associated surface dipole will decrease accordingly.

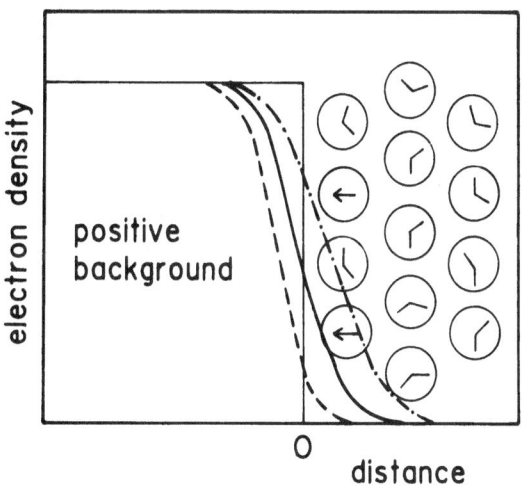

Figure 4. Electron spill over at a metal/water interphase.

On the contrary, a shift of the centre of symmetry outwards will generate a negative charge density; in this case the weaker electrostatic interaction between the jellium and the inhomogeneous electron gas will make the latter more deformable, with a consequent decrease in the slope of the electron density profile and an increase in the associated surface dipole. In conclusion, the surface dipole potential difference χ_{el} between the metal and the solution becomes progressively more positive with an increase in the negative value of σ, and hence tends to oppose the effect of the external electric field $4\pi\sigma$, just as the surface dipole potential difference due to water reorientation.

It is interesting to attempt an estimate of the contribution to the entropy of formation of the interphase from electron spill over [7]. We will start once again from the pseudothermodynamic relationship between this quantity and the potential difference $\Delta\varphi$ across the interface:

$$(d\varphi^{M\text{-}S}/dT)_{\sigma,m} = -(\partial\Delta S/\partial\sigma)_{T,m} \tag{1}$$

where the entropy of formation of the interface ΔS is given by the equation

$$\Delta S = S_{ex} - \Gamma_M S_M - \Gamma_{el} S_{el} - \Gamma_w S_w - \Gamma_+ S_+ - \Gamma_- S_- \tag{2}$$

Here S_{ex} is the *entropy surface excess*, namely the entropy of the real thermodynamic system incorporating the interphase minus the entropy of the corresponding ideal (or Gibbs) system. Now, since the interphase contains an excess (either positive or negative) of the various constituents with respect to the ideal system, in order to obtain ΔS from S_{ex} we must correct S_{ex} by subtracting the sum of the surface excesses of the various constituents (namely the metal atoms M, the electrons el, the water w and the cations and anions of the electrolyte), each excess being multiplied by the corresponding partial molar entropy in the bulk of the respective phase. It should be noted that the electronic term $\Gamma_{el} S_{el}$ is constantly ignored in the literature, often unintentionally, and sporadically because it is explicitly regarded as negligibly small. At any rate, the entropy of formation of the interphase deprived of the electronic term, which we shall denote by ΔS^*, can be determined experimentally, for example through the equation [3]:

$$\Delta S^* = \Delta S + \Gamma_{el} S_{el} = -(\partial\gamma/\partial T)_{E^+,m} + \sigma(S_+ + S^{ref})/F \tag{3}$$

where $(\partial\gamma/\partial T)_{E^+,m}$ is the temperature coefficient of the interfacial tension at a constant applied potential as measured relative to a reference electrode reversible to the electrolyte cation; moreover, S^{ref} is the molar entropy of the pure components of the reference electrode and S_+ is the partial molar entropy of the electrolyte cation. This is clearly the only quantity in eqn.(3) whose estimate involves some reasonable extrathermodynamic assumptions. If we differentiate the first two members of eqn.(3) with respect to σ and we note that $-F\Gamma_{el}$ is just equal to σ, substitution of $\partial\Delta S/\partial\sigma$ from the resulting equation into eqn.(1) yields:

$$(\partial \varphi^{M-S}/\partial T)_{\sigma,m} = -S_{el}/F - (\partial \Delta S^*/\partial \sigma)_{T,m}$$

Upon separating φ^{M-S} into the sum of its three contributions, namely the surface dipole potential χ_{el} due to electron spill over, the contribution φ_w due to water structuring and the GC contribution φ_2 due to point charges and to the dielectric continuum in which they are immersed, we obtain :

$$\left(\frac{\partial \Delta S^*}{\partial \sigma}\right)_{T,m} = -\frac{S_{el}}{F} - \left(\frac{\partial \chi_{el}}{\partial T}\right)_\sigma - \left(\frac{\partial \varphi_w}{\partial T}\right)_\sigma - \left(\frac{\partial \varphi_2}{\partial T}\right)_{\sigma,m} \qquad (4)$$

If we now consider the first two terms of this equation, we may note that $S_{el} = -(d\mu_{el}/dT)$, where μ_{el} is the chemical potential of electrons in the bulk metal. Hence, the sum of these two terms is just the temperature coefficient of the sum of the energy required to overcome the chemical interactions which bind the electron in the bulk metal and of the energy $F\chi_{el}$ required to overcome the potential difference χ_{el} due to electron spill over. This is just the temperature coefficient of the *work function* of the metal, namely the work required to bring the electron from the Fermi level within the metal just outside the metal surface. Upon integrating eqn. (4) over the charge σ starting from $\sigma=0$ we get:

$$\Delta S^*(\sigma) = \Delta S^*(\sigma=0) - \int_0^\sigma \left(\frac{\partial \varphi_2}{\partial T}\right)_{\sigma,m} d\sigma - \int_0^\sigma \left(\frac{\partial \varphi_w}{\partial T}\right)_\sigma d\sigma - \frac{1}{F}\int_0^\sigma \left(\frac{\partial \Phi}{\partial T}\right)_\sigma d\sigma \qquad (5)$$

which shows that ΔS^* may be regarded as consisting of three contributions, namely a contribution from the ions embedded in a dielectric continuum, a contribution from water reorganization and a contribution from electron spill over. We have already seen that, once we subtract from ΔS^* the GC contribution, $-\int_0^\sigma (\partial \varphi_2/\partial T)_{\sigma,m} d\sigma$, from the ions embedded in a dielectric continuum, we obtain a quantity ΔS_i called the entropy of formation of the inner layer, which is roughly independent of the electrolyte concentration [3]. However, to single out the contribution from solvent structuring, we must also subtract the contribution from electron spill over, namely $-(1/F)\int_0^\sigma (\partial \Phi/\partial T)_\sigma d\sigma$.

In principle, both the work function Φ at zero charge and its temperature coefficient are experimentally accessible. Unfortunately the temperature coefficient of Φ at Hg was estimated only indirectly by Farrell and Mc Tigue [8] from measurements of the temperature coefficient of the compensating potential difference of a voltaic cell on the basis of a number of extrathermodynamic assumptions. Assuming that the temperature coefficient of Φ is approximately independent of σ and using the pseudothermodynamic value by Farrell and Mc Tigue, we obtain a correction term for the electronic gas which is obviously proportional to σ. Upon subtracting this contribution from the entropy of formation of the inner layer we obtain the contribution from water structuring, which shows a maximum at a more negative charge, namely at about -9 μC cm^{-2} (curve b in Fig. 5).

The temperature coefficient of the work function of Hg on the basis of a jellium model with pseudopotentials was also estimated [7]. Jellium models do not include the temperature dependence explicitly. Nonetheless, this can be introduced by considering that the electron density in the bulk of mercury is twice as high as the density of the mercury atoms, whose temperature dependence is well known from the literature. The correction term so calculated shows a curvature when plotted against σ (curve c' in Fig. 5), but its behaviour is comparable with that obtained from Farrell and Mc Tigue's pseudothermodynamic value, and gives rise to a contribution to the entropy of formation of the interphase from solvent structuring which shows a maximum at ~-8 μC cm^{-2} (curve c in Fig. 5). Note that this entropy maximum is now close to the minimum of the inner layer differential capacity C_i, as shown in Fig.1. This indicates that the maximum disorder of the interfacial water molecules roughly corresponds to a minimum in their capability of undergoing a reorientation under the effect of a change in the external electric field [7].

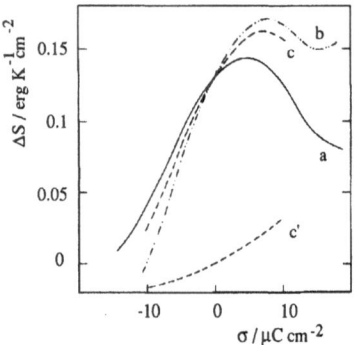

Figure 5.(a): plot of ΔS_i vs. the charge σ at the Hg / 0.1 M NaF solution interphase. (b) and (c): plots of $\Delta S_i + (1/F)\int(\partial\Phi/\partial T)d\sigma$ as obtained from the Φ value of Farrel and Mc Tigue (b) and from the jellium model (c). (c') $(1/F)\int(\partial\Phi/\partial T)d\sigma$ as obtained from the jellium model [7].

We may summarize the above experimental results as follows:
1) The differential capacity C_i corrected for the diffuse layer on the basis of the GC theory, which can be ascribed to water reorganization near the electrode surface and to electron spill

over, shows a maximum at a positive charge.

2) The entropy of formation of the interphase as corrected for the diffuse layer on the basis of the GC theory shows a maximum at a negative charge.

We may further add that several pieces of experimental evidence combined with a few moderate extrathermodynamic assumtions indicate that the potential difference φ^{M-S} across the mercury/water interphase at zero charge is negative.

The above three features can be explained by water reorganization close to the electrode surface, by electron spillover or by both these causes. In the early literature electron spillover was completely disregarded. However, during the last decade a new school of thought has taken up an opposite position, by trying to explain the properties of C_i almost exclusively on the basis of electron spillover. We will briefly examine this position later on. For the moment we will try to predict the behaviour of water molecules against a charged wall and how this behaviour may affect interfacial properties.

3. Models of Water Molecules Against a Charged Wall

3.1. MONOLAYER MODELS

Monolayer models, which represent the solution side of the interphase by a two-dimensional array of solvent molecules, are not in a condition of explaining the above experimental features unless a high number of adjustable parameters is introduced into the model [4]. When this is the case, however, it is difficult to establish how much of the interpretative potentialities of these models is to be ascribed to their capability of mimicking reality and how much to the high number of adjustable parameters. Consider, for instance, the negative value of φ^{M-S} at zero charge. The two-dimensional lattice of solvent molecules is actually situated in vacuo, and hence it cannot distinguish the bulk metal from the bulk liquid at zero charge, when the external electric field is also equal to zero. Hence the average orientation of the solvent molecules in the direction normal to the array, and consequently φ^{M-S}, will be necessarily equal to zero unless we introduce a fictitious external *adsorption potential* favouring the water orientations with the negative end of the dipole pointing towards the metal. This involves at least one adjustable parameter. If we now assume a random distribution of the various water orientations over the adsorption sites of the two-dimensional lattice, in spite of the different energies of lateral interaction between the different orientations, then the calculated capacity maximum turns out to lie at a negative charge, rather than at a positive one as observed experimentally. To predict a positive capacitive maximum one is forced to complicate the physical picture, for instance by introducing both water monomers and H-bonded water clusters and by ascribing different dipole moments and different energies of chemical interaction to these various

318

species. This involves further adjustable parameters, whose number is no less than 5 and may easily be appreciably higher [4].

An appreciable improvement is attained by considering explicitly H-bond formation between the adsorbed water molecules and by accounting for the *local order* which any given adsorbed water orientation tends to create around its own adsorption site by favouring the occupation of the neighbouring sites by those water orientations which interact more attractively with it. (We will return on the importance of accounting for local order in connection with organic adsorption on metals from water.) In this way the total number of adjustable parameters is reduced to three [9]. According to this model water molecules are assumed to occupy a square lattice and to take any of 25 orientations which are obtained, via a slight rotation, starting from the ideal orientations having the four H-bonding directions pointing towards four of the eight vertices of the elementary cubic cell (see Fig. 6).

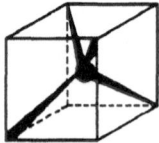 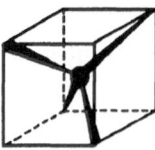

Figure 6. Allowed orientations of water molecules with H-bonds directed towards the vertices of the unit cell.

Clearly, all these orientations have the same a priori probability of forming a maximum of two H-bonds with the adjacent water molecules of the solution phase. The slight rotations imparted to the above ideal orientations are such as to lift one of the two H-bonding directions pointing towards the metal so as to permit the formation of an almost horizontal, slightly bent H-bond between two adsorbed water molecules occuping cubic cells having a lateral edge in common (Fig. 7).

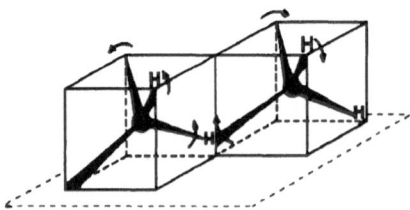

Figure 7. Rotations required to allow H-bonding between two water molecules of the monolayer [9].

To account for the different tendency of different metals to polarize the atomic orbitals of the adsorbed water molecules with the *electron tail* due to electron spill over (or, putting it in a somewhat different way, to interact chemically with the oxygen atom of the adsorbed water molecules), a *chemisorbed water orientation* is also introduced in the model. This orientation is one of the ideal orientations of Fig. 6 having the dipole moment fully aligned in the direction towards the solution, and hence it is not allowed to form H-bonds within the adsorbed monolayer. In this respect this orientation interacts more attractively not only with the adjacent metal phase but also, and mainly, with the adjacent water phase, due to its higher tendency to form H-bonds with the overhanging water molecules of the latter phase (Fig. 8).

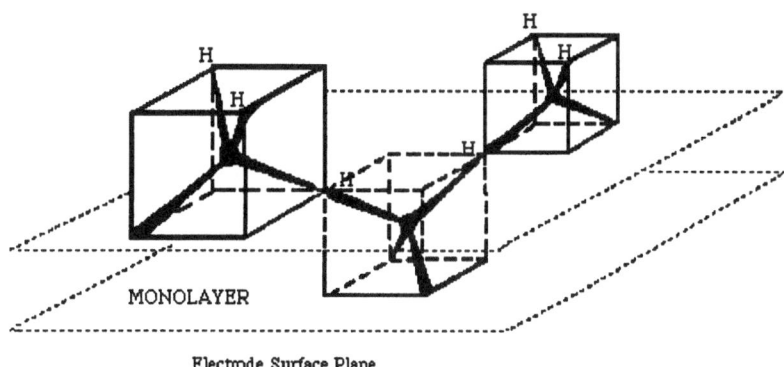

Figure 8. Sketch of H-bonding of a chemisorbed water molecule in the monolayer with two *bulk* water molecules.

The energy of *chemical interaction* of the chemisorbed water orientation with the two contacting phases is therefore assumed to be more negative, namely more attractive, than that of the other water orientations by an amount which we shall denote by ΔU_c.

This model involves only three adjustable parameters, namely (1) the distance of closest approach between two water molecules, l, which is also regarded as a measure of the thickness of the monolayer, (2) a dielectric constant ε accounting for distortional polarization, and (3) ΔU_c. In practice, ε and l can be varied only within relatively narrow ranges. In fact, physically significant values of l and ε cannot differ too much from the minimum O-O distance in ice ($l = 0.278$ nm) and from the square of the refractive index of water ($n^2=4.5$). Hence, the only widely adjustable parameter is ΔU_c, which depends on the nature of the metal. By treating this model of differently oriented and H-bonded water molecules on the basis of an approximate statistical mechanical procedure accounting for local order called the *quasi-chemical approximation*, calculated differential capacity curves are obtained which agree satisfactorily with the experimental ones [9]. Even this model, however, fails to predict the position of the entropy maximum at negative charges. In fact, this maximum is generally predicted to lie at a slightly positive charge, in the proximity of the maximum of the calculated differential capacity curve [10].

To correctly predict the position of the entropy maximum, those models which use solvent monomers and clusters in the random mixing approximation are forced to introduce a number of additional adjustable parameters ascribing an *ad hoc* temperature dependence, say, to the dipole moments and/or to the energy of chemical interaction of the water clusters [4]. With these models the number of adjustable parameters ranges from 9 to 12.

The need is therefore felt to interpret the salient features of water reorganization at an electrified interphase by a model which makes use exclusively of molecular parameters determined by independent means from the behaviour of bulk water. Let's see how we can proceed in this direction. First of all the pseudothermodynamic quantities corrected on the basis of the GC theory and for electron spill over are independent of the electrolyte concentration, and hence it should be possible to interpret them on the basis of a model of water molecules extending from the charged wall up to the bulk liquid in the absence of ions. This seems to be an attractive procedure because the simultaneous consideration of the short-ranged but strong H-bonds and of the long-ranged ion-ion and ion-wall interactions represents a stumbling block from a statistical-mechanical point of view.

3.2. A MONTE CARLO SIMULATION OF WATER MOLECULES NEAR A CHARGED WALL

Before estimating the contribution from water reorganization to the potential difference across an electrified interphase on the basis of a molecular model excluding ions, we must ask ourselves which is the limit of the GC theory when the electrolyte concentration in a dielectric continuum tends to zero. To this end it is convenient to consider a model of point charges embedded in a dielectric continuum between two oppositely charged parallel walls, adopting as a boundary condition for the solution of the PB equation the electroneutrality of the whole solution enclosed between the two walls [11]. We can see from Fig. 9 that at sufficiently high electrolyte concentrations the electric potential over the intermediate region between the two walls is constant, and hence separates the whole system into two independent electrified interphases. As the electrolyte concentration is gradually decreased, however, the two interphases tend to merge and the electric field tends to penetrate into the middle of the solution, so that it is no longer possible to refer to a bulk electric potential or to a potential difference between any of the two walls and the bulk liquid. Ultimately, when the electrolyte concentration becomes practically equal to zero, the electric field, which is measured by the slope of the potential-distance profile, tends to a maximum constant value throughout the solution. This is equal to $4\pi\sigma/\varepsilon$, where ε is the dielectric constant of the continuum. This is indeed the limiting behaviour of the GC theory when the electrolyte concentration tends to zero, and hence the GC corrected quantities measure the deviations

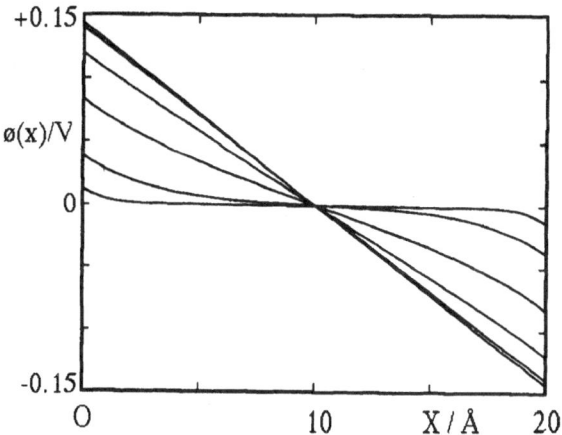

Figure 9. Potential-distance profiles for an aqueous 1-1 valent electrolyte of concentration 10, 1, 0.1, 10^{-2}, 10^{-3} and 10^{-4} M (when proceeding upwards on the left-hand side of the figure) between two planar, oppositely charged electrodes at a distance of 20Å. The charge on the left electrode equals $10\mu C$ cm^{-2} [11].

from this behaviour as exhibited by a system of discrete water molecules enclosed between two charged walls in the absence of ions.

This problem has been tackled by a Monte Carlo (MC) simulation of a system of soft water spheres enclosed between two soft parallel charged walls at a distance of 30 Å from each other [12]. The water-water interactions are expressed by the TIP4P intermolecular potential (Fig. 10). This consists of a Lennard-Jones 12-6 spherical potential and of three partial charges incorporated in the spherical molecule, which simulate H-bonding. Two equal positive charges $+\delta$ are in a peripheral position whereas a negative charge -2δ is close to the centre of the sphere, albeit slightly shifted towards the two positive charges.

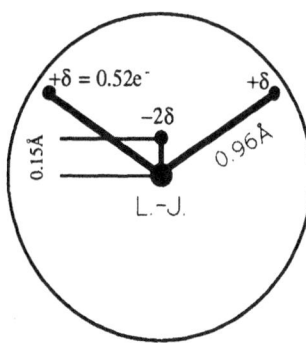

Figure 10. Sketch of the TIP4P water model.

The distribution of these partial charges was estimated in such a way as to account for as many thermodynamic properties of bulk water as possible [13]. The potential-distance profile between the two walls is shown in Fig. 11 for different values of the charge density σ on the left wall.

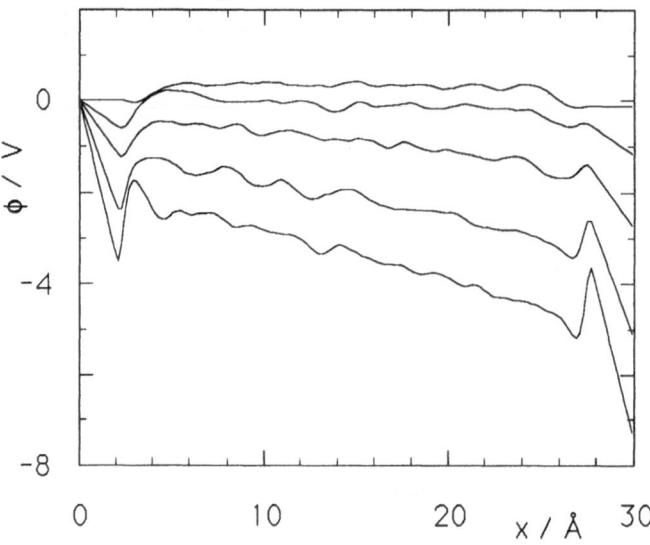

Figure 11. Potential-distance profiles between two planar, oppositely charged walls at a distance of 30Å, with TIP4P water molecules in between. The charge density on the left wall equals 0, 2.5, 5, 10 and 15 μC cm^{-2}, starting from the upper curve [12].

It should be noted that at zero charge the potential difference between any of the two walls and the liquid is negative. This negative potential difference, which correctly accounts for experimental behaviour, stems in a natural way from the asymmetrical distribution of the partial charges inside the water molecule, without having to introduce a fictitious external adsorption potential such as to favour water orientations with the negative end of the dipole pointing towards the metal. At zero charge the adsorbed water molecules tend to assume compromise orientations which favour H-bond formation parallel to the electrode surface, due to the impossibility of H-bonding in the direction towards the metal, while still retaining a sufficiently high H-bonding capability in the direction towards the bulk liquid. On the average these compromise orientations have the more peripheral partial charges (namely the positive ones) slightly directed towards the bulk liquid. Note that a symmetrical point-charge water-water potential such as the Ben Naim-Stillinger potential, with two positive and two negative charges distributed symmetrically at the corners of a tetrahedron, would lead necessarily to a zero value of the potential difference across the interface at zero charge for obvious symmetry reasons. It is now generally recognized, however, that asymmetrical point-charge water-water potentials, such as the ST2 or the TIP4P potential adopted by us, are superior to symmetrical ones in predicting the bulk properties of water.

At charges different from zero the electric potential shows a rapid change within the *vacuum* region enclosed between the charged wall and the equilibrium distance of closest approach of the centres of the water molecules. Here the potential gradient is equal to the electric field *in vacuo* between the two charged walls, i.e. $4\pi\sigma$. At larger distances from the wall we first have an abrupt decrease in the absolute value of the electric potential due to the orientation polarization of the water molecules in the immediate vicinity of the electrode, and then small oscillations which are mainly due to the noise in the numerical calculations involved in the MC simulation. Apart from this noise, we may observe that in the intermediate region between the two walls the electric potential varies almost linearly with distance, exhibiting a slope which is proportional to the charge density σ. Such a slope is clearly equal to the electric field $4\pi\sigma/\varepsilon$, and hence provides a value for the bulk dielectric constant ε which is consistent with the choice of the TIP4P potential as the water-water intermolecular potential. The potential-distance profile also shows that the wall-water correlation tends to vanish at distances greater than one, or at most two, molecular diameters; beyond this distance the wall practically "sees" the water molecules as a dielectric continuum.

If the water behaved as a dielectric continuum throughout the whole region between the walls, as predicted by the GC theory at infinite dilution, then the potential profile would show the slope $4\pi\sigma/\varepsilon$ starting from the wall surface, as shown by curve *a* in Fig. 12. It is

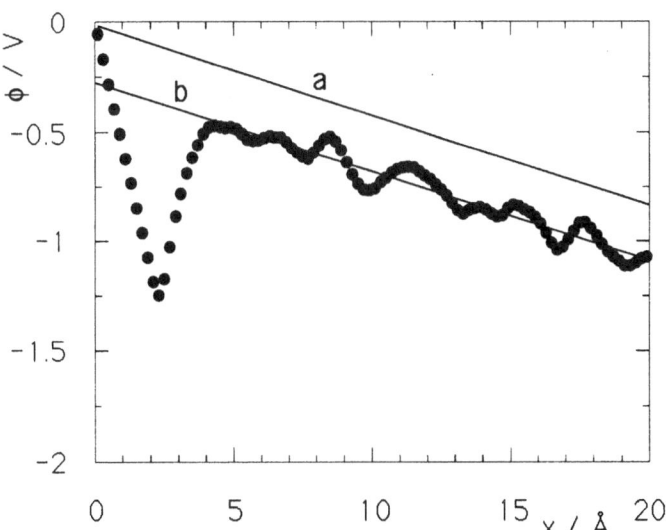

Figure 12. Potential-distance profile between two parallel, oppositely charged walls at a distance of 30 Å, for a charge of 5 µC cm^{-2} on the left wall: MC results (dots), GC behaviour (**a**) and extrapolation of the bulk potential-distance profile to the left wall (**b**).

therefore evident that the vertical separation between curves *a* and *b* in Fig. 12 measures the contribution to the wall-liquid potential difference due to the deviations of the solvent behaviour from that of a dielectric continuum with the same bulk dielectric constant, namely the contribution φ_w from water reorganization.

3.3. THREE-DIMENSIONAL LATTICE MODEL OF TIP4P WATER MOLECULES NEAR A CHARGED WALL

In order to reduce the noise of the MC simulation as well as the computational time to a notable extent without losing the main advantages of the model, we have simplified the model by adopting a three-dimensional lattice model of water molecules interacting according to the TIP4P point-charge intermolecular potential [14]. The statistical-mechanical treatment of this regular lattice was carried out by accounting for local order on the basis of the pair correlation approximation of Kirkwood's variation method [15]. According to this method the probability density $P(\{i_\lambda\})$ of a given configuration, which consists of a well defined water orientation i_λ for each lattice site λ, is first set equal to the product of the singlet probability densities $P(i_\lambda)$ over all sites, where $P(i_\lambda)$ expresses the probability of occupation of a single site λ by the corresponding water orientation i_λ. If we confined ourselves to this product, we would assume that the occupation of a given site by a given orientation does not affect the probability of occupation of the neighbouring sites, namely that the different orientations are distributed randomly over the sites, in spite of the large differences in the energies of lateral interaction between the different pairs of orientations. This is indeed a very rough assumption, because the energy of a H-bond is about 6 times greater than the energy kT of thermal agitation. To account for the local order created by H-bonding, a further product is introduced into the expression for the probability $P(\{i_\lambda\})$ of the given configuration $\{i_\lambda\}$. This is a product of pair correlation functions $g(i_\lambda,j_\mu)$ over all possible pairs (λ,μ) of nearest neighbour sites; $g(i_\lambda,j_\mu)$ expresses the probability that the two pairs of nearest neighbour sites λ and μ are occupied by the water orientations i_λ and j_μ with respect to the probability which we would expect in the case of a random distribution of the various orientations over the sites. In other words, all pair correlation functions $g(i_\lambda,j_\mu)$ would tend to unity in the case of a random distribution over the sites. Naturally, all pair correlation functions must be determined by an iterative numerical procedure through the minimization of the Helmholtz free energy of the system, upon taking into account the geometrical constraints imposed by the lattice structure *via* a set of Lagrangian undetermined multipliers.

The notable decrease in the computational time attained by the use of a three-dimensional lattice model of water molecules is due to the freezing of the translational degrees of freedom as well as to the ready singling out of the "most probable" orientations, namely the orientations which make by far the major contribution to the partition function of the whole system. These are indeed the orientations which point at least two of their four H-bond forming directions towards any two of the first nearest-neighbour sites. The symmetrical location of the nearest-neighbour sites around the central site ensures that the selected water orientations are distributed symmetrically along the space directions. An apparent drawback of lattice models is represented by the fact that they underestimate entropy and, in fact, they entirely neglect the communal entropy which gas molecules possess with respect to molecules in a crystal. Fortunately, this seems to be a minor drawback when estimating surface excess quantities and their dependence upon charge. Thus, under the reasonable assumption that the communal entropy varies only slightly in passing from the interfacial region to the bulk, its additive contribution cancels out from the entropy surface excess. A further justification for the use of three-dimensional lattice models stems from the observation that the choice of the lattice geometry does not affect the qualitative conclusions which can be drawn from the model. Thus, a body-centered cubic lattice and a face-centered cubic lattice yield qualitatively analogous results. An experimental justification for the use of lattice models comes from the observation that the interfacial properties vary in a gradual way when passing across the freezing point. Thus, the differential capacity at the interphase between gold and $HClO_4 \cdot 5.5\ H_2O$ shows the typical capacity hump commonly ascribed to water reorientation; this hump does not undergo critical changes in passing from the liquid to the solid state, in spite of the resulting suppression of translational degrees of freedom [16].

Here we will report the results obtained with a body-centered cubic lattice. The two groups of allowed water orientations are shown in Fig. 13. Twelve orientations have the hydrogens directed towards the ends of any of the twelve diagonals of the six faces of the elementary cubic cell (Fig.13a). The other twelve orientations have the hydrogens approximately directed towards the ends of any of the twelve edges of the cubic cell (Fig. 13b). The statistical mechanical calculation was carried out on the pairs formed by the central site not only with the eight first nearest-neighbour pairs, but also with the six second nearest-neighbour pairs.

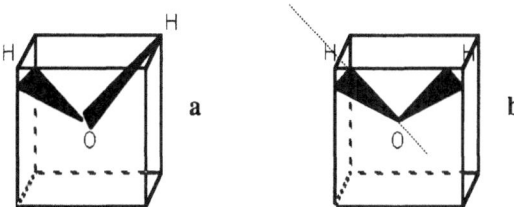

Figure 13. The two groups of allowed orientations used in the model.

It should be noted that the TIP4P distribution of partial charges within a water molecule has a dipole moment of 2.18 D, which is greater than that, 1.85 D, of an isolated water molecule, since it incorporates the effect of distortional polarization in bulk water. Since at an electrified interphase distortional polarization is due primarily to the anisotropic external electric field $4\pi\sigma$, it was deemed it convenient to scale down the partial charges of the TIP4P water molecule by the factor (1.85/2.18), which yields a dipole moment of 1.84 D; distortional polarization due both to the external electric field and to the other water molecules of the three-dimensional lattice was then accounted for in the random-mixing approximation by using the scalar polarizability of bulk water, which amounts to ≈ 1.5 Å^3.

According to the present model both the water polarization and the entropy per unit surface of the lattice attain constant "bulk" values after two, or at most three, lattice planes of water molecules as counted from the charged wall. Surface excess quantities were therefore readily calculated by subtracting the bulk values of these quantities from the corresponding values on the first three lattice planes adjacent to the charged wall. It must be noted that the surface excess of the electric potential, which measures the contribution φ_w from solvent reorganization to the potential difference across the interface, also includes the potential difference *in vacuo*, $4\pi\sigma d$, across the distance d separating the charged wall from the first lattice plane minus the corresponding potential difference in the bulk, $4\pi\sigma d/\varepsilon$, across the same distance. Since the bulk dielectric constant ε as estimated from the model is >>1, as expected, the second contribution to the potential difference is negligible with respect to the first one, and hence the estimated value of φ_w is notably affected by the choice of d. Note that d is generally different from the Wan der Waals radius of water molecules. This is because water-metal dispersion interactions, which affect the distance of separation d, are generally different from water-water dispersion interactions. A purely theoretical calculation of the potential energy curve for metal-water dispersion interactions is a formidable task. In the absence of such a calculation the d value must be regarded as an adjustable parameter, since it comes into play when the metal and the water phase come into contact, and hence its value cannot be inferred by independent means from the behaviour of the two separate phases. Fortunately enough the d value, while affecting the magnitude of the calculated values of φ_w and of the corresponding differential capacity C_w, has a negligible effect upon the position of the extreme points of these quantities on the charge axis.

Figures 14 and 15 show plots of φ_w vs. σ and of $C_w = d\sigma/d\varphi_w$ vs. σ for two different values of d. It can be seen that φ_w is negative at zero charge, in agreement with experiment. The φ_w value at zero charge, ≈ -0.3 V, is in good agreement with the value estimated by the MC simulation [12,17]. The curve of the differential capacity $C_w = d\sigma/d\varphi_w$ against σ shows a maximum value at a positive charge $\sigma \approx +5$ μC cm^{-2}, once again in agreement with experiment. Finally, Fig. 16 shows the calculated curve of the entropy surface excess ΔS_w against σ, which exhibits a maximum at a negative charge $\sigma \approx -6$ μC cm^{-2}.

It is evident that the present model accounts satisfactorily for all salient features of the experimental behaviour without the introduction of adjustable parameters, apart from the distance of separation d, whose value, however, does not effect the qualitative conclusions. The fact that the entropy maximum lies at negative charges is readily

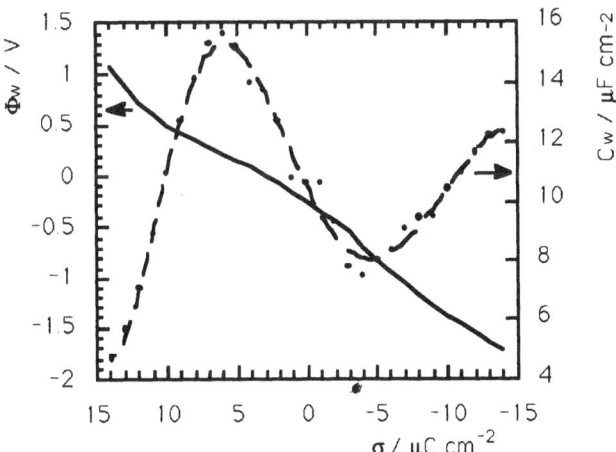

Figure 14. Calculated plots of φ_w and C_w vs. σ for $d = 2.5$Å.

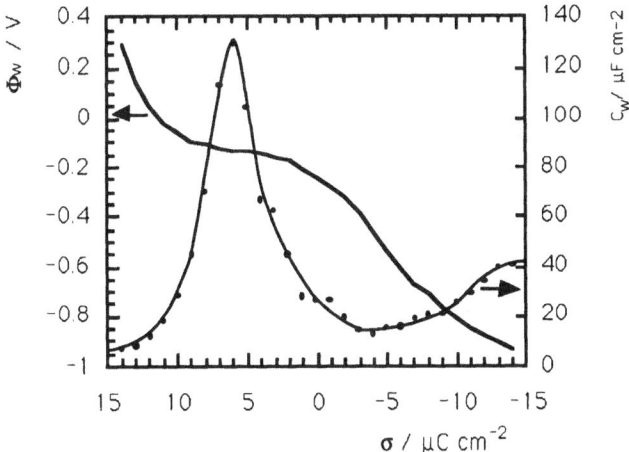

Figure 15. Calculated plots of φ_w and C_w vs. σ for $d = 3$Å.

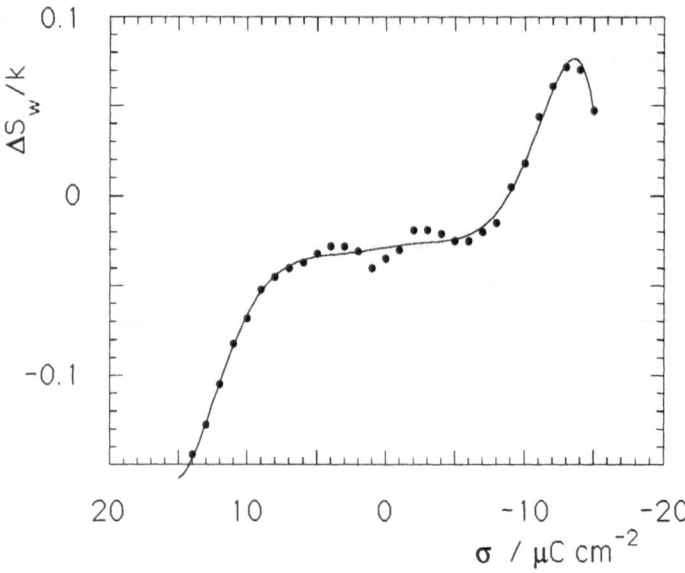

Figure 16. Calculated plot of ΔS_w vs. σ for $d = 2.5$ Å.

understood if we consider that the orientations which are generated from each other by rotation around the axis normal to the wall are equivalent, in that they are characterized by the same probability density. Now, for any given lattice structure, including the one adopted by us, the number of equivalent orientations is higher the more horizontal is their dipole moment, since the chosen orientations are distributed rather uniformly along the directions of space (see Fig. 17a). It follows that the more horizontal are the water dipoles, the higher will be the number of equivalent orientations on which the water molecules will scatter uniformly, with the resulting attainment of a maximum entropy. In view of the average orientation of the adsorbed water dipoles at zero charge with the negative end slightly turned towards the metal, these dipoles will become more horizontal as the charge density σ is shifted towards negative values, and hence the entropy surface excess will attain its maximum value there.

On the other hand, the differential capacity will attain its maximum value where the availability of the adsorbed water molecules to undergo a change in orientation with a change in σ is a maximum. This situation is encountered when the water molecules start to align to an appreciable extent in the direction normal to the wall. Such an alignment is self-enhancing because, once one adsorbed water molecule is strongly aligned, it will no longer be in a condition of forming H bonds laterally with the surrounding adsorbed water molecules. Hence, the latter molecules will also tend to undergo a rapid alignement with

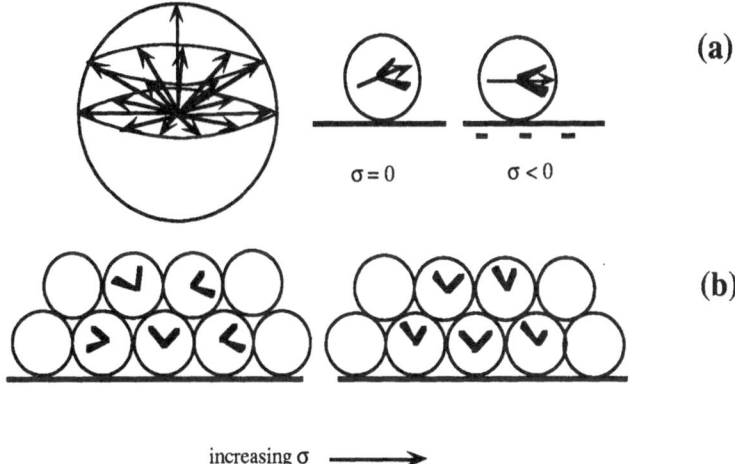

increasing σ ⟶

Figure 17. (a) Uniform distribution of the directions of water dipoles in space (only the upper half-space is shown) and the expected change in the average orientation of water dipoles in passing from zero charge to negative charge values. (b) Rapid increase in the alignment of water dipoles along the direction of the external electric field as σ becomes progressively more positive.

increasing the electric field, so as to be able to form stronger H bonds with the overhanging water molecules (see Fig.17b). In view of the average orientation of the adsorbed molecules at zero charge, an increase in their partial alignment can only be realized by shifting the charge density towards positive values. It is therefore in this direction that the capacity shows a maximum.

4. Effect of Electron Spill over Upon Modelistic Predictions

After having examined some models of water molecules against a charged wall and their comparison with experiment, let us consider how the inclusion of electron spill over can modify this picture. We have seen that a gradual shift of the charge density σ towards more negative values causes a concomitant shift of the electron density profile outwards and a progressive decrease in its slope, namely a spreading out of the electron tail. If the water molecules approach the jellium edge, $x=0$, up to a minimum equilibrium distance d, it is possible to show that the potential difference φ^{M-S} across the interface due to the overall charge of the uniform background of positive charges (*jellium*) and of the nonuniform electron gas, with the only exclusion of the contribution from water polarization, can be simply represented as the potential difference across the plates of an ideal capacitor:

$$\phi^{M\text{-}S}(\sigma) - \text{contribution from water polarization} = \phi_j^{M\text{-}S}(\sigma) = 4\pi\sigma(d - x_\sigma) \qquad (6)$$

While the plate of this capacitor on the water side is located at $x=d$, the plate on the metal side is located at the "center of mass" of the overall charge of the (jellium + electron gas):

$$x_\sigma = \int x\,[n_+\theta(-x) - n_\sigma(x)]dx \ / \int [n_+\theta(-x) - n_\sigma(x)]dx =$$

$$= \int x[n_+\theta(-x) - n_\sigma(x)]dx \ / \ \sigma \qquad (7)$$

where n_+ is the constant charge density of the uniform positive background, $\theta(-x)$ is the Heaviside step function, and $n_\sigma(x)$ is the density of the electron gas at the charge density σ and at a distance x from the jellium edge. In practice, x_σ may also be regarded as a weighted average of the distance x from the jellium edge, the weighting factor being represented by the overall charge density at that distance. It is more convenient to refer both $\phi_j^{M\text{-}S}(\sigma)$ and x_σ to the corresponding values at zero charge, by writing:

$$\phi_j^{M\text{-}S}(\sigma) - \phi_j^{M\text{-}S}(0) = 4\pi\sigma \ (d - x'_\sigma) \qquad (8)$$

where

$$x'_\sigma = x_\sigma - x_0 = -\int x\,[n_\sigma(x) - n_0(x)] \ dx \ / \int [n_\sigma(x) - n_0(x)] \ dx =$$

$$= -\int x[n_\sigma(x) - n_0(x)] \ dx \ / \ \sigma \qquad (9)$$

where $n_0(x)$ is the electron density at zero charge. Note that the *jellium* charge density n_+ cancels out from the expression of x'_σ, since it does not change with varying σ. The position of x'_σ can be qualitatively localized by considering the electron density profile both at positive and negative charge densities with respect to that at zero charge (see Fig. 18). The progressive decrease in slope of this profile with increasing $-\sigma$ causes the difference $[n_\sigma(x)-n_0(x)]$ to attain a maximum absolute value at a positive x value, namely outside the *jellium*, such that the location $x=x'_\sigma$ of the centre of charge is also positive. Differently stated, the metal-side plate of our ideal capacitor stands in front of the jellium edge and is expected to move outwards as σ becomes progressively more negative.

From eqn.(8) it is apparent that the separation between the $x=x'_\sigma$ and $x=d$ planes is a measure of the reciprocal of the integral capacity of the inner layer, with the exclusion of the contribution from water polarization:

$$[\phi_j^{M\text{-}S}(\sigma) - \phi_j^{M\text{-}S}(0)] \ / \ \sigma \ = \ 1 \ / \ K_j = 4\pi \ (d - x'_\sigma) \qquad (10)$$

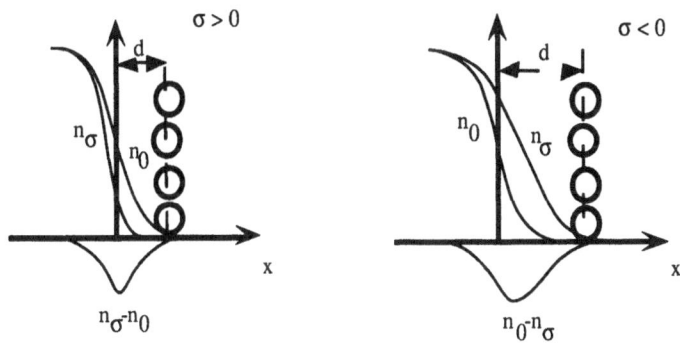

Figure 18.Electron density profile n_σ both at a positive and at a negative charge density with respect to the corresponding profile at zero charge, n_0.

Hence, if the position $x=d$ of closest approach of the water molecules to the jellium edge is kept constant, then the gradual shift outwards of the centre of mass of the overall charge, $x=x'_\sigma$ (also referred to as the *image plane*), which takes place as σ becomes progressively more negative, causes the two plates of our ideal capacitor to approach each other with a consequent increase in the integral capacity K_j (see Fig. 19). Since the experimental integral capacity in the proximity of zero charge increases in the opposite direction, namely towards positive charges, this experimental behaviour can only be explained on the basis of water polarization and of its dependence upon σ [18].

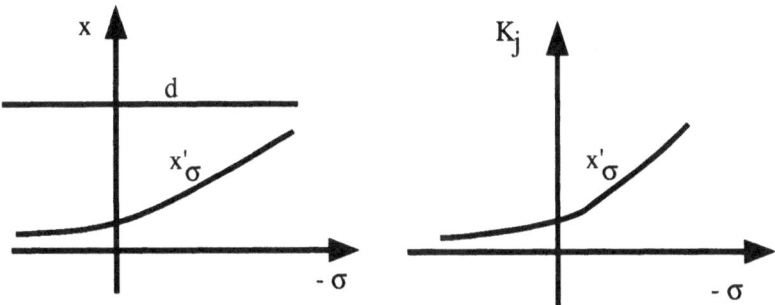

Figure 19. Positions of the $x=d$ and $x=x'_\sigma$ planes and integral capacity K_j as a function of the charge density σ.

The modelistic predictions can be notably modified if the $x=d$ plane is allowed to move with varying the charge density σ on the metal [18,19]. Indeed, its equilibrium position can be regarded as determined by a balance of opposing pressures, namely:

332

(I) a Van der Waals pressure attracting the water molecules towards the metal;

(II) a pressure resulting from the interaction of the electron tail with the electron shells of the water molecules, which repels them away from the metal;

(III) an electrostatic pressure resulting from the interaction of the electron tail with the water dipoles, and which may be either attractive or repulsive depending on the average orientation of the latter dipoles;

(IV) an electrostatic pressure resulting from the attraction between the diffuse layer ions and the metal charge σ, and which all theoretical approaches regard as proportional to the square σ^2 of the charge density.

Pressure (II) tends to produce a progressive repulsion of the water molecules, and hence of the $x=d$ plane, as the image plane $x=x'_\sigma$ moves outwards, thus causing the $x=d$ plane to move roughly parallel to the image plane. On the other hand, the electrostatic pressure (IV), provided it is of comparable magnitude, tends to push the $x=d$ plane progressively towards the jellium edge as σ^2 increases. As a result of these two pressures, the $x=d$ plane is expected to attain a maximum distance from the *jellium* edge at a negative charge and to approach gradually the *jellium* as we depart from this charge (see Fig. 20). At sufficiently positive σ values the $x=d$ plane, as distinct from the image plane $x=x'_\sigma$, may be predicted not to approach the jellium edge beyond a minimum distance, for obvious physical reasons. Upon focusing our attention on the separation between these two planes and on its physical significance as expressed by eqn.(10), we may therefore predict an integral capacity showing a maximum at a positive charge and a minimum at a negative charge, as shown in Fig. 20.

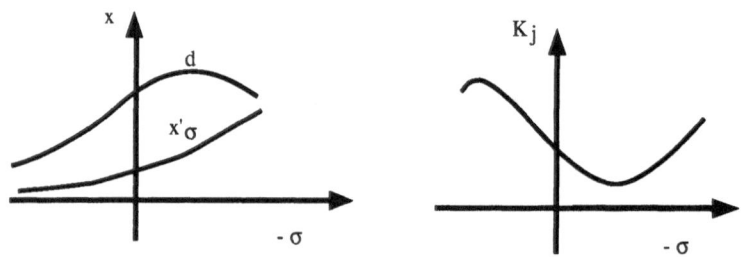

Figure 20. Effect of the balance of the different pressures upon the positions of the $x=d$ and $x=x'_\sigma$ planes and upon the integral capacity K_j as a function of the charge density σ.

The positive maximum shown by the experimental curves of the inner-layer differential capacity C_i vs. σ (see Fig.1) may thus be qualitatively justified on the basis of the *jellium* model and of a variable distance of closest approach of the solvent molecules, without having recourse to water polarization [18].

The above arguments seem, however, exceedingly optimistic. Thus, the Van der Waals attractive pressure (I) between metals and water molecules, as estimated on the basis of electrodynamical considerations [20], turns out to be relatively high. Hence, in order to attain an equilibrium distance of closest approach $x=d$ which is not unphysically short, we must oppose this attractive pressure with a high repulsive pressure (II), which can be generated, for instance, by a potential barrier fixed with respect to the water molecules and repelling the electron tail. As concerns the electrostatic pressure (IV) due to the diffuse-layer ions, it can be shown on general grounds [21] that it is equal to $2\pi\sigma^2/\varepsilon$, where ε is a dielectric constant. If ε is ascribed a value close to unity, as done by Badiali et al.[19], then this pressure is comparable with the other pressures (I) and (II) for reasonable values of σ, and hence has an effect similar to that schematically depicted in Fig. 20. However, the expression $2\pi\sigma^2/\varepsilon$ can also be directly derived from the Gouy-Chapman theory; hence, if we accept the validity of this theory and we compare theoretical predictions with GC corrected quantities, then the ε value cannot be a matter of choice, since the GC correction for the diffuse layer yields satisfactory results only provided ε is given the value, ≈ 80, for bulk water. With this ε value, the electrostatic pressure (IV) is more than one order of magnitude lower than the two opposing pressures (I) and (II), and therefore exerts no appreciable effect upon the location of the $x=d$ plane.

Figure 21 shows plots of $x=x'_\sigma$ and $x=d$ vs. σ as obtained upon combining the three-dimensional lattice model of TIP4P water molecules previously described with a simple one-parameter jellium model, and by determining the equilibrium distance $x=d$ by an iterative self-consistent procedure based on a balancing of the previously listed contributions to the pressure, from (I) to (IV). In practice, once we set $\varepsilon=80$, only the two opposite pressures I and II bear a significant contribution, and both planes $x=x'_\sigma$ and $x=d$ move progressively away from the jellium edge with increasing negative charge. In view of eqn.(10) the contribution from the (jellium + electron gas) to the dependence of the integral capacity upon σ is small, because the separation between the plates of our ideal capacitor varies only slightly with varying σ. The contributions to the overall potential difference φ^{M-S} across the interface as obtained from this model are shown in Fig. 22. It can be seen that the potential difference $4\pi\sigma d$ created by σ in the absence of polarization effects does not show a perfect linear dependence upon σ because of the charge dependence of d ; as opposed to this potential difference, the surface dipole potential differences χ_{el} and χ_w, due to the polarization of the electron gas and of water molecules respectively, become progressively more positive as σ shifts towards negative values. The algebraic sum of these three contributions yields the whole potential difference φ^{M-S} across the interface, whose plot against σ shows a minimum slope at positive charge densities; this corresponds to a maximum in the differential capacity $C_i=\partial\sigma/\partial\varphi^{M-S}$. It is readily seen that such a maximum is determined by water polarization, which opposes the external electric field $4\pi\sigma$ more effectively just at these positive charge densities.

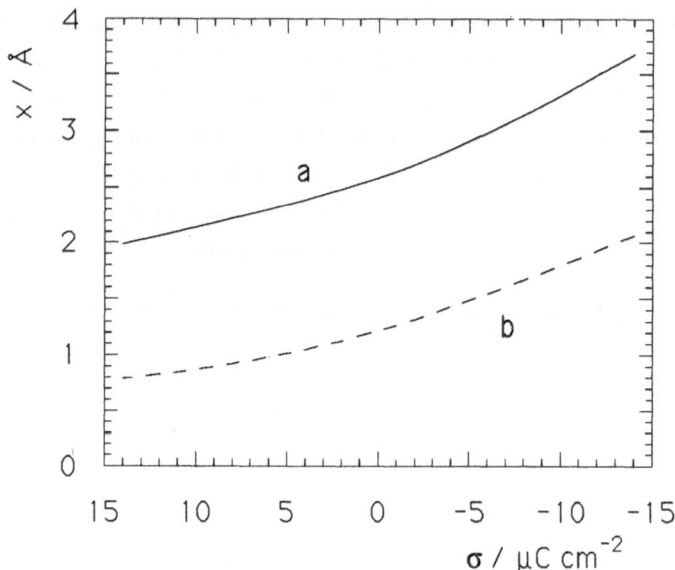

Figure 21. Calculated positions of the x=d (a) and x=x'$_\sigma$ (b) planes as a function of σ for ε = 80.

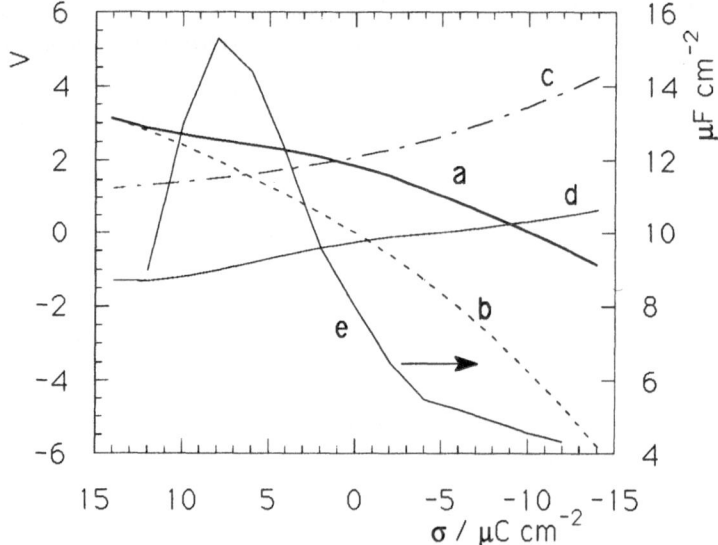

Figure 22. Curves of the overall potential difference across the interphase (**a**), of $4\pi d\sigma$ (**b**), of χ_{el}(**c**) and of χ_w(**d**) against the charge density σ, as calculated for ε = 80 . Curve (**e**) shows the differental capacity vs. σ.

For comparison, Figs. 23 and 24 show plots analogous to those in Figs. 21 and 22, obtained on the basis of the same model upon setting ε=1 in place of ε=80 in the expression for the electrostatic pressure due to diffuse-layer ions. It can be seen from the vertical separation between the curves of x=d and x=x'$_\sigma$ vs. σ that in the present case the integral capacity due to the (jellium + electron gas) increases appreciably towards positive charges. Even in this case, however, by far the major contribution to the positive maximum in the differential capacity C_i of Fig. 24 is still made by water polarization, as can be seen by comparing the curves of χ_{el} and χ_w in the same figure.

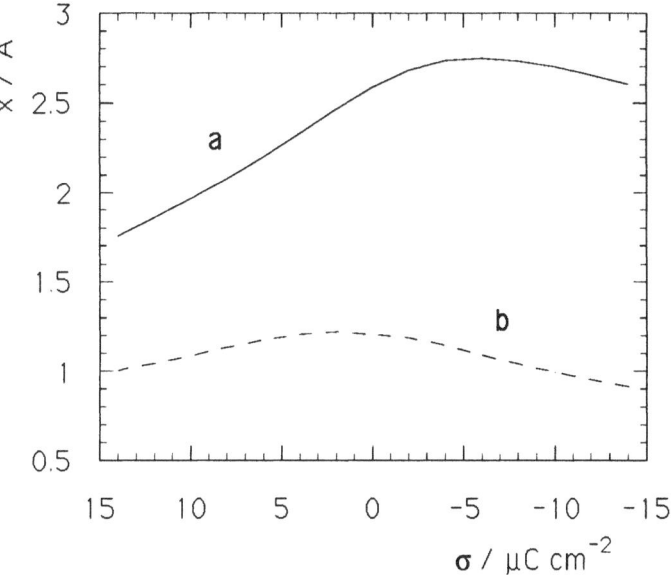

Figure 23.Calculated positions of the x=d (a) and x=x'$_\sigma$ (b) planes as a function of σ for ε = 1.

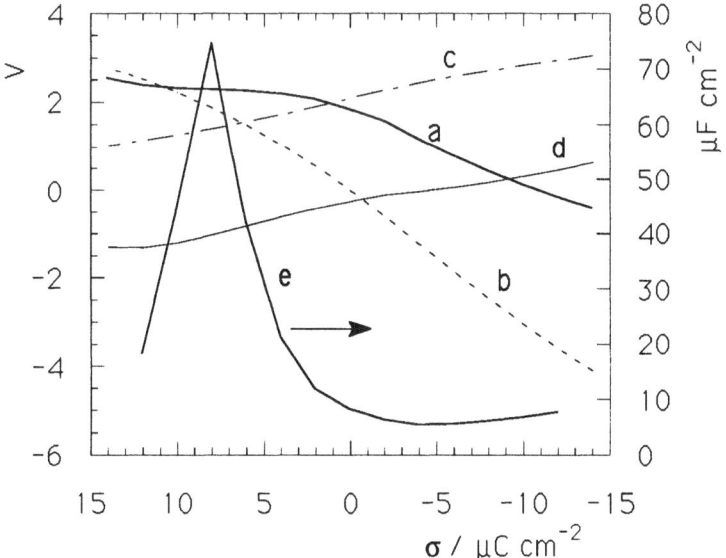

Figure 24. Curves of the overall potential difference across the interphase (**a**), of $4\pi d\sigma$ (**b**), of χ_{el}(**c**) and of χ_w(**d**) against the charge density σ, as calculated for ε = 1. Curve (**e**) shows the differential capacity vs. σ

References

(1) D.C.Grahame, J.Am.Chem.Soc., 79(1957)2093.

(2) G.Aloisi and M.Carlà, Chem.Phys.Lett., 131(1986)103.

(3) J.A.Harrison, J.E.B. Randles and D.J.Shiffrin, J.Electroanal.Chem., 48(1973)359.

(4) For a recent review see R.Guidelli in A.F.Silva (Ed.), Trends in Interfacial Electrochemistry, Reidel, Dordrecht, 1986, pp.387-452.

(5) S.L.Carnie and D.Y.C.Chan, J.Chem.Phys., 73(1980)2949; Adv.Colloid Interface Sci.,16(1982)81; L.Blum and D.Henderson, J.Chem.Phys., 74(1981)1902; D.Henderson and L.Blum, J.Electroanal.Chem., 132(1982)1.

(6) W.Schmikler and D.Henderson, J.Chem.Phys., 85(1986)1650; W.Schmikler and D.Henderson, Prog.Surf.Sci., 22(1986)323.

(7) R.Guidelli, G.Aloisi, E.Leiva and W.Schmickler, J.Phys.Chem., 92(1988)6671.

(8) J.R.Farrell and P.J.Mc Tigue, J.Electroanal.Chem.,82(1977)391.

(9) R.Guidelli, J.Electroanal.Chem., 197(1986)77.

(10) R.Guidelli and M.L.Foresti, J.Electroanal.Chem., 250(1988)23.

(11) G.Aloisi, M.L.Foresti and R.Guidelli, J.Chem.Phys., 89(1988)1141.

(12) G. Aloisi, M.L. Foresti, R. Guidelli and P. Barnes, J. Chem. Phys., 91(1989)5592.

(13) W.L.Jorgensen, J.Chandrasekhar and J.D.Madura, J.Chem.Phys., 79(1983)926.

(14) G. Aloisi and R. Guidelli, J. Electroanal. Chem., 260(1989)259.

(15) J.A.Barker, J.Chem.Phys., 44(1966)4212; A.R.Allnatt, Mol.Phys., 14(1968)289.

(16) A.Hamelin, S.Röttgermann and W.Schmickler, J.Electroanal.Chem., 230(1987)281.

(17) G.Aloisi, R.Guidelli, R.A.Jackson, S.M.Clark and P.Barnes, J.Electroanal.Chem., 206(1986)131.

(18) A.A. Kornyshev, Electrochim. Acta, 34(1989)1829.

(19) S. Amokrane, V. Russier and J.P. Badiali, Surface Sci.,217(1989)425.

(20) J.F. Annett and P.M. Echenique, Phys. Rev. B 32 (1986) 6853.

(21) D. Henderson, L. Blum and J.L. Lebowitz, J. Electroanal. Chem., 102 (1979) 315.

MOLECULAR MODELS OF ORGANIC ADSORPTION FROM WATER AT CHARGED INTERFACES

ROLANDO GUIDELLI and GIOVANNI ALOISI

Dipartimento di Chimica - Università di Firenze
Via G.Capponi,9 - 50121 FIRENZE - Italy

ABSTRACT.   Molecular models of organic adsorption from water at charged interfaces are briefly reviewed, and their predictions are critically compared with the experimental electrosorption behaviour of simple aliphatic compounds on *sp* metals. The fundamental role played by H-bonding between interfacial water molecules in determining the main distinguishing features of organic adsorption at electrodes is stressed. Since the energy of H-bonds is almost one order of magnitude greater than kT at room temperature, the importance of accounting for local order is emphasized. A three-dimensional lattice model of H-bonded water molecules and of mono- or dimeric solute molecules extending from the charged wall into the bulk liquid is outlined, in which local order is accounted for on the basis of a first-order statistical mechanical approximation. Some effects of the nature of the metal on organic adsorption are tentatively explained on the basis of a difference in the extent of electron spillover at different metals.

1. Introduction

It is well known that adsorption of a solute from an aqueous solution on a metal is a competition between solute and solvent molecules for adsorption sites. In the case of neutral organic solutes, solute-solute interactions are usually appreciably weaker than the strong water-water interactions responsible for H-bonding, so that the latter are expected to exert a profound influence upon the electrosorption behaviour of organic surfactants. Hence molecular models of solvent and solute molecules against a charged wall which aim at interpreting the electrosorption behaviour of the solute must also be able to account satisfactorily for the experimental behaviour of metal-water interphases, when the solute molecules are excluded from the model.

The first attempt to account simultaneously for the behaviour of a metal-water interphase in the absence of solutes and for electrosorption of solute molecules on the basis of the same model was made by Bockris and coworkers [1], who represented both solvent and solute molecules by dipolar spheres confined to a monomolecular layer adjacent to a charged wall. This model was subsequently complicated by the introduction of H-bonded water clusters in order to predict an increasing number of experimental features [2].

R. Guidelli (ed.), Electrified Interfaces in Physics, Chemistry and Biology, 337–367.
© 1992 *Kluwer Academic Publishers.*

2. The Adsorption Behaviour of Aliphatic Compounds

The adsorption behaviour of a large number of aliphatic compounds on sp metals from aqueous solutions shows appreciable similarities in spite of the presence of different functional groups. Some typical features are [3,4]:

I) Adsorption satisfies the Frumkin isotherm:

$$c \exp(-\Delta G^\circ_{ads}/kT) = \theta \exp(a\theta)/(1-\theta) \qquad (1)$$

where θ is the surface coverage by the organic surfactant S, c is its bulk concentration, ΔG°_{ads} is the standard Gibbs energy of adsorption at zero coverage and a is the so called Frumkin interaction factor. Roughly linear plots of $ln\ [c\ (1-\theta)/\theta]$ vs θ at constant applied potential E or at constant charge σ (henceforth referred to as Frumkin-isotherm plots) have been reported countless times in the literature. Usually the slope of these plots, which measures the interaction factor a, assumes negative values ranging from -1 to -3.

II) The standard Gibbs energy of adsorption, ΔG°_{ads}, is a quadratic function of the charge density σ as well as of the applied potential E. The minimum value of ΔG°_{ads}, which corresponds to a maximum in the adsorptivity of the organic surfactant, is attained for a charge value, σ_m, which frequently lies in the range from -1 to -3 $\mu C\ cm^{-2}$ [5].

3. The Frumkin Isotherm

Frumkin derived the isotherm of eqn.(1) on the basis of a model which lacks a molecular character. However, this isotherm can be readily derived on the basis of a statistical-mechanical procedure for the case of non-electrochemical adsorption of a gas on a solid [6]. The Frumkin isotherm can also be derived for the case of adsorption at an electrified interface from an electrolytic solution [7], provided the following rough assumptions are made:

i) The adsorbed monolayer is approximated by a two-dimensional lattice, whose adsorption sites are occupied either by water or by surfactant molecules.

ii) Water and surfactant molecules have the same size and are adsorbed in a single orientation, namely are unable to rotate.

iii) The adsorbing water and solute molecules are assumed to be distributed randomly over adsorption sites (random-mixing approximation). In other words, the "local order" which an adsorbed particle tends to create about its own adsorption site by favouring the occupation of the nearest-neighbour sites by adsorbing particles which interact with it attractively over those which interact repulsively, is not accounted for.

With these assumptions, the following expression for the Frumkin isotherm is obtained [7]:

$$c = \frac{q_w}{q_s} \frac{\theta}{1-\theta} \exp(a\theta) + \text{constant} \tag{2}$$

where

$$a = \frac{z}{kT} (w_{ww} + w_{ss} - 2w_{ws})$$

Here the subscript W denotes the water molecules and S the surfactant molecules; z is the coordination number of the lattice, namely the number of nearest-neighbour sites which surround any given site of the two-dimensional lattice; the w's are energies of lateral interaction between pairs of nearest-neighbour particles; q_w and q_s are single-molecule partition functions, namely the Boltzmann factors accounting for the energies of interaction of the adsorbed molecules with the two contacting phases, i.e. the metal phase and the solution phase.

In general the q_i's have the form:

$$q_i = \exp[-(U_i^c - X\mu_i)/kT] \; ; \; i = W,S \tag{3}$$

with $X = 4\pi\sigma/\varepsilon_\infty$.

Here U^c_i is the energy of *chemical* interaction of the i-th adsorbed particle with the two contacting phases, and $-X\mu_i$ is the energy of electrostatic interaction of this particle with the external electric field X created by the charge density σ on the metal within the adsorbed monolayer. The latter energy is obtained by approximating the adsorbed molecules to nonpolarizable point dipoles with a dipole moment normal component μ_i in the direction towards the solution, and by accounting for distortional polarizability through the fictitious dielectric constant ε_∞.

If we combine eqns. (2) and (3) as follows:

$$ln \frac{c (1- \theta)}{\theta} = \text{const} + a\theta - \frac{4\pi\sigma}{\varepsilon_\infty kT} (\mu_s - \mu_w) \tag{4}$$

we immediately realize that this equation predicts a linear Frumkin-isotherm plot of slope a. However, the intercept of this plot on the $\theta=0$ axis, which measures $(\Delta G°_{ads}/kT)$ apart from an additive constant, is a linear function of σ. This prediction contrasts with the quadratic dependence of $\Delta G°_{ads}$ upon charge as observed experimentally. Such a prediction is a direct consequence of the assumption that solvent and solute molecules are both adsorbed in a single orientation; hence, the replacement of a solvent molecule by a solute molecule will produce a charge-independent change (μ_s-μ_w) in the dipole-moment normal component, which will be constantly favoured by a shift of the external electric field X in one direction and constantly opposed by a shift in the opposite direction.

To justify the experimental observation that the slope a of Frumkin-isotherm plots is generally negative, we must assume, in view of eqn. (2), that:

$$(w_{ww} + w_{ss}) < 2w_{ws},$$

namely that lateral interactions between like particles are more negative (namely more attractive) than those between unlike particles. This prediction can be intuitively understood by considering that incipient adsorption of surfactant molecules on a solvent-filled adsorbed monolayer will involve the establishing of solvent-solute interactions at the expense of solvent-solvent interactions, and hence it will be somewhat hindered by solvent-solvent interactions which are more attractive than solvent-solute interactions. However, as θ increases progressively, the number of WW pairs per unit surface will decrease and that of SS pairs will increase, so that the displacement of solvent molecules by the adsorbing molecules of the surfactant will become gradually easier with respect to the ideal situation in which the energies of lateral interaction are all vanishingly small (or, which is practically the same, all identical). Incidentally, this ideal situation corresponds to the Langmuir isotherm behaviour, and yields a horizontal plot of ln [c(1-θ)/θ] vs θ. Hence, θ will increase with increasing c faster than predicted by the Langmuir isotherm, and consequently ln [c(1-θ)/θ] will decrease with increasing θ. Frumkin ascribed the negative experimental value of a to strong attractive interactions of the Van der Waals type between solute molecules. As a matter of fact, the strongest lateral interactions are undoubtedly the hydrogen bonds (H-bonds) which can be formed between the adsorbed water molecules, but this simple model cannot account for H-bond formation.

In conclusion, the statistical-mechanical treatment leading to the Frumkin isotherm is not in a condition of accounting for the parabolic dependence of ΔG°_{ads} upon σ.

4. Consideration of the Different Size of Solvent and Surfactant Molecules

The physical picture can be apparently improved by considering that organic surfactant molecules are usually more bulky than water molecules. We can, therefore, assume as a first approximation that a solute molecule S occupies n contiguous adsorption sites, and we can account for the different size between solvent and surfactant molecules on the basis of Flory-Huggins statistics. Upon assuming once again that solvent and solute molecules are adsorbed in a single orientation and retaining the simplifying assumptions (i) and (iii) reported for the Frumkin isotherm, the statistical-mechanical treatment of the resulting model [7] yields an adsorption isotherm which has the form of a *generalized Frumkin isotherm* [8]:

$$c = \frac{q_w^n}{q_s} \frac{\theta}{(1-\theta)^n} \exp(a\theta) + \text{constant} \tag{5}$$

where

$$a = \frac{zn}{kT}(w_{ww} + v^2 w_{ss} - 2w_{ws}) \; ; \; v = 1 - \frac{2(n-1)}{zn}$$

Here v is a number close to unity which accounts for the geometrical constraints imposed by the fact that the n *monomeric* segments composing a solute molecule must necessarily occupy contiguous adsorption sites. Equation (5) reduces to the simple Frumkin isotherm when $n=1$.

The main difference with respect to the Frumkin isotherm is represented by the presence of the entropic factor $(1-\theta)^n$, instead of $(1-\theta)$, in the denominator of eqn. (5). In practice, this factor expresses the probability of an adsorbing solute molecule finding n contiguous adsorbed solvent molecules to displace from the monolayer. In the random-mixing approximation, the probability of finding one adsorbed solvent molecule is expressed by $(1-\theta)$, that of finding two contiguous solvent molecules is expressed by $(1-\theta)^2$, and so on. The presence of the $(1-\theta)^{-n}$ factor in the generalized Frumkin isotherm of eqn. (5) is such that, if we plot this isotherm on a $ln\,[c(1-\theta)/\theta]$ vs θ graph, we no longer obtain a straight line, but rather a curved line with an upward concavity which is the more pronounced the higher is n (see Fig. 1).

In view of eqns. (3) and (5), extrapolation of the generalized Frumkin isotherm to the $\theta=0$ axis yields a standard Gibbs energy of adsorption ΔG°_{ads} which varies linearly with charge. In conclusion, consideration of the different size of solvent and solute molecules does not improve agreement with experiment as concerns the charge dependence of ΔG°_{ads}, while it worsens it as concerns the dependence of θ on c at constant charge.

5. Consideration of Reorientation of the Solvent Molecules

To account for the quadratic dependence of ΔG°_{ads} upon σ, the physical picture must be further improved by allowing for the reorientation of the adsorbed molecules, in particular the adsorbed water molecules. In the preceding chapter we have seen that the presence of a hump in the differential-capacity curves at metal-water interphases in the absence of specific adsorption points to a reorientation of the solvent molecules with varying the electrical

342

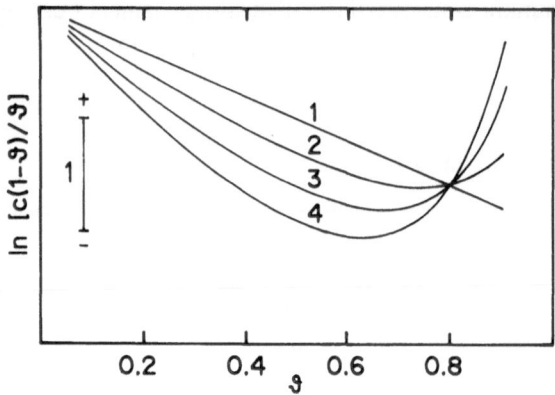

Figure 1. Plots of $ln[c(1-\theta)/\theta]$ vs. θ as obtained from eqn.(5) for $zw_{WW}/kT=-2$ and $w_{SS}=w_{WS}=0$. Numbers on each curve denote n values.

state. The simplest way to account for water reorientation consists in assuming that water molecules may take any of two different orientations, one with the dipole moment pointing towards the bulk solution (up orientation, henceforth denoted by u) and the other with the dipole moment pointing towards the metal (down orientation, henceforth denoted by d). In practice, the model consists of a two-dimensional lattice populated by *up* and *down* *solvent molecules* and by solute molecules adsorbed in a fixed orientation; the random-mixing approximation is once again retained. For simplicity, we shall assume that the up and down water molecules have the same energy of chemical interaction with the two contacting phases and that their dipole-moment normal components are equal in magnitude and opposite in sign ($\mu_u = -\mu_d$). Moreover, we shall assume that the solute molecules have the same size as the solvent molecules and that their dipole-moment normal component is negligibly small. More general conditions could be easily introduced [7], but their consideration would be of no use for the present discussion. The statistical-mechanical treatment of the present model yields the following isotherm:

$$c = \frac{\theta}{1-\theta} \sqrt{\frac{4}{1-R^2}} + \text{constant} \qquad (6)$$

where

$$R = \tanh\left(\frac{4\pi\sigma\mu_u}{kT\varepsilon_\infty} - \frac{z\,w_{uu}(1-\theta)R}{kT}\right) \qquad (7)$$

Here μ_u is the positive dipole-moment normal component of the up water dipoles and w_{uu} is the positive (i.e. repulsive) energy of lateral interaction between contiguous up dipoles. Note that eqn.(6) has not the form of a Langmuir isotherm, because the R term in this equation depends upon θ. Figure 2 shows plots of the isotherm of eqn. (6) on a $ln[c(1-\theta)/\theta]$ vs. θ graph at different negative charges. Plots at positive charges are identical to those at the corresponding negative charges. Extrapolation of these plots to the $\theta=0$ axis yields a standard Gibbs energy of adsorption which exhibits the correct parabolic

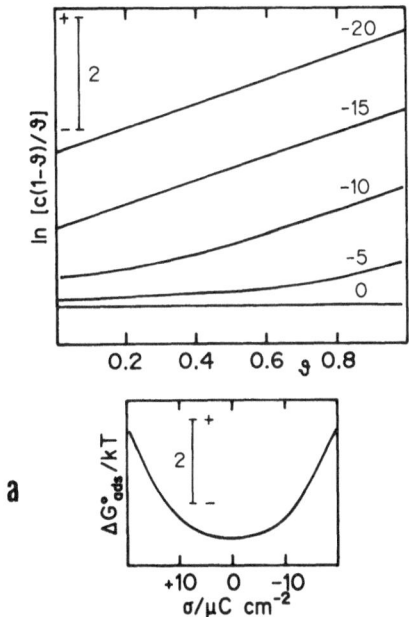

Figure 2. Plots of $ln[c(1-\theta)/\theta]$ vs. θ as obtained from eqns.(6) and (7) for $\mu_u=1.84D$, $T=298°K$, $\varepsilon_\infty=6$, $z=11$ and $w_{uu}/kT=0.2$. Numbers on each curve denote the charge σ, in $\mu C\ cm^{-2}$. (a) Plot of $\Delta G°_{ads}/kT$ vs. σ [7].

dependence upon charge (see Fig. 2a). This satisfactory result is due to the assumed reorientability of water molecules, which allows the water dipoles to take the electrostatically more favourable orientation with varying the magnitude and sign of the external electric field, compatible with the randomizing forces of thermal agitation. Hence,the solute molecules will be more easily adsorbed in the proximity of zero charge, where the electrostatic interactions between the external electric field due to σ and the water dipoles are weaker. In spite of this favourable feature of the present model, the Frumkin-isotherm plots in Fig. 2 are horizontal at zero charge, then they become curved with an increasingly positive average slope as the absolute value of the charge increases, and finally they become straight lines with a maximum positive slope at extreme charges. This behaviour contrasts with the experimental negative slope of Frumkin isotherm plots [3-5]. In other words, consideration of water orientability explains the quadratic dependence of ΔG°_{ads} upon charge, but causes more pronounced deviations from the Frumkin-isotherm behaviour.

At this point it is important to realize that such deviations are a direct consequence of the random-mixing approximation. In fact, when one orientation of the water molecules (say, the up orientation) prevails over the other, then the nearest neighbours of the up molecules, which are the majority, are also predominantly up molecules (see Fig. 3a).

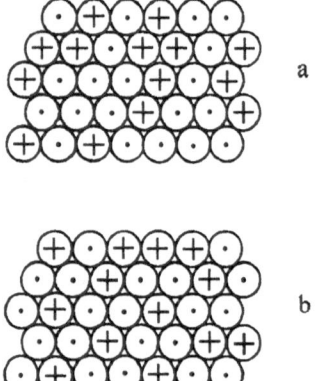

Figure 3. Typical distribution of up (.) and down (+) dipoles over a hexagonal lattice for a ratio of up and down dipoles equal to 2/3. In **a** random mixing is assumed, while in **b** local order is accounted for.

Hence, on the average, repulsive interactions between contiguous water molecules prevail over attractive interactions other than at zero charge; here, up and down water molecules are equal in number, and consequently average water-water interactions are vanishingly small. At charges different from zero the average water-water interactions are repulsive, and hence favour the displacement of adsorbed water molecules from the monolayer by adsorbing solute molecules. However, as θ increases, an increasing number of "non-interacting" adsorbed solute molecules surrounds the remaining adsorbed solvent molecules. Hence the latter molecules are less strongly repelled by their nearest neighbours, which makes their removal by further adsorbing solute molecules more difficult. In other words, θ increases with c more slowly than in the absence of lateral interaction forces, namely more slowly than predicted by the Langmuir isotherm, and hence $ln[c(1-\theta)/\theta]$ increases with increasing θ as shown in Fig. 2. To remove this unsatisfactory feature of the model, it is, therefore, necessary to abandon the random-mixing approximation and to use a statistical-mechanical procedure accounting for the local order which any adsorbed particle tends to create about its own adsorption site.

6. Consideration of Local Order

Computer simulation techniques, such as the Monte Carlo method [9], are the most rigorous statistical-mechanical techniques accounting for local order. An approximate procedure yielding results which are often in excellent agreement with the Monte Carlo method is the *quasi-chemical approximation* [10]. Upon retaining all the physical features of the model and using the quasi-chemical approximation in place of the random-mixing approximation, it is possible to reverse the unsatisfactory trend of Fig. 2 and to obtain

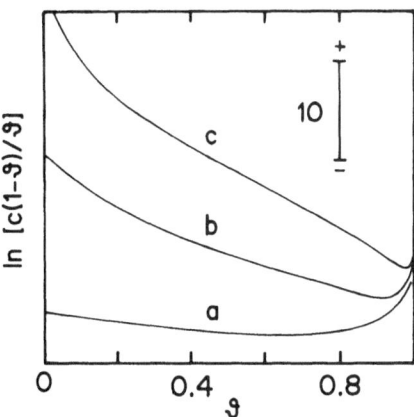

Figure 4. Plots of $ln[c(1-\theta)/\theta]$ vs. θ as obtained by the quasi-chemical approximation for T=298°K, $z=6$, $n=3$, $\varepsilon_\infty=6$, $\sigma=+5$ μC cm^{-2}, $\mu_u=1.84$D and for $w_{uu}=1$ (a), 2 (b) and 3kT (c).

$ln[c(1-\theta)/\theta]$ vs. θ plots with an average negative slope. This is shown in Fig. 4, which refers to a hexagonal lattice populated by up and down water dipoles and by non-interacting solute molecules consisting of three monomeric segments [7]. As the energy w_{uu} of lateral interaction between contiguous water dipoles increases, the negative slope of the calculated Frumkin-isotherm plots increases. Extrapolation of these plots to the $\theta=0$ axis yields a standard Gibbs energy of adsorption which once again exhibits the correct quadratic dependence upon σ. It should be noted that the passage from a positive to a negative slope in the calculated Frumkin isotherm plots is not due to an improvement in the physical modeling, but rather to an improvement in the statistical-mechanical treatment which now accounts for local order. Once local order is accounted for, water dipoles in a given orientation will tend to be surrounded by the maximum number of water dipoles in the opposite orientation (see Fig. 3b), compatible with the randomizing forces of thermal agitation, because the resulting configurations favour attractive interactions between contiguous solvent dipoles. These attractive solvent-solvent interactions tend to oppose water displacement by the adsorbing solute molecules, at least at low surface coverages. However, as θ increases, the increasing number of solute molecules interposing between the residual water molecules decreases the number of water-water nearest-neighbour pairs, and hence makes the removal of water molecules progressively easier. In other words, θ increases with c faster than in the absence of lateral interactions, and hence $ln[c(1-\theta)/\theta]$ decreases with increasing θ.

7. Consideration of Hydrogen Bonding

To predict a negative slope of the Frumkin isotherm plot in the framework of a model of up and down water dipoles, the energy $w_{uu}=w_{dd}$ of lateral interaction between neighbouring dipoles must be given unrealistically high values (see Fig. 4). Since the strongest water-water interactions are due to H-bonding, monolayer models can be further improved by taking H-bonding into account. This can be done by ascribing to the water molecules a relatively high number of orientations, so as to allow for the possibility that two contiguous water molecules have a proton-donating direction and a proton-accepting direction pointing towards each other; when this requirement is met, the usual dipole-dipole interaction energy is increased by an amount equal to the experimental value of the energy of H-bond formation [11-13]. The monolayer model of H-bonded water molecules which we shall use is described in the preceding chapter (see Fig. 5 for a schematic picture of the allowed water orientations). In the absence of solute molecules this model involves three adjustable parameters, which are estimated by fitting the differential capacity curve calculated from the model to the corresponding experimental curve. For the mercury/water interphase the best

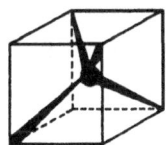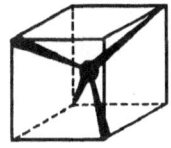

Figure 5. Orientations of water molecules with H-bonds directed towards the vertices of the cubic cell.

fit is obtained for $l = 3.2$ Å, $\varepsilon_\infty = 5.5$ and $\Delta U_c = 14.5$ kJ/mol [12]. If these values are retained when generalizing the model so as to include the presence of solute molecules, it is evident that these parameters should not be regarded as adjustable as far as the predicted adsorption behaviour is concerned, because they were derived by independent means. In the generalized model the solute molecules are assumed to interact both between themselves and with water molecules via dipole-dipole interactions [13]. This generalization involves the introduction of two adjustable parameters, namely the number n of water molecules displaced by one adsorbing solute molecule and the normal component μ_S of the dipole moment of the solute molecule. Figure 6 shows a series of isotherms calculated at a charge

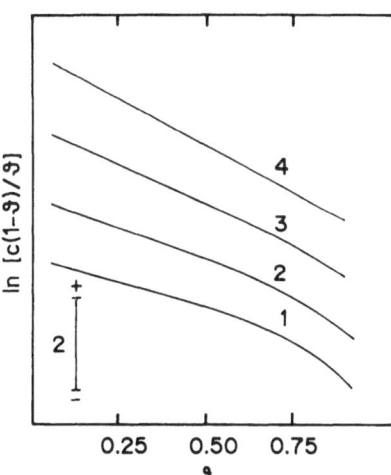

Figure 6. Plots of $ln[c(1-\theta)/\theta]$ vs. θ as obtained by the quasi-chemical approximation for T=298°K, l=3.2Å, ε_∞=5.5, ΔU_c=14.5 kJ mol^{-1} and μ_S=-0.4n D. Numbers on each curve denote n values.

348

density σ=-4 μC cm^{-2} on the basis of this monolayer model using the quasi-chemical approximation to account for local order. The isotherms, which are plotted on a Frumkin-isotherm graph, were obtained by using the l, ε_∞, and ΔU_c values which provide the best fit between calculated and experimental differential-capacity curves on mercury, and by setting $\mu_S/n = -0.4$ D and $n = 1,2,3$ and 4. It can be seen that these plots are almost linear over a wide range of θ values, independent of the number n of water molecules displaced by one adsorbing solute molecule, and exhibit a negative slope in accordance with experiment. The linearity of these plots, which contrasts with the predictions of the generalized Frumkin isotherm for $n > 1$ (see Fig. 1), can be explained by the strong attractive water-water interactions. Taking local order into account, these attractive interactions cause water molecules to aggregate, namely to occupy a maximum of adjacent sites compatible with the randomizing forces of thermal agitation. Hence, the probability of an adsorbing polymeric solute molecule finding n adjacent water molecules to displace from the adsorbed monolayer is greater than expressed by the $(1-\theta)^{-n}$ factor in the generalized Frumkin isotherm. Loosely speaking, for θ values which are not too high, this probability tends to be expressed by the $(1-\theta)$ factor in accordance with the simple Frumkin-isotherm behaviour [7,11,13].

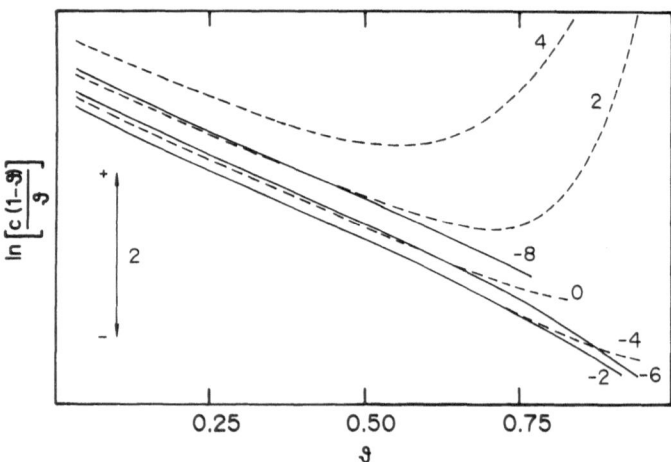

Figure 7. Plots of $ln[c(1-\theta)/\theta]$ vs. θ as obtained by the quasi-chemical approximation for T=298°K, l=3.2Å, ε_∞=5.5, ΔU_c=14.5 kJ mol^{-1}, n=3 and μ_S=-1.2D. Numbers on each curve denote charge values in μC cm^{-2} [13].

Figure 7 shows a number of isotherms calculated for different σ values, $n = 3$ and $\mu_S = -1.2$ D, as plotted on a Frumkin-isotherm graph. By least-squares fitting of these plots to a straight line, values of $\Delta G^{\circ}_{ads}/kT$ and of the Frumkin interaction factor a are immediately obtained from the corresponding intercepts and slopes. Figure 8 shows $\Delta G^{\circ}_{ads}/kT$ vs. σ and a vs. σ plots as calculated for $\mu_S = -0.6$ and -1.2 D. It is apparent that ΔG°_{ads} depends almost parabolically upon σ and that the charge σ_m of maximum adsorption, corresponding to the minimum of the parabola, shifts towards more negative values with a gradual increase in the negative value of μ_S, namely the more the positive end of the solute dipoles points towards the metal; for $\mu_S = 0$, the model predicts a charge σ_m of maximum adsorption of about +1 μC cm^{-2}. The calculated Frumkin interaction factor a in Fig. 8 assumes negative values ranging from -3 to -3.5, the minimum value being attained at an intermediate charge close to σ_m.

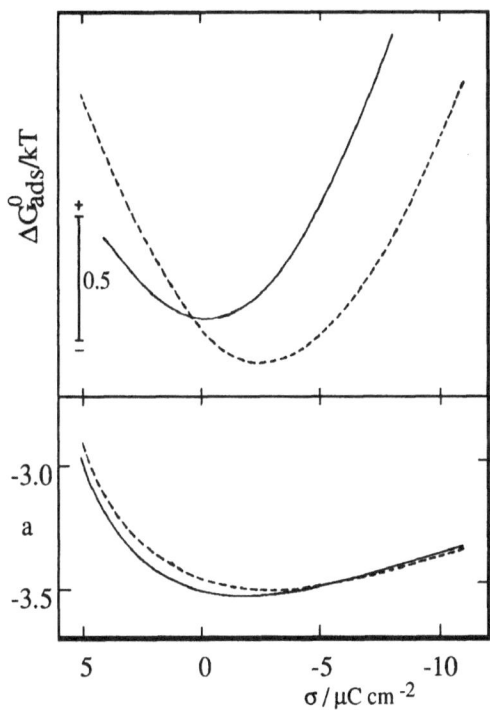

Figure 8. Plots of $\Delta G^{\circ}_{ads}/kT$ and of a vs. the charge σ, as calculated for μ_S=-0.6D (solid curve) and μ_S=-1.2D (dashed curve). All other parameters as in Fig. 7 [13].

The above predictions account satisfactorily for the adsorption behaviour of the majority of aliphatic compounds. Thus, we have already stated that these compounds satisfy the Frumkin isotherm with interaction factors ranging from -1 to -3. Even the roughly parabolic dependence of a upon charge with a minimum negative value in the proximity of σ_m, as predicted in Fig. 8, was reported for several aliphatic amines and alcohols [14-18]. The model predicts a slope, b_-, of the $\Delta G^\circ_{ads}/kT$ vs $(\sigma-\sigma_m)^2$ plot for negative $(\sigma-\sigma_m)$ values which decreases with a negative shift of σ_m, and a corresponding slope, b_+, for positive $(\sigma-\sigma_m)$ values which increases in the same direction (see the calculated plots of b vs. σ_m in Fig. 9). The experimental b values for a number of aliphatic

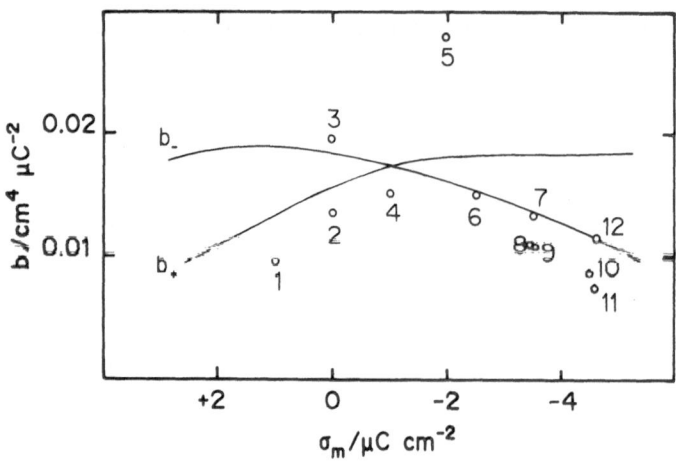

Figure 9. The solid curve is a plot of b vs. the charge of maximum adsorption σ_m, as calculated for different μ_s values, with other parameters as in Fig.7. Circles are experimental values for 2-methylthioethanol (1), sucrose (2=b_+, 3=b_-), 2-buthyne-1,4-diol (4), n-butanol (5), butane-1,4-diol (6), ethylene glycol (7), propionitrile (8), acetonitrile (9), diethyl ether (10) and succinonitrile (11=b_-, 12=b_+) [13].

compounds, denoted by circles in Fig. 9, are in fairly good agreement with these predictions. With the majority of aliphatic compounds, the experimental plots of the potential drop φ^{M-2} across the inner layer vs. θ at constant charge exhibit a typical behaviour. Thus, the average slope of these plots passes from negative to positive values with increasing charge, becoming approximately zero in the proximity of σ_m [13-27]. Figure 10 shows this typical behaviour for the case of n-hexylamine [15], whose maximum adsorption is attained at a charge $\sigma_m = -3.2$ μC cm^{-2}. Even this typical feature is predicted by the present model, as shown by the calculated φ^{M-2} vs. θ curves at constant charge in Fig. 11. The curvature exhibited by these calculated curves is due to a reorientation of water molecules at constant charge with increasing θ.

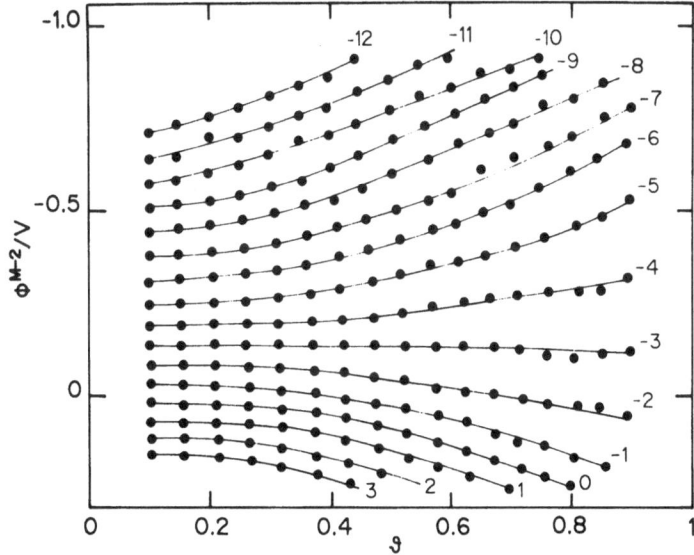

Figure 10. Experimental ϕ^{M-2} vs. θ plots for triethylamine adsorption on mercury from aquous 0.1M NaOH at 25°C. Numbers on each curve denote charge values in μC cm^{-2} [15].

8. A Three-Dimensional Lattice Model of Water and Solute Molecules Extending into the Bulk Liquid

In the preceding chapter we already pointed out that monolayer models completely disregard direct correlations between the water molecules of the monolayer and the adjacent water molecules of the solution phase, only averaged interactions being incorporated into the model under the form of one or more adsorption potentials. Direct correlations may be included by using a three-dimensional lattice model of water molecules extending into the bulk water phase, as described in the preceding chapter. An improvement in the quality of the predictions concerning organic adsorption can therefore be achieved by generalizing the latter model so as to include the presence of solute molecules [28].

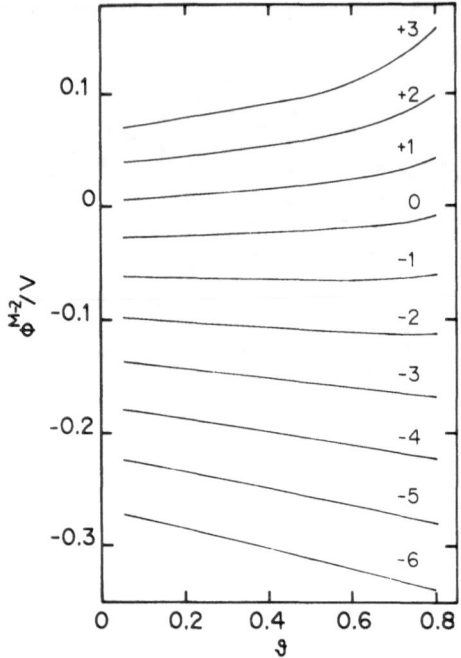

Figure 11. ϕ^{M-2} vs. θ plots calculated for $T=298°K$, $l=3.2Å$, $\varepsilon_\infty=5.5$, $\Delta U_c=14.5$ kJ mol^{-1}, $n=2$ and $\mu_s=$ -0.2 D. Numbers on each curve denote charge values in μC cm^{-2} [13].

8.1. MONOMERIC SOLUTE MOLECULES

We will first assume that the solute molecules have the same size as the water molecules and that they interact neither between themselves nor with the water molecules. In this way direct solute-solute interactions will be excluded, and only the solute-solute interactions induced by water structuring, namely the so called *hydrophobic interactions*, will be left. The calculated adsorption behaviour of this hypothetical solute will thus serve to single out the role played by hydrophobic interactions. The adsorption isotherm was obtained by equilibrating the water and solute molecules in the first lattice plane adjacent to the wall with those in the remaining lattice planes. To this end the chemical potential of the water molecules was regarded as uniform throughout all lattice planes for a given charge density on the wall. As concerns the solute, it was assumed to be present in the bulk water phase at such a low concentration as to permit us to disregard its presence in all lattice planes except for that immediately adjacent to the electrode surface plane. This assumption implies that the solute behaves ideally in the bulk, and hence its chemical potential was assumed to increase linearly with kT *ln c*. Adsorption isotherms at constant charge were then obtained by increasing gradually the chemical potential of the solute in the bulk, kT *ln c* + constant, and by estimating the corresponding increase in the solute concentration in the first lattice plane, namely the *solute surface concentration*. This estimate was made by forcing the chamical potential of the solute in the first lattice plane to be constantly equal to its bulk value.

In comparing the adsorption isotherms calculated at different charge values, a correction was made for the change in the chemical potential of water with varying charge; this is because in our model the bulk lattice planes are subject to the external electric field created by the charge density on the wall, whereas under experimental conditions the electric field in the bulk liquid equals zero. This correction was made by calculating how the chemical potential of water has to be varied with varying the charge density σ in order to maintain the solute concentration in the bulk lattice planes at a fixed low level. Now, when the solute is adsorbed from a bulk liquid which is not subject to an electric field, the chemical potential of water remains constant not only with varying the distance from the electrode surface but also with varying σ, Hence, the chemical potential of water at the various charge values, as estimated by this procedure, was subtracted from that at σ=0 whenever comparing adsorption isotherms at different charge values.

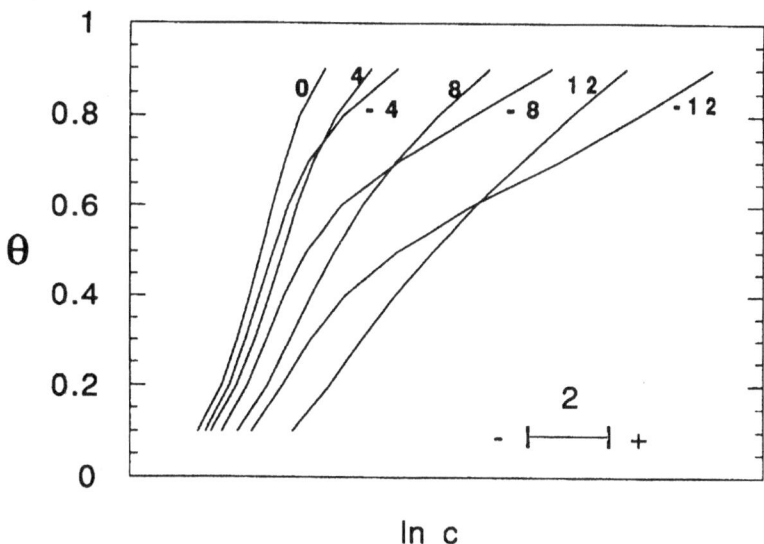

Figure 12. Calculated θ vs. ln *c* plots for the adsorption of a nonpolar monomeric solute. Numbers on each curve denote σ values in μC cm^{-2}.

354

Figure 12 shows plots of the surface coverage θ vs. *ln c* as calculated for a nonpolar monomeric solute at different charge values. It can be seen that the surface coverage tends to an apparent limiting value less than unity, which is lower the more the charge departs from its value of maximum adsorption. Such a behaviour is actually shown by the adsorption isotherms at constant charge of the majority of aliphatic compounds [14-16,22,29-33]. As an example, Fig. 13 shows the experimental adsorption isotherms of

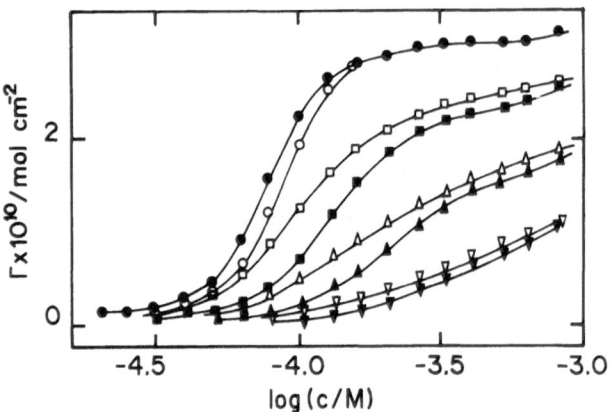

Figure 13. Γ vs. log c plots at constant charge for 4-octanol adsorption on Hg from aqueous 0.1M NaF at σ=-2 ($\bullet$), 0 (o), -4 ($\square$), +2 ($\blacksquare$), -6 ($\triangle$), +4 ($\blacktriangle$), -8 ($\triangledown$) and +6 μC cm^{-2} ($\blacktriangledown$) [16].

4-octanol at different charges [16]. According to the model, this behaviour is explained by considering that the removal of a water molecule from the adsorbed state by one solute molecule involves both the breaking of H-bonds and the overcoming of the electric field strength which anchors the water molecules to the adsorbed state. In the proximity of zero charge, the external electric field $4\pi\sigma$ is small and the breaking of H-bonds plays a major role. Under these circumstances, as the surface coverage θ increases, the progressive interposition of noninteracting solute molecules between adsorbed water molecules decreases the probability of H-bond formation between the latter molecules. Hence, the residual water molecules will be more easily displaced from the adsorbed state by the adsorbing solute molecules the higher is θ, and θ will increase with $ln\ c$ faster than predicted by the Langmuir isotherm, in rough agreement with the behaviour of a Frumkin isotherm with an attractive interaction factor a. As the absolute value of the charge density σ increases, however, water molecules tend to reorient under the electric field in order to minimize their electrostatic potential energy compatible with the randomizing forces of thermal agitation. Hence, they will be more strongly anchored to the electrode surface, and the slope of the plot of θ vs. $ln\ c$ at constant charge will decrease accordingly. It should also be noted that the electric field which anchors the water molecules to the electrode surface is not exclusively the external field $4\pi\sigma$, but rather the whole *local field* consisting of $4\pi\sigma$ plus the additional polarization field created by the neighbouring water molecules, which acts in the direction opposite to the external field. As the surface coverage θ increases, the polarization field decreases accordingly, and hence the local field at constant charge increases in absolute value, approaching $4\pi\sigma$. Hence, the residual adsorbed water molecules will be held in the adsorbed state by the local field more firmly the higher is θ, and this explains why we reach an apparent limiting value of the surface coverage, albeit ill defined, which decreases the higher is the absolute value of the charge density.

The standard Gibbs energy of adsorption at zero coverage as calculated from this model shows the usual parabolic dependence upon the charge density σ (see curve a in Fig. 14). The minimum of the ΔG°_{ads} vs. σ curve, which corresponds to a maximum of adsorptivity, lies at a small negative charge of about $-1 - -2\ \mu C\ cm^{-2}$. If the monomeric solute molecule is *polar* and is adsorbed in a fixed orientation with the positive end of the dipole pointing towards the metal, then the model predicts an increase in adsorptivity, and hence a decrease in the standard Gibbs energy of adsorption, at negative charge values (see curve b in Fig. 14). Conversely, if the monomeric polar solute molecule is adsorbed with the positive end of the dipole pointing towards the bulk solution, an increase in adsorptivity is predicted at positive charge densities, as expected (see curve c in Fig. 14).

The prediction that the maximum adsorptivity of a nonpolar, noninteracting solute molecule lies at a small negative charge is particularly significant, because we have seen that the charge σ_m of maximum adsorption for the majority of aliphatic compounds takes values ranging from -1 to $-3\ \mu C\ cm^{-2}$, independent of the particular nature of the functional group. It is therefore useful to point out the rationale behind such a modelistic prediction. Since a nonpolar solute molecule interacts neither with its neighbours nor with the external electric field, its behaviour is induced by water molecules; in particular, maximum adsorptivity corresponds to minimum resistance of adsorbed water molecules to displacement. A measure of such a resistance is provided by the surface excess of the Helmholtz free energy of water in the absence of solute as a function of σ. The entropic contribution to this surface excess is practically negligible with respect to the internal energy contribution. Hence, in Fig. 15 only the contribution to the surface excess of internal energy due to the electrostatic energy of the water molecules within the external

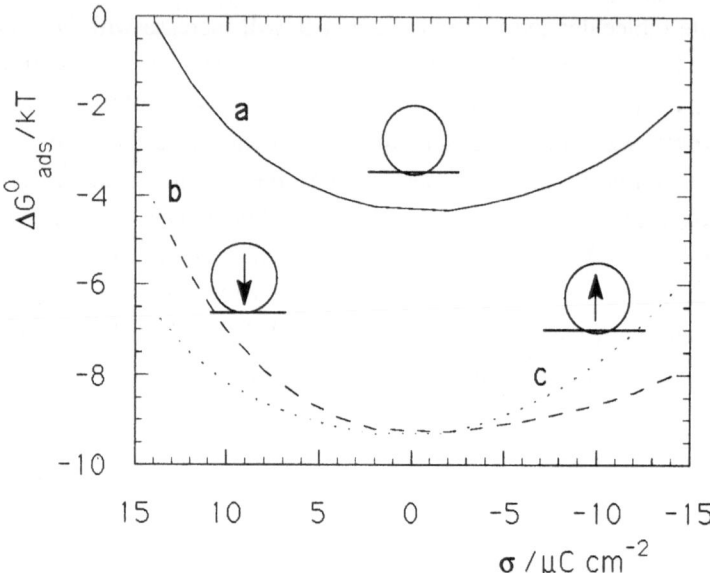

Figure 14. Calculated ($\Delta G^o_{ads}/kT$) vs. σ plots for the adsorption of a nonpolar solute monomer (a), a monomer with a 0.5 D dipole pointing towards the metal (b), and a monomer with a 0.5 D dipole pointing towards the bulk solution.

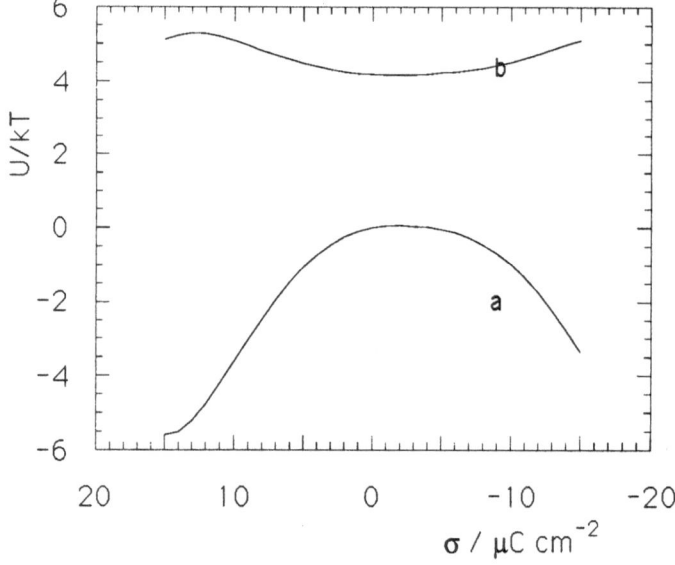

Figure 15. Internal-energy surface excess of water molecules as a function of σ: (a) contribution from their electrostatic energy in the external electric field $4\pi\sigma$; (b) contribution from H-bonding.

electric field and the further contribution due to H-binding are plotted against σ. In can be seen that the electrostatic energy contribution attains a maximum value of zero over a charge range from 0 to -5 μC cm^{-2}. This is due to the fact that at zero charge the external electric field $4\pi\sigma$ is obviously equal to zero, whereas at a charge of about -5 μC cm^{-2} it is the average value of the dipole-moment normal component of the adsorbed water molecules which equals zero. The asymmetry in the maximum of this curve with respect to the zero charge is practically compensated for by an analogous asymmetry in the minimum of the internal energy due to H-binding among the water molecules, so that the sum of these two contributions provides a curve of the internal energy surface excess vs. σ which is almost symmetrical with respect to the zero charge. At this point, however, we must consider that when a solute molecule displaces a water molecule from the adsorbed state, a part of the energy spent in breaking the H-bonds which bind it to the neighbouring water molecules is recovered, since the water molecules surrounding the adsorbed solute molecule will tend to rearrange so as to increase H-bond formation between themselves. Hence, a major role in solute adsorption with water displacement will be played by the electrostatic energy of the desorbing water molecule in the external electric field, which is a maximum at a small negative charge density σ, thus favouring water desorption there.

8.2. DIMERIC SOLUTE MOLECULES

A monomeric solute molecule is a poor model for an aliphatic molecule. The model can be improved by considering a dimeric solute molecule consisting of a nonpolar segment, which simulates a very short hydrocarbon chain, and of a polar segment, which simulates the polar head. For simplicity, the dipole moment of the polar head will be regarded as directed either towards the corresponding tail or else in the opposite direction. At first, this dimeric solute will be allowed to assume any of four equivalent horizontal orientations as well as any of the two opposite vertical orientations.

The curves of the surface coverage θ against the natural logarithm of the solute concentration calculated by this model are shown in Fig. 16 at different charge values. In this figure the surface coverage is defined in such a way as to attain the unit value when the first monolayer is completely occupied by solute molecules in any of the two vertical orientations. It can be seen that the adsorption isotherms show an inflection at an intermediate surface coverage. The inflection is determined by a relatively rapid passage of the solute molecules from a horizontal to a vertical orientation, as is shown by the plots of the percent values of the number densities of the three different orientations (the horizontal and the two vertical ones) against surface coverage at three different charge values (Fig. 17). At zero charge, the vertical orientation with the tail towards the metal prevails decidedly over the opposite vertical orientation. This prediction, which stems from the

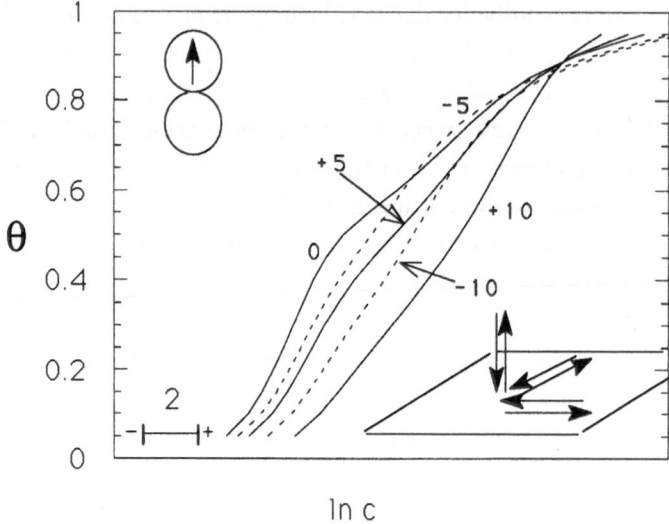

Figure 16. Calculated θ vs. *ln c* plots at constant charge for the adsorption of a dimeric solute whose polar segment has a 1 D dipole moment of 0.2 Å length, and points away from the corresponding nonpolar segment. Numbers on each curve denote σ values in μC cm⁻². The inset shows the allowed orientations of the dimer.

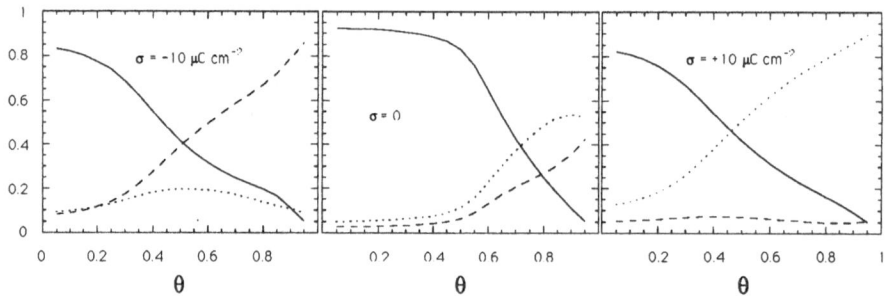

Figure 17. Percent values of the number densities of the horizontal orientations (solid curves), the vertical orientation with the polar head towards bulk water (dotted curves) and the vertical orientation with the polar head towards the metal (dashed curves) as a function of θ, for σ=-10, 0 and +10 μC cm⁻².

greater ability of the polar head to interact with the water molecules, agrees with what is commonly believed to be the behaviour of adsorbed aliphatic compounds from water. If the dipole moment of the polar head points away from the corresponding tail, then the predominance of the vertical orientation with the tail towards the metal becomes still more evident at positive charge values, since it is also favoured by the external electric field. However, even at a negative charge density of -10 μC cm^{-2}, the two opposite vertical orientations maintain a comparable weight up to a surface coverage of about 0.5. Obviously, the trend with respect to the charge density is reversed if the dipole moment of the polar head points towards the corresponding tail. That the vertical orientation with the polar head pointing towards the bulk water is favoured with respect to the opposite orientation by attractive interactions with water molecules is confirmed by using a more flexible model, in which the dipole moment of the polar head is allowed to point towards any of the four vertices of the elementary cubic cell enclosing the head, on the opposite side with respect to the corresponding tail. These orientations favour H-bonding with the neighbouring water molecules to such an extent that at zero charge vertical dimers with the polar head towards the metal are practically absent, as shown in Fig. 18.

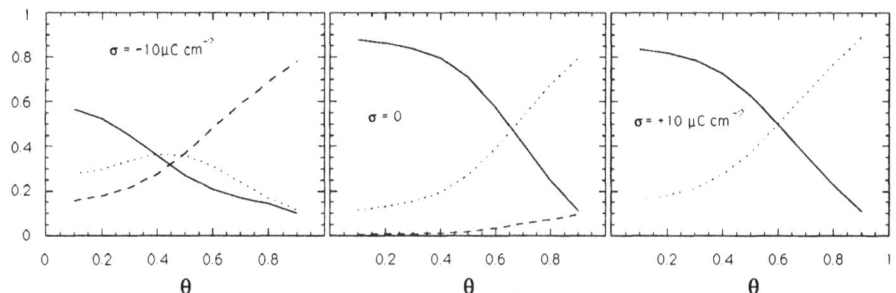

Figure 18. Curves analogous to those in Fig. 17, but referring to a dimeric solute molecule with a polar head assuming four different orientations relative to the corresponding tail (see the text).

In view of the wide range of bulk solute concentrations covered by the horizontal axis in Fig. 16, only the first portion of the adsorption isotherms is expected to be experimentally accessible. If we enlarge this portion (Fig. 19), we see that the shape of the isotherms is analogous to that already calculated for a nonpolar monomeric solute molecule. Thus, the isotherms show an ill defined limiting surface coverage which decreases as we depart from the charge of maximum adsorption. However, in this case the limiting surface coverage is close to 0.5, and is followed by a second rise due to an incipient change from a horizontal to a vertical orientation. This prediction is in agreement with the experimental behaviour of practically all n-aliphatic compounds, which show indeed a second rise.

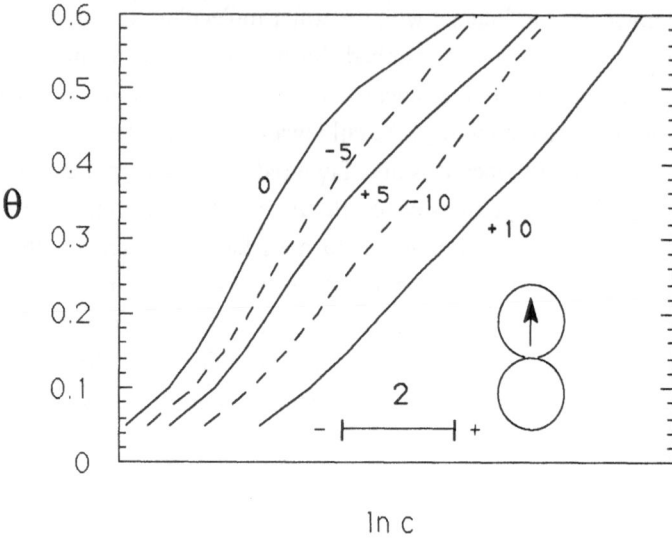

Figure 19. Enlarged lower portion of the adsorption isotherms in Fig. 16. Numbers on each curve denote σ values in μC cm^{-2}.

Moreover, simple geometrical considerations indicate that the experimental limiting coverage attained by n-aliphatic compounds ranges from 40 to 70% of that corresponding to a compact monomolecular film of surfactant molecules in a vertical orientation [15-18]. If we consider the first portion of the isotherms in Fig. 19 at charges close to σ_m and we plot $ln[c(1-\theta')/\theta']$ against θ', where θ' is now the ratio of the surface concentration to its apparent limiting value, we obtain a roughly straight line with a negative slope of about -1, in agreement with the predictions of the Frumkin isotherm with a negative (namely attractive) Frumkin interaction factor a. This further prediction is also in agreement with the experimental behaviour of the majority of aliphatic compounds.

The standard Gibbs energy of adsorption at zero coverage as calculated for dimeric solute molecules shows a minimum at a slightly negative charge of about -1 – -2 μC cm^{-2} (see Fig. 20) independent of whether the dipole moment of the polar head points towards the corresponding tail or in the opposite direction. This is not surprising, because in both cases the orientations of the dimer which predominate decidedly at very low coverages are the horizontal ones. The reason for the minimum standard Gibbs energy of adsorption lying at a negative charge is the same already described for a nonpolar monomeric solute molecule.

If we allow the dimeric solute molecules to assume a number of tilted orientations in addition to the horizontal and vertical orientations, thus imparting them a greater flexibility, we may predict a gradual passage from the horizontal to the vertical orientation

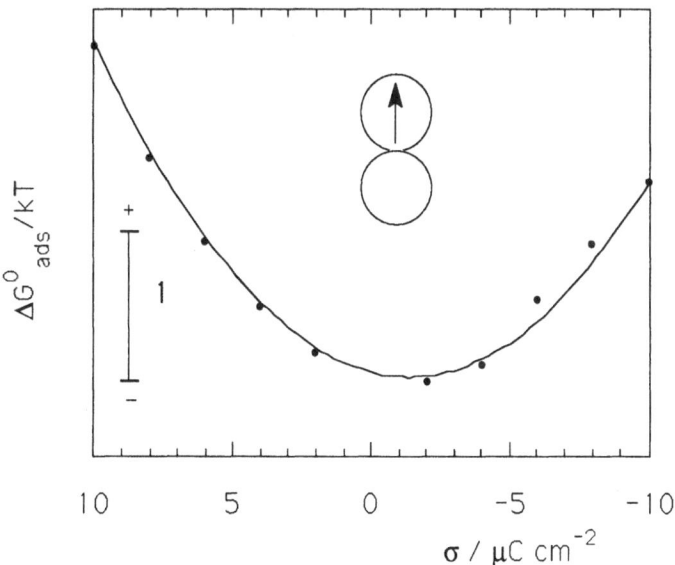

Figure 20. Calculated (ΔG^o_{ads}/kT) vs. σ plot for the adsorption of the dimeric solute of Fig. 16. The solid curve is a least-squares fitted quadratic.

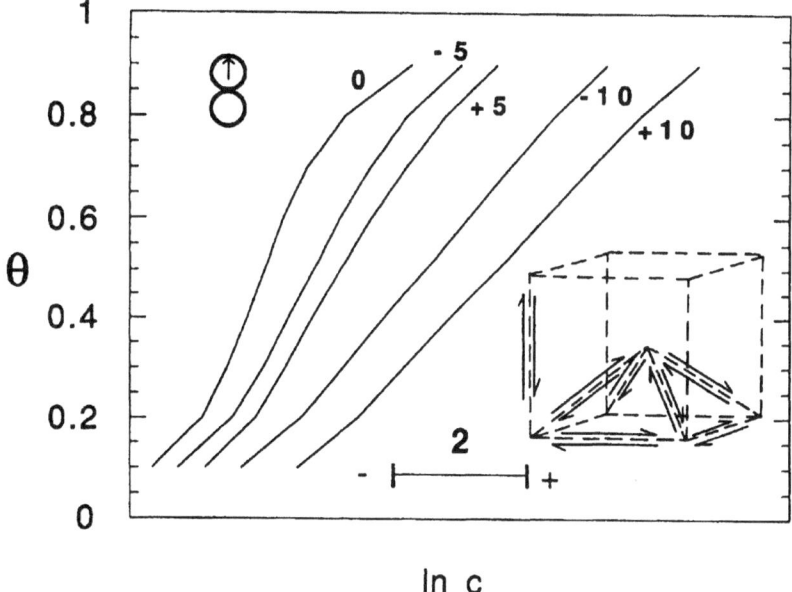

Figure 21. Calculated θ vs. *ln c* plots at constant charge for the adsorption of the dimeric solute of Fig. 16, which is now allowed to assume tilted orientations. The inset shows all allowed orientations. Numbers on each curve denote σ values in μC cm^{-2}.

with increasing the surface coverage θ, without any intermediate inflection in the corresponding adsorption isotherm. This is shown in Fig. 21, together with a schematic representation of all allowed orientations of the dimer. The isotherms so derived have once again a general behaviour which is analogous to that exhibited by the first portion of the isotherms in Fig. 19. The gradual passage from the horizontal to the vertical orientations with increasing θ at zero charge is shown in Fig. 22, which reports the percent values of the number densities of the horizontal orientations and of the tilted and vertical orientations in their two opposite directions. It can be seen that, even in this case, the vertical and tilted orientations with the tail towards the metal prevail over the corresponding opposite orientations, as expected.

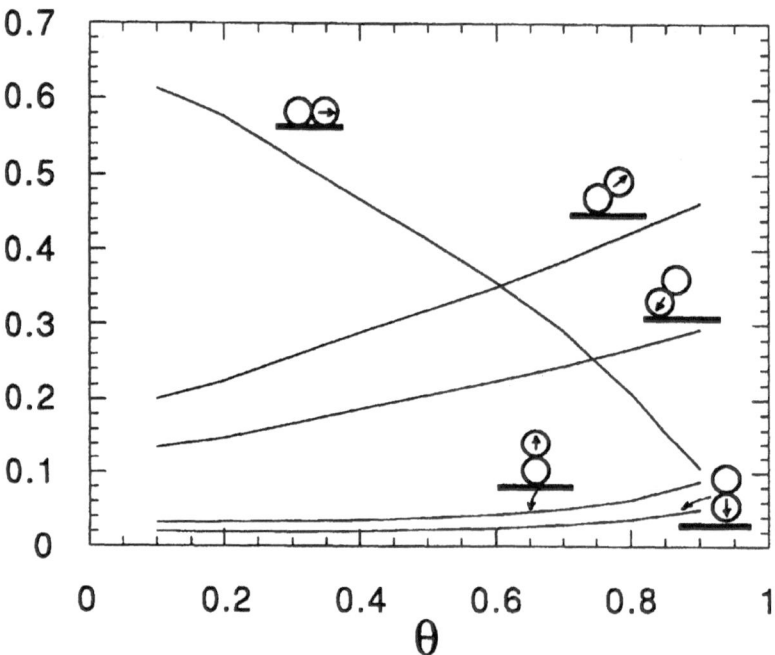

Figure 22. Percent values of the number densities of the horizontal, tilted and vertical orientations of the dimeric solute of Fig.21 as a function of θ, at zero charge.

8.3. EFFECT OF THE NATURE OF THE METAL

So far we have disregarded the effect of the nature of the metal, even though it is well known that the adsorptivity of a given surfactant from water is somewhat different on different *sp* metals even at far negative charges, where anionic specific adsorption and metal surface oxidation can be ruled out. Such a difference in behaviour can be ascribed, at least partially, to a different extent of electron spillover from the metal into the water phase. This electron spillover creates an electric field which tends to oppose the spillover itself and which is therefore constantly directed from the metal towards the solution. This electric field is superimposed on the uniform electric field $4\pi\sigma$ created by the charge density σ on the metal surface but, as distinct from the latter, it becomes vanishingly small ad distances greater than 1-2 Å from the metal surface plane. Hence it will only act upon the first lattice plane of water and solute molecules, or, more precisely, upon the most periferal partial charges of these water and solute molecules. The electric field produced by electron spillover as well as the distance of closest approach of the first lattice plane to the jellium edge were estimated by the self-consistent procedure outlined in the preceding chapter on the basis of a simple one-parameter jellium model and of a pressure balance. Figure 23 shows isotherms calculated for a monomeric polar solute adsorbed with the dipole moment constantly directed towards the metal at a charge density of -12 μC cm^2, both in the presence and in the absence of electron spillover. It can be seen that at the same negative

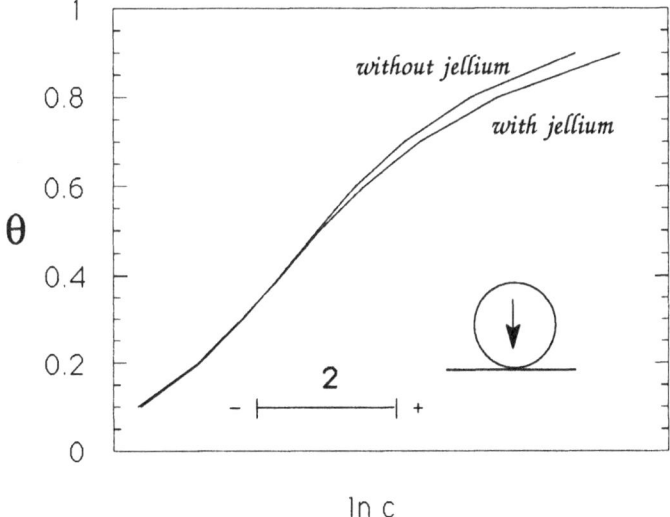

Figure 23. Calculated θ vs. ln c plots at $\sigma = -12$ μC cm^{-2} for the adsorption of a monomeric dipolar solute molecule pointing towards the metal and having a 2 D dipole moment of 2 Å length, both in the absence and in the presence of jellium. In the latter case, the centre of the first lattice plane of solvent and solute molecules was positioned at 2.3 Å from the jellium edge.

charge the adsorptivity is lower when electron spillover is accounted for, because the same surface coverage is attained at higher bulk solute concentrations. This result is readily explained by noting that the electrostatic potential energy of a solute dipole with the positive end towards the metal in the electric field produced by electron spillover is positive and hence favours the expulsion of the solute dipole from the adsorbed state. Obviously, an opposite effect is predicted in the case of a solute dipole with the negative end pointing towards the metal, as shown in Fig. 24.

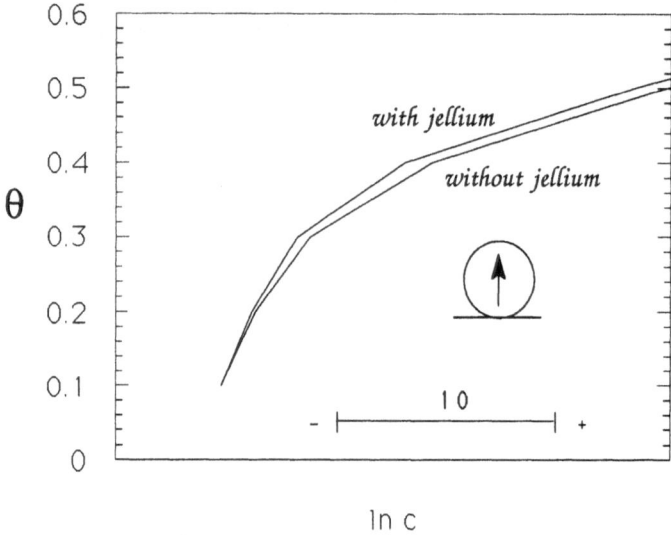

Figure 24. Calculated plots analogous to those in Fig. 23, but with the monomeric dipolar solute molecule pointing away from the metal.

These predictions can explain the increase in the adsorptivity of aliphatic alcohols such as n-butanol, n-pentanol [32], *iso*-propanol and *tert*-pentanol [33] when passing from mercury to gallium, even at far negative charges, where water chemisorption on Ga can be ruled out (see Fig. 25). In fact, independent of whether these alcohols are adsorbed in a flat or in a vertical orientation, the positive ends of the dipoles of the H-C bonds, which lie on the H atoms, will be closer to the jellium edge than the corresponding negative ends, which lie on the C atom. This behaviour can be explained by noting that electron spillover at a given charge density is more pronounced the higher is the electron density in the bulk metal, and this is twice as high in Ga than in Hg. Hence, the more pronounced electron

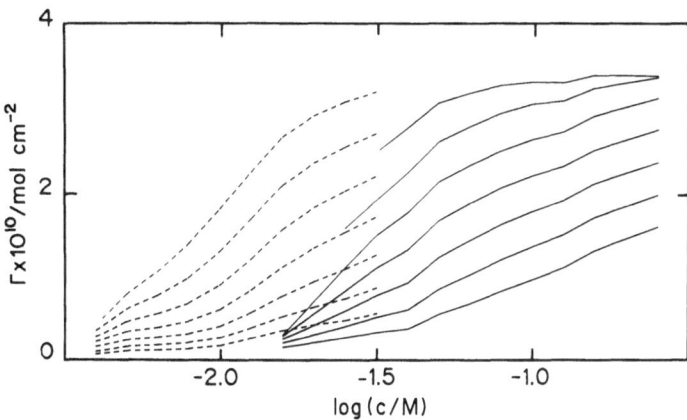

Figure 25. Γ vs. log c plots at constant charge for *tert*-pentanol adsorption at 32° C from aqueous 0.5 M Na$_2$SO$_4$ on Ga (solid curves) and Hg (dashed curves). For both sets of curves, σ increases from -12 to -6 μC cm^{-2} by 1 μC cm^{-2} steps when proceeding upwards [33].

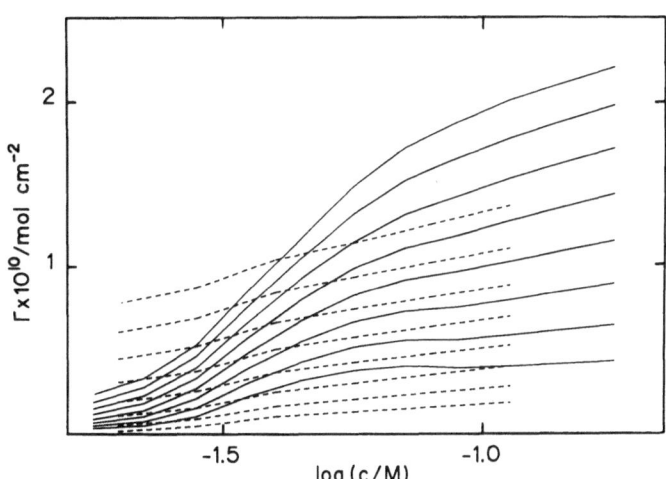

Figure 26. Γ vs. log c plots at constant charge for thiourea adsorption at 32°C from aqueous 0.5M Na$_2$SO$_4$ on Ga (solid curves) and Hg (dashed curves). For both sets of curves, σ increases from -13 to -6 μC cm^{-2} by 1 μC cm^{-2} steps when proceeding upwards [34].

spillover on Ga will exert a stronger repulsive effect on the adsorbed alcohol molecules, thus decreasing their adsorptivity with respect to mercury. According to predictions, thiourea, which is adsorbed with the negative end of the dipole towards the metal, is more strongly adsorbed on Ga than on Hg at far negative charges [33], as shown in Fig. 26.

References

(1)   J. O'M. Bockris, M.A.V. Devanathan and K. Müller, Proc. Roy. Soc., **A274**, 55 (1963); J. O'M. Bockris, E. Gileadi and K. Müller, Electrochim. Acta **1 2**, 1301 (1967).

(2)   For a review, see R. Guidelli, *Trends in Interfacial Electrochemistry*, edited by A.F. Silva (Reidel, Dordrecht, 1986).

(3)   A.N. Frumkin and B.B. Damaskin, *Modern Aspects of Electrochemistry*, Vol. 3, edited by J. O'M. Bockris and B. Conway, Butterworths, London, 1964, pp. 149-223.

(4)   B.B. Damaskin, O.A. Petrii and V.V. Batrakov, *Adsorption of Organic Compounds on Electrodes*, Plenum, New York, 1971.

(5)   H. Raous, D. Schuhmann, E. Tronel-Peyroz and P. Vanel, J. Electronal. Chem. **1 3**7,393(1982).

(6)   R. Fowler and E.A. Guggenheim, *Statistical Thermodynamics*, Cambridge University Press, Cambridge, 1965, p.431.

(7)   R. Guidelli, J. Electroanal. Chem. **1 1 0**,205(1980).

(8)   R. Parsons, J. Electroanal. Chem. **8**,93(1964).

(9)   G. Aloisi, M.L. Foresti, R. Guidelli and P. Barnes, J. Chem. Phys. **9 1**,5592(1989).

(10) R. Guidelli, G. Aloisi, M. Carla' and M.R. Moncelli, J. Electroanal. Chem. **1 9**7,143(1986).

(11) R. Guidelli, J. Electroanal. Chem. **1 2 3**,59(1981).

(12) R. Guidelli, J. Electroanal. Chem. **1 9**7,77(1986).

(13) R. Guidelli and M.L. Foresti, J. Electroanal. Chem. **1 9**7,103(1986).

(14) M. Carla', G. Aloisi, M.L. Foresti and R. Guidelli, J. Electroanal. Chem. **1 9**7,123(1986).

(15) M.L. Foresti, M.R. Moncelli, G. Aloisi and M. Carla', J. Electroanal. Chem. **2 5**5,267(1988).

(16) M. Carla', G. Aloisi, M.R. Moncelli and M.L. Foresti, J. Colloid Interface Sci. **1 3**2,72(1989).

(17) M.R. Moncelli, G. Pezzatini and R. Guidelli, J. Electroanal. Chem. **2 7**2,217(1989).

(18) M.R. Moncelli, M.L. Foresti and R. Guidelli, J. Electroanal. Chem. **2 9**5,225(1990).

(19) E. Dutkiewicz, J.D. Garnish and R. Parsons, J. Electroanal. Chem. **16**,505(1968).

(20) A. De Battisti and S. Trasatti, J. Electroanal. Chem. **48**,213(1973).

(21) B.A. Abd-El-Nabey and S. Trasatti, J. Chem. Soc. Faraday Trans.1 **71**,1230(1975).

(22) F. Pulidori, G. Borghesani, R. Pedriali and C. Bighi, J. Electroanal. Chem. **72**,65(1976).

(23) A. De Battisti, B.A. Abd-El-Nabey and S. Trasatti, J. Chem. Soc. Faraday Trans.1 **72**,2076(1976).

(24) F. Pulidori, G. Borghesani, R. Pedriali, A. De Battisti and S. Trasatti, J. Chem. Soc., Faraday Trans.1 **74**,79(1978).

(25) R. Amadelli, A. Daghetti, L. Vergano, A. De Battisti and S. Trasatti, J. Electroanal. Chem. **100**,379(1979).

(26) O. Ikeda, K. Kogo and H. Tamura, J. Electroanal. Chem. **151**,163(1983).

(27) M. Goledzinowski, J. Dojlido and J. Lipkowski, J. Electroanal. Chem. **185**,131(1985).

(28) G. Aloisi and R. Guidelli, J. Chem. Phys., in press.

(29) H. Nakadomari, D.M. Mohilner and P.R. Mohilner, J. Phys. Chem. **80**,1761(1976).

(30) B.B. Damaskin and S.L. Dyatkina, Elektrokhimiya **14**,152(1978).

(31) M.R. Moncelli and R. Guidelli, J. Electroanal. Chem. **295**,239(1990).

(32) G. Pezzatini, M.R. Moncelli, M. Innocenti and R. Guidelli, J. Electroanal. Chem. **295**,275(1990).

(33) G. Pezzatini, M.L. Foresti, M.R. Moncelli and R. Guidelli, J. Colloid Interface Sci., in press.

(34) G. Pezzatini, M.R. Moncelli and R. Guidelli, J. Electroanal. Chem. **301**,227(1991).

THE INTERFACE BETWEEN A METAL AND A SOLUTION IN THE ABSENCE OF SPECIFIC ADSORPTION

WOLFGANG SCHMICKLER

Physics Department, Utah State University, Logan, UT 84322, USA

1. INTRODUCTION

The theory of the double layer at the interface between a metal and an electrolyte solution dates back to the work of Helmholtz [1] in 1881. He pointed out that a difference in the electrostatic poten- tial between the metal and the solution implies a charge separation at the interface, and that the resulting charge configuration re- sembles that of a parallel plate condenser. Gouy [2] in 1910 and Chapman [3] in 1913 developed the theory of the diffuse double lay- er, based on a model in which the solution consists of point ions in a dielectric continuum (the solvent), and the metal is a perfect conductor. In 1924, Stern [4] introduced a charge-free layer, named the *Stern layer* or the *inner layer*, into the model of Gouy and Chapman, and the resulting Gouy-Chapman-Stern theory dominated double layer theory for more than five decades.

During the last decade two new lines of ideas have evolved, which are not consistent with the conventional view. Firstly, the metal is no longer regarded as a region of constant potential. Quantum mechanical models of the metal surface have been introduced to account for the penetration of electric fields into the metal surface, and for the *spillover* of electrons, effects that are well known from the physics of metal surfaces. Secondly, the existence of an *inner layer* of solvent molecules with special properties has been disputed; molecular models of electrolytes have been advanced in which the distribution of the particles and the potential can be calculated in the interfacial region, and in which the notion of a local dielectric constant is not used. In the major part of this work, we shall review and discuss these new ideas, and present what we think is the best double layer model at this time: the jellium / hard sphere electrolyte model.

Double layer theory has generally been based on an analysis of the differential capacity, and till now this is still the major

R. Guidelli (ed.), Electrified Interfaces in Physics, Chemistry and Biology, 369–398.
© 1992 *Kluwer Academic Publishers.*

source of information about the structure of the interface. Therefore the theories which we discuss will be compared to double layer
capacity data. However, several of the new in situ techniques that
have been developed during the last few years, in particular optical techniques and scanning tunneling microscopy (STM), can be applied to the interface in the absence of adsorption, and these data
can be interpreted within the same model. As an example we will
discuss the operation of the STM.

2. The Helmholtz Capacity

2.1 Gouy - Chapman Theory

This is the simplest reasonable theory for the distribution of ions
and of the electrostatic potential at the interface. It is based on
the same ideas as the well known Debye - Hückel theory, which it
preceded by one decade. Just as the latter it is correct in the
limit of infinite dilution of the ions, and thus serves as a reference point for the behavior at higher concentrations, and also for
more elaborate theories.

In the Gouy - Chapman theory the metal is considered to be a
perfect conductor, and the solution is modeled as point ions in a
dielectric continuum. Since this theory is extensively reviewed in
the literature, we only give the results for the case of a simple
z-z electrolyte with a concentration c_0. The capacity of the interface is:

$$C_{GC} = \left[\frac{2z^2 e_0^2 \varepsilon c_0}{kT} \right] \cosh \left[\frac{z e_0 \varphi_0}{2kT} \right] \qquad (2.1)$$

where φ_0 is the potential on the metal surface taken with
respect to its value at zero charge. For small excess charges q
this can be simplified to:

$$C_{GC} = \frac{\varepsilon \kappa}{4\pi} \qquad (2.2)$$

where $\kappa = \left[\dfrac{8\pi z^2 e_0^2 c_0}{\varepsilon kT} \right]^{1/2}$ is the inverse Debye screening length of

the solution.

2.2 Definition of the Helmholtz Capacity

The Gouy Chapman theory is the simplest quantitative model for the charge and potential distribution near the interface. It gives good results for very dilute electrolyte solutions, and thus serves as a convenient reference framework. Very early experiments [6] already showed that even at intermediate concentrations, where the Gouy-Chapman theory would still be expected to hold, there are systematic deviations. In this region the inverse double layer capacity can be written as the sum of two terms:

$$1/C = 1/C_H + 1/C_{GC} \qquad (2.3)$$

where C_{GC} is the usual Gouy-Chapman expression, and C_H is independent of the ionic concentration. C_H is called the *inner layer* or *Helmholtz capacity*; we shall use the latter terminology, since it does not imply an interpretation. C_H is conveniently determined from a Parsons and Zobel [5] plot: the measured inverse capacity $1/C$ is plotted versus the calculated Gouy-Chapman expression $1/C_{GC}$. At low and intermediate concentrations one obtains a straight line, and the Helmholtz capacity is obtained from the intercept; the inverse slope gives the roughness of the electrode.

A typical plot of this type is shown in fig.1 for various surface charge densities q on the electrode. If there is no specific adsorption of the ions, one obtains a set of parallel straight lines. At high electrolyte concentrations there are small but systematic deviations, indicating that eq.(2.3) is no longer valid in this region. We shall return to this point later.

In this way one can determine capacity - charge characteristics $C_H(q)$ for a particular metal/electrolyte interface. Experimentally one finds that, at least near the potential of zero charge (pzc), the Helmholtz capacity is almost independent of the nature of the ions, provided they are not specifically adsorbed.

It is important to realize that the Helmholtz capacity as defined here is the first order correction to the Gouy and Chapman theory. Since the latter is really a very simple theory, it is to be hoped that a model of the interface which represents the main features, but neglects the finer details, may be sufficient to explain the experimental data.

372

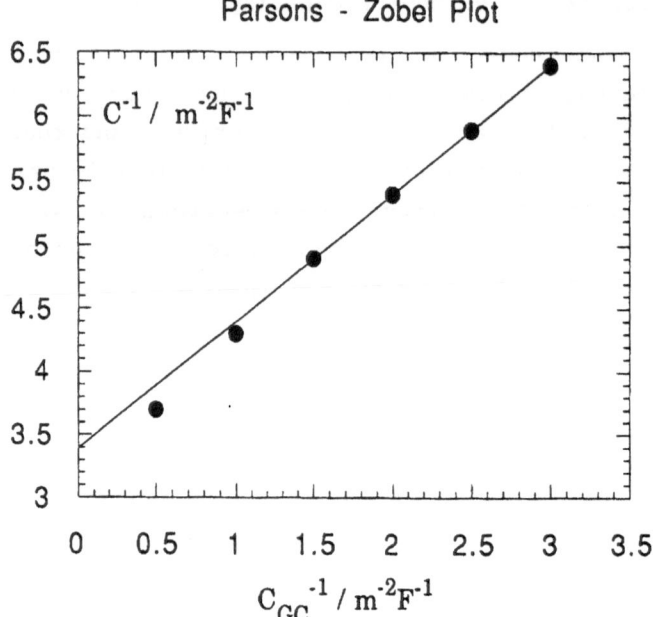

Fig.1: Parsons and Zobel plot (schematic).

2.3 Examples of capacity-charge characteristics

The Gouy-Chapman capacity depends only on the trivial system parameters: temperature T, electrode surface charge density q, and ionic concentration. In contrast, the Helmholtz capacity depends on
the nature of the metal and the solvent, and for a given interface
it also depends on T and q. It thus contains important information
about the double layer, and a good model should explain the main
features of these dependences.

Early investigations of the double layer capacity focused on
the interface between mercury and an aqueous solution. Precise measurements of the Helmholtz capacity $C_H(q)$ for this system were made
by Grahame [6]. More recently single crystal surfaces of several
metals, in particular gold and silver, were investigated. Figure 2
shows the capacity / charge characteristics of mercury and of an
Ag(100) surface in contact with aqueous solutions [7]; the great
influence of the metal is immediately apparent. Note that both curves have a maximum, the so called *capacity hump*, near the potential
of zero charge.

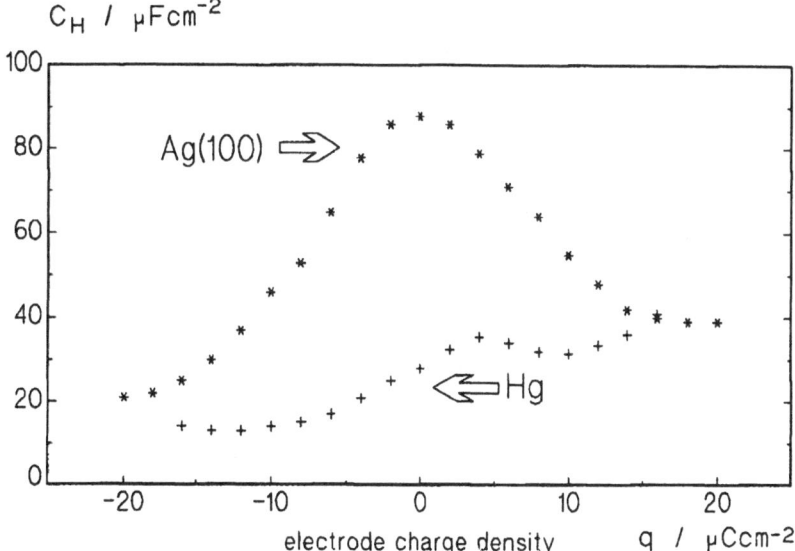

Fig.2: Experimental capacity - charge characteristics for an Ag(100) surface and for Hg in contact with an aqueous solution.

2.4 Physical interpretation

Obviously, the Helmholtz capacity contains information about the structure of the interface, but it can be interpreted only within a physical model. The quality of such a model has to be judged by three criteria: (1) Is it internally consistent? (2) How well does it explain the experimental data? (3) Does it comply with our knowledge of metals, solutions, and interfaces from other branches of science?

The simplest interpretation of eqn.(2.3) is to regard it as representing two capacitors in series, and accordingly to divide the interfacial region into two parts. Since C_H is independent of the ionic concentration, the existence of an inner layer of solvent molecules adjacent to the metal surface was postulated by Stern; C_H is then attributed to this inner layer, and C_{GC} to the diffuse layer.

While this interpretation is suggestive, it is by no means unique. The Gouy-Chapman theory holds only at very low concentrations. In the construction of more elaborate theories it is natural to use it as a starting point, and to seek expressions for the capacity in the form of a power series in $\kappa=1/L_D$, where L_D is the Debye length of the solution:

$$\frac{1}{C} = a\kappa^{-1}+b+c\kappa+... \qquad (2.4)$$

the first term dominates at low concentrations and must hence reduce to the Gouy - Chapman expression. The second term is independent of the electrolyte concentration, can be obtained from the intercept of a Parsons and Zobel plot, and should be identified with the Helmholtz capacity. Higher order terms contribute at rather high concentrations, and give rise to deviations from the straight line in the Parsons and Zobel plot [8].

The main difficulty with the old interpetration by Stern is that is cannot be developed into a consistent theory. If the Helmholtz capacity were caused solely by a special inner layer of solvent molecules, we would expect the intrinsic properties of the solvent, such as its dipole moment and its polarizability, to be independent of the metal. But then it is not possible to explain the strong dependence of the capacity on the metal which is evident from fig.2. On the other hand, the new models, in particular the jellium / hard sphere model of the interface which we will describe below, postulate a direct contribution of the metal to the interfacial capacity, and provide reasonable estimates for its magnitude. Also, they are more in line with our knowledge from other branches of surface science than the older theories.

3. The Jellium / Hard Sphere Electrolyte Model

This model has been worked out by Badiali et al [9] and by Schmickler and Henderson [10,11]. Both versions are similar, but differ in some important details. We will mailny follow the latter version here, which has been worked out in greater detail. First we shall describe the jellium model, then the ensemble of hard sphere ions and dipoles; finally, we shall specify the interaction of the two subsytems, and combine them into a model for the interface.

3.1. The Jellium Model

In the simple version of the model the positive charge on the metal ions is smeared out into a constant positive background charge, which drops abruptly to zero at the metal surface; so, in a suitable coordinate system in which the metal surface is at the plane $x = 0$, the positive charge distribution is:

$$n_+(x) = \begin{cases} n & \text{for } x < 0 \\ 0 & \text{for } x > 0 \end{cases} \tag{3.1}$$

where n is the bulk density of the positive charge and, because of electroneutrality, also of the electronic charge. The electrons are treated as an inhomogeneous electron gas interacting with the positive background charge, with itself, and with any other external field that may be present. Due to its small mass the electron can tunnel a short distance into the region outside the metal surface, and the electronic density $n_-(x)$ drops from its bulk value n to zero over a region of a few Ångstrom thickness (see fig. 3). Methods for its calculation will be described below.

Since the ionic distribution has been smeared out, the jellium model contains no information on the lattice structure, and is therefore applicable to polycrystalline metals only. It can, however, be modified by introducing a lattice of pseudopotentials, so that single crystal surfaces can be studied [12,13]. Also, the electron gas model is not suited for electrons in d-bands, which are narrow and have a more localized character. Application of the simple jellium model is therefore limited to sp-metals with a wide conduction band; however, Russier and Badiali [14] have recently suggested a way to include the low-lying d-bands of copper and silver, which seems to be quite successful.

The electronic density profile $n_-(x)$ is generally calculated within the density functional formalism; according to a theorem established by Hohenberg, Kohn, and Sham [15,16], the total energy of the inhomogeneous electron gas can be written as a functional of the electronic density only:

$$E[n_-(\mathbf{r})] = T_s + E_{xc} + \int v(\mathbf{r})n_-(\mathbf{r})d\mathbf{r} + \int \frac{n_-(\mathbf{r})n_-(\mathbf{r}')}{|\mathbf{r}-\mathbf{r}'|} d\mathbf{r}\ d\mathbf{r}' \tag{3.2}$$

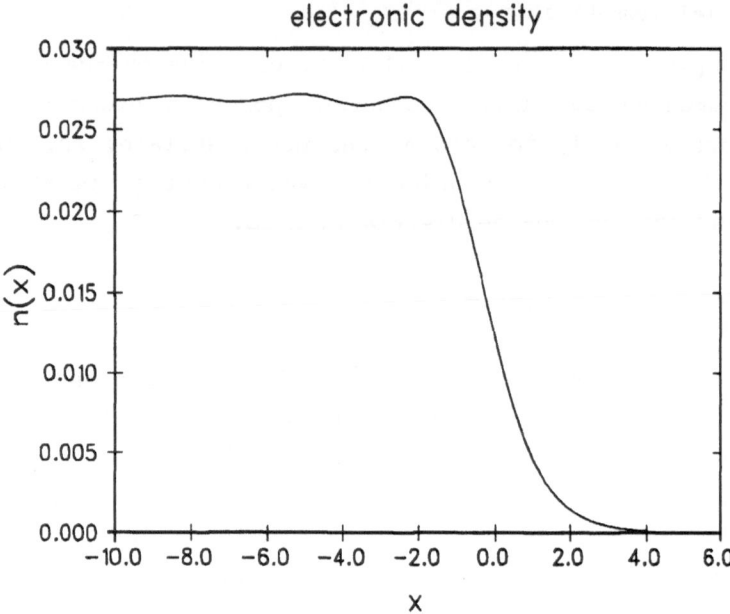

Fig.3: Relative electronic density $n_-(x)/n$ in the surface region of jellium; all quantities in atomic units, note: 1 a.u. = 0.529 Å.

where T_s is the kinetic energy, and E_{xc} the sum of the exchange and correlation energies; the third term represents the interaction with any external potentials such as that generated by the positive background charge, and the fourth is the electrostatic self-interaction. The exchange and correlation energies have a non-local character; however, useful local approximations have been derived which give good results even for surface properties. The following formula, due to Wigner, is often employed:

$$E_{xc}[n_-(\mathbf{r})] = -\frac{3}{4}\left\{\frac{3}{\pi}\right\}^{1/3}\int n_-(\mathbf{r})\ d\mathbf{r} -0.056 \int \frac{n_-(\mathbf{r})^{4/3}}{0.079+n_-(\mathbf{r})^{1/3}}\ d\mathbf{r} \quad (3.3)$$

where atomic units have been used.

The energy functional of eq.(3.2) attains its minimum for the true electronic density profile. Exact solutions can be obtained by solving the corresponding Euler equations [17]. For this purpose, electronic wave functions have to be introduced, so that some of

the simplicity of the density functional formalism is lost, and a
fair amount of numerical calculations is involved. Alternatively,
one can use local approximations for the kinetic energy based on a
gradient expansion [18,19], and minimize the energy within a suitable family of trial functions. Schmickler and Henderson [20] have
used the following functions, which incorporate the so-called *Friedel oscillations*, density oscillations known to exist near the surface, and which are particularly pronounced for low bulk electronic
densities:

$$n_-(x) = n \begin{cases} 1 - A \cos(\gamma x + \delta) \exp(\alpha x) & \text{for } x < 0 \\ B \exp(-\beta x) & \text{for } x > 0 \end{cases} \tag{3.4}$$

This family of functions contains six parameters, which are related
by the conditions that the density and its first derivative must be
continuous at $x = 0$, and by charge balance. Straightforward inser-
tion into the density functional and minimization would yield a
vanishing oscillation parameter γ. However, by using the pressure
balance relation established by Budd and Vannimenus [21]:

$$n [\phi(0) - \phi(-\infty)] = - p(-\infty) - 2\pi q^2 \tag{3.5}$$

as a subsidiary condition, where ϕ denotes the electrostatic potential, and p is the electronic pressure, the number of free parameters is reduced to two, and a minimum is found with a non-zero
value of γ giving electronic density profiles close to the exact
solution.

Within the jellium model all electronic properties are a function of the bulk electronic density n only; this is calculated from
the density under the assumption that all valence electrons of the
metal are free. Instead of n, the *Wigner - Seitz radius* $r_s = [3/4\pi n]^{1/3}$ is often given, which is the radius of a sphere containing one electron. As an example for the application of the jellium
model, we consider the work function Φ of simple metals. This can
be decomposed into contributions: $\Phi = D_e - \mu$, where μ is the bulk
chemical potential due to the kinetic, exchange, and correlation
energies, and D_e is the surface dipole potential due to the charge
distribution at the interface: $D_e = \phi(\infty) - \phi(-\infty)$. Figure 4 shows
the work function of jellium and experimental values for selected
polycrystalline sp-metals. Considering the conceptual simplicity of

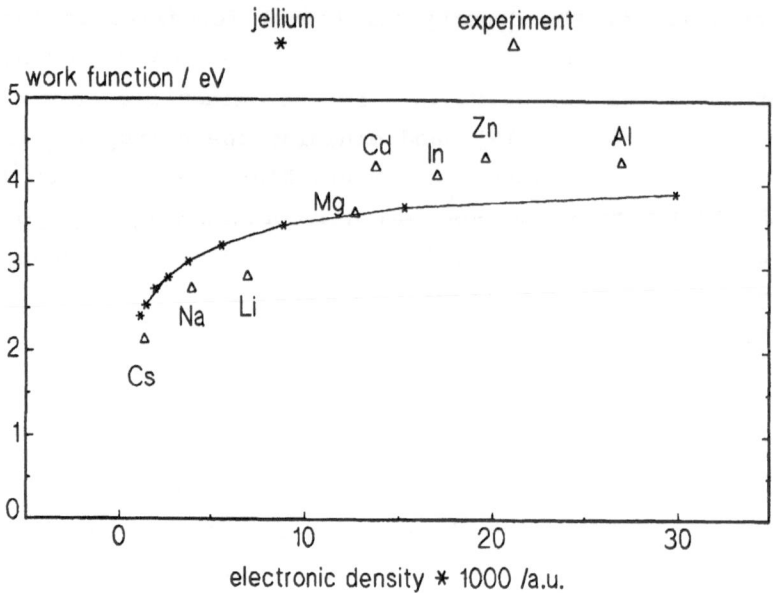

Fig.4: Work function of jellium and selected experimental values for polycrystalline sp-metals.

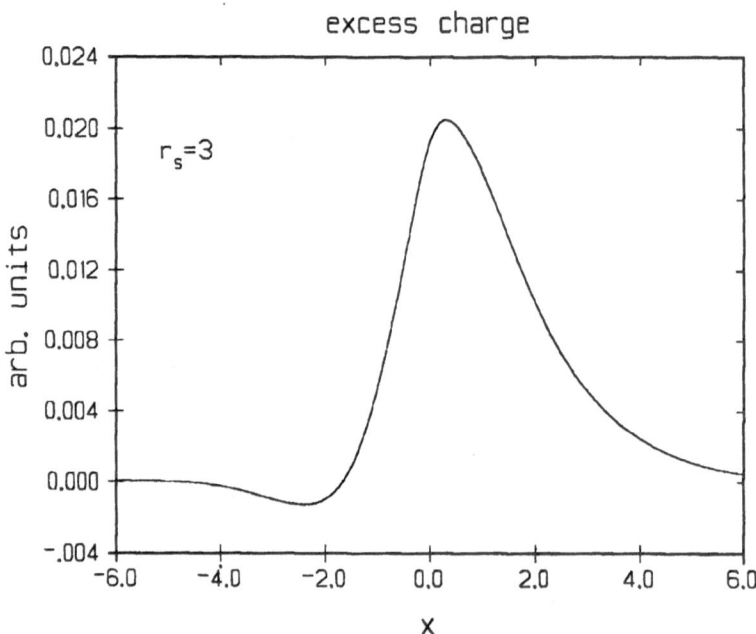

Fig.5: Distribution of a small excess charge at the surface of jellium; distances in a.u., (1 a.u. = 0.529 Å).

the model, the agreement is surprisingly good. This gives us confidence that jellium may be also a good model for the interfacial properties of metals in the electrochemical environment.

We are particularly interested in charged surfaces; fig. 5 shows the distribution of a small excess charge at the jellium / vacuum interface. Note that most of the induced charge is in front of the jellium edge. This is a consequence of the pressure balance condition; by differentiation one can show that the variation of the left half moment:

$$\delta h = \int_\infty^0 x\, \delta n_-(x)\, dx = \frac{1}{4\pi}\, \delta[\phi(0) - \phi(-\infty)] = \frac{-4\pi q}{n} \qquad (3.6)$$

vanishes to first order in q. Consequently, the center of mass x_{im} of the induced charge lies in front of the jellium edge. In addition, x_{im} is the position of the effective image plane, i.e. a small test charge q_t placed at a distance $d > 0$ from the metal surface will experience an image potential $V_{im} = -q_t/[4(d-x_{im})]$; we shall show later that x_{im} also determines the contribution of the metal to the interfacial capacity. Values for x_{im} for small excess charges are given in tab.1; it increases with the electronic density.

r_s /a.u.	x_{im} / Å
2	0.85
3	0.75
4	0.65

Tab.1. Effective position of the image plane for various electronic densities of jellium.

3.1.1. Jellium with Pseudopotentials

To introduce structure into the jellium model, one has to replace the constant positive background charge with a lattice of pseudopotentials. The simplest choice is the Ashcroft Potential [22], which has the form:

380

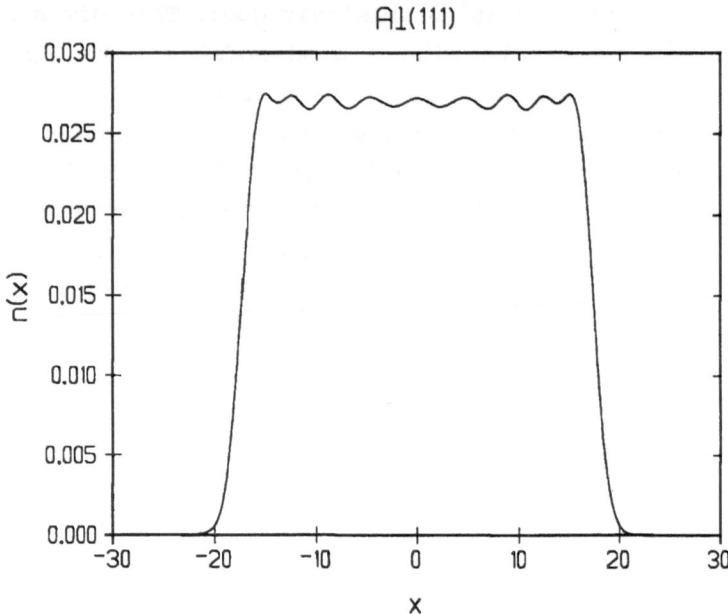

Fig.6: Electronic density at an Al(111) surface calculated in the
jellium model with pseudopotentials; all quantities in
atomic units

$$V_{ps}(r) = \begin{cases} 0 & \text{for } r < r_c \\ z/r & \text{for } r > r_c \end{cases} \qquad (3.7)$$

where z is the valence of the ion, and the *pseudopotential radius*
r_c is chosen to give good bulk properties for the metal under consideration. Several lists of recommended values for r_c exist.

To keep the model one-dimensional the pseudopotentials are usually averaged parallel to the metal surface. Various ways have been
devised to solve for the electronic density. Figure 6 shows an
exact solution for the electronic density of a finite slab of jellium with parameters corresponding to the (111) surface of aluminium [21]. Note the oscillations in the bulk caused by the periodic
lattice structure. The distribution of an excess charge is similar
to that in ordinary jellium, but the position of the effective
image plane depends strongly on the single crystal plane under consideration. For an fcc crystal its distance from the jellium edge,
which is half a lattice spacing in front of the first lattice
plane, is shortest for the compact (111) plane, and largest for the

open (110) plane. Thus, it is generally largest on the plane with the lowest work function.

3.2 An ensemble of hard sphere ions and dipoles

A good model for the solution must describe both the ions and the solvents on a molecular level. Further, it must be amenable to the mathematical techniques of statistical mechanics, since a model is only useful if we can perform calculations on it. As an alternative to doing calculations one can perform computer simulations, but due to the long range nature of the Coulomb force simulations of an ensemble of ions and dipoles at an interface are not yet feasible - they are beyond the power even of modern supercomputers. However, simulations of pure water at interfaces have been performed, and are reported elsewhere in this volume.

The mathematical difficulties of the statistical mechanics of solutions severely limit the complexity of the models that can be used. So far, the best model for which calculations have been performed is an ensemble of hard sphere ions and dipoles in contact with a hard wall, and even these have been done for small excess charges q on the wall only within the so-called *mean spherical approximation* [24,25]. This gives for the potential drop V between a charged hard wall and the bulk of the electrolyte:

$$V = \frac{4\pi q}{\varepsilon\kappa} + \frac{2\pi q}{\varepsilon}\left(\sigma_i + \frac{\varepsilon-1}{\lambda}\sigma_d\right) + \ldots \tag{3.8}$$

where ε is the bulk dielectric constant of the solvent, σ_i and σ_d are the diameters of the ions and the solvent, respectively, and λ is a constant characterizing the dielectric behaviour of the solvent near a boundary, and is determined from ε via the implicit relation:

$$\lambda^2(1+\lambda)^4 = 16\varepsilon \tag{3.9}$$

Typical values for λ are of the order of 2 - 3, so that it is much smaller than ε.

Hence the ensemble of hard spheres makes the following contribution to the inverse interfacial capacity:

$$\frac{1}{C_s} = \frac{4\pi}{\varepsilon\kappa} + \frac{2\pi}{\varepsilon} \left(\sigma_i + \frac{\varepsilon-1}{\lambda} \sigma_d\right) + \dots \tag{3.10}$$

which has precisely the form of the expansion given in eq.(2.2): the first term is the Gouy - Chapman capacity for small excess charges, and the second term can be identified with the Helmholtz capacity, more precisely with the contribution of the solution to the Helmholtz capacity, since in a complete model there will also be a contribution from the metal. Note that the contribution of the solvent to the Helmholtz capacity is considerably larger than that of the ions, since $(\varepsilon-1)/\lambda \gg 1$. This is in line with experimental data, which indicate that the interfacial capacity depends only weakly on the nature of the ions as long as they are not specifically adsorbed.

The finite size of the ions gives rise to oscillations in their distribution functions, which can be understood as a packing effect. These give rise to oscillations in the electrostatic potential which are illustrated in fig.7. Thus there exists a boundary region at the interface, which extends over several solvent diameters, where the structure of the solution differs significantly from that in the bulk, and which is characterized by oscillations for the distribution functions. In the limit $\sigma_i \to 0$ and $\sigma_d \to 0$ these oscillations disappear, the theory reduces to the Gouy - Chapman theory, and the Helmholtz capacity in eq.(3.10) vanishes.

3.2.1. Heuristic Extension to Higher Charge Densities

For high charge densities on the metal, the equations for an ensemble of hard sphere ions and dipoles have not yet been solved. Schmickler and Henderson [10] have devised a heuristic formula for the contribution of the solvent to the Helmholtz capacity, which interpolates between the value at the pzc and the value for complete dielectric saturation at very high fields. The interpolation is uniquely determined if one stipulates that it is effected by the Langevin function:

$$\frac{1}{4\pi C_s} = \frac{\sigma_i}{2} - \frac{1}{2} c_d\mu \ \sigma_i \ \frac{d}{dq} \ \mathcal{L}\left(4\pi q\mu\frac{9(\sigma_i-\sigma_s/\lambda)}{\lambda^2(\lambda+2)^2\sigma_i kT}\right) \tag{3.11}$$

where $\mathcal{L}(x) = \coth x - 1/x$

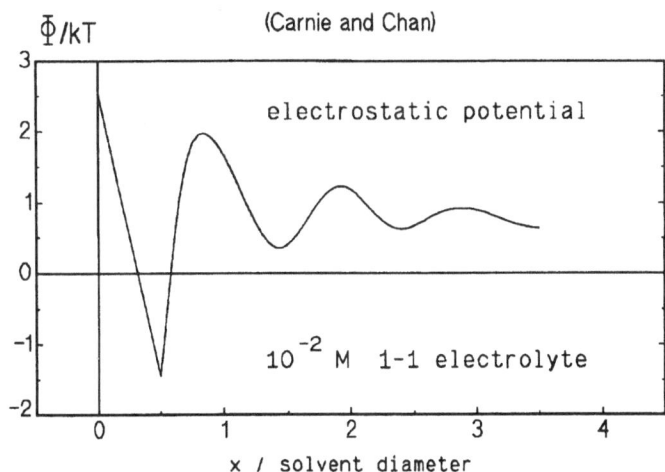

Fig.7: Electrostatic potential for a 0.01 M model 1-1 electrolyte near a charged hard wall; $\sigma_i = \sigma_d = \sigma = 3$ Å. The solid curve gives the result of the mean spherical approximation, the broken curve the result of the linearized Gouy - Chapman theory. Adapted with permission from S.L. Carnie and D.Y.C. Chan [25].

denotes the Langevin function, which describes the average orientation of a freely orientable dipole in an external field.

3.3 The Jellium / Hard Sphere Electrolyte Model

The ensemble of hard sphere ions and dipoles can be combined with the jellium model to obtain a model for the interface. The metal surface forms a hard wall for the ensemble of hard spheres; the two phases interact through the electrostatic fields generated by the electrons and the positive background on the one hand, and by the ions and dipoles on the other. This interaction modifies in principle both the electron distribution on the metal surface and the distribution functions of the particles in the solution. The dominant term is the classical electrostatic interaction due to the total excess charge; minor effects caused by the interpenetration of the electron tail and the first layer of the solvent have been treated by various approximations. In addition there can be a repulsive interaction of the metal electron tail with the water molecules. Since the details of this interaction are not known, a simple ad-hoc form is usually chosen. In particular, the following heuristic potentials have been used [10]:

$$V_{rep}(x) = A \; \delta(x-a) \tag{3.12a}$$

$$V_{rep}(x) = V_b \; \theta(x) \tag{3.12b}$$

$$V_{rep}(x) = V_s \; x \; \theta(x) \tag{3.12c}$$

all of which introduce unknown constants into the model; the forms of eqs. (3.12b) and (3.12c) are preferable, since they introduce only one such constant.

Due to the limitations of the hard sphere model calculations are limited to the case of small excess charges on the metal. Usually the modification of the distribution functions of the ions and dipoles due to the interpenetration of the metal electrons are neglected, since they are small and difficult to account for. In this case one obtains a particularly transparent expression for the Helmholtz capacity of the interface:

$$C_H^{-1} = \frac{2\pi}{\varepsilon} \left(\sigma_i + \frac{\varepsilon-1}{\lambda} \sigma_d \right) - 4\pi x_{im} \tag{3.13}$$

where x_{im} is the effective position of the metal image plane discussed in section 3.1.2 above; of course, x_{im} has to be calculated with the electron distribution modified by the interaction with the solution. Eq.(3.13) has a simple interpretation: the effect of the electronic spill-over is simply a shift of the position of the image plane towards the solution side, so that the effective charge separation is decreased, and the capacity gets larger. A different view of the same physical effect is illustrated in fig.8: the electronic density shifts with the surface charge density; it thus forms a highly polarizable substance, whose presence enhances the interfacial capacity. The latter interpretation allows a prediction of the dependence of the capacity on the electronic density of the metal: the higher the latter, the larger the polarizability of the electronic tail, and the higher the capacity. This is indeed borne

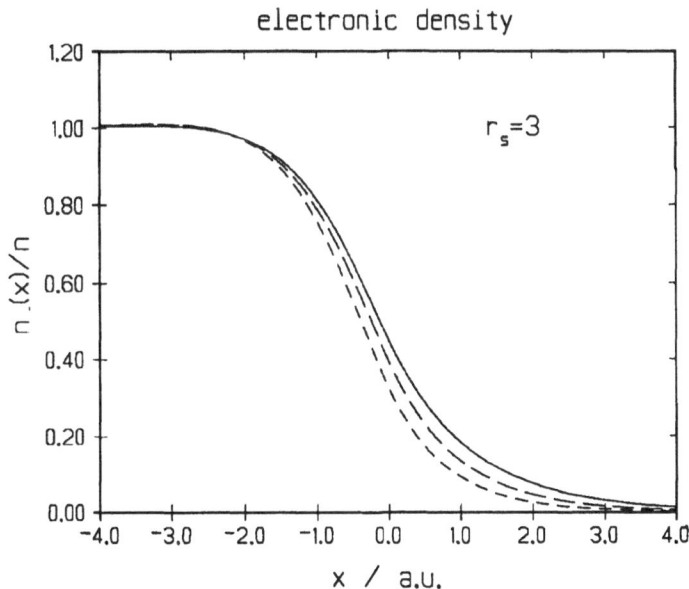

Fig.8: Electronic density profile for uncharged jellium (curve in the middle), jellium with a small negative (upper curve) and a small positive (lower curve) excess charge.

out by model calculations which will be discussed below, and also by the values of x_{im} for the vacuum situation given in tab.1 above.

4 Comparison of the Jellium / Hard Sphere Model with Experiment

The model of jellium in contact with an ensemble of hard sphere ions and dipoles allows quantitative calculations of a number of interfacial properties. The obvious quantity to compare with experiment is the Helmholtz capacity, for which a large set of data exists. Schmickler and Henderson [11] have performed model calculations for a number of systems, and we shall report on them below.

In the simple jellium model a metal is uniquely characterized by its electronic density; if pseudopotentials are included, the corresponding radius r_c is also required, which can be obtained from recommendations in the literature. In the hard sphere model the solution is characterized by its bulk dielectric constant ε and by the radii of the solvent molecules and of the ions. The van der Waals radius can be taken for the former; the radii of the ions is not so important, and usually values for hydrated ions are taken, which gives values of $\sigma_i \approx 6\text{-}8$ Å. The only unknown parameter is introduced by the repulsive potential; this has to be chosen for each solvent.

4.1 The Helmholtz Capacity at the pzc

From eq.(3.13) the Helmholtz capacity at the pzc can be calculated. Its dependence on the nature of the metal is of particular interest, since this was left completely unexplained by older theories. Figure 9 shows the results of such model calculations for the Helmholtz capacity of polycrystalline sp-metals in contact with aqueous solutions at the pzc; these calculations were performed with the two-parameter family of trial functions of eq.(3.4). The experimental points for sp-metals of the second and third column, and to some extent also those of the fourth column, lie close to the theoretical curve, while the semi-metals bismuth and antimony (not shown) lie rather far away. This is not unexpected, since the metals of the second and third column conform better to the free electron model than those of the fourth, while the semi-metals are known to behave rather differently.

Within the Schmickler-Henderson version of the model, there is one unknown parameter, which characterizes the short range repulsion of the metal electrons from the electronic shell of the solvent molecules. This parameter is fixed for each solvent. Changing its value within reasonable limits produces curves which run parallel to the curve shown in fig.9, so that the variation of the Helmholtz capacity with the electronic density is independent of the specific value given to this parameter. In particular, any reasonable jellium calculation predicts the increase of the Helmholtz capacity with increasing electronic density which is observed for the sp-metals of the second and third column.

Somewhat better results can be obtained for the metals of the fourth and fifth columns of the periodic table by introducing pseudopotentials into the jellium. Leiva and Schmickler [26] have modeled polycrystalline sp-metals by averaging the pseudopotentials over the principal crystal planes. Since in this model each metal is characterized by two parameters, the Helmholtz capacity can no longer be plotted as a function of the electronic density. However,

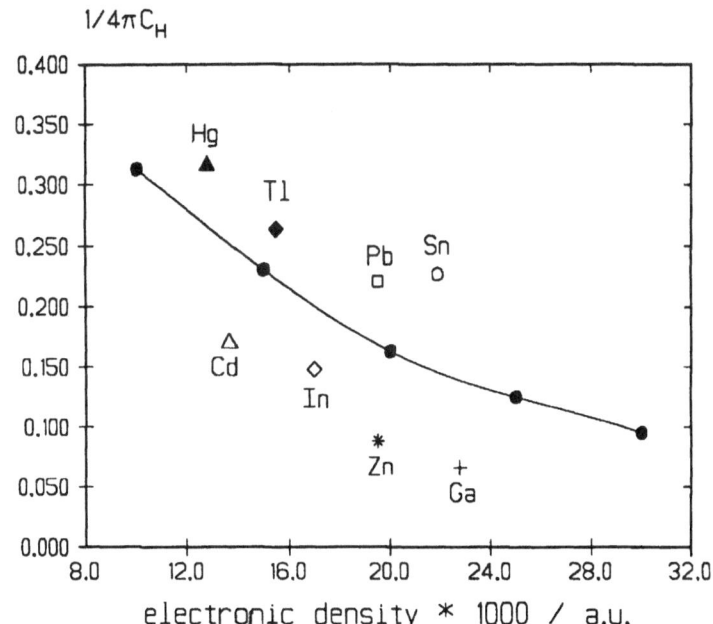

Fig.9: Inverse Helmholtz capacity $1/4\pi C_H$ / Å at the pzc of poly crystalline sp-metals in contact with an aqueous solution; The solid line is the result of a jellium calculation with a repulsive barrier of the type of eq.[37c] with V_b = 2 eV/Å.

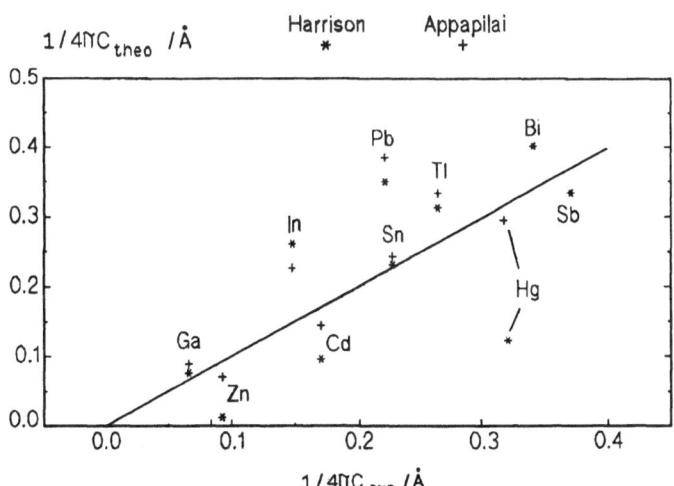

Fig.10: Theoretical inverse Helmholtz capacity at the pzc of simple metals in contact with an aqueous solution versus experimental values.

if one plots the metal contribution $-4\pi x_{im}$ to the inverse Helmholtz capacity versus the inverse of the experimental capacity, one should obtain a straight line with slope unity, and the intercept should be the negative of the solvent contribution. The calculated values fall indeed close to this line (see fig.10), and the metals of the forth and fifth column are also well represented by this model. These calculations were performed with a one parameter family of trial functions and without a repulsive interaction potential between the electrons and the solvent, and are thus free of any adjustable parameters. It is not clear at the present time whether V_{rep} is really negligible, or whether it will have to be reintroduced in more exact treatments of the problem.

4.2 Capacity - Charge Characteristics

Using the heuristic extension of the hard sphere model presented in section 3.2.2 capacity-charge characteristics can be calculated. Figure 11 shows two curves calculated for different electronic densities of the metal, which have to be compared with the experimantal curves of fig.2. Both sets of curves show a maximum near the pzc, but the theoretical curves have theirs at slightly negative charge densities, while in real systems it is usually found at small positive charges. Within this model, the decrease of the capacity on both sides of the maximum is caused by dielectric saturation of the solvent.

4.3 Discussion

The jellium / hard sphere electrolyte model presented above provides us with a detailed microscopic picture of the metal /solution interface, which gives at least a qualitative understanding of such properties as the differential capacitance. Indeed, the model is even quantitative in the sense that it gives the correct trends and variations for these properties without any adjustable parameters. The same model also gives good values for the surface dipole potential [11, 26] - as one might expect, since the capacity is mainly determined by the variation of the dipole potential. Very recently similar models have been successfully applied to second harmonic generation spectroscopy [27], which we will not consider here, and

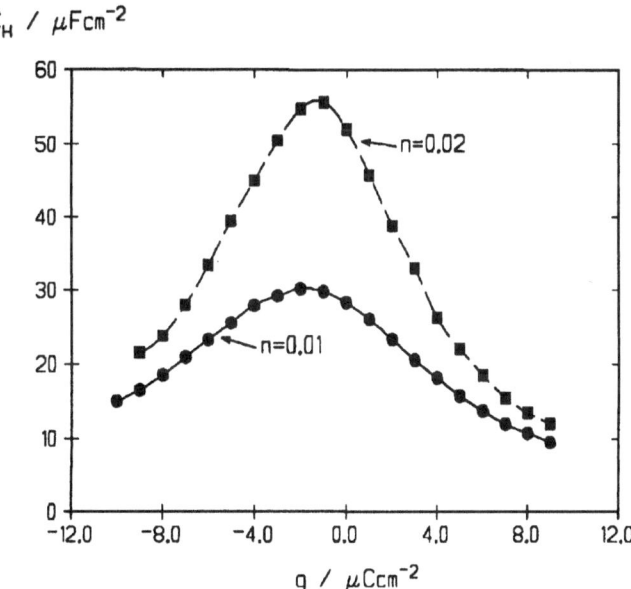

Fig.11: Calculated capacity - charge characteristics for jellium
with two different electronic densities.

to the theory of the scanning tunneling microscope operating in a
solution, which will be elaborated below.

The relative success of this model may at first seem somewhat
surprising considering its simplicity. However, most experimental
techniques give surface properties averaged over the electrode surface, so that the one-dimensional nature of the model does not matter. To represent a hydrogen bonded liquid like water as dipolar
hard spheres would seem to be a more serious approximation, but by
using the bulk dielectric constant of water rather than the molecular dipole moment, which in liquid water is badly defined, as input
parameter for numerical estimates, one makes, in a sense, an extrapolation from bulk to surface properties, so that the defects of
the model are not so important.

No doubt, more detailed and quantitative theories for the interface will be developed in the future. While a simple model may
have drawbacks, simplicity can also be a virtue in that provides a
direct physical understanding of the mechanisms underlying a phenomenon, which may be obscured in a more complex model, particularly
in one that is based on extensive numerical calculations.

5. Application to the Scanning Tunneling Microscope

Since its invention the scanning tunneling microscope (STM) and its variants have been widely applied to the study of surfaces in vacuo and in air. Recently, several groups have begun to use this technique for the in-situ investigation of the metal / solution interface [28]. This poses a number of sizable problems, such as the insulation of the tip, the separation of the tunneling current from residual Faradaic currents, and the difficulty of finding experimental conditions under which both the metal electrode and the tip are inert. Nevertheless, the STM is a potentially extremely useful probe of the interfacial structure, and in a few cases images with near atomic resolution have been obtained [29].

The STM in solutions has an advantage in principle over the same instrument in vacuo: the potentials of the substrate and of the tip can be varied independently with respect to a reference electrode, so that, for example, one can investigate a charged substrate with a similarly charged tip. This control over one additional parameter may open new possibilities for the investigation of the interfacial structure.

While the theory of the STM in vacuo has been intensively investigated, very little work has been done on the metal solution interface. This has prompted us to extend the ideas of double layer theory presented above to the electrochemical interface in the STM configuration.

We represent the metal electrode as a semi-infinite jellium occupying the region $x < 0$. The solution is treated on the Debye Hückel level, i.e. as point ions in a dielectric continuum - the equations for an ensemble of hard sphere ions and dipoles cannot be solved with the complicated boundary conditions imposed by the presence of the tip. The STM tip is modeled as a jellium sphere of radius R centered on the z-axis at $z = L$. On account of mathematical difficulties related mainly to the solution side of the model, our present treatment is limited to small charge densities on the tip and the substrate, and the current will be calculated in the linear response regime.

5.1 The potential energy surfaces for the transferring electron

In order to calculate the potential energy surface experienced by a tunneling electron we have modified the expression for the electrostatic energy on the solution side of the interface by including the optical dielectric constant ε_∞ of the solvent; a similar modification was made by Hirabayashi [30] within a different context. The reason for doing this is the following: the transferring electron will interact with the polarization of the solvent. However, both the librational and the vibrational modes of the solvent are slow compared to the speed of the tunneling electron, and cannot respond to its transfer; only the electronic polarization, which gives rise to ε_∞, is expected to be sufficiently fast to follow the electronic motion.

For the semi-infinite jellium representing the metal electrode, we have used the same two-parameter family of trial functions as for the double layer calculations above (see eq. 3.4).For the electronic density of the jellium sphere representing the tip we have used trial functions suggested by Cini [31]:

$$n_-(r) = \begin{cases} a + br + c \ exp[\alpha(r-R_0)] \text{ for } r < R_0 \\ g \cdot exp(-\beta(r-R_0)) \end{cases} \tag{5.1}$$

where a, b, c, g, α, β, R_0 are variational parameters; these are related by the conditions that $n_-(r)$ and its derivative are continuous, by charge balance, and by the condition that $dn_-(r)/dr$ vanishes at the origin. For an uncharged tip one may set $R_0 = R$, as one might expect, leaving two parameters to be determined by energy minimization. While this family does not give very good results for the electronic density inside the sphere, it works well on the outside. We need only calculate the potential energy profile between the sphere and the substrate, so this approximation is good enough for our purpose. Indeed, since a jellium sphere is only a crude representation of a metal tip, there is no point in obtaining exact solutions for the sphere.

The potential energy surface is easily obtained from the electronic density profile; for the semi-infinite jellium the potential energy experienced by an electron is (in atomic units):

392

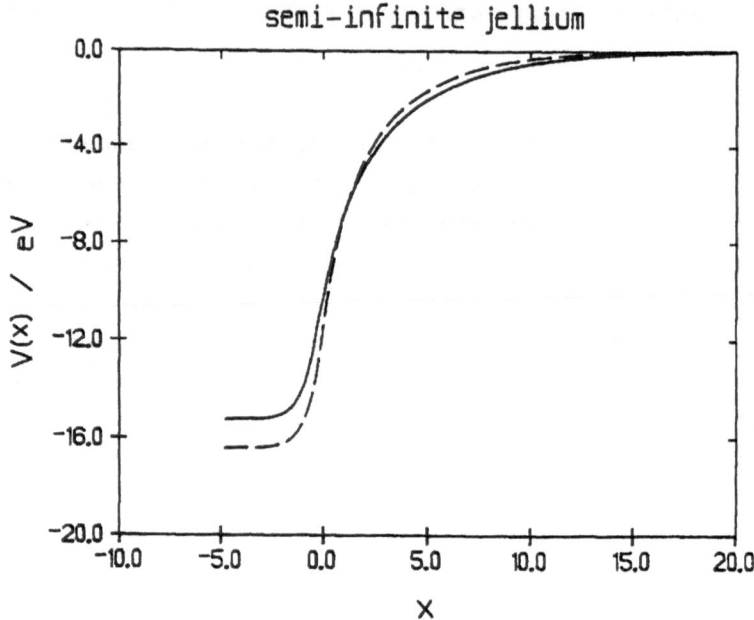

semi-infinite jellium

Fig.12: Electronic potential energy profile for a semi-infinite
jellium in contact with a solution of dielectric constant
ε_∞ = 1.88 (full line) and with the vacuum (broken line).

$$V(x) = \phi(x) - \left(\frac{3}{\pi} n_-(x)\right)^{1/3} - \frac{0.056 \cdot n_-(x)^{2/3} + 0.0059 \cdot n_-(x)^{1/3}}{(0.079 + n_-(x)^{1/3})^2} \qquad (5.2)$$

where $\phi(x)$ is the electrostatic potential generated by the charge
distribution in the boundary region; a corresponding relation holds
for the jellium sphere. Within the present model, the effect of the
solution is mainly a lowering of the potential energy barrier due
to the screening of the electrostatic interaction by ε_∞. This is
illustrated in fig.12, which shows the potential energy $V(x)$ for a
semi-infinite jellium both in vacuo and in contact with a solution.
Far from the metal surface, this potential should reduce to the
classical image law: as in all jellium calculations, we do not obtain the correct asymptotic behavior, if a local approximation to
the exchange and correlation energies is used. However, this does
not affect our results, since the image potential at long distances
makes a negligible contribution to the overall barrier. In the real
system the work function is lowered further by a small net orientation of the water in the first layer, which tends to have the oxy-

gen end a little closer to the metal surface. We shall include this effect, which has been estimated to be of the order of a few tenths of an eV [11], in future work.

5.2 Tunneling current

In the limit of small voltages and temperatures, the tunneling current is given to first order by [32]:

$$I = \frac{2\pi}{\hbar} e_0^2 U \sum_{\mu,\nu} |M_{\mu\varphi}|^2 \delta(E_\nu - E_\mu) \delta(E_\mu - E_F) \tag{5.3}$$

where U is the applied voltage, M is the tunneling matrix element between states ϕ_μ of the electrode and ϕ_φ of the tip, which have energies E_μ and E_ν, respectively, and E_F is the Fermi level.

The unperturbed wave functions of the semi-infinite jellium are characterized by $\mu = (k_\parallel, k_z)$, where $k_\parallel$ is the momentum parallel to the surface, and have the form:

$$\Psi_\mu = exp(ik_\parallel x_\parallel) \phi_{k_x}(x) \tag{5.4}$$

where $x_\parallel = (0, y, z)$, and $\phi_{k_x}(x)$ is a solution of:

$$\left[-\frac{\hbar^2}{2m} \frac{d^2}{dx^2} + V(x) + \frac{\hbar^2 k_\parallel^2}{2m} - E \right] \phi_{k_x}(x) = 0 \tag{5.5}$$

where $V(x)$ is given by eq.5.2. In the bulk of the jellium, $\phi_{k_x}(x)$ takes on the form:

$$\phi_{k_x}(x) \simeq A \, cbs(k_x x + \delta) \qquad k_x^2 = 2m[E - V(-\infty)] - k_\parallel^2 \tag{5.6}$$

where we have normalized A to unity.

The Schrödinger equation for a jellium sphere has a discrete spectrum for $E < 0$, and need not have a solution with energy E_F. However, the real system is a tip linked to a bulk metal, and has a continuous spectrum near the Fermi level. We have therefore calculated the s-like wave function of energy E_F for the one-electron potential of the jellium sphere, and not imposed the condition $d\Psi/dr = 0$ at $r = 0$, which leads to a discrete spectrum. Thus, the jellium sphere is merely a device to obtain the potential energy surface in the region between the tip and the substrate.

The matrix elements were calculated using Oppenheimer's version of first order perturbation theory; the integrals over x and the momentum $k_\parallel$ were performed numerically, making use of the fact that $M_{\mu\nu}$ depends on the absolute value of $k_\parallel$ only.

5.3 Results and Discussion

Model calculations were performed for a semi-infinite jellium and a jellium sphere with the same Wigner-Seitz radius $r_s = 2$ for a small applied voltage U. The optical dielectric constant for the solvent was taken as $\varepsilon_\infty = 1.88$ corresponding to water. The tip radius was taken as $R = 5$ a.u., the substrate - tip separation $l = L-R$ was varied in the region 6 - 35 a.u..

The variation of the conductance $S = I/U$ with the separation l is shown in fig.13 both for operation in the liquid and in vacuo. Within our model, the work function for electron emission into the solvent (2.68 eV) is substantially lower than the corresponding vacuum value (3.88 eV); this is in line with experimental data for photoemission into solutions [33]. For this reason the decrease of the conductivity with distance is faster in the solution, and the current is higher at all separations. Differentiating the conductivity with respect to the separation gives the effective barrier height:

$$W = \frac{1}{2m} \left(\frac{\hbar}{2} \frac{d \ln(S)}{dl} \right) \tag{5.7}$$

which is shown in fig.14. It increases with the separation, and tends asymptotically to the corresponding work function, but even at $l = 35$ a.u. it is still a little lower. This relatively slow rise is due to the fact that the potential energy $V(x)$ for the isolated jellium approaches zero only at a sizable distance from the jellium edge (see fig.12).

So, within our model, the effective barrier height for the tunneling electron is substantially lowered due to the presence of the solution; in an aqueous solution, it would be further reduced by the small preferential orientation of the water molecules at the metal surface, which we have mentioned above. This may explain the

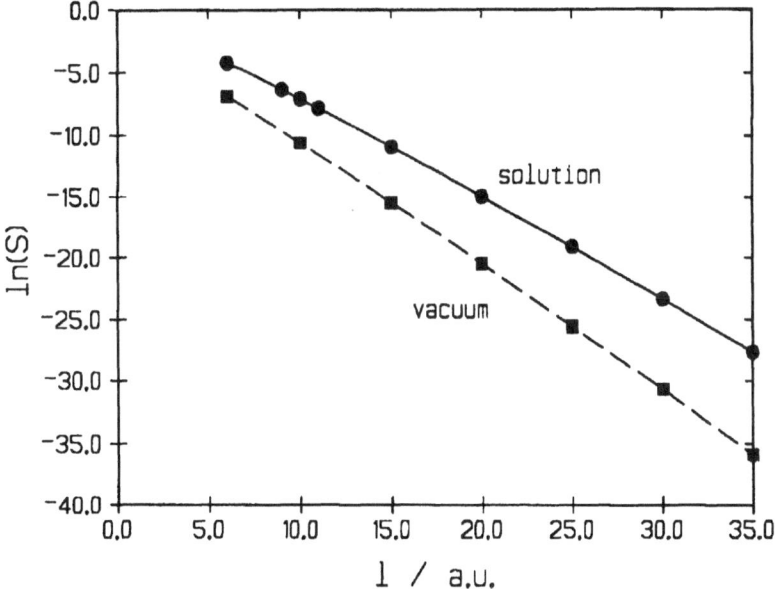

Fig.13: Conductance S = I/U for various metal - tip separations;
——•—— solution, - - ■- - vacuum.

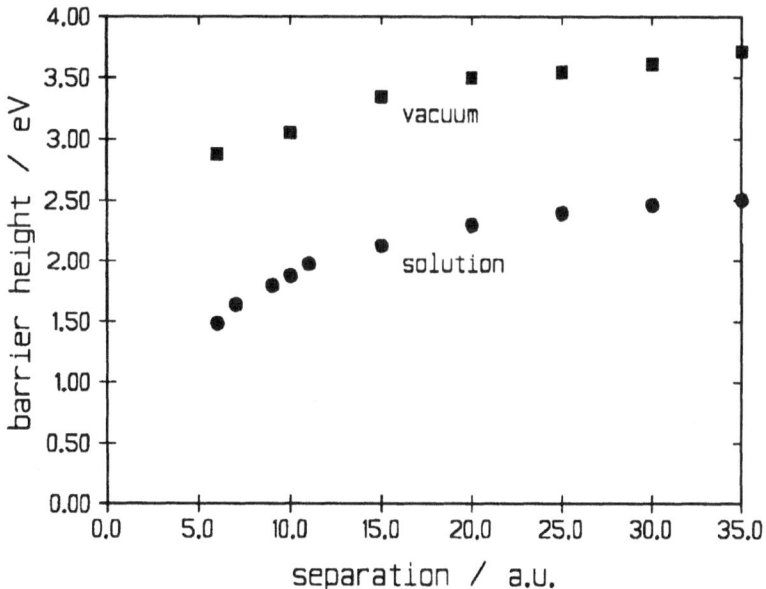

Fig.14: Barrier height as a function of the metal - tip
separation; • solution, ■ vacuum.

small barrier heights which have been observed in a number of ex
periments performed in aqueous solutions [34].

It follows from the work of Tersoff and Hamann [32] that under
the usual operating conditions the resolution of the STM is the
greater, the lower the barrier height. Thus, within our present
model, the theoretical resolution is greater in the solution than
in vacuo, though the experimental difficulties mentioned in the
introduction may make this difficult to achieve in practice. Also,
it is not clear whether this feature will persist in better models
accounting for the discrete nature of the solution.

Acknowledgement

It is a pleasure to thank my colleagues with whom I have cooperated
in this area of research: Drs. D. Henderson, A. Kornyshev, and E.
Leiva. Financial support through a start up fund from Utah State
University and by the NATO Science committee (Collaborative Research Grant No. RG-86/0068) is gratefully acknowledged.

References

1) H.von Helmholtz, Monatsber. Preuss. Akad. Wiss. Nov. 1881

2) G. Gouy, J. Phys. **9** (1910) 457

3) D.L. Chapman, Philos. Mag. **25** (1913) 475

4) O. Stern, Z. Elektrochem. **56** (1924) 508

5) L. Blum, D. Hemderson, and R. Parsons, J. Electroanal. Chem.
   161 (1984) 389

6) D.C. Grahame, J. Am. Chem. Soc. **80** (1958) 4201; D.C. Grahame
   and R. Parsons, J. Am. Chem. Soc. **83** (1961) 1291

7) G. Valette, J. Electroanal. Chem. **122** (1981) 265

8) R. Parsons and F.G.R. Zobel, J. Electroanal. Chem. **9** (1965) 333

9) J.P. Badiali, M.L. Rosinberg, F. Vericat, and L. Blum, J.
   Electroanal. Chem. **158** (1983) 253

10) W. Schmickler and D. Henderson, J. Chem. Phys. **80** (1984) 3381

11) W. Schmickler and D. Henderson, J. Phys. Chem. **82** (1985) 2925

12) N.D. Lang and W. Kohn, Phys. Rev. B **3** (1971) 1215; Phys. Rev. B**7** (1973) 2541

13) R. Monnier and J.P. Perdew, Phys. Rev. B **17** (1975) 2595

14) V. Russier and J.P. Badiali, Phys. Rev. B **39** (1989) 13193

15) P. Hohenberg and W. Kohn, Phys. Rev. **136** (1964) B864

16) W. Kohn and L.J. Sham, Phys. Rev. **140** (1965) A1133

17) N.D. Lang and W. Kohn, Phys. Rev. B **1** (1970) 4555

18) J.R. Smith, Phys. Rev. **181** (1969) 522

19) C.Q. Ma and V. Sahni, Phys. Rev. B **19** (1979) 1290

20) W. Schmickler and D.J. Henderson, Phys. Rev. B **30** (1984) 3081

21) H.F. Budd and J. Vannimenus, Phys. Rev. Lett. **31** (1973) 1212, 1439E

22) N.W. Ashcroft, Phys. Rev. Lett. **23** (1966) 48

23) E. Leiva, unpublished results, personal communication

24) S.L. Carnie and D.Y.C. Chan, Adv. Coll. Interface Sci. **16** (1982) 81

25) L. Blum and D. Henderson, J. Chem. Phys. **74** (1981) 1902

26) E. Leiva and W. Schmickler, J. Electroanal. Chem. **205** (1986) 323

27) P. Guyot-Sionnest, A. Tadjeddine, and A. Liebsch, Phys. Rev. Lett. **64** (1990) 1678; P.G. Dzhavakhidze, A.A. Kornyshev, A. Liebsch, and M. Urbakh, in press; W. Schmickler and M. Urbakh, unpublished results

28) early papers are: R. Sonnenfeld and B.C. Schardt, Appl. Phys. Lett. **49** (1986) 1172; O. Lev, F.R. Fau, and A.J. Bard, J. Electrochem. Soc. **135** (1988) 783; D.H. Craston, C.W. Lin, and A.J. Bard, J. Electrochem. Soc. **135** (1988) 785; K. Itaya, K. Higaki, and S. Sugawara, Chem. Lett. (Japan) (1988) 421; I. Otsuka and T. Iwasaki, J. Microsc. **152** (1988) 289; P. Lustenberger, H. Rohrer, R. Christoph, and H. Siegenthaler, J. Electroanal. Chem. **243** (1988) 225; J. Wiechers, T. Twomey, D.M. Kolb, and R.J. Behm, J. Electroanal. Chem. **248** (1988) 451

398

29) O.M. Magnussen, J. Hotlos, R.J. Nichols, D.M. Kolb, and R.J. Behm, Phys. Rev. Lett. **64** (1990) 2929

30) K. Hirabayashi, Phys. Rev. B **3** (1971) 4023

31) M. Cini, J. Catalysis **37** (1975) 187

32) J. Tersoff and D.R. Hamann, Phys. Rev. B **31** (1985) 805

33) Yu.V. Pleskov and Z.H. Rotenberg, in: Advances in Electrochemistry and Electrochemical Engineering, Vol. XI, John Wiley & Sons, New York, Chichester, Brisbane, Toronto, 1978

34) R. Christoph, H. Siegenthaler, H. Rohrer, and H. Wiese, Electrochim. Acta **34** (1989) 1011; also private communication.

ADSORPTION AT THE METAL / SOLUTION INTERFACE

Wolfgang Schmickler

Physics Department, Utah State University, Logan, UT 84322, USA

1 Introduction

While we would like to use similar concepts and models for the double layer in the presence and in the absence of specific adsorption, it is obvious that the adsorption of ions on the metal surface introduces complications such as the chemical interaction of the ions with the metal, their interaction with the solvent, and adsorbate - adsorbate interactions. Much work is still to be done in this area, but some progress has been made in two relatively simple cases: the adsorption of simple ions at low coverage, and the deposition of a monolayer of metal ions on a foreign metal, which is also known as *underpotential deposition* (upd). We shall first consider simple ions like the alkali and halide ions, and then upd.

2. Adsorption of Small Amounts of Ions on Metal Electrodes

Carnie and Chan [1] have extended the hard sphere electrolyte model to the case where a small amount of ions is adsorbed on the metal surface. When their work is combined with the jellium model, the *dipole moment induced by the adsorbate* can be calculated within linear response theory. We shall first explain the notion of an adsorbate dipole moment [2,3], which has not been used extensively in the literature.

2.1 The Dipole Moment Induced by an Adsorbate

The concept of a surface dipole moment is commonly used in surface science, so let us briefly look at adsorption in the ultra high vacuum (uhv) to point out the analogies. Adsorption of a species generally gives rise to a dipole moment μ per adsorbate directed perpendicularly to the metal surface, which modifies the electronic work function Φ of the substrate:

$$\Phi = \Phi_0 - 4\pi s\mu \tag{2.1}$$

R. Guidelli (ed.), *Electrified Interfaces in Physics, Chemistry and Biology*, 399–425.
© 1992 *Kluwer Academic Publishers*.

where Φ_0 is the work function of the bare metal surface, and s is the surface concentration of the adsorbate in particles per unit area. Note that the work function is measured under conditions where the surface is uncharged; hence any charge on the adatom is balanced by the image charge on the metal. The simplest interpretation of the dipole moment μ is that it is caused by a charge q_a on the adsorbate situated at a distance d from the metal surface, so that $\mu = q_a d$. It has, however, been pointed out [4] that this view is too simplistic: if the adatom is charged, the corresponding image charge is in front of the metal surface, and may partially surround the adatom, so that it is difficult to infer the charge on the adsorbate from the dipole moment μ. As we shall see below, this is even more difficult for adsorption from solution.

In a typical uhv experiment, the metal surface is electrically neutral before and after the adsorption process. Thus, any excess charge on the adsorbate is balanced by the corresponding image charge on the metal. The corresponding electrochemical experiment would be the following: Before the adsorption, the electrode is kept at the pzc; in the elementary act, an ion is transferred from the bulk of the solution to the metal surface, while simultaneously the image charge flows onto the electrode. If the ion is partially or totally discharged, a fraction or all of the image charge is transferred to the ion. After the experiment, the total charge at the interface is zero; any excess charge on the adsorbates is balanced by the image charge on the metal. Hence, there is no excess charge in the adjacent diffuse layer region. But note that the metal surface is generally not at the pzc, since it carries the image charge. The change $\Delta\varphi$ in the electrode potential during this experiment is equivalent to the change in the work function during adsorption from the gas phase, and $\mu = \Delta\varphi/4\pi s$ gives the surface dipole moment per adsorbed particle.

Naturally, the real experiment does not have to be performed in this way. In the classical works of Grahame and Parsons [5], the electrode potential is plotted against the formal charge q_i on the adsorbate layer (assuming that the ions still carry their full charge) for various values of the total formal thermodynamic charge q on the metal. The electrode potential for $q_i = -q$, minus the potential of zero charge without adsorption, gives the desired surface potential change $\Delta\varphi$.

Unfortunately, this procedure sometimes requires an extrapolation of the experimental data, which limits its accuracy. A much more commonly used concept is the *electrosorption valency* [6]. Fortunately the one can be calculated from the other, as we shall proceed to show. We shall restrict our consideration to the adsorption of a small amount of simple ions near the pzc, since this is the case for which we have a model. Even at the pzc, and in the absence of specific adsorption, there is a difference in the inner (and outer) potential between the metal and the bulk of the solution. Let $\Delta\varphi$ denote the variation of this potential difference due to a charge density q on the metal or to a charge density $q_i = sze_0$ on the adsorbate layers. We may write:

$$\Delta\varphi = Aq + Bq_i + \frac{1}{C_{GC}}(q+q_i) \qquad (2.2)$$

where C_{GC} denotes the Gouy - Chapman capacity, the term Aq gives the first order deviations from the Gouy - Chapman theory in the absence of specific adsorption, and Bq_i is the potential drop caused by the adsorbate layer. Eq.(2.2) is simply a first order correction to the Gouy - Chapman equation, and should hold for small q_i and q independent of any model; in particular, it is also correct if part or all of the adsorbate charge has passed to the metal. Setting $q_i = 0$, we obtain $A = 1/C_H$. Setting $q = -q_i$ we obtain for the surface dipole moment of the adsorbate:

$$\mu = -\Delta\phi/4\pi s = -ze_0(B-A)/4\pi, \qquad (2.3)$$

where z is the charge number of the adsorbed ion. The electrosorption valency $\tilde{\gamma}$, not corrected for the diffuse layer potential, is defined as:

$$\tilde{\gamma} = -z\left(\frac{\partial q}{\partial q_i}\right)_{\Delta\varphi} \qquad (2.4a)$$

and thus characterizes the adsorption at constant potential. Generally, the electrosorption valency is corrected for the Gouy - Chapman potential; from eq.(2.2) the corrected value γ at the pzc

402

is:

$$\gamma = \frac{zB}{A} \tag{2.4b}$$

which is the same as the *thickness ratio* defined earlier by Grahame [7]. We obtain:

$$\mu = \frac{ze_0}{4\pi C_H} \, (1-\gamma/z) \tag{2.5}$$

which is the desired relation between dipole moment and electrosorption valency; a more general proof of this relation is given in ref.[8]. Schultze and Koppitz [9] have collected data for γ at the pzc and for small coverages, so we may immediately convert these to a table for the dipole moment μ.

2.2 The Dipole Moment of a Hard Sphere Adsorbed on Jellium

Carnie and Chan [1] have extended the hard sphere electrolyte model to the situation in which one kind of ion can be adsorbed at the metal surface through an adsorption potential; they represented the associated Boltzmann factor by a Dirac delta function centered at a distance $\sigma_i/2$ from the metal, where σ_i is the diameter of the adsorbed ion. When both the charge density q on the metal and the charge density q_i due to the specifically adsorbed ions are small, application of the mean spherical approximation yields the following expression for the potential drop between the hard wall and the bulk of the solution:

$$\Delta\varphi = \frac{4\pi}{\varepsilon\kappa} \, (q+q_i) + \left[\frac{2\pi\sigma_i}{\varepsilon} + \frac{2\pi\sigma_s(\varepsilon-1)}{\lambda\varepsilon} \right] q + \frac{2\pi\sigma_s(\varepsilon-1)}{\lambda\varepsilon[1+\lambda\sigma_i/\sigma_s]} \, q_i \tag{2.6}$$

Here σ_s is the diameter of the solvent, ε is the dielectric constant, κ is the inverse Debye length of the solution, and λ is given by $\lambda^2(1+\lambda)^4 = 16 \, \varepsilon$. Just as in eq. (2.2) the first term is the familiar Gouy - Chapman term; the other terms are the first order deviations from the Gouy and Chapman theory, and are due to the structure of the solution in a boundary region extending over several solvent diameters.

When the hard wall is replaced by jellium, we must account for the surface dipole potential of the metal, and its variation with the external field. When the interaction between the metal and the adsorbed ions is relatively weak, the response of the metal will be mainly determined by the electrostatic field, and the corresponding change in the dipole potential will be $q/C_m = -4\pi q x_{im}$, where C_m is the contribution of the metal to the capacity at the pzc. Then the total change in the potential drop between the metal and the solution due to the charges q and q_i is [2,3]:

$$\Delta\varphi = \frac{4\pi}{\varepsilon\kappa}(q+q_i) + 4\pi\left[\frac{\sigma_i}{2\varepsilon} + \frac{\sigma_s(\varepsilon-1)}{2\lambda\varepsilon} - x_{im}\right]q + \frac{2\pi\sigma_s(\varepsilon-1)q_i}{\lambda\varepsilon[1+\lambda\sigma_i/\sigma_s]} \quad (2.7)$$

The following comments are in order: (1) Both the charges on the metal and on the adsorbate must be so small that the response of the system is linear; this implies a low coverage with the adsorbate, and excludes island formation. (2) It has been assumed that there is no charge exchange between the ions and the metal; indeed, the assumption of a relatively weak interaction between the ions and the metal implies that the charge transfer coefficient l is small. Formally, a small charge transfer can be incorporated into the model by writing: $q_i = (z_i+l)e_0\Gamma_i$, $q = -e_0(\Gamma_e+l\Gamma_i)$, where Γ_i is the surface concentration of the adsorbed species, and Γ_e the surface excess of electrons. (3) In the absence of specific adsorption, i.e. for $q_i = 0$, we recover the result for the interfacial capacity given in the previous chapter.

Setting $q_i = -q$ in eq.(2.7) gives for the adsorbate dipole moment:

$$\mu = ze_0\left[\frac{\sigma_i}{2\varepsilon} + \frac{\sigma_i\sigma_s(\varepsilon-1)}{2\varepsilon(\sigma_s+\lambda\sigma_i)} - x_{im}\right] \quad (2.8)$$

If a partial charge $l\cdot e_0$ has been transferred onto the adsorbate, z must be replaced by $(z+l)$.

3 The Dipole Moments of Simple Adsorbates

We have calculated adsorbate dipole moments for a number of systems, mostly from literature data for the electrsorption valency. Table 1 gives the values for alkali, halide, and similar ions on a

Metal $(1/4\pi C_H)$	Ion	γ	μ/D	$r_i/Å$	$e_0 r_i/D$	μ_{theo}/D
Hg	K^+	0.16	1.21	1.33	6.38	0.72
(0.3Å)	Rb^+	0.15	1.22	1.48	7.1	0.78
	Cs^+	0.2	1.18	1.68	5.66	0.86
	Cl^-	-0.2	-1.15	1.81	8.69	-0.91
	Br^-	-0.34	-0.95	1.95	9.36	-0.95
	I^-	-0.45	-0.79	2.16	10.37	-1.01
	ClO_4^-	-0.3	-1.01	3.2	15.36	-1.23
	NO_3^-	-0.25	-1.08			
	N_3^-		-1.01	1.57	7.54	-0.82
	SCN^-	-0.3	-1.01	1.74	8.35	0.88
Bi	Cs^+	0.17	1.36	1.68	5.66	1.01
(0.34Å)						
Ga	Rb^+	0.2	0.27	1.48	7.1	0
(0.07Å)	Cs^+	0.2	0.27	1.68	5.66	0
Ag(110)	Cl^-	0	-0.43	1.81	8.69	0
(0.089Å)	Br^-	-0.18	-0.35	1.95	9.36	0
Au(311)	Br^-	-0.49	-0.36	1.95	9.36	-0.17
(0.147Å)						

Table 1: Adsorption characteristics of simple ions from aqueous solution at the p.z.c; electrosorption valencies γ from Schultze and Koppitz [9]; the dipole moments μ are from Schmickler [3]. r_i is the ionic radius.

number of metals. For energetic reasons, these ions are expected to keep almost their total charge when they are adsorbed [10]; note that the interaction with the solvent favours a charged state of the adsorbates, so that the adsorbate charge should be greater than for adsorption from the gas phase.

In all systems considered, the absolute value $|\mu|$ of the adsorbate dipole moment is substantially smaller than $e_0 \sigma_i/2$, which would be the value for the adsorption of an ion of diameter σ_i on a perfect conductor in a medium of dielectric constant $\varepsilon=1$. So the charge on the ion is strongly screened. The dipole moment also depends on the metal substrate. The five metal surfaces for which data exist can be divided into two groups: on Hg and Bi the dipole moments are relatively large, on Ga, Ag(110), Au(311) they are much smaller. Note that the former group has small Helmholtz capacities C_H at the pzc in the absence of specific adsorption, while the latter has large capacities C_H. This parallelism indicates that μ and C_H both involve the screening properties of the metal: if a metal screens an external field well, it should have a high double layer capacity C_H, and a small dipole moment $|\mu|$; in the former case, the surface electrons respond to the field generated by the diffuse double layer, in the latter case to the field of the adsorbed ions.

A schematic diagram illustrating the screening of the adsorbate charge is shown in fig.1. The adsorbed ion polarizes the solvent molecules in its vicinity, so that they are preferentially oriented with their dipole moment antiparallel to the adsorbate dipole moment. The image charge has its centre outside the metal surface, and partially surrounds the adsorbate.

Theoretical values for the dipole moment μ calculated from eq.(2.8) under the assumption that no charge has been transferred onto the ions, are given in the last column of tab. 1. For these calculations x_{im} was estimated from experimental data for the Helmholtz capacity, using the hard sphere value of $1/4\pi C_s \approx 0.57$ Å for the solvent contribution. It is gratifying to note, that it gives the correct order of magnitude, and explains the parallelism between the double layer capacity C_H and the dipole moment discussed above. Second order trends like the systematic variation of μ for halide ions adsorbed on Hg are, however, not reproduced; this may be due to a small charge transfer onto these ions, which is expected to increase in the series Cl^-, Br^-, I^- [9,10].

Eqation (2.8) is based on a linear response formalism, in which negative and positive charges are screened equally well. For higher fields as they occur in the adsorption of ions, one would expect that positive charges are shielded better than negative

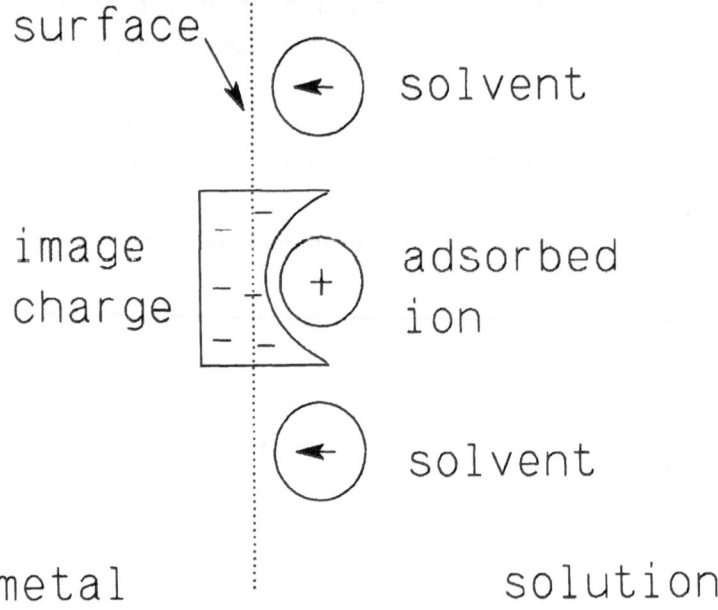

Fig.1)   Distribution of the image charge (schematic)

ones, since there is a suffecient number of electrons at the inter face to screen any external positive charge, while a negative charge can only push back the tail of electrons extending into the solution, while the rest of the screening charge is inside the metal. Indeed, explicit calculations based on the jellium model [11,12] indicate that x_{im} is larger for fields directed towards the metal. We should thus expect that cations are shielded better than anions; there is, however, no evidence for this effect in the data of tab.1; possibly, it is masked by the greater polarizability of the anions, which also decreases the dipole moment.

4 Underpotential Deposition on Polycrystaline Surfaces

The adsorption of metal ions on a foreign metal substrate is a case of particular importance. In many cases up to a monolayer can be electrochemically deposited at potentials substantially higher than the potential for the deposition of the bulk metal. This phe-nomenon, known as *underpotential deposition (upd)*, indicates that the binding energy of the adsorbed monolayer to the foreign metal is greater than the corresponding binding energy to the same metal.

4.1 Phenomenological Relation between Adsorption Potential and Work Function

Kolb, Przasnyski, and Gerischer [13] have shown that underpotential deposition onto polycrystalline substrates shows a surprisingly simple correlation with the work function. They introduced the *underpotential shift* ΔE_p which they defined as the difference in the potential of the desorption peak for a monolayer of metal A adsorbed on the foreign metal substrate B, and the potential of the dissolution peak of the bulk metal A; the latter is practically identical to the Nernst potential for the bulk deposition of the metal A. Note, that ΔE_p is independent of the concentration of the metal ion A^{z+} in the solution, and that $ze_0\Delta E_p$ is a measure of the difference in the free adsorption energies. Investigating a large number of systems both in aqueous and non-aqueous solutions, Kolb et al. obtained the correlation:

$$e_0\Delta E_p \approx (\Phi_A - \Phi_B)/2 \qquad (4.1)$$

where Φ_A, Φ_B denote the work functions of the metals A and B, and e_0 is the unit of charge. Note that the correlation does not involve the difference of the average energy of adsorption $ze_0\Delta E_p$. This correlation also implies that generally underpotential deposition occurs only on a bulk metal with a work function higher than that of the deposited metal, although there are a few exceptions to this rule.

This simple correlation suggests that the adsorption of a layer of metal atoms is mainly determined by the electronic properties of the two metals involved. This is in accord with theoretical work by Lang [14], who has shown that the change in work function due to the adsorption of alkali ions can be explained in terms of a model, in which both the substrate and the adsorbate are modeled as *jellium*, i.e. an electronic plasma embedded in a positive background charge. Lang's work is for adsorption from the vacuum, but

408

the same principles should also operate in electrochemical adsorption. In the following, we shall therefore apply this model to the underpotential deposition of metal ions.

4.2 A Model for upd on Polycrystalline Surfaces

Following the work by Lang [14], we represent the metal substrate as a semiinfinite jellium with a positive background charge n_1, and the adsorbate layer as a jellium slab of thickness d and background charge density n_2. Using the coordinate system depicted in fig.2, the positive background charge is:

$$n_+(x) = n_1\theta(-x) + n_2\theta(x)\theta(d-x) \tag{4.2}$$

where $\theta(x)$ denotes the Heaviside function. Since for underpotential deposition the work function Φ_1 of the substrate must be greater than the work function Φ_2 of the adsorbate, and since the work function correlates with the electronic density n, we may limit ourselves to the case $n_1 > n_2$.

We shall show below that for an estimate of the underpotential shift ΔE_p we require the surface energy of the system. Unfortunately, the jellium model only gives good values for the surface energies of metals with a bulk electronic density lower than about $9\cdot10^{-3}$ a.u. [15], or with a Wigner-Seitz radius $r_s \geq 3$, where $r_s = (3/4\pi n)^{1/3}$. We have therefore taken a value of n_1 corresponding to a Wigner-Seitz radius of $r_{s,1} = 3$ for the substrate, and considered lower electronic densities for the adsorbate. Since we are interested in a qualitative understanding of the adsorption process rather than in calculating quantitative data for specific systems, this limitation in the electronic densities investigated is not severe. By including pseudopotentials into the jellium, we can extend the calculations to higher densities and to single crystals (see below).

We have calculated the electronic density profile using the variational trial function method explained in the previous chapter. For the problem at hand, we have taken functions of the form:

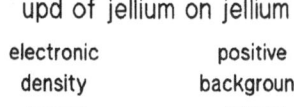

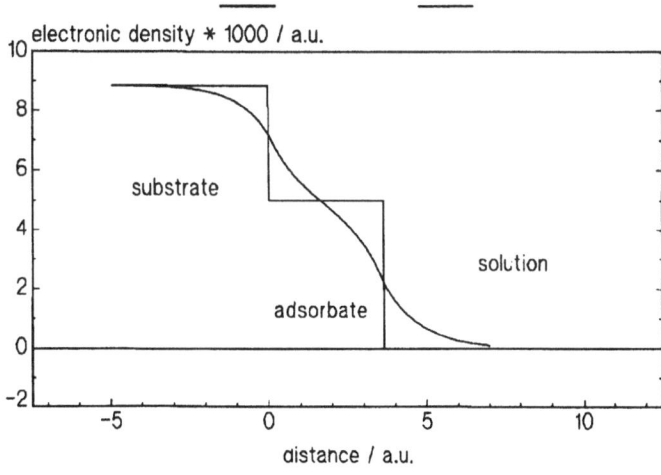

Fig.2) Positive background charge and electronic density profile in our model for monolayer adsorption. The calculations correspond to a substrate with Wigner Seitz radius of $r_s = 3$, for the adsorbate, $r_s = 4$; the thickness d corresponds to an open plane and a charge number of $z = 1$.

$$n_-(x) = \begin{cases} n_1 [1 - A \cdot e^{\alpha x}] & \text{for } x \le 0 \\ n_1 \cdot B \cdot e^{-\beta x} + n_2 [1 - C \cdot e^{\gamma(x-d)}] & \text{for } 0 \le x \le d \quad (4.3) \\ n_1 \cdot B \cdot e^{-\beta x} + n_2 \cdot D \cdot e^{-\delta(x-d)} & \text{for } d \le x \end{cases}$$

Continuity of $n_-(x)$ and its first derivative leads to the relations:

$$1-A = B + n_2(1-C \cdot e^{-\gamma d})/n_1 \qquad\qquad 1-C = D$$

$$\alpha A = \beta B + n_2 \gamma C \cdot e^{-\gamma d}/n_1 \qquad\qquad \gamma C = \delta D \qquad (4.4)$$

while charge balance gives:

$$n_1\left(\frac{A}{\alpha} - \frac{B}{\beta}\right) + n_2\left[\frac{C}{\gamma}(1-e^{-\gamma d}) - \frac{D}{\delta}\right] = q \qquad (4.5)$$

where q is the surface charge density of the system; we consider uncharged surfaces only. The set of equations (4.4) and (4.5) reduces the number of unknown parameters A, B, C, D, α, β, γ, δ to three independent ones, for which we have taken C, α, and γ; from these, the others can be determined by straightforward calculations.

The three free parameters are determined by energy minimization. When the adsorbate layer is left out ($d \to 0$), our trial functions reduce to those employed by Smith [16]. It was shown in ref.17 that, with the inclusion of the second order gradient term for the kinetic energy, these functions give excellent results for the surface energy of a semiinfinite jellium , so there is good reason to believe that our extended set of functions gives good values for the surface energy in presence of an adsorbate.

4.3 The adsorption potential

To estimate the adsorption energy and the corresponding electrode potential, we have calculated the surface energy of the system substrate + adsorbate, which we define as:

$$E_s^{ad} = E[n_-(x)] - E[n_+(x)] \qquad (4.6)$$

where the first term is the total electronic energy of the system with the correct electronic density profile, and the second term is the corresponding energy for a hypothetical system where the electronic density would be equal to the background charge density $n_+(x)$ given in eq.(4.2). In addition, we require the surface energy E_s^{sub} of the pure substrate, for which we obtained the same values as Ma and Sahni [17]. Further, we define the energy E_{slab} of a hypothetical slab of jellium with a thickness d and constant background charge density n_2, where the electronic density would be equal to the background charge density. Since our energy functional is local, E_{slab} is the same, independent of the surroundings of the slab. This quantity need not be calculated since it will cancel when we consider relative adsorption energies.

During the deposition of a monolayer of jellium with density n_2 onto a semiinfinite jellium of the same density, the substrate simply gains an amount d in thickness, and its energy increases by E_{slab}. So the corresponding adsorption energy is:

$$E_{dep} = E_{slab} - N \cdot G_{sol} \qquad (4.7)$$

where G_{sol} is the solvation energy of the deposited ion when it is in the solution, and N is the number of atoms per unit area of the deposited monolayer.

For the deposition of the adsorbate on the foreign metal substrate, the corresponding energy change is:

$$E_{upd} = E_s^{ad} + E_{slab} - N \cdot G_{sol} - E_s^{sub} \qquad (4.8)$$

So the energy difference between adsorption on the foreign substrate and the same substrate is simply:

$$\Delta E_{upd} = E_s^{ad} - E_s^{sub} \qquad (4.9)$$

To convert this to an average difference in potential between underpotential deposition and bulk deposition, we have to divide ΔE_{upd} by the total number of adsorbed atoms per unit area of the monolayer, which is $N = n_2 \cdot d/z$, and divide by the charge number z to convert energies into potentials - this means that we suppose that the ions are in their highest oxidation state in the solution. This gives:

$$\Delta\phi_{upd} = \frac{E_s^{ad} - E_s^{sub}}{n_2 \cdot d} \qquad (4.10)$$

Finally, we have to specify the thickness d of the monolayer slab. We shall assume that the adsorbate layer is incommensurate with the substrate; this is probably true for most experimentally investigated systems, since the adsorbate atoms are typically larger than those of the substrate. We considerthe two cases in which the adsorbate layer forms either the most compact plane (i.e. the (111)- plane) or the most open principal plane (i.e. the (110)-plane) of an fcc lattice. So we have:

$$
\begin{aligned}
d = 1.4774 r_s \cdot z^{1/3} \qquad &\text{compact layer} \\
d = 0.9047 r_s \cdot z^{1/3} \qquad &\text{open layer}
\end{aligned}
\qquad (4.11)
$$

These two limiting cases should bracket the situation for real systems.

In the above considerations we have neglected any specific interaction of the metal surfaces with the solvent; more precisely, we have assumed that the difference in the energies of adsorption of the solvent on the adsorbate and on the substrate is negligible. This assumption should at least be good for water as a solvent: experiments in the uhv indicate that the interaction of a water molecule with a metal substrate is typically weaker than the water - water interaction [18].

4. 4. Results and Discussion

Explicit model calculations were performed [19] for adsorbates with electronic densities in the range $6 \geq r_s \geq 3.5$ for a compact adsorbate layer and a charge number of $z = 1$ for the adsorbed ion, and for an open adsorbate surface and charge numbers of $z = 1, 2, 3$. When the calculated upd-shifts $\Delta\phi_{upd}$ are plotted against the difference in the work function correlations similar to eq.(4.1) result (see fig.3). The slopes are a little different for the four cases investigated: it is highest for the open surface and $z = 1$, it is smaller for higher values of z, and lowest for the compact adsorbate layer.

Our calculated $\Delta\phi_{ad}$ values are of the same order of magnitude as the experimental ΔE_p values, but the slopes of our plots are lower than that of the experimental correlation in all cases. This could be due to the following fact: The peak potential for underpotential deposition occurs typically at a coverage of about half a monolayer, so the experimental ΔE_p values correspond to adsorbate layers that are even more open than the (110) face of an fcc lattice. It must also be said that our calculated $\Delta\phi_{upd}$ and the experimental ΔE_p do not have quite the same meaning: $\Delta\phi_{upd}$ is the average deposition potential for a monolayer, while ΔE_p corresponds to the potential at which the desorption rate of the layer attains its maximum.

Our model explains why the experimental correlation involves the potential difference ΔE_p rather than the corresponding difference $z\Delta E_p$ in adsorption energy. The surface energy E_s^{ad} of the

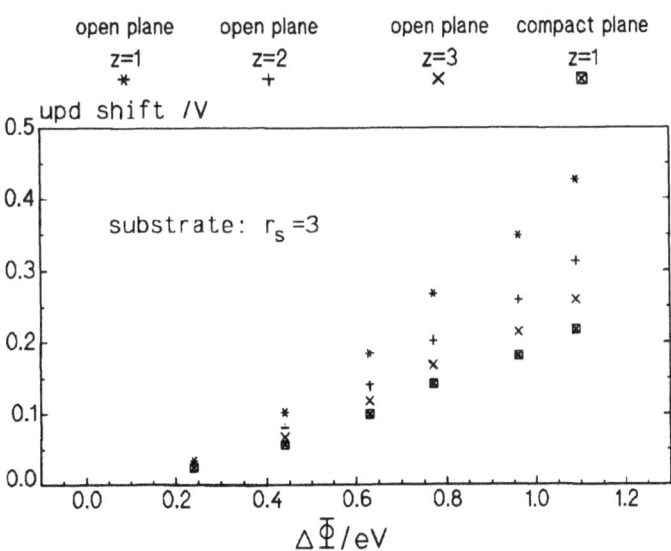

Fig.3)   Calculated upd-shift versus work function difference for
various situations.

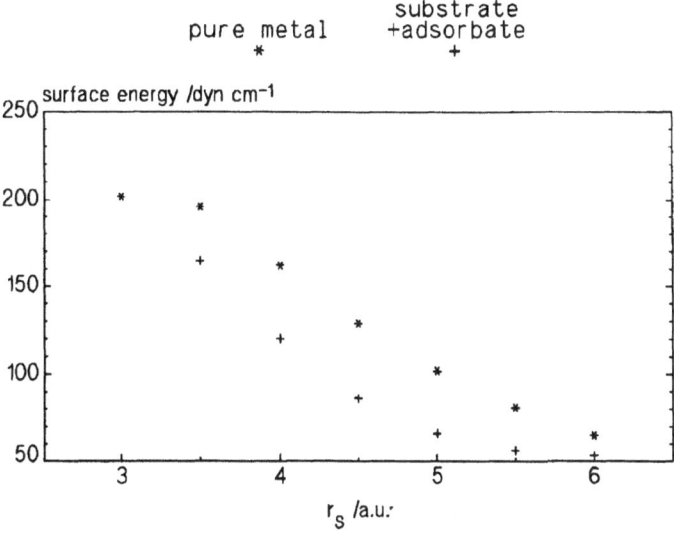

Fig.4)   Surface energy E_s^{ad} of the metal / adsorbate system and
surface energy of a pure metal.

substrate / adsorbate system depends only weakly on the thickness d of the adsorbate layer, so that, to a first approximation, $\Delta\phi_{upd}$ is proportional to $z^{-1/3}$. Thus the dependence of the upd-shift on the charge number z is much smaller than one might expect. In the experimental plot, this dependence may not be observable due to the scatter of the data. This may at least partially be due to the fact that the ΔE_p values of different systems often correspond to different coverages - it can be seen from fig.3 that the density of the surface layer may have a larger effect than the charge number z of the deposited ion. Also, since our present model for underpotential deposition is quite simple, we should be satisfied that it correctly reproduces the general trend.

What is the physical origin of this correlation between the upd-shift and the work function difference? According to our calculations, two mechanisms operate. The first one has to do with the surface energy of the system. Metals with high work functions tend to have high surface energies. Within the jellium model, this trend is reproduced for electronic densities corresponding to a Wigner-Seitz radius $r_s \geq 3$ for reasons mentioned above. So the substrate, which has the higher work function, also has the higher surface energy, and energy is gained by covering the substrate with an adsorbate with a lower surface energy. In this context, it is of interest to note that the metals Pt, Au, and Ag, which are usually used as substrates for upd, have particularly high surface energies. The second mechanism operates at the substrate / adsorbate interface: electrons flow from the adsorbate layer, which has the lower work function, to the substrate, thus creating a double layer of charges. That the corresponding energy is negative can be seen from the fact that the surface energy E_s^{ad} of the metal / adsorbate system is generally lower than the surface energy of a bulk metal composed of the adsorbate, as is shown in fig.4. While the latter mechanism has been discussed before [13], the role of the surface energy has so far been ignored; this is surprising, since it is the influence of the surface energy which prevents the underpotential deposition of a metal with a high work function on a substrate with a low work function.

Figure 4 also explains why our points in fig.3 fall on a curved line at small work function differences: at a the Wigner -
Seitz radius of $r_s = 3$, which we have chosen for the substrate,
the surface energy begins to level off.

5 Underpotential Deposition on Single Crystal Surfaces

While on polycrystalline electrodes structural effects average
out, the situation is more complex for adsorption on single crystal surfaces. The energy of adsorption depends on the orientation
of the crystal, and often several desorption peaks are observed
corresponding to different structures of the adsorbate. A particularly nice example is the adsorption of Pb^{++} on gold, which has
been examined over a whole series of single crystal surfaces [20].

Obviously, the simple jellium model is not suitable for single
crystals, since all information on the lattice structure is lost
when the ionic charge is smeared out into a constant background
charge. So a discrete ionic lattice has to be introduced.

5.1 Model for upd on Single Crystal Surfaces

As before, we suppose the metal substrate to be semi-infinite and
to occupy the half-space $x \leq 0$. Its ion lattice is now represented
by a lattice of pseudopotentials, which we take to be of the Ashcroft type [21]:

$$
w(r) = \begin{cases} 0 & \text{for } r < r_c \\ z/r & \text{for } r > r_c \end{cases} \tag{5.1}
$$

The choice of the pseudopotential radius r_c will be discussed below. The adsorbed ions are supposed to form a two-dimensional lattice in the yz-plane at $x = d$, which is represented by a corresponding lattice of pseudopotentials. The electrons of the system
are assumed to form an electronic plasma, which interacts with the
lattices of pseudopotentials and with itself through the usual
electrostatic and quantum-mechanical interactions. The substrate
and the adsorbate are thus represented as *jellium with pseudopo-*
tentials, a good model for sp-metals, from which Lang and Kohn

[15] have successfully calculated the work function and the surface energies of simple metals.

Within the density functional formalism [22], the ground state energy of our system can be written as a functional of the electronic density $n(\vec{r})$ in the following way:

$$E[n(\vec{r})] = T_s[n] + E_{xc}[n] + \frac{e^2}{2} \int \frac{e(\vec{r})e(\vec{r}')}{|\vec{r} - \vec{r}'|} \, d\vec{r} \, dr' + E_{ion} +$$

$$\int d\vec{r} \sum_{\vec{i}} v_i\left(\vec{r} - \vec{i}\right) n(\vec{r})$$

(5.2)

where the first term denotes the kinetic energy of the electrons, the second the exchange and correlation energy, the third and forth the electrostatic self-interactions of the electrons and of the ion lattice, respectively, and the last term the interaction of the electrons with the pseudopotentials; $\vec{i}$ denotes the positions of the ion cores, and v_i takes on different values depending on whether $\vec{i}$ corresponds to a lattice point of the substrate or of the adsorbate.

As is usually done in jellium calculations, we assume that the electronic density is constant within the bulk of a metal, and that in the surface region it is a function of the x-coordinate only. Accordingly, the pseudopotential terms are averaged parallel to the surface as prescribed by Lang and Kohn [15] and Monnier and Perdew [22]. The electronic density profile is then calculated with the variational method using the same set of trial functions as for the polycrystalline case discussed above; this also gives the electronic part of the surface energy of the system.

5.2 Classical Electrostatic Interactions

A constant positive background charge and a constant electronic density do not give rise to an interaction energy; hence there is no ionic term in the simple jellium model. However, an ionic lattice and a constant electronic density interact electrostatically, and this gives rise to a term in the binding energy. When we apply the model *jellium with pseudopotentials* to metallic adsorption, we require two such electrostatic terms: the electrostatic self-interaction of the adsorbate layer, and that of the adsorbate layer with the substrate.

The former is the electrostatic self-interaction of a two-dimensional lattice of ions of charge number z embedded in a uniform layer of compensating negative charge density of thickness 2d. The electrostatic energy per ion is one half of the electrostatic energy of a positive ion due to the electrons and the other ions. The two-dimensional case, in which the negative charge is confined to the lattice plane, has been treated by Bonsall and Maradudin [23]. It was shown in ref. 24 how to generalize their result to a slab with a finite width. For a square lattice with lattice constant a, the energy per ion is:

$$E_{el}^{2d} = - \frac{(ze)^2}{2a} (3.900265 - \pi \cdot d/a) \qquad (5.3)$$

where the first term in the brackets is for the two-dimensional case, and the second corrects for the finite thickness of the electron layer.

Concerning the electrostatic interaction of the adsorbate layer with the substrate, the interaction of the electron layer with the substrate averages to zero, so we only need to consider the interaction of an ion with the substrate. This depends strongly on the position of the ion with respect to the substrate lattice, and on the ratio s of the distance from the first lattice plane to the lattice constant. The interaction energy is attractive for a hollow site, and repulsive for the a-top position. It can be written in the form:

$$E_{el}^{int} = -z \cdot z_{sub} \cdot f(s) / a_{sub} \qquad (5.4)$$

where a_{sub} is the lattice constant of the substrate, and z_{sub} its charge number; the function $f(s)$ depends on the relative position of the ion, its dimenionless distance s from the substrate surface, and the structure of the substrate. It can be calculated by the same procedure used for the classical contribution to the surface energy [15].

For the binding energy $E_{bind}(A)$ of the adsorbate we require the electrostatic self energy of the three-dimensional lattice.

For this we have taken the usual literature values [25] and added the appropriate correction terms derived by Plaskett and Hall [26]. These correction terms are quite important for our model; without them, calculations containing both two- and three-dimensional lattice sums are inconsistent.

5.3 Choice of Pseudopotentials

Several tables of pseudopotentials, based on different criteria, exist in the literature. For our problem the pseudopotentials must be chosen in such a way that the bulk metal is at equilibrium:

$$\frac{dE[n]}{dn} = 0 \qquad (5.5)$$

where n is the bulk electronic density, and $E[n]$ is the functional given in eq.5.2. The same criterion was used by Paasch and Hietschold [27]; our values differ, since we include the correction term of Plaskett and Hall [26] to the electrostatic energy.

With the energy functional specified above this gives for the pseudopotential radius:

$$r_c = \frac{1}{3}\left[-4.42\ r_s + 0.9162\ r_s^2 + \frac{0.878\ r_s^4}{(r_s+7/.8)^2} - M \cdot z^{2/3} \cdot r_s^2\right]^{1/2} \qquad (5.6)$$

Here $r_s = (3/4\pi n)^{1/3}$ is the Wigner-Seitz radius associated with the electronic density n, and M is the Madelung constant appropriate for the lattice, which is negative; atomic units are used.

The reason for this particular choice of the pseudopotentials is the following: the density of the adsorbate A is generally different from that of the bulk material A; since the bulk metal is in equilibrium, the lower electronic energy should tend to lower the self-energy of the adsorbate. Only if we choose the pseudopotentials in such a way that our model for the bulk metal is also in equilibrium, can we be sure that we reproduce this trend correctly.

Table 3 shows values for the pseudopotential radii r_c of a few simple metals calculated by this procedure using the data in tab.2, and contributions to the binding energies per electron. We have not accounted for the band structure energy, which is typi-

metal	r_c/au	E_{kin}	E_{xc}	E_{ps}	E_{el}	E_{bind}^{cal}	E_{bind}^{exp}
Al	0.95	7.02	-7.23	4.18	-20.30	-16.3	-18.9
Li	1.16	2.80	-4.88	1.56	-6.16	-7.0	-6.7
Na	1.57	1.89	-3.12	1.59	-5.07	-5.7	-6.2
Zn	0.85	5.68	-6.60	2.44	-12.99	-11.5	-14.3
Cd	1.04	4.48	-5.96	2.54	-11.49	-10.4	-13.5
Tl	1.21	4.89	-6.19	3.92	-16.73	-13.6	-14.6

Tab.2: pseudopotential radii and contributions to the binding energy E_{bind} per electron of a few simple metals. E_{kin}: kinetic energy, E_{xc}: exchange and correlation energy, E_{ps}: pseudopotential energy, E_{el}: electrostatic energy. All energies in electron volts.

metal	E_{kin}	E_{xc}	E_{ps}	E_{el}^{2d}	E_{el}^{int}	ΔE_s	$\Delta\phi_{upd}$/V
Li	1.98	-4.22	0.93	-5.39	-0.50	-0.14	0.64
Na	1.60	-3.85	1.23	-4.80	-0.38	-0.21	0.76
Zn	3.74	-5.52	1.30	-11.48	-0.28	0.03	0.82
Cd	3.20	-5.26	1.53	-10.85	-0.35	-0.01	1.21
Tl	3.51	-5.37	2.39	-14.93	-0.27	0.02	1.02

Tab.3: contributions to the adsorbate binding energy per electron and calculated upd-shifts. E_{el}^{2d}: electrostatic self energy, E_{el}^{int}: electrostatic interaction energy with substrate, ΔE_s: electronic relaxation energy of system minus that of the pure aluminium substrate. All energies in eV.

cally less than half an electron volt. A comparison between the calculated and the experimental binding energies gives an idea of the order of accuracy that can be expected from this kind of calculation.

6 Model Calculations

As an illustration we have performed model calculations of the underpotential shift for a number of systems. In real systems the substrate is typically a d- or an sd- metal with a relatively high work function such as platinum or gold. Since we are limited to sp- metals we had to compromise on this point, and have taken aluminium, which has a high electronic density and work function, as the substrate; specifically, we have taken the (100)-plane as the electrode surface. The electrostatic interaction between the substrate and the adsorbate favors adsorption on the fourfold hollow sites of the (100)- plane, so we have assumed that the adatoms form a $\sqrt{2}$ x $\sqrt{2}$ (45°) superlattice. As adsorbates we have taken a few sp-metals which show underpotential deposition on gold and platinum, and which are small enough to fit into this superlattice: Li, Na, Zn, Cd, Tl.

The thickness of the adsorbate layer was determined by the following considerations: The substrate atoms and the adatoms are each assigned a hard sphere radius equal to half the nearest neighbor distance of the corresponding atom when it is in its own lattice. The positions of the adatom ion cores were then calculated from billiard-ball geometry; this gives the distance d_{as} between the adatom plane and the first lattice plane of the substrate. In the absence of an adsorbate the first lattice plane of an fcc(100) surface is situated at $x = -a/4$, where a is the lattice constant (4.05 Å for Al), with respect to the jellium edge at $x = 0$; so we have taken the positions of the adsorbate plane at: $x = d = d_{as}-a/4$. This procedure determines the geometry and the electronic density of the adsorbate layer.

For all systems investigated the electronic density of the adsorbate layer is substantially less than that of the corresponding bulk metal; the jellium width $2d$ is generally smaller than half the lattice constant of the adsorbate, so that the structure of the adsorbate layer is quite different from that of a lattice plane in the bulk material. Calculated values for the upd-shift $\Delta\phi_{upd}$ and the contributing energy terms are given in tab.4. For Zn, Cd, and Tl this lattice distortion leads to a noticeable decrease of the self energy per electron, which is surprising in view of the fact that the pseudopotentials were chosen such that the bulk crystal obtains its minimal energy at the experimentally observed density. On the other hand, one would not be able to

metal	E_{kin}	E_{xc}	E_{ps}	E_{el}^{2d}	E_{el}^{int}	ΔE_s	$\Delta\phi_{upd}/V$
Li	1.98	-4.22	0.93	-5.39	-0.50	-0.14	0.64
Na	1.60	-3.85	1.23	-4.80	-0.38	-0.21	0.76
Zn	3.74	-5.52	1.30	-11.48	-0.28	0.03	0.82
Cd	3.20	-5.26	1.53	-10.85	-0.35	-0.01	1.21
Tl	3.51	-5.37	2.39	-14.93	-0.27	0.02	1.02

Tab.4: contributions to the adsorbate binding energy per electron and calculated upd-shifts. E_{el}^{2d}: electrostatic self energy, E_{el}^{int}: electrostatic interaction energy with substrate, ΔE_s: electronic relaxation energy of system minus that of the pure aluminium substrate. All energies in eV.

build a three dimensional lattice from the configurations of the two-dimensional adsorbate structures. It must also be said that for the hexagonal structure the procedure of determining the pseudopotential is less satisfactory then for the fcc or bcc lattices, since the minimization has to be done at constant c/a ratio. Apart from this the individual terms show the expected trends when one compares the bulk values in tab.3 with the adsorbate values: the kinetic and the pseudopotential energies are lower in the adsorbate, since the electronic density is lower; conversely, the electrostatic self energies and the exchange-correlation energies are higher in the bulk crystal. The electrostatic interaction of the adsorbate with the substrate makes a sizable contribution to the the upd-shift $\Delta\phi_{upd}$. This shows that the formation of structurally commensurate adsorbate layers, with the adatoms sitting at the hollow sites, is favored. The electronic relaxation energy is important only for the alkali adsorbates, whose work functions are much lower than those of the substrate.

The calculated upd-shifts are of a reasonable order of magnitude; there are not sufficient experimental data for underpotential deposition on fcc(100) faces, but values for the upd-shift of Zn, Cd, Tl on polycrystalline gold and platinum are of the order of 0.5 - 1.3 V, and the same order of magnitude is expected for open single crystal surfaces. Experimental values for $\Delta\phi_{upd}$ of Li and Na on polycrystalline platinum and gold are in the range 1.2 - 1.6 V and are thus higher than our calculated value. However, these experimental data were obtained in non-aqueous solutions and may be inaccurate due to contamination with water. In any case, at this stage we cannot expect exact agreement between theory and experiment.

From our calculations we obtain another quantity of interest: the dipole potential ϕ_{dip} caused by the adsorbate, which is the difference in work function between the pure substrate and the adsorbate covered substrate. We have calculated the work function from the change in energy of the system due to the removal of an electron. As expected, ϕ_{dip} follows the difference in work function between the adsorbate and the substrate (see tab.5).

metal	Φ_{ad}	Φ_{poly}	ϕ_{dip}/V
Li	2.96	3.36	−0.84
Na	2.88	3.06	−0.92
Zn	3.45	3.79	−0.35
Cd	3.43	3.68	−0.38
Tl	3.71	3.73	−0.10

Tab.5: work function Φ_{ad} of the adsorbate covered substrate, jellium work function Φ_{poly} of the corresponding bulk system calculated within the jellium model, and dipole potential ϕ_{dip} induced by the adsorbate. Φ_{poly} was interpolated from the values of Lang and Kohn [15]. Work functions in eV.

Recently [29] these calculations have been extended to the other principal planes of aluminium. A few representative results are given in tab.6. In each case the calculations were performed for a dense commensurate adsorbate layer; for the (111) surface two different adsorbate structures were considered: the hexagonal (hex) and the honeycomb (hc) structures.

Generally the calculated upd shifts are of a reasonable order of magnitude; what is more important: in line with experimental

metal	Li	Na	Tl	Pb
(110)	1.78	1.19	1.58	1.00
(100)	0.64	0.50	1.05	0.54
(111)hex	no upd	0.29	no upd	no upd
(111)hc	0.15	-	0.11	-

Tab.6: upd shifts (in V) of several adorbates on single crystal planes of aluminium [28]. When no entry is given the adsorbate does not fit into the structure.

observations [19] they predict that adsorption should be stronger on the open (110) plane than on the compact (111) plane.

7 Conclusion

Adsorption at the metal / solution interface is a very important, and also a very complex phenomenon. While it is gratifying to see that the simple models presented help to explain a number of observations in a qualitative, and sometimes in a semi-quantitative way, there is still plenty of room for improvement. In particular, adsorption at medium coverage, and adsorbate - adsorbate interactions, are but poorly understood.

Acknowledgement

It is a pleasure to thank my colleagues with whom I have cooperated in this area: Drs. A. Kornyshev, E. Leiva, and R. Guidelli. Financial support through a start up fund from Utah State University and by the NATO Science committee (Collaborative Research Grant No. RG-86/0068) is gratefully acknowledged.

References

1) S.L. Carnie and D.Y.C. Chan, J. Chem. Phys. **73** (1980) 2949

2) W. Schmickler and R. Guidelli, J. Electroanal. Chem. **235** (1987) 387

424

3) W. Schmickler, J. Electroanal. Chem. **249** (1988) 25;

4) N.D. Lang, Surf. Science **127** (1983) L 118

5) D.C. Grahame and R. Parsons, J. Am. Chem. Soc. **83** (1961) 1291

6) K.J. Vetter and J.W. Schultze, Ber. Bunsenges. Phys. Chem. **76** (1972) 920, 927

7) D.C. Grahame, J. Am. Chem. Soc. **80** (1958) 4201;

8) K. Bange, B. Strachler, J.K. Sass, and R. Parsons, J. Electroanal. Chem. **229** (1987) 87

9) J.W. Schultze and F.D. Koppitz, Electrochim. Acta **21** (1976) 327

10) A.Kornyshev and W. Schmickler, J. Electroanal. Chem. **185** (1985) 253

11) W. Schmickler and D.J. Henderson, Phys. Rev. B **30** (1984) 3081

12) P. Gies and R.R. Gerhardts, Phys. Rev. B, **33** (1986) 982

13) D.M. Kolb, M. Przasnyski, and H. Gerischer, J. Electroanal. Chem. **54** (1974) 25; H. Gerischer, D.M. Kolb, and M. Przasnyski, Surface Sci. **43** (1974) 282

14) N. Lang, Phys. Rev. B **12** (1971) 4234

15) N. Lang and W. Kohn, Phys. Rev. B **1** (1970) 4555; **3** (1971) 1215

16) J.R. Smith, Phys. Rev. **181** (1969) 522

17) C.Q. Ma and V. Sahni, Phys. Rev. B **19** (1979) 1290

18) K. Bange, D.E. Grider, T.E. Madey, and J.K. Sass, Surface Sci. **136** (1984) 38; E. Stuve, R.J. Madix, and B.A. Sexton, Surface Sci. **111** (1981) 11

19) W. Schmickler and E. Leiva, Chem. Phys. Lett. **160** (1989) 75

20) A. Hamelin, J. Electroanal. Chem. **165** (1984) 167; A. Hamelin and J. Lipkowski, J. Electroanal. Chem. **171** (1984) 317

21) N.W. Ashcroft, Phys. Rev. Lett. **23** (1966) 48

21) R. Monnier and J.P. Perdew, Phys. Rev. B **17** (1975) 2595

23) L. Bonsall and A.A. Maradudin, Phys. Rev. B **15** (1977) 1959

24) W. Schmickler, Chem. Phys. **141** (1990) 95

25) R.A. Coldwell-Horsfall and A.A. Maradudin, J. Math. Phys. 1 (1960) 395

26) G.L. Hall, Phys. Rev. **B 19** (1979) 3921

27) G. Paasch and M. Hietschold, phys. stat. solid. b) **67** (1975) 753, ibid. **83** (1977) 209

28) N.W. Ashcroft and N.D. Mermin, *Solid State Physics,* Holt, Rinehart and Winston, 1976

29) W. Lehnert and W. Schmickler, J. Electroanal. Chem., in press

ELECTRON-TRANSFER REACTIONS AT METAL-SOLUTION INTERFACES:
AN INTRODUCTION TO SOME CONTEMPORARY ISSUES

MICHAEL J. WEAVER
Department of Chemistry
Purdue University
West Lafayette, Indiana 47907

ABSTRACT. Some pivotal aspects of the contemporary treatment of electrochemical kinetics for one–electron redox couples are outlined, and some commonalities with homogeneous–phase electron transfers are pointed out. Two illustrative examples of the interplay between theory and experiment are briefly discussed; specifically to the relation between the rates of electrochemical and homogeneous–phase processes, and to the recognition and elucidation of dynamical solvent effects.

I. INTRODUCTION

Examinations of the kinetics as well as thermodynamics of electron–transfer reactions at metal electrodes have long been of concern not only to electrochemists but also to others concerned with redox processes in a variety of homogeneous as well as heterogeneous chemical environments. This interest stems in part from the unique control of the electrical variable afforded at metal–solution interfaces by the electrode potential. In addition, such electrochemical processes involve inherently only a *single* redox couple

$$Ox + ne^- \rightleftharpoons Red \qquad (1)$$

In contrast, electron–transfer reactions in homogeneous solution and at most liquid–liquid interfaces involve *pairs* of redox couples. The thermodynamic and kinetic parameters that describe electrochemical charge–transfer process therefore comprise fundamental characteristics of such individual redox couples (so–called "half–reactions"), of value to unraveling the behavior of redox processes on a more general basis.
While the characterization of electrochemical charge–transfer reactions, often involving proton– and coupled atom–, as well as electron–, transfers has a long and venerable history, understanding of most kinetic and mechanistic aspects of electrode kinetics has tended to lag behind that for analogous homogeneous–phase systems. This is especially the case for the simplest type of reaction, involving one–electron transfer between stable solution (Ox and Red) species. Nevertheless, there are close parallels in the treatment of hetero-

R. Guidelli (ed.), Electrified Interfaces in Physics, Chemistry and Biology, 427–442.
© 1992 *Kluwer Academic Publishers.*

geneous- and homogeneous-phase processes of this type that allow a largely unified description to be developed.

Presented here is a brief overview of some fundamental features of electrode kinetics in relation to electron transfers in general. Discussion is also provided of two topics of research interest to the author, namely the relationship between the rates of heterogeneous- and homogeneous-phase processes, and the role of solvent dynamics in electron transfer. Rather than attempting a broadbased overview, then, the present account is written from an admittedly personal perspective. Nevertheless, the text is intended to illustrate some of the flavor of contemporary research in this area. A more comprehensive review of fundamental aspects, along with expanded citations, is to be found in ref. 1.

II. ELECTROCHEMICAL RATE PARAMETERS AND REACTION MECHANISMS

The basic characteristic of electrochemical reaction rates is their sensitivity to the applied electrode potential, E. In the absence of mass-transport limitations, the dependence of the current density for a redox process as in Eq. (1) upon E can be expressed as

$$i = i_o \{\exp[-\alpha nf(E-E_{eq})] - \exp[(1-\alpha)nf(E-E_{eq})]\} \qquad (2)$$

Here α is the cathodic transfer coefficient, $f = F/RT$, E_{eq} is the equilibrium potential for a solution containing given concentrations of Ox and Red, and i_o is the exchange current density. The last quantity denotes the equal (and counterbalancing) cathodic and anodic currents, i_c and i_a, respectively, that necessarily flow at $E = E_{eq}$, corresponding to the first and second exponential terms in Eq. (2). At least for electrode reactions displaying first-order kinetics, it is useful to define cathodic and anodic rate constants k_c and k_a, respectively. For one-electron first-order reactions, these are given by

$$k_c = i_c/FC_{ox}, \quad k_a = i_a/FC_{red} \qquad (3)$$

where C_{ox} and C_{red} are the bulk-phase (solution) concentrations of Ox and Red. From Eqs. (2) and (3),

$$k_c = k_{ex} \exp[-\alpha f(E-E°)] \qquad (4)$$

The "standard" (or exchange) rate constant, k_{ex}, equals k_c (and k_a) measured at the standard electrode potential, $E°$; note that $E° = E_{eq}$ when $C_{ox} = C_{red}$. The standard potential, and hence k_{ex}, are of particular significance since $E°$ corresponds to the point where the interfacial potential experienced by the transferring electron equals the chemical-potential difference between Ox and Red, so that the electrochemical free-energy driving force for the overall electron-transfer reaction, $\Delta G°_{rc}$, equals zero.

Even for such ostensibly "single-step" electrochemical processes for solution-phase reactants, it is useful to separate the overall reaction into components associated with the formation of a "precursor state" with the reactant in a suitable geometry within the interfacial region, and the "elementary electron–transfer step" where the activation energy barrier is surmounted with the reactant located within the precursor state [2]. This "preequilibrium" model differs from earlier treatments which presume (albeit vaguely) that the reaction proceeds via reactant "collisions" with the metal surface [2]. Provided that the formation of the precursor state involves a markedly smaller barrier than for electron transfer itself, we can decompose the overall observed rate constant, k_{ob} (cm s^{-1}), into a precursor equilibrium constant, K_p (cm), and a *unimolecular* rate constant, k_{et} (s^{-1}), characterizing the elementary electron–transfer step, according to [2]

$$k_{ob} = K_p k_{et} \qquad (5)$$

In addition to its simplicity, this treatment has the virtue of enabling the influence of the interfacial environment upon the reactant thermodynamic stability to be separated from the vagaries of the electron–transfer step itself. Most importantly, the preequilibrium model is applicable to processes where the electrode–reactant interactions are weak and nonspecific as well as those where surface–reactant bonds are formed [2]. The former type includes most so-called "outer–sphere" processes, defined usually for electrochemical processes as those where the reactant is separated from the electrode surface by a layer of solvent or other molecules in the transition state. The latter reaction type can be considered to be "inner–sphere" processes, defined most generally as those featuring transition states where the reactant, or a coordinated ligand(s), binds to (or otherwise contacts) the metal surface [2]. For many (but by no means all) inner–sphere reactions, the electrode–reactant interactions are sufficiently favorable to enable the interfacial (precursor–state) reactant concentrations to be determined analytically, allowing K_p to be evaluated [3].

A similar situation can also be achieved in a few cases for outer–sphere electrochemical processes [4]. More commonly, however, diffuse–layer theory needs to be utilized to yield approximate K_p values for outer–sphere processes. In general, K_p can be related to the work of forming the precursor state from the bulk solution reactant, w_p, by [2]

$$K_p = K_o \exp(-w_p/RT) \qquad (6)$$

where K_o (cm) is the "statistical" component of K_p. The latter can be approximated as the "reaction zone thickness", δr, denoting the effective range of precursor–state geometries (as separations from the metal surface) that contribute importantly to electron transfer; δr is typically of the order of 10^{-8} cm [5]. For reaction sites close to the outer Helmholtz plane (oHp), $w_p \approx ZF\phi_d$, where Z is the reactant charge number and ϕ_d is the diffuse–layer potential as deduced from the Gouy–Chapman model.

III. THEORETICAL RATE FORMULATIONS FOR THE ELECTRON-TRANSFER STEP

Of primary interest in electron-transfer kinetics is the understanding and rationalization of the observed rate constants, especially k_{et}, in terms of dynamical and energetic factors. The unimolecular rate constant can usefully be expressed in this fashion as [1,2,6]

$$k_{et} = \kappa_{el} \Gamma_n \nu_n \exp(-\Delta G^*/RT) \tag{7}$$

Here κ_{el} is the electronic transmission coefficient (≤ 1), Γ_n is the nuclear tunneling factor (≥ 1), ν_n is the nuclear frequency factor, and ΔG^* is the free energy of activation for the electron-transfer step. The first three terms together constitute the "preexponential factor" A_{et}, that describes the net dynamics of surmounting the classical free-energy barrier ΔG^*. It is important to recognize that this contemporary treatment of the preexponential factor, involving unimolecular activation within a preequilibriated precursor state, differs substantially from the classical (and flawed) description which emphasizes reactant-surface collisions as a prerequisite for electron transfer.

The ν_n term in Eq. (7) describes the net velocity along the reaction coordinate associated with the various motions which together constitute the nuclear reorganization barrier ΔG^*. The Γ_n term corrects the electron-transfer rate for the occurrence of sufficiently high-frequency motions so that significant nuclear-tunneling occurs *through* the classical free-energy barrier. In cases where the "inner-shell" barrier component, ΔG^*_{is}, arising from reactant bond distortions, is large, ν_n is often approximated by an appropriately weighted average of the relevant vibrational frequencies. In circumstances where the barrier is associated primarily with solvent reorganization, ΔG^*_{os}, the net dynamics of the solvent repolarization process can provide the predominant contribution to ν_n [7]. The role of solvent dynamics in electrochemical kinetics is considered in more detail below.

Complete control of the rate of electron-transfer barrier crossing by such nuclear dynamics will only be achieved when the electronic coupling between the donor and acceptor sites (the redox center and the electrode surface for electrochemical reactions) is sufficiently strong so to maintain so-called "reaction adiabaticity". Physically, this corresponds to the occurrence of sufficient resonance splitting between the lower and upper potential-energy surfaces in the vicinity of the intersection region (saddle point) so that the system stays primarily on the lower, reactive, surface. Such an "adiabatic" passage through the transition-state region corresponds to $\kappa_{el} \sim 1$ in Eq. (7). On the other hand, if the donor-acceptor electronic coupling is relatively weak, then a large fraction of passages through the intersection region will involve nonreactive transitions to the upper surface. For such "nonadiabatic" pathways, $\kappa_{el} \ll 1$, reflecting the relatively small fraction of reactive passages over the barrier.

It is important to recognize that for such nonadiabatic pathways, the net barrier-crossing frequency will be essentially *independent* of the nuclear dynamics, depending instead chiefly upon the degree of electronic coupling. This point can be discerned readily by considering the following simplified, yet illustrative, Landau–Zener expression for the transmission coefficient κ_{el} [6]:

$$\kappa_{el} = 2[1 - \exp(-\nu_{el}/2\nu_n)]/[2 - \exp(-\nu_{el}/2\nu_n)] \qquad (8)$$

The "electronic frequency factor" ν_{el} is given by

$$\nu_{el} = H_{12}^2(\pi^3/\Delta G^* h^2 k_B T)^{\frac{1}{2}} \qquad (8a)$$

where h is Planck's constant, k_B is Boltzmann's constant, and H_{12} is the electronic coupling matrix element. While $\kappa_{el} \to 1$ for sufficiently strong electronic coupling (i.e. large H_{12}) so that $\nu_{el} \gg \nu_n$, Eq. (8) reduces to $\kappa_{el} \approx \nu_{el}/\nu_n$ for weaker coupling ($\kappa_{el} \ll 1$) so that $\nu_{el} \ll \nu_n$. Combining the preexponential factor A_{et} ($=\kappa_{el}\Gamma_n\nu_n$) from Eq. (7) with the latter nonadiabatic limit, yields $A_{et} = \nu_{el}\Gamma_n$, i.e., the net preexponential factor is proportional to H_{12}^2 [Eq. (8a)] yet *independent* of ν_n.

While some small variations of A_{et} can also be anticipated, the dependence of k_{et} (and hence k_{ob}) upon the electrode potential arises primarily from variations in ΔG^*. Generally, for one–electron reactions we can separate ΔG^* into "intrinsic" and "thermodynamic" (driving–force) contributions [1]:

$$\Delta G^* = \Delta G_{int}^* + \alpha_{et}\Delta G_{et}^o \qquad (9)$$

where α_{et} is the (cathodic) transfer coefficient (symmetry factor) for the electron–transfer step. The corresponding driving force, ΔG_{et}^o, can be related to the "standard overpotential" (E–E°) by [1]

$$\Delta G_{et}^o = F(E–E^o) + (w_s - w_p) \qquad (10)$$

where w_s is the work of assembling the successor state (i.e., the state immediately following electron transfer) from the bulk–phase product. [Note that the combined term $(w_s - w_p)$ in Eq. (10) accounts for the difference in the potential dependence of the thermodynamics for the electron–transfer step and for the overall solution–phase reaction (1)].

As already noted, it is convenient to separate the overall free-energy barrier into inner–shell (reactant distortional) and outer–shell (solvent reorganizational) components. For calculational purposes, it is convenient to apply this distinction to the intrinsic barrier, ΔG_{int}^*, yielding more manageable formulae. Thus the inner–shell component, ΔG_{is}^*, of ΔG_{int}^* can be estimated from the simple harmonic oscillator formula [6]:

$$\Delta G_{is}^* = 0.5 \Sigma f_i(\Delta a/2)^2 \qquad (11)$$

where f_i is the force constant of each bond undergoing distortion by Δa as a result of electron transfer.

The outer–shell component, ΔG_{os}^*, is usually estimated by means of the well-known formulae derived from dielectric-continuum theory, which for one-electron electrochemical reactions can be expressed as [8]

$$\Delta G^*_{os,e} = (e^2/8)(a^{-1} - R_e^{-1})(\epsilon_{op}^{-1} - \epsilon_s^{-1}) \qquad (12)$$

where a is the radius of the (presumed spherical) reactant, R_e is twice the reactant–metal surface distance (i.e., the reactant–image distance) in the transition state, and ϵ_{op} and ϵ_s are the so-called optical and static (zero-frequency) solvent dielectric constants, respectively. A related treatment for one-electron transfer between identical pairs of redox couples in homogeneous solution yields [8]

$$\Delta G^*_{os,h} = (e^2/4)(a^{-1} - R_h^{-1})(\epsilon_p^{-1} - \epsilon_s^{-1}) \qquad (13)$$

where R_h is the internuclear distance between the reactants (having equal radius a).

The extent to which these dielectric–continuum formulae provide reliable estimates of the solvent reorganization energy has engendered considerable discussion over the years [9]. A number of analytic treatments appearing recently provide somewhat modified descriptions of $\Delta G^*_{os,e}$ and/or $\Delta G^*_{os,h}$, most straightforwardly for the limit where R_e, $R_h \to \infty$, i.e., activation for "isolated" spherical reactants. In this limit, several improved treatments, such as that employing the "mean spherical approximation" (MSA) [10] and "nonlocal electrostatic" models [11], yield ΔG^*_{os} values that are significantly (ca 10–30%) smaller than in the dielectric continuum limit [12]. Moreover, a modified treatment of imaging effects suggests that the R_e^{-1} term in Eq. (12) is inappropriate in most circumstances [13]. Unfortunately, *direct* experimental tests of these models at metal–solution interfaces are absent; rate measurements by themselves provide only indirect comparisons since estimation of the preexponential component is not straightforward (vide infra) [13b]. Nevertheless, the applicability of the dielectric continuum approach for homogeneous–phase electron transfer can be tested more directly from the energies, E_{op}, of optical electron transfer within binuclear complexes, since E_{op} can be related simply to activation free energies for thermal electron transfer [14]. While the dielectric–continuum treatment has proved to be semiquantitatively reliable on this basis, somewhat improved agreement with experiment can be achieved by using more sophisticated models [12].

IV. INTERPLAY BETWEEN THEORY AND EXPERIMENT – GENERAL

At least for one-electron processes, there is much scope for utilizing such theoretical treatments to explore the rich diversity of experimental kinetics both at electrode surfaces and in solution. Indeed, a major application of kinetic theory in this vein is to collate, with an eye to rationalizing, sequences of rate parameters involving systematic variations in the system physical or chemical state [1]. Common examples of such experimental variables include electrode potential, temperature, electrode material, solvent composition, or reactant structure. Specific theoretical relationships, or formalisms, can be derived for this purpose, to confront with corresponding experimental data. Such tactics can be viewed as utilizing "relative" theory-experiment comparisons [1].

An ultimate objective of chemical kinetic theory, however, is to predict rates of *individual* reactions in a given reaction environment. Such "absolute" theory-experiment comparisons can provide a demanding test of the underlying theoretical models. Their application, however, has been relatively limited in practice, even for outer-sphere one-electron processes, by the extensive structural and redox thermodynamic data that are required [15]. "Relative" theory-experiment comparisons are often less likely to exhibit deviations from expectations, due to the cancellation of terms in the theoretical expressions when the kinetics of closely related processes, and/or for a given process in different reaction environments, are compared. On the other hand, examination of relative rate parameters for judiciously chosen variations of physical and/or chemical state can yield much insight into *particular* factors that influence experimental systems since unwanted (and often unknown) components of the observed reactivities can thereby be held constant [1].

Outlined briefly below are two illustrative examples of such "relative" theory-experiment comparisons, specifically involving the rates of related electrochemical and homogeneous-phase reactions, and solvent effects upon the reaction dynamics. Emphasis is restricted here to a summary of the underlying concepts; numerical and other details can be found in the cited literature.

V. REACTIVITIES OF RELATED ELECTROCHEMICAL AND HOMOGENEOUS-PHASE PROCESSES

An issue of longstanding interest in electrochemical reactions concerns the relationship of the observed kinetics with those for the same redox couples involved in homogeneous-phase electron transfers. Given that the latter processes involve inevitably *pairs* of redox couples, several different types of comparisons can be envisaged. The simplest involve homogeneous self-exchange processes, since these reactions feature identical redox couple partners, thereby yielding the same reactants and products, so that $\Delta G_{et}^{o} = 0$.

The most common relationship used to examine the rates of outer-sphere electrochemical exchange, k_{ex}^{e} [Eq. (4)], and homogeneous self-exchange, k_{ex}^{h}, for a given redox couple is [16]

$$k_{ex}^{e}/A^{e} = (k_{ex}^{h}/A^{h})^{\frac{1}{2}} \tag{14}$$

where A^{e} and A^{h} are the preexponential factors for the overall electrochemical and homogeneous-phase reactions, respectively (equal to $A_{et}K_{o}$) [1]. Equation (14) is derived for the special case where $R_{e} = R_{h}$ [Eqs. (12), (13)], predicted when the reaction partners (or the reactant-electrode pair) are in contact, so that $\Delta G_{os,e}^{*} = 0.5\Delta G_{os,h}^{*}$. The rate constants need to be corrected for electrostatic work terms, so that they reflect the kinetics of the elementary electron-transfer step [other than the inclusion of the statistical term K_{o}, Eq. (6)]. A related, yet distinct, formula results from the assumption that reactant-electrode imaging, and the inner-shell component of ΔG_{int}^{*}, can both be neglected, whereupon simply [17]

$$k_{ex}^e/A^e - k_{ex}^h/A^h \qquad (15)$$

In appropriate circumstances, both Eqs. (14) and (15) can provide acceptable fits to experimental data [1].

Nonetheless, it is desirable to provide formalisms that are applicable in the more commonly encountered cases that involve homogeneous—phase cross reactions. A formulation designed for this purpose, which we have discussed recently [18], involves comparisons between a given electrochemical reaction and a corresponding homogeneous—phase reduction (or oxidation) that utilizes a reversible redox reagent Ox_2/Red_2, having a standard potential E_2^o. This treatment emphasizes the nature of metal surfaces as a special type of coreactant, having infinite radius and zero inner—shell barrier, yet a continuously variable "redox potential" equal to the applied electrode potential. Provided that the inner—shell barrier associated with Ox_2/Red_2 is small (or can be corrected for), and reactant—electrode imaging is unimportant, then the (work—term corrected) electrochemical rate constant measured at E_2^o, k_2^e, is predicted on this basis to be related to the corresponding homogeneous—phase rate constant involving Ox_2/Red_2, k_2^h, by [18]

$$k_2^e/A^e - k_2^h/A^h \qquad (16)$$

Note that the form of Eqs. (15) and (16) are similar; indeed, the former is a special case of the latter. In the particular case where both the electrochemical and homogeneous—phase reactions are adiabatic ($\kappa_{el} \approx 1$) or at least feature comparable transmission coefficients, then[18] $A^h/A^e - 4\pi N r_h^2$, where r_h is the reactant internuclear distance for the homogeneous—phase process, and N is Avogadro's number. Equation (16) therefore becomes

$$4\pi N r_h^2 k_2^e - k_2^h \qquad (17)$$

The $4\pi N r_h^2$ term in Eq. (17) accounts for the difference between the planar and spherical reactant geometries characteristic of the electrode and homogeneous—phase coreactants, respectively, thereby converting the usual heterogeneous (cm s^{-1}) rate units into those ($M^{-1}s^{-1}$) appropriate for second—order solution—phase processes.

Given the various assumptions involved in deriving Eq. (17), widespread agreement with experiment is not anticipated. The virtues of this and related expressions lie instead in their treatment of related electrochemical and homogeneous—phase processes in an equivalent, if rather idealized, fashion. The manner and extent of the observed deviations of experimental rate data from Eq. (17) provide a useful measure of the degree to which additional, especially "specific" factors, such as electron tunneling, solvation dynamics, imaging interactions, etc., affect the kinetics of a given redox couple differently in the chosen heterogeneous— and homogeneous—phase reaction environments. Since the formal potential of only the homogeneous *coreactant* is required for the analysis, it is applicable to chemically irreversible and even multielectron electrochemical reactions.

In an exploratory comparison with experimental data [18], a number of outer–sphere electrochemical reactions at mercury–aqueous interfaces, including the reduction of various metal complexes and dioxygen, were found to exhibit rates that are significantly (up to ca 10^4 fold) greater than anticipated from homogeneous–phase data on the basis of Eq. (17). These differences were ascribed primarily to the occurrence of larger transmission coefficients in the electrochemical reaction environment. However, other factors, such as the occurrence of lower electrochemical free–energy barriers from metal imaging or other factors, may contribute to the observed behavior. Closer correspondence to the predictions of Eq. (17) are obtained, however, for some processes in nonaqueous media [18].

VI. DYNAMICAL SOLVENT EFFECTS IN ELECTROCHEMICAL KINETICS

Examining the kinetic as well as thermodynamic consequences of altering the solvent medium is of obvious fundamental interest in redox chemistry given the central role of solvation in all types of liquid–phase electron–transfer processes. The latter free–energy component (ΔG^o_{et}) often provides the predominant contribution to the observed rate–solvent dependencies. However, this term can be held fixed by examining the solvent dependence of the *standard* rate constant, k_{ex}, (i.e., at E^o in *each solvent*) thereby altering the electrode potential and therefore the free energy of the transferring electron(s), so to cancel the inevitable alterations in the solvation energies of Ox versus Red. Aside from solvent–dependent work terms, the remaining anticipated solvent effects upon k_{ex} arise at least from variations in the outer–shell reorganization energy, ΔG^*_{os}, and possibly also in the nuclear frequency factor, ν_n [Eq. (7)]. The presence of the former component has long been recognized, especially from the work of Marcus in the USA, Hush in England, and from Levich and Doganadze in the Soviet Union, dating back to the 1950's.

Perhaps surprisingly, the realization that electron–transfer rates can be influenced additionally by the *dynamics* of solvent reorganization, affecting ν_n as noted above, is of much more recent origin. This recognition arose from, and indeed forms an integral part of, the remarkable upsurge of interest in dynamical solvent effects in condensed–phase chemical processes in general that has occurred during the last decade [19]. We now outline in general terms the physical origins and likely consequences of such solvent dynamical effects in electrochemical kinetics. (See ref. 7 for further details and more complete literature citations).

In broad outline, the physical origin of such effects in electron transfer can be understood simply from the need to reorient collectively solvent dipoles in the vicinity of a redox center in order to achieve an appropriate nonequilibrium configuration so that electron transfer can occur. Even for simple dipolar fluids, however, it is important to distinguish between two distinct solvent dynamical regimes. In so–called low–dielectric friction media (associated loosely with low fluid viscosities), the rate of the necessarily *collective* dipolar motion will be limited only by the moment of inertia of the individual dipoles together with the dielectric properties of the surrounding medium [20,21]. Reaction dynamics controlled by solvent motion under these "underdamped" circumstances correspond to the so-called transition-state theory (TST) limit, where the system passes smoothly along the reaction coordinate from reactants to products. More commonly, however, it is expected that the frequency of collective dipole motion will fall significantly, or substantially, below the TST value, ω_o, by irreversible energy transfer from a given rotating dipole to its surroundings. This latter occurrence, known as solvent "dielectric friction", corresponds to slower "overdamped" (often-termed "diffusive") motion along the reaction coordinate.

A key feature of such phenomena is that while the underdamped (TST) dynamics are only mildly dependent on the solvent structure [21], the overdamped frequencies are markedly solvent-sensitive. At least for Debye solvents in the dielectric-continuum limit, the relevant overdamped dynamics are described by the longitudinal solvent relaxation time, τ_L [7,19,20]. This quantity is usually extracted from the "Debye" relaxation time, τ_D, obtained from solvent dielectric-loss spectra, by using [19]

$$\tau_L = (\epsilon_\infty/\epsilon_s)\tau_D \qquad (18)$$

where ϵ_∞ is the so-called "infinite-"(microwave) frequency dielectric constant. [The observation of a single dielectric dispersion (τ_D) in the dielectric loss spectrum is often referred to as "Debye-like" behavior.] Interestingly, τ_L varies by at least 50 fold in common polar solvents at ambient temperatures, from ca 0.2 to 10ps [22,23]. Measurements of longitudinal relaxation dynamics can be made in some cases more directly from time-dependent fluorescence Stokes shifts (TDFS) following ultrafast laser-induced creation of charge-transfer excited states within suitable chromophore solutes [24]. The time resolution of most published measurements of this type is strictly inadequate for a reliable description of the solvent dynamics relevant to electron-transfer kinetics. However, recent subpicosecond TDFS measurements by Barbara and coworkers have enabled solvation relaxation times, τ_s, to be extracted even in low-friction media [24a,b]. In most cases, $\tau_s \sim (1.5-2)\tau_L$ [24b], although relaxation components substantially shorter than τ_L are also often observed (vide infra).

Of central interest are the relationships between the various dielectric relaxation parameters and the resultant adiabatic barrier-crossing frequency, ν_n, in Eq. (7). We consider first the situation where the reaction coordinate is dominated by solvent, rather than additionally by inner-shell, reorganization. In this case, in the TST limit simply $\nu_n = \omega_o/2\pi$ [7,19]. The situation in the presence of dielectric friction is somewhat more complicated, since the reaction dynamics depend on the shape of the barrier top. For cusp-like barriers (i.e., for small resonance splitting, H_{12}, of the barrier top), $\nu_n \sim \tau_L^{-1}$ [7,20]. As the barrier top becomes more "rounded" (i.e., H_{12} increases) ν_n is predicted to diminish somewhat. A simple rationalization of the latter is that barrier recrossings, yielding a lower net frequency of successful diffusive passages through the intersection region and hence a smaller ν_n, should become more prevalent as the barrier top becomes broader. For the degree of barrier-top roundedness required typically to yield adiabatic reaction pathways, ν_n is anticipated to be ca 2-5 fold smaller than τ_L^{-1} [25b]. One might therefore expect that at least approximate correlations between τ_L^{-1} and ν_n would often be obtained.

At least two further factors, however, are anticipated on theoretical grounds to complicate the nature of solvent dynamical effects on electron transfer. Firstly, as noted above, solvent relaxation components substantially faster than τ_L have been observed, even in "simple" dipolar fluids, from TDFS measurements. Additional faster relaxation times can also be extracted in some cases from dielectric loss spectra, such as in primary alcohols [25]. (Solvents exhibiting the latter behavior are referred to as "non-Debye" media.) Such higher-frequency components are predicted to exert disproportionately large contributions to the barrier-crossing frequencies, so that ν_n values substantially higher than τ_L^{-1} are often anticipated under these conditions [25].

Secondly, some deviations of the effective net relaxation time from τ_L are generally expected, even in Debye media, as a result of "solvent molecularity" effects, whereby solvent molecules in the immediate vicinity of the reacting solute behave differently from those in the "bulk" liquid [7,20]. An additional source of divergence from the simple expectation that $\nu_n \sim \tau_L^{-1}$ is expected for systems where the reaction coordinate contains substantial contributions from inner-shell (i.e., reactant vibrational) distortions as well as solvent reorganization. This situation has been discussed in detail by Marcus and coworkers [26]. Their treatment predicts that the dependence of ν_n upon solvent dynamics becomes increasingly muted as the frequency of inner-shell motion and/or the ratio $\Delta G_{is}^*/\Delta G_{os}^*$ increases; however, a substantial contribution from solvent dynamics can remain in the overdamped case even when $\Delta G_{is}^* \sim \Delta G_{os}^*$.

In the last five years or so, a sizable number of experimental studies have been published that aim to test the manner and extent to which such solvent dynamical factors influence activated electron-transfer reactions, both at electrode surfaces and in homogeneous solution [7]. The primary tactic employed has been to explore the solvent dependence of the rate constants, k_{ex}, for electrochemical exchange or homogeneous self-exchange of suitable one-electron redox couples that are anticipated or known to follow outer-sphere pathways. It is appropriate here to comment briefly on some of the limitations of, as well as key results from, these experimental studies, especially since insufficiently critical analyses have appeared on occasion in the recent literature. A more detailed discussion of these and related issues will be available elsewhere [27].

A key element in all these studies is the attempted separation of the observed k_{ex}-solvent dependencies into energetic and dynamical components, associated with variations in ΔG^* and ν_n, respectively. The reliable partition of these factors is often problematical, especially for electrochemical reactions, since the solvent-induced variations in ΔG^* typically have been estimated by using a theoretical formula, usually from Eqs. (12) or (13), rather than having a firm experimental basis. Admittedly, the *functional form* of Eq. (13) is typically roughly in

accord with some experimental E_{op} data for photoinduced electron transfers
(vide supra) [14]. However, the uncertainties in the values of the
"geometric factors" $(a^{-1} - R^{-1})$ which act to scale the resulting ΔG^*
values, as well as in the details of the underlying model itself, can
lead to substantial mistrust of the resulting estimated ΔG^*-solvent
variations. This difficulty is heightened by the common expectation that
the influence of the energetic and dynamical components upon the observed
k_{ex}-solvent dependencies are often expected to be roughly comparable
[22,23].

Fortunately, however, at least the *qualitative recognition* of the
presence of solvent dynamical effects on this basis is aided by the
roughly linear correlation between ΔG^*_{os} and $\log \tau_L^{-1}$ (and hence $\log \nu_n$)
expected for a number of solvents [22]. [This correlation is not
unexpected since the outer-shell barrier ΔG^*_{os} in Eqs. (12) and (13) is
determined primarily by the inverse optical dielectric constant ϵ_{op}^{-1}, and
longer τ_L values are commonly associated with larger, more polarizable,
solvents that also tend to exhibit greater ϵ_{op} values.] In a series of
solvents with progressively decreasing ϵ_{op} values, then, in the absence
of solvent dynamical effects the k_{ex} values will tend to decrease since
the $\exp(-\Delta G^*/RT)$ term in Eq. (7) will diminish in this sequence. In the
presence of solvent dynamical effects upon ν_n, however, this trend will
be offset systematically since $\log \nu_n$ tends to increase under these
conditions. In some cases, the latter dynamical effect appears to
dominate since net *increases* in k_{ex} are often observed in a given solvent
sequence as τ_L^{-1} (and ϵ_{op}^{-1}) increase [22,23]. In this circumstance, strong
evidence for the qualitative presence of solvent dynamical effects upon
k_{ex} is at hand even in the *absence* of reliable corrections for the
solvent-dependent barrier since the latter effect acts in the opposite
direction [22].

Additional difficulties that can (and sometimes have) plagued the
interpretation of solvent-dependent electrochemical kinetics in this vein
include uncertainties in the work terms ("double-layer" effects) as the
solvent is varied, and the erroneous neglect of (or inadequate correction
for) solution resistance and related artifacts during the experimental
evaluation of k_{ex}. The latter problem is particularly insidious since
many reactions otherwise suitable for such studies are sufficiently rapid
so to make corrections for solution resistance fraught with difficulty.
Moreover, the effect of residual uncompensated resistance is often to
depress the apparent k_{ex} values in a solvent-dependent manner that
unfortunately mimics closely the anticipated influence of overdamped
solvent dynamics. Unfortunately, then, the latter artifactual effect can
easily be mistakenly identified as signaling a dominant role of solvent
dynamics in electrochemical kinetics. For these and other reasons,
multiparametric dielectric-continuum analyses of solvent effects in
electrochemical kinetics, espoused by some authors [28], should be used
only with extreme caution as an indicator of dynamical factors.

In spite of these pitfalls, the reliable quantitative diagnosis of
dynamical solvent effects has been achieved in several systems. In our
laboratory, we have preferred to explore most quantitative details of
such effects by utilizing homogeneous self-exchange rather than solely

electrochemical processes [7,23]. This was brought about in part by the identification of a suitable series of metallocenium–metallocene reactions, of the general form $Cp_2M^{+/o}$, where Cp is a cyclopentadienyl ligand and M = Co or Fe [23]. Reliable estimates of the solvent-dependent barriers for these systems can be obtained from E_{op} values for closely related biferrocene cations [12]. This enables the solvent-dependent k_{ex} values to be corrected directly for the variations in barrier height. The substantial (up to 10^2 fold) variations in k_{ex} in a given solvent which are induced by altering Cp and/or M have been traced primarily to differences in the degree of electronic coupling (i.e., the magnitude of H_{12}) and consequently in the reaction adiabaticity [i.e., the κ_{el} value in Eq. (7)], reflecting the alterations in electronic structure and the orbital symmetry for the transferring electron.

Interestingly, the logarithmic dependence of the barrier–corrected rate constants, k'_{ex}, upon τ_L^{-1} in Debye–like solvents varies systematically with the magnitude of k'_{ex} in a given solvent [23a]. For the least facile redox couples (e.g., ferrocenium–ferrocene), k'_{ex} is virtually independent of τ_L^{-1}, whereas for the most facile couples (e.g., $Cp'_2Co^{+/o}$, where Cp' = pentamethylcyclopentadienyl) the log k'_{ex} – log τ_L^{-1} slopes approach unity at small τ_L^{-1} values. This systematic behavior is consistent with a progression from virtually nonadiabatic to near–adiabatic electron transfer, reflected in the increasing emergence of solvent dynamical effects. Indeed, a detailed analysis of the solvent–dependent kinetic data yields H_{12} values for $Cp_2Co^{+/o}$ and $Cp_2Fe^{+/o}$ (ca 0.5 and 0.1 kcal. mol^{-1}, respectively) that are in accordance with independent theoretical estimates [23a].

In contrast, the corresponding solvent–dependent k_{ex} values for $Cp_2M^{+/o}$ electrochemical exchange are uniformly facile, and exhibit marked effects from solvent dynamics, apparently reflecting a greater degree of donor–acceptor orbital overlap engendered at the metal–solution interface [23b,c]. Indeed, most electrochemical reactions examined so far in this context exhibit comparable behavior. An interesting exception, however, is provided by the $Ru(hfac)_3^{o/-}$ couple (hfac = hexafluoroacetylacetonato) in that the solvent–dependent electrochemical exchange kinetics of this system are both sluggish and essentially independent of the solvent dynamics [29]. By comparison with the corresponding self–exchange kinetics and with related electrochemical systems, this behavior has been traced to the occurrence of nonadiabatic electrochemical reaction pathways [29].

Given the largely satisfactory picture noted above for metallocene self–exchange reactions in Debye–like solvents, we have been interested in exploring their behavior in "non–Debye" media (e.g., propylene carbonate, primary alcohols, water) that as noted above are characterized by additional higher–frequency solvent relaxations [25b,30]. Broadly speaking, the substantial (ca 10 fold) rate accelerations predicted from theoretical models in such media [25] are borne out by the experimental findings [25b,30]. Similar results are also obtained for some electrochemical systems [22a,31]

One issue that has received little experimental attention so far is the interplay between nuclear dynamical effects arising from reactant vibrational and overdamped solvent motion. This is probably due to the

440

paucity of experimental systems suitable for solvent-dependent studies for which the magnitude of the inner-shell barrier is known and (preferably) can be altered. Given the prevalence of electrochemical systems for which the inner-shell barrier is substantial, however, such studies would be of significant practical interest.

VII. CONCLUDING REMARKS

While the understanding, as well as interpretation, of electron-transfer kinetics at metal-solution interfaces has been placed on an increasingly molecular-level basis in recent years, manifold challenges remain. Even for the simplest class of reaction considered here: outer-sphere one-electron exchange processes, there are substantial uncertainties in both the activation dynamics and energetics associated with solvent reorganization. Nevertheless, the increased emphasis placed on the elucidation of preexponential factors in electrochemical kinetics, especially involving electron tunneling and solvent dynamical factors, has brought some fresh insight into this previously neglected topic. The increasing tendency to consider the properties of related electrochemical and homogeneous-phase processes in concert, rather than in isolation, is also having a beneficial effect on the level of interpretation applied to the former type of reactions.

A longstanding, yet continuing, problem concerns the reliable measurement of rate parameters for rapid reactions. The advent of "ultramicroelectrodes", with the consequent diminution of the deleterious effects of solution resistance, can contribute significantly in this regard [32], but the straightforward use of such electrodes for the evaluation of fast rate constants has apparently yet to be demonstrated.

Probably the greatest and the most significant challenges in electrochemical, as in homogeneous-phase kinetics, concern multielectron and other multistep reactions, such as coupled electron/proton and electron/atom transfers. Even though many of these reactions probably involve rate-determining single-electron steps, a major difficulty remains the estimation of the thermodynamic and structural reorganization parameters associated with electron transfer in the face of such additional steps. Nevertheless, some significant progress is being made [33], suggesting that the future of electrochemical kinetics may well include an increased breadth in the type of systems amenable to molecular-level interpretation.

ACKNOWLEDGMENT

The continued support of our research in this area by the U.S. Office of Naval Research is gratefully acknowledged.

REFERENCES

1.   M.J. Weaver, in "Comprehensive Chemical Kinetics", Vol. 27, R.G. Compton, ed, Elsevier, Amsterdam, 1987, Chapter 1.

2.   J.T. Hupp, M.J. Weaver, J. Electroanal. Chem., 152, 1 (1983).

3.   For example: (a) S.W. Barr, M.J. Weaver, Inorg. Chem., 23, 1657 (1984); (b) T.T-T. Li, M.J. Weaver, J. Am. Chem. Soc., 106, 1233 (1984).

4.   M.A. Tadayyoni, M.J. Weaver, J. Electroanal. Chem., 187, 283 (1985)

5.   J.T. Hupp, M.J. Weaver, J. Phys. Chem., 88, 1463 (1984).

6.   N. Sutin, Prog. Inorg. Chem., 30, 441 (1983).

7.   For a recent conceptually based overview, see: M.J. Weaver, G.E. McManis, Acc. Chem. Res., 23, 294 (1990).

8.   R.A. Marcus, J. Chem. Phys., 43, 679 (1965); 24, 966, 979 (1956).

9.   For an extreme view, see: J.O'M. Bockris, S.U.M. Khan, "Quantum Electrochemistry", Plenum Press, New York, 1979.

10.   (a) P.G. Wolynes, J. Chem. Phys., 86, 5133 (1987); (b)   G.E. McManis, M.J. Weaver, J. Chem. Phys., 90, 1720 (1989).

11.   A.A. Kornyshev, J. Ulstrup, Chem. Phys. Lett., 126, 74 (1986).

12.   G.E. McManis, A. Gochev, R.M. Nielson, M.J. Weaver, J. Phys. Chem., 93, 7733 (1989).

13.   (a) P.G. Dzhavakhidze, A.A. Kornyshev, L.I. Krishtalik, J. Electroanal. Chem., 228, 329 (1987); (b) D.K. Phelps, A.A. Kornyshev, M.J. Weaver, J. Phys. Chem., 94, 1454 (1990).

14.   For a review, see: C. Creutz, Prog. Inorg. Chem., 30, 1 (1983).

15.   (a) B.S. Brunschwig, C. Creutz, D.H. MaCartney, T-K. Sham, N. Sutin, Disc. Far. Soc., 74, 113 (1982); (b)   J.T. Hupp, M.J. Weaver, J. Phys. Chem., 89, 2795 (1985).

16.   R.A. Marcus, J. Phys. Chem., 67, 853 (1963).

17.   N.S. Hush, Electrochim. Acta., 13, 1005 (1968).

18.   M.J. Weaver, J. Phys. Chem., 94, 8608 (1990).

19.   For concise general overviews of solvent dynamical effects in chemical kinetics, see: (a) J.T. Hynes, J. Stat. Phys., 42, 149

442

(1986); (b) G.R. Fleming, P.G. Wolynes, Physics Today, <u>43</u> (May), 36 (1990).

20. D.F. Calef, P.G. Wolynes, J. Phys. Chem., <u>87</u>, 3387 (1983).

21. G.E. McManis, A. Gochev, M.J. Weaver, Chem. Phys., in press.

22. (a) G.E. McManis, M.N. Golovin, M.J. Weaver, J. Phys. Chem., <u>90</u>, 6563 (1986); (b) T. Gennett, D.F. Milner, M.J. Weaver, J. Phys. Chem., <u>89</u>, 2787 (1985).

23. (a) G.E. McManis, R.M. Nielson, A. Gochev, M.J. Weaver, J. Am. Chem. Soc., <u>111</u>, 5533 (1989); (b) R.M. Nielson, G.E. McManis, M.N. Golovin, M.J. Weaver, J. Phys. Chem., <u>92</u>, 3441 (1988); (c) R.M. Nielson, M.N. Golovin, G.E. McManis, M.J. Weaver, J. Am. Chem. Soc., <u>110</u>, 1745 (1988).

24. See for example: (a) M.A. Kahlow, W. Jarzeba, T.J. Kang, P.F. Barbara, J. Chem. Phys., <u>90</u>, 151 (1989); (b) P.F. Barbara, W. Jarzeba, Adv. Photochem., <u>15</u>, 1 (1990); (c) M. Maroncelli, J. MacInnis, G.R. Fleming, Science, <u>243</u>, 1674 (1989).

25. (a) J.T. Hynes, J. Phys. Chem., <u>90</u>, 3701 (1986); (b) G.E. McManis, M.J. Weaver, J. Chem. Phys., <u>90</u>, 912 (1989).

26. (a) H. Sumi, R.A. Marcus, J. Chem. Phys., <u>84</u>, 4894 (1986); (b) W. Nadler, R.A. Marcus, J. Chem. Phys., <u>86</u>, 3906 (1987).

27. M.J. Weaver, D.K. Phelps, Chem. Rev., to be published.

28. For example, see: W.R. Fawcett, C.A. Foss, Jr., J. Electroanal. Chem., <u>270</u>, 103 (1989).

29. M.J. Weaver, D.K. Phelps, R.M. Nielson, M.N. Golovin, G.E. McManis, J. Phys. Chem., <u>94</u>, 2949 (1990).

30. (a) M.J. Weaver, G.E. McManis, W. Jarzeba, P.F. Barbara, J. Phys. Chem., <u>94</u>, 1715 (1990); (b) R.M. Nielson, G.E. McManis, M.J. Weaver, J. Phys. Chem., <u>93</u>, 4703 (1989).

31. M. Opallo, J. Chem. Soc. Far. Trans. I, <u>82</u>, 339 (1986).

32. R.M. Penner, M.J. Heben, T.L. Longin, N.S. Lewis, Science, <u>250</u>, 1118 (1990).

33. For informative overviews, see: (a) W.E. Geiger, Prog. Inorg. Chem., <u>33</u>, 275 (1985); (b) T.J. Meyer, J. Electrochem. Soc., <u>131</u>, 221C (1984).

THE SOLID-ELECTROLYTE INTERFACE AS EXEMPLIFIED BY HYDROUS OXIDES; SURFACE CHEMISTRY AND SURFACE REACTIVITY

WERNER STUMM
Swiss Federal Institute of Technology (ETH Zürich)
Institute for Water Resources and Water Pollution Control (EAWAG)
Switzerland

ABSTRACT

The interaction of solutes (H^+, OH^-, metal ions and ligands) with hydrous oxides can be described in terms of surface complex formation. Equilibrium constants can be corrected for electrostatic effects by considering the Gouy Chapman theory. The surface reactivity depends on surface structure (surface speciation). A general rate law on the surface controlled dissolution of oxide minerals can be derived by combining concepts of surface coordination chemistry with established models of lattice statistics and activated complex theory.

Introduction

The interaction of a solute with a surface – be it in terms of adsorption or surface complexation – requires a characterization of the physical and chemical properties of the solvent (electrolyte), the solute and the sorbent. While the surface in the electric double layer model is assumed to be a structureless continuum which interferes with the solution only by its electric charge, the basic concept in the surface coordination model are the surface functional groups formed on all natural inorganic and organic hydrous solids; they are responsible for the surface reactivity (and mechanism of adsorption). These functional groups contain the same donor atoms as found in functional groups of soluble ligands; e.g. the surface hydroxyl group on a hydrous oxide or an organic solid has similar donor properties as the corresponding counterparts in dissolved molecules such as hydroxide ions and carboxylates. We need to consider, however, that the functional groups are bound into a solid framework and their reactivity is, in essence, a cooperative property.

A hydrous oxide particle can be treated like a polymeric oxoacid (on base) which tends to undergo protolysis and to coordinate with metal ions. The central ions in the surface layer of the oxide – acting as Lewis acids – can replace their coordinated OH^- ions by ligand exchange reactions with anions or weak acids (Sigg and Stumm, 1981).

Table 1 summarizes some of the features of the coordination chemistry of the oxide water interface (Stumm et al., 1990).

In this chapter we will review 1) the adsorption equilibria of hydrous oxide surfaces with H^+, OH^-, metal ions and ligands; and 2) illustrate on the basis of a few examples how the surface reactivity (dissolution, redox reactions) depends on surface structure (surface speciation).

R. Guidelli (ed.), Electrified Interfaces in Physics, Chemistry and Biology, 443–472.
© 1992 *Kluwer Academic Publishers.*

Table:1 Coordination Chemistry of the Oxide-Water Interface; Concepts and Important Applications in Natural and Technical Systems
from Stumm et al., 1990

Surface Complex Formation

Interaction with
– H⁺, OH⁻
– Metal ions
– Ligands (ligand exchange)

Thermodynamics of Surface Complex Formation
– K (mass law constants, corrected for electrostatic effects)
– DG, DH

Kinetics of Surface Complex Formation
Rates of sorption and desorption

Structure of Surface Compounds (Surface Speciation)
– Inner-sphere versus outer-sphere
– Mononuclear versus binuclear
– Monodentate versus bidentate

Structure of Lattice
– Defect sites
– Adatoms, kinks, steps, ledges
– Lattice statistics

Thermodynamic Applications: Distribution of Solutes between Water and Solid Surface

Binding of Reactive Elements to Aquatic Particles in Natural Systems
– Regulation of metals in soil, sediment, and water systems
– Regulation of Oxyanions of P, As, Se, Si in water and soil systems
– Interaction with phenols, carboxylates and humic acids

Binding of Cations, Anions and Weak Acids to Hydrous Oxides in Technical Systems
– Corrosion; passive films
– Processing of ores, flotation
– Coagulation, flocculation, filtration
– Ceramics, cements
– (Photo)electrochemistry (electrodes, oxide electrodes and semiconductors

Kinetic Applications: Rates depend on Surface Speciation

Natural Systems
Dissolution of Oxides and Sicilates
– Weathering of minerals
– Proton and ligand promoted dissolution
– Reductive dissolution of Fe(III) and Mn(III,IV) oxides

Formation of Oxide Phases
– Heterogeneous nucleation
– Surface precipitation, crystal growth
– Biomineralization

Surface Catalyzed Processes
– (Photo)redox processes
– Hydrolysis of esters
– Transformations of organic matter by Fe and Mn (photo)redox-cycles
– Oxygenation of Fe(II), Mn (II), Cu(I) and V(IV)

Technical Systems
– Passive films (corrosion)
– Photoredox processes with colloidal semiconductor particles as photocatalyst, e.g. degradation of refractory organic substances
– Photoelectrochemistry, e.g. photoredox processes of semiconductor electrodes

Generic and Specific Aspects of the Oxide-Electrolyte Interface

In the Gouy-Chapman (Stern) double layer model the surface is treated as if it were an electrode; the distribution of ions in solution is governed by thermal motion and electrostatic interaction. The (Stern) Gouy-Chapman model has been useful in describing the distribution of charges on the solution side and predicting many phenomena of colloid stability. However, the selectivity of interactions of hydrous oxide surfaces with many solute species can be accounted for only by considering specific chemical interactions at the solid surface. In Figure 1a the surface complex formation model is represented schematically. Hydrous surfaces contain functional groups, e.g., hydroxo groups >Me-OH, at the surfaces of oxides of Si, Al and Fe. These surface functional groups represent enormous facilities for the specific adsorption of cations and anions.

Figure 1

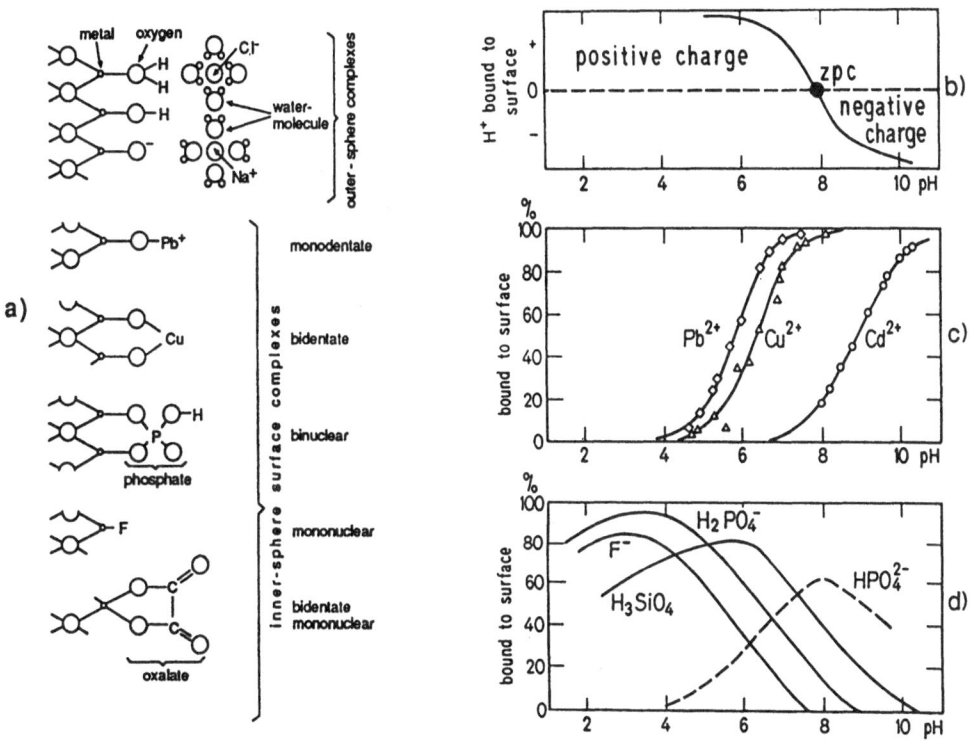

Figure 1: (a) An oxide surface, covered in the presence of water with amphoteric surface hydroxyl groups, >M-OH, can be looked at as a polymeric oxyacid or base. The surface OH group has a complex-forming 0-donor atom that coordinates with H^+ (b) and metal ions (c). The H^+ bound to the surface can be determined experimentally by alkalimetric titration of an oxide suspension in a given electrolyte solution; ZPC is the point of zero proton condition ("zero point charge"). The underlying central ion in the surface layer of the oxide acting as a Lewis acid can exchange its structural OH ions against other ligands (anions or weak acids) (d). The extent of surface coordination and its pH dependence can be quantified by mass-action equations and can be explained by considering the affinity of the surface sites for metal ions or ligands and the pH dependence of the activity of surface sites and ligands. The tendency to form surface complexes may be compared with the tendency to form corresponding solute complexes.

446

Figure 1 e

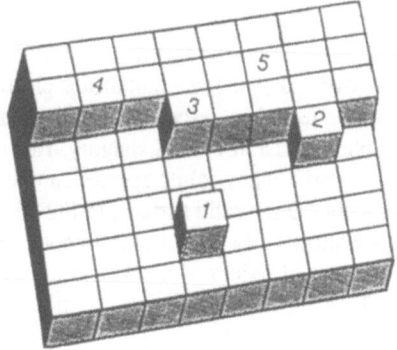

(e)The geometry of a square lattice surface model. The five different surface sites are 1) adatom, 2) ledge, 3) kink, 4) step and 5) face. The five types of octahedral surface complexes are bonded to 1, 2, 3, 4 and 5 neighboring surface links (ligands).From a point of view of surface reactivity (e.g., dissolution rate), obviously the various surface sites have different activation energies, the adatom-site (1) is most reactive and the face-site (5) (linked to 5 neighboring sites) is least reactive. The overall dissolution rate is based on the parallel dissolution reactions of all sites, but the overall dissolution kinetics is dictated by the fastest individual reaction rates. The latter is essentially given by the product of the first order reaction rate specific for each type of site and the relative concentration of surface sites of each category. Monte Carlo methods, where individual activation energies were assigned to the distinct sites, were able to show that a steady state distribution of the various surface sites can be maintained during the dissolution and that on type of thesurface sites essentially accounts for the overall dissolution rate. The model (Wehrli, 1989) suggests that the kink-sites (3) – although reacting much slower than ledge and adatom sites, but being present at much higher relative concentrations than these less-linked surface sites – control the overall dissolution rate. (Fig a is modified from Sposito, 1984).

The pH dependent charge of a hydrous oxide surface results from proton transfers at the surface. The surface OH groups represent σ-donor groups and are, like their counterparts in solution, able to form complexes with metal ions. The central ion acting as a Lewis acid can exchange its structural OH^- ion against other ligands such as anions or weak acids (ligand exchange) (Stumm et al., 1980; Sigg and Stumm, 1981). The concept of surface complex formation has been extensively documented in recent reviews (Schindler and Stumm, 1987; Sposito, 1989). The extent of surface coordination and its pH dependence can be quantified by mass action equations. The equilibrium constants are conditional stability constants - at constant temperature, pressure and ionic strength - the values of which can be corrected for electrostatic interaction. The nature of surface complexes and the nature of its bonding has far-reaching implications for the mechanism of interfacial processes and their kinetics.

Lyklema (1987) has recently proposed to separate adsorption phenomena on hydrous oxides into a specific part, determined by the chemical nature of the surface and a generic part, solely determined by the solution site. A simple example is given by Figure 2, illustrating the proton binding at various surfaces (Wieland et al., 1988). Different oxide surfaces, because of specific chemical interaction, have different affinities for protons. On the other hand if normal behaviour is fulfilled (the surface potential shifts by 59 mV for each pH unit) the pH-axis, corresponding to the potential axis, can be normalized with respect to the point of zero charge (ZPC=point of zero proton condition). If the concentration of the protons bound to the surface of the various oxides is plotted as a function of $\Delta pH = pH_{ZPC} - pH$ (Figure 2b), the curves describing the adsorption of protons become congruent; i.e. the same protonation curve describes various particle surfaces at constant ionic strength; this illustrates the generic property of the double layer which produces the same surface proton charge on different surfaces. As Lyklema (1987) points out the charge formation is determined from the solution side. On rutile and hematite, and possibly on other oxide surfaces the H^+-dependent surface charge is not only congruent on ΔpH but also temperature-congruent (Lyklema, 1987).

Figure 2

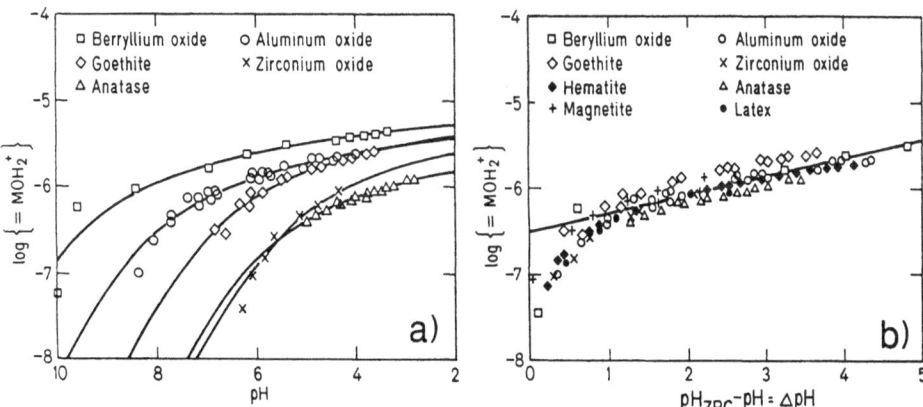

Figure 2: Surface protonation isotherm. Dots represent experimental data from titration curves at ionic strength I = 0.1 (Hematite, I = 0.2). The concentration of protonated sites $\{=MOH_2^+\}$ is given in moles m^{-2}. BET surface data were used to calculate the surface concentrations. (a) Frumkin isotherms. (b) Surface concentration as a function of $pH_{ZPC} - pH = \Delta pH$. The adsorption isotherm at $\Delta pH > 1$ can be interpreted as a Freundlich master isotherm (from Wieland, Wehrli and Stumm, 1988).

Figure 3

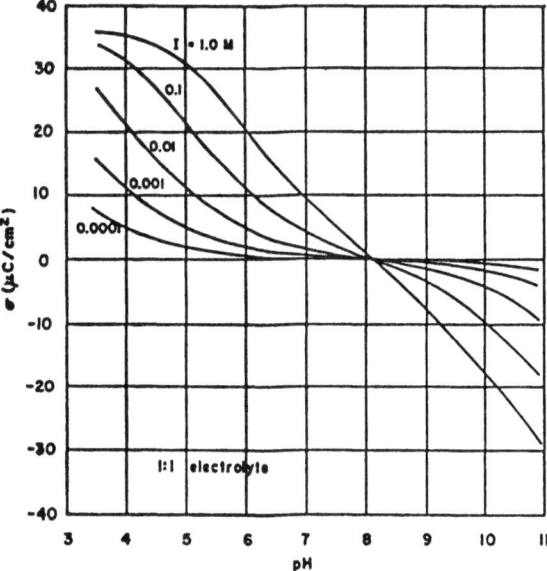

Figure 3: Predicted variation of surface charge density σ with pH and ionic strength (1:1 electrolyte) for cases in which surface charge is determined only by surface acid-base reactions (i.e. low cation/ anion sorption density). Example: Hydrous ferric oxide surface (from Dzombak and Morel, 1990).

pH$_{ZPC}$ reflects the affinity of the surface for protons. Its value can be estimated on the basis of a simple electrostatic model (Parks, 1967). Table 2 gives a few representative values for pH$_{ZPC}$.

Table 2: <u>ZERO POINT OF CHARGE</u> [a] (from Stumm and Morgan, 1981)

Material	pH$_{ZPC}$
α-Al$_2$O$_3$	9.1
α-Al(OH)$_3$	5.0
γ-AlOOH	8.2
CuO	9.5
Fe$_3$O$_4$	6.5
α-FeOOH	7.8
γ-Fe$_2$O$_3$	6.7
"Fe(OH)$_3$"(amorph)	8.5
MgO	12.4
δ-MnO$_2$	2.8
β-MnO$_2$	7.2
SiO$_2$	2.0
ZrSiO$_4$	5
Feldspars	2-2.4
Kaolinite	4.6
Montmorillonite	2.5
Albite	2.0
Chrysotile	>12

[a] The values are from different investigators who have used different methods and are not necessarily comparable. They are given here for illustration

The specific properties, the chemical interaction with the functional groups of the surface (Figure 1a) reflect the possibilities for the formation of a great diversity of surface complexes. Geometrical considerations and chemical measurements indicate an average surface density of 5 (typical range 2-12) hydroxyls per square nanometer of an oxide mineral. The various surface hydroxyls formed may not be structurally and chemically fully equivalent, but to facilitate the formulation of thermodynamic equilibria one usually considers the chemical reactions of "a" surface hydroxyl group S-OH. These functional groups contain the same donor atoms as found in functional groups of soluble ligands; e.g., the surface hydroxyl group on a hydrous oxide or an organic solid has similar donor properties as the corresponding counterparts in dissolved molecules such as hydroxide ions or carboxylates. We need to consider, however, that the functional groups are bound into a solid framework and their reactivity is, in essence, a cooperative property (Sposito, 1984). In a mean field approach, a hydrous oxide particle can be treated statistically like a polymeric oxoacid (or base) which tends to undergo protolysis and to coordinate with metal ions and ligands. An inner-sphere complex, as indicated in Figure 1a involves a chemical (often a covalent) bond; no H$_2$O molecule is interposed between the functional group or the central ion and the bond species. Outer-sphere surface complexes involve electrostatic binding mechanisms; at least one H_2O molecule is interposed between the complex partners.

The ideas developed here are largely based on the concept of the coordination at the (hydr)oxide interface; the ideas apply equally well to silicates. Somewhat modified concepts for the surface chemistry of carbonate, phosphate, sulfides and disulfide minerals can be developed.

The Coordination Chemistry of the Oxide-Water Interface

The presence of two lone electron pairs and a dissociable hydrogen ion at surface hydroxyl groups S-OH (we will also use >M-OH if the central ion is a metal ion) indicates that these groups are ampholytes (i.e., they can act as acids or bases). Adsorption of H^+ and OH^- ions is thus based on protonation and deprotonation of surface hydroxyls.

Energies of interaction of surface functional groups include electrostatic and chemical contributions. The selectivity of interaction of hydrous oxides and silicates with many solute species (metal ions and ligands) can be accounted for only by considering specific interactions (Stumm and Wollast, 1990).

$$S\text{-}OH + H^+ \rightleftharpoons S\text{-}OH_2^+ \tag{1}$$

$$S\text{-}OH(+OH^-) \rightleftharpoons S\text{-}O^- + H^+ (+H_2O) \tag{2}$$

Deprotonated surface hydroxyls exhibit Lewis base behavior; that is, the adsoprtion of metal ions is therefore understood as competitive complex formation involving one or two surface hydroxyls.

$$S\text{-}OH + M^{2+} \rightleftharpoons S\text{-}OM^{(z-1)+} + H^+ \tag{3}$$

$$2S\text{-}OH + M^{2+} \rightleftharpoons (S\text{-}OH)_2M^{(z-2)+} + 2H^+ \tag{4}$$

The central ion, acting as a Lewis acid (i.e., a metal center that can bind covalently with a ligand), can exchange its structural OH^- ions against other ligands (complex-forming anions or weak acids; ligand exchange) (Sigg and Stumm, 1981):

$$S\text{-}OH + L \rightleftharpoons S\text{-}L^+ + OH^- \tag{5}$$

$$2S\text{-}OH + L \rightleftharpoons S_2L^{2+} + 2OH^- \tag{6}$$

For the case where L is a polydentate ligand, under certain circumstances ternary surface complexes may be formed:

$$S\text{-}OH + L \rightleftharpoons S\text{-}L\text{-}M^{(z-1)+} + OH^- \tag{7}$$

EQUILIBRIUM CONSTANTS

The concept of surface complexation permits us to handle adsorption equilibria in the same way as equilibria in solutions. Hence uptake and release of H^+ ions in solutions of constant ionic strength (equations (1) and (2)) can be described by the acidity constants

$$K_{a1}^{s} = \frac{\{SOH\}[H^+]}{\{SOH_2^+\}} \text{ (mol dm}^{-3}) \tag{8}$$

$$K_{a2}^{s} = \frac{\{SO^-\}[H^+]}{\{SOH\}} \text{ (mol dm}^{-3}) \tag{9}$$

where braces denote the concentrations of surface species in moles per square meter of adsorbing solid. (Frequently, the concentration of surface species is given in moles per kilogram of adsorbing solids. Conversion into surface densities is easily accomplished if the specific surface area of the adsorbing solid is known.)

The quotients K_{a1}^{s}, K_{a2}^{s} introduced above, and similar expressions related to the metal complexation and to ligand exchange, e.g.

$$K_{M}^{s} = \frac{\{S\text{-}OM^+\}[H^+]}{\{S\text{-}OH\}[M^{2+}]} \tag{10}$$

$$K_{L}^{s} = \frac{\{S\text{-}L\}[OH^-]}{\{S\text{-}OH\}[L^-]} \tag{11}$$

are experimentally accessible. These constants have the rank of conditional stability constants. For exact considerations we need to correct by a coulombic term for electrostatic interaction, because the surface is charged. This coulombic term is in fact a surface activity coefficient which is calculated from the Gouy Chapman theory of the electric double layer. Before considering this correction, we may illustrate the coordinating interactions given and the corresponding mass law expressions (Figures 4-5). The mass law expressions permit to calculate the extent of adsorption (or surface speciation) and its pH-dependence. For every given pH, the equilibrium expression permits to calculate an adsorption isotherm (surface bound species, i.e. sorption density vs residual (equilibrium) concentration of adsorbate). There is no difference between a Langmuir equation and the simple mass law expression given for surface complexation, e.g., for (10)

$$\frac{K_{M}^{s}}{[H^+]} = K_{M}^{s} \text{ (pH)} = \frac{\{S\text{-}OM^+\}}{\{S\text{-}OH\}[M^{2+}]} = K_{ads} = \exp\left(\frac{\Delta G}{RT}\right) \tag{12}$$

where K_{M}^{s} (pH) is the conditional surface complex formation (or adsorption) constant valid for a given pH.

Figure 4

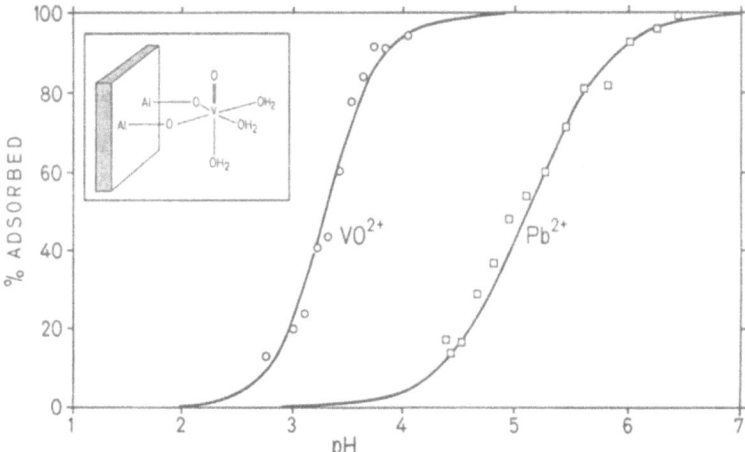

Figure 4: Adsorption of VO^{2+} to Al_2O_3 (10g/l oxide, $[V]_T = 2.5 \times 10^{-4}$ M) [from Wehrli and Stumm, 1988] and of Pb^{2+} at similar conditions (Hohl and Stumm, 1976). One may note that the binding of these cations occurs on a positively charged surface; obviously the chemical coordinative interaction outweighs electrostatic repulsion. The inset presents a postulated structure for the vanadyl surface complex (Motschi and Rudin, 1984).

Assuming a total constant number of surface sites

$$\{S_T\} = \{S\text{-}OH\} + \{S\text{-}OM^+\} \tag{13}$$

we obtain

$$\{S\text{-}OM\} = \{S_T\} \ \frac{K_{ads}\ [M^{2+}]}{1 + K_{ads}[M^{2+}]} \tag{14}$$

if we use the terms, generally used in the Langmuir equation, $\{SO, M\}$ corresponds to Γ_M, the surface density, and $\{S_T\}$ is equivalent to $\Gamma_{M(max)}$, the maximum surface density.

$$\Gamma_M = \Gamma_{M(max)} \ \frac{K_{ads}\ [M^{2+}]}{1 + K_{ads}\ [M^{2+}]} \tag{15a}$$

One may note that K_{ads} corresponds to the surface complex formation equilibrium constant valid for a given pH. A fit of equation (15) is often obtained by plotting it as a linear function in the reciprocal form (Γ_M^{-1} vs $[M^{2+}]^{-1}$) which is particularly convenient to obtain from experimental data K_{ads} and Γ_{Max} or S_T.

$$\Gamma_M^{-1} = \Gamma_{M\ (max)}^{-1} + \Gamma_{M\ (max)}^{-1}\ K_{ads}^{-1}\ [M^{2+}]^{-1} \tag{15b}$$

452

Figure 5

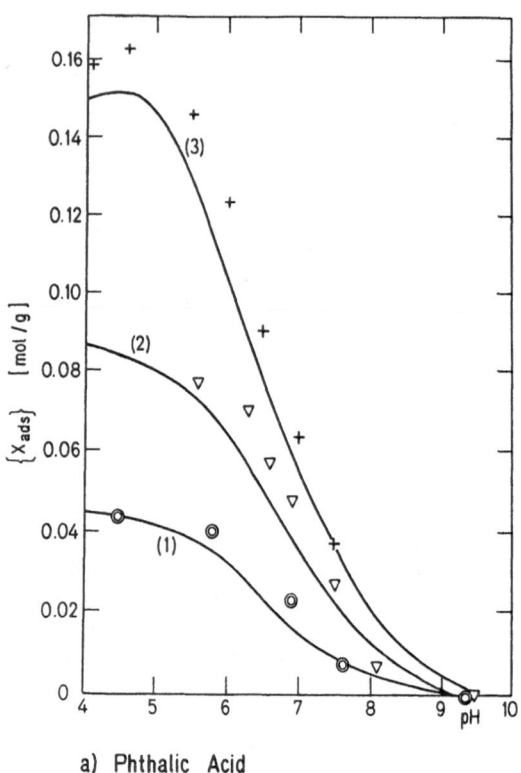

a) Phthalic Acid

Figure 5: pH dependence of adsorption of phthalic acid on δ-Al$_2$O$_3$. Concentration of δ-Al$_2$O$_3$ was 2.21 g/l. Total phthalic acid concentrations from top to the bottom are 4×10^{-4} M, 2×10^{-4} M, 1×10^{-4} M, respectively. Curves are calculated from surface complex formation equilibrium constants, points are experimental.

Linear free energy relations. Treating surface complex formation in a similar way as complex formation in solution is justified because the energy of interaction of metal ion with a oxygen donor atom of a deprotonated surface hydroxo-group is of similar tendency as with a deprotonated hydroxo-group of a solute ligand.

$$Cu^{2+} + HOH \rightleftarrows Cu(OH)^+ + H^+; *K_1$$

$$Cu^{2+} + \equiv AlOH \rightleftarrows \equiv AlOCu^+ + H^+; *K_1^s$$

Similarly for ligand exchange we may compare the reactions

$$Fe(OH)^{2+} + HL \rightleftarrows FeL + H_2O ; K_{FeL}$$

$$\equiv Fe(OH) + HL \rightleftarrows \equiv FeL^+ + H_2O ; K_{FeL}^s$$

Linear free energy relationships can be established between surface complex formation constants and corresponding constants in solution and can be used to predict surface equilibrium constants from solute data (Sigg and Stumm, 1981; Schindler and Stumm, 1987; Dzombak and Morel, 1990).

THE COULOMBIC CORRECTION FACTOR

Returning to Eq. (10) we can consider the electrostatic interaction of the metal ion M^{2+} with the surface in the following way

$$\frac{\{S\text{-}OM^+\}\,[H^+]}{\{S\text{-}OH\}\,[M^{2+}]} = K_M^s = K_{M(intr)}^s \exp(-zF\psi/RT) \qquad (16)^*$$

In other words K_M^s the <u>apparent</u> constant is a product of an <u>intrinsic</u> constant, $K_{(intr)}$, (a constant valid for a hypothetical uncharged surface) and a Boltzman factor. ψ is the surface potential, F the Faraday and z=charge of the surface complex. $K_{(intr)}^s$ is experimentally accessible by extrapolating experimental data to a surface charge $\sigma_P = 0$ where $\psi = 0$. (σ_P is the net total particle charge; for definition see Eq. (20)

The correction as applied in (16) (Stumm et al., 1970) assumes the classical diffuse double layer model (a planar surface layer and a diffuse layer) of counter ions. The constant capacitance model (Schindler and Kamber, 1968; Hohl et al., 1980; Schindler and Stumm, 1987) is characterized by a planar surface layer and a plane of counter ions; this double layer model is especially applicable at relatively high ionic strength (because under these conditions the surface charge σ_P is directly proportional to the surface potential $\sigma_P = C\psi$, where C = integral capacitance of the double layer). The correction factor of the constant capacitance model is

$$K_M^s = K_{M(intr.)}^s \exp(-zF\sigma_P/CRT) \qquad (17)$$

(here σ_P and C have the units, respectively, coulomb m^{-2} and coulombs $Volt^{-1}\,m^{-2}$).
The two models (diffuse double layer and constant capacitance) are closely related and do not differ significantly. They both fit the experimental data equally well.

The <u>surface charge</u> is experimentally accessible (e.g. by measuring cations H^+ and OH^- and anions that have been bound to the surface), e.g., in case of adsorption of M^{2+}:

$$Q_P = (\{S\text{-}OH_2^+\} + \{SO\text{-}M^+\} - \{S\text{-}O^-\}) \qquad (18)$$

where Q_P is the surface charge accumulated at the interface in moles kg^{-1}. Q_P can be converted in σ_P (coulombs m^{-2}) ($\sigma_P = QF/s$) where s is the specific surface area in $m^{-2}\,kg^{-1}$.
If sufficient amount of M^{2+} sorbs, the surface will have a more positive charge than if proton exchange reactions alone were governing the surface charge.
The charge can be introduced into Eq. (15) or we can estimate ψ from σ on the basis of Gouy Chapman diffuse layer model

$$\text{at } 25^\circ C \quad \sigma \cong 0.1174 \sqrt{c} \ (\sinh Z\psi \times 19.46), \text{ or} \qquad (19)$$

$$\sigma \approx 2.5 \sqrt{I}\,\psi \qquad (19a)$$

---

* Equation (16) corresponds to a subdivision of the total free energy of adsorption, ΔG_{tot}^o, into a chemical and electrical contribution: $\Delta G_{tot}^o = \Delta G_{(intr.)}^o + zF\psi$.

where σ has the units coulomb m^{-2}, ψ is in volt, c the electrolyte concentration, and I the ionic strength in M. The value of ψ calculated from (19) is used in (16). Computer programs are available to make these equlibrium calculations routinely. For details see Dzombak and Morel, 1990.

The Stern layer. The number of functional groups (binding sites) limits the amount of surface charge that can be developed on an oxide surface. As has been pointed out by Dzombak and Morel (1990) this constraint is not inherently part of the Gouy-Chapman theory. The modifications of this theory proposed by Stern and Grahame, who considered specific sorption at the surface of the Hg electrode were developed in part to avoid predicting physically impossible ion densities very close to the surface. Hence, consideration of surface complexation reactions represents a modification of the Gouy-Chapman theory similar to that introduced by Stern and Grahame.

The following concepts are characteristic for all surface complexation models (Dzombak and Morel, 1990):
 i) Sorption on oxides takes place at specific coordination sites;
 ii) sorption reactions (on oxide surfaces) can be described by mass law equations;
 iii) surface charge results from the sorption reaction itself; and
 iv) the effect of surface charge on sorption can be taken into account by applying a correction factor derived from the electric double layer theory to the mass law constants for surface reactions.

SURFACE CHARGE

Solid particle surfaces can develop electrical charge in two principal ways:
either from isomorphic substitutions in minerals or from the reactions of surface functional groups with ions in aqueous solution (Eq.18). Thus different type of surface charge contribute to the net total particle charge on a colloid, denoted σ_P.

$$\sigma_P = \sigma_O + \sigma_H + \sigma_{IS} + \sigma_{OS} \tag{20}$$

where σ_P = total net surface charge

$\quad \sigma_O$ = permanent structural charge (usually for a mineral) caused by isomorphic substitutions in minerals. Significant charge is produced primarily in the 2:1 phyllosilicates.

$\quad \sigma_H$ = net proton charge, i.e., the charge due to the binding of protons or the binding of OH^- ions (equivalent to the dissociation of H^+). Protons in the diffuse layer are not included in σ_H

$\quad \sigma_{IS}$ = inner-sphere complex charge,

$\quad \sigma_{OS}$ = outer-sphere complex charge.

The unit of σ is usually coulomb m^{-2}, it can be derived from mol kg^{-1} or mol m^{-2} (1 mol of charge units equals 1 Faraday or 96490 coulombs).

The mechanism of sorption of some ions e.g., fulvates or humates, is not known. Ionic species carrying a hydrophobic moiety may bind inner-spherically or outer-spherically depending whether the surface-coordinative or the hydrophobic interaction prevails.

Although aquatic particles may bear electric charge, this charge is balanced by the charges in the diffuse swarm which move about freely in solution while remaining near enough to colloid surfaces to create the effective (counter) charge σ_D that balances σ_P

$$\sigma_P + \sigma_D = 0 \tag{21}$$

The following <u>points of zero charge</u> can be distinguished:

PZC: Point of zero charge: $\sigma_P = 0$

This is often referred to as isoelectric point. It is the condition where particles do not move in an applied electric field.

PZPC: Point of zero proton condition: $\sigma_H = 0$

This is often referred to in connection with pH as a variable as pH_{ZPC} (pH of the zero point of charge; it is preferable to speak of the pH of zero net proton charge or of zero proton condition).

STRUCTURAL IDENTITY

Unfortunately, spectroscopic methods are seldom sufficiently sensitive to reveal the structure of surface complexes. Motschi (1987) used electron spin resonance spectroscopy to study Cu(II) surface complexes. Additional studies were carried out with electron nuclear double resonance spectroscopy (ENDOR) and electron spin echo envelope modulation (ESEEM) in order to elucidate structural aspects of surface-bound Cu(II), of ternary copper complexes (in which coordinated water is replaced by ligands), and of vanadyl ions on δ-Al_2O_3. Application of ENDOR spectroscopy allows the resolution of weak interactions between the unpaired electron and nuclei within a distance of about 5 Å. From these so-called hyperfine data, structural parameters can be derived, e.g., bond distances of the paramagnetic center to the coupling nuclei or ligands. In the ENDOR spectrum of adsorbed VO^{2+} on δ-Al_2O_3, signals caused by the coupling with the surface Lewis center (^{27}Al) are more strongly split than is calculated from molecular modeling. The existence of an inner-sphere coordination between the hydrated oxide and the metal is confirmed experientially (Motschi, 1987).

Similarly, studies by Zeltner et al. (1986) with cylindrical internal reflection Fourier transform infrared spectroscopy suggest that salicylate adsorbs on goethite by forming a chelate structure in which each salicylate ion replaces two hydroxyls attached to a single ion atom at the surface.

Direct in situ X ray (from synchroton radiation) adsorption measurements (EXAFS) (Hayes et al., 1987; Brown et al., 1989) permit the determination of adsorbed species to neighboring ions and to central ions on oxide surfaces in the presence of water. Such investigations showed, for example, that selenite is inner-spherically and selenate is outer-spherically bound to the central Fe(III) ions of a goethite surface. It was also shown by this technique that Pb(II) is inner-spherically bound to δ-Al_2O_3 (Chisholm-Brause et al., 1989).

Structure and Reactivity at Solid-Solution Interfaces

It is not sufficient to describe thermodynamically how solutes are distributed between the surface and the solution. We need to know the structural identity of the surface species. Many heterogeneous processes (formation and dissolution of solid phases, redox and photochemical processes) at the solid-water interface are kinetically controlled by a reaction step at the surface (and not by a transport step).

Obviously the surface reactivity depends on the surface species and their structural identity, which – in turn – depend on the coordination chemical interactions that occur at the solid water interface, i.e., the geometry of the coordination shell of surface sites and of reactants at surfaces need to be known.

From thermodynamic data alone we cannot unequivocally predict the nature of surface species. Such data together with information on linear-free energy comparisons between solute and surface complex stability constants (Schindler and Stumm, 1987) or ionic strength dependence of these constants (Hayes and Leckie, 1987), and correlations with reactivity usually permit us to assess whether inner-sphere or outer-sphere complexes prevail and may give some hints whether monodentate or bidentate, mononuclear or binuclear complexes are formed.

A GENERAL RATE LAW FOR SURFACE-CONTROLLED DISSOLUTION

We would like to provide the reader first with a qualitative understanding of the subject of dissolution kinetics. If the reactions at the surface are slow in comparison with diffusion or other reaction steps, the dissolution processes are controlled by the processes at the surface. In this case the concentrations of solutes adjacent to the surface will be the same as in the bulk solution. The dissolution kinetics follows a zero-order rate law if the steady state conditions at the surface prevail:

$$r = \frac{dC}{dt} = ka \ [M \ s^{-1}] \tag{22}$$

where the dissolution rate $r \ [M \ s^{-1}]$ is proportional to the surface area of the mineral, $a \ [m^2]$; k is the reaction rate constant $[M \ m^{-2} \ s^{-1}]$.

In the dissolution reaction of an oxide mineral, the coordinative environment of the metal changes; for example, in dissolving an aluminum oxide layer, the Al^{3+} in the crystalline lattice exchanges its O^{2-} ligand for H_2O or another ligand L. The most important reactants participating in the dissolution of a solid mineral are H_2O, H^+, OH^-, ligands (surface complex building), and reductants and oxidants (in the case of reducible or oxidizable minerals)

Thus the reaction occurs schematically in two sequences:

$$\text{surface sites + reactants } (H^+, OH^-, \text{ or ligands}) \xrightarrow{\text{fast}} \text{surface species} \tag{23a}$$

$$\text{surface sepcies} \xrightarrow[\text{detachment of Me}]{\text{slow}} Me(aq) \tag{23b}$$

where Me stands for metal. Although each sequence may consist of a series of smaller reaction steps, the rate law of surface-controlled dissolution is based on the idea (1) that the attachment of reactants to the surface sites is fast and (2) that the subsequent detachment of the metal species from the surface of the crystalline lattice into the solution is slow and thus rate limiting. In the first sequence the dissolution reaction is initiated by the surface coordination with H^+, $OH^{-,}$ and ligand which polarize, weaken, and tend to break the metal-oxygen bonds in the lattice of the surface. Since reaction (23b) is is rate-limiting , the rate law on the dissolution reaction will show a dependence on the concentration (activity) of the particular surface species, $C_j \ [mol \ m^{-2}]$:

$$\text{dissolution rate} \propto \{\text{surface species}\} \tag{23c}$$

We reach the same conclusion (equation 23c) if we treat the reaction sequence according to the activated complex theory (ACT), often also called the transition state theory. The particular surface species that has formed from the interaction of H^+, OH^-, or ligands with surface sites is the precursor of the activated complex (Figure 6):

$$\text{dissolution rate} \propto \{\text{precursor of the activated complex}\} \qquad (23d)$$

Figure 6:

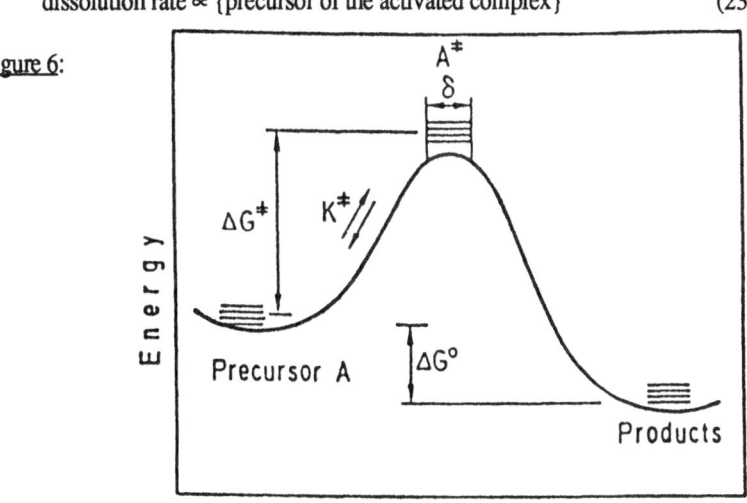

Figure 6: Activated complex theory for the surface-controlled dissolution of a mineral far from equilibrium. A is the precursor, i.e., a surface site that can be activated to $A^{\ddagger}$.

The surface concentration of the particular surface species, C_j, corresponds to the concentration of the precursor of the activated complex. Note that we use braces and brackets to indicate surface concentrations [mol m^{-2}] and solute concentrations [M], respectively. Equation (23b) can usually be determined from the knowledge of the number of surface sites and the extent of surface protonation or surface deprotonation or the surface concentration of ligands. Surface protonation or deprotonation can be measured from alkalimetric or acidimetric surface titrations, and ligands bound to the surface sites can be determined analytically, from the change in the concentration of ligands in solution. Most generally, the dissolution rate R [mol m^{-2} s^{-1}] is, in line with (23c) and (23d), proportional to the surface concentration (activity) of the precursor species [mol m^{-2}]:

$$R = kC_j \qquad (23e)$$

C_j is proportional to the density of surface sites, S, the mole fraction χ_a of these surface sites which are dissolution active, and the probability P_j of finding a site in the suitable coordinative arrangement of the precursor complex:

$$C_j = \chi_a P_j S \qquad (24)$$

Thus the dissolution rate can be generalized (Wieland et al., 1988) into

$$R = k\chi_a P_j S \qquad (25)$$

where   R   dissolution rate[mol $m^{-2} s^{-1}$];

         k   appropriate rate constant $[s^{-1}]$;

         S   surface concentration of sites [mol m^{-2}];

         χ_a mole fraction of dissolution active sites;

         P_j probability of finding a specific site in the coordinative
                arrangement of the precursor complex

We shall provide more information on the chemistry of the mineral-water interface to describe χ_a and P_j and to interpret k. But it may be convenient first to give rate laws for the ligand-promoted and the proton-promoted dissolution rate.

The ligand-promoted dissolution rate R_L is proportional to the density of sites occupied by adsorbed (surface bound) ligands, Pj $\propto$ {>ML}

$$R_L = k_L' \{>ML\} \tag{26}$$

The acid-promoted (proton-promoted) dissolution rate R_H is proportional to the density of protonated sites to the power of j, $P_j \propto \{>MOH_2^+\}^j$:

$$R_H = k_H' \{>MOH_2^+\}j = k'(C_H^S)^j \tag{27}$$

where {>ML} is the concentration of the surface complex of the ligand L, C_L^S [mol m^{-2}], {>MOH$_2^+$} is the concentration of protonated surface hydroxo group, ($C_H^S = (C_H^S$ [mol m^{-2}], which is equivalent to the surface protonation; and j is the exponent, which in ideal cases corresponds to the oxidation state of the central metal ion in the crytalline lattice (e.g., j= 3 for Al and Fe(III); j= 2 for Be(II) and Mg(II); j= 4 for Si(IV).

The densitiy of surface sites occupied with L, {>ML}, or with protons, {>MOH$_2^+$}, is related to the concentration of L in solution, [L], by a complex formation equilibrium) or by a surface protonation equilibrium, respectively. The rate laws should preferably be written in terms of surface species. Since the concentration of surface species at equilibrium is usually not linearly dependent on the concentration of the species in solution, fractional order dependence of solution rates on [L] and [H$^+$] result if rate laws are formulated in terms of solutes:

$$R_L \propto [L]^m \text{ or } RH \propto [H^+]^n$$

where m<1 and n<1.

SURFACE REACTIVITY

As Figure 7 illustrates schematically, a high degree of surface protonation of the surface ligands accelerates the dissolution because it leads to highly polarized interatomic bonds in the immediate proximity of the surface central ions and thus facilitates the detachment of a cationic surface group into the solution. Similarly, binding of OH$^-$ to the surface groups (which experimentally cannot be distinguished from deprotonation of the surface ligands -SO(OH) $\rightarrow$ -SO$^-$) occurring at higher pH values facilitates the detachment of an anionic surface group into the solution.

Ligands that form surface complexes by ligand exchange with surface hydroxyl groups bring negative charge into the coordination sphere of the Lewis acid center of the hydrous oxide surface and can polarize the critical Me-oxygen bonds, thus enabling the detachment of the central ion into the adjacvent solution. Most effective in enhancing the dissolution rate are bidentate ligands that form mononuclear complexes. The same ligands that form surface complexes usually also form complexes with these ions (e.g., Al(III), Fe(III), Mg(II) in solution and thus increase the solubility and,in turn, the activity gradient at the mineral solution interface. This enhancement of the solubility alone, however, has no effect on the dissolution rate if the dissolution is surface-controlled.

Blocking of surface functional groups by cations, e.g., VO^{2+}, Cr(III), and Al(III), tends to retard dissolution. Inhibitory effects may also result from binuclear or multinuclear binding of ligands (Figure 7). Obviously, the formation of surface films (heterogeneous nucleation, surface precipitation) and subsequent phase transformations at the surface as it may, for example, occur with phosphates, or silicate on Fe(III) (hydr)oxide surfaces, modifies surface reactivity. These effects are also of significance in metall corrosion (passivity of oxide films).

Figure 7

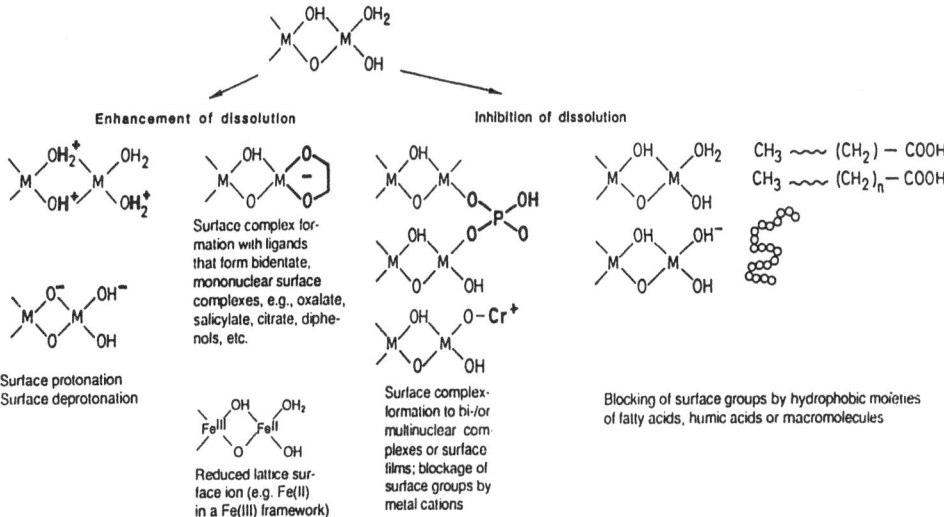

Figure 7: Effect of protonation, complex formation with ligands and metal ions and reduction on dissolution rate. The structures given here are schematic short hand notations to illustrate the principal features they do not reveal the structural properties nor the coordination numbers of the oxides under consideration; charges given are relative).

DISSOLUTION SCENARIOS

We will first describe a relatively simple scenario for the enhancement of the dissolution of Al_2O by a complex-forming ligand. As we have seen ligands tend to become adsorbed specifically and to form surface complexes with the Al(III) Lewis acid centers of the hydrous oxide surface. They also usually form complexes with Al(III) in solution. Complex formation in solution increases the solubility. This has no direct effect on the dissolution rate, however, since the dissolution is surface-controlled.

460

The enhancement of the dissolution rate by a ligand in a surface-controlled reaction implies that surface complex formation facilitates the release of ions from the surface to the adjacent solution. These ligands bring negative charge into the coordination sphere of the surface Al species, lowering their Lewis acidity. This may polarize the critical Al-oxygen bonds, thus facilitating the detachment of the Al from the surface. It has been shown that bidentate ligands (i.e., ligands with two donor atoms) such as dicarboxylates and hydroxy-carboxylates, e.g., oxalate (see Figure 8a), can form relatively strong surface chelates, i.e., ring-type surface complexes.

In Figure 8 a simple scheme of reaction steps is proposed. Some of the assumptions of our model are summarized in Table 3. The short-hand representation of a surface site is a simplification that does not take into account either detailed structural

Figure 8

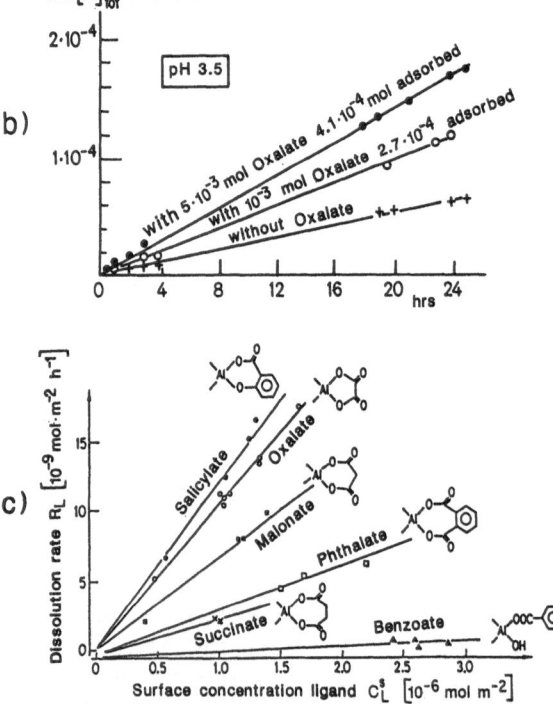

Figure 8: (a) The ligand-catalyzed dissolution reaction of a M_2O_3 can be described by three elementary steps: a fast ligand adsorption step (equilibrium), a slow detachment process, and fast protonation

subsequent to detachment restoring the incipient surface configuration. (b) In accordance with the reaction scheme of (a) the rate of ligand-catalyzed dissolution of δ-Al_2O_3 by the aliphatic ligands oxalate, citrate, and succinate, R_L (nmol m^{-2} h^{-1}), can be interpreted as a linear dependence on the surface concentrations of chelate complexes C_L^s (from Furrer and Stumm, 1986).

aspects of the oxide surface or the oxidation state of the metal ion and its coordination number. It implies (model assumption 2 in Table 3) that all functional surface groups, such as those in a cross-linked polyhydroxo-oxo acid, are identical.

The scheme in Figure 8 indicates that the ligand, for example, oxalate, is adsorbed very fast in comparison to the dissolution reaction (Hachiya et al., 1984; Ikeda et al., 1982); thus adsorption equilibrium may be assumed. The surface chelate formed is able to weaken the original Al-oxygen bonds on the surface of the crystal lattice. The detachment of the oxalato-aluminum species is the slow and rate-determining step. The negative charge of the surface site after detachment is neutralized by two subsequent fast protonation steps. When initial sites are regenerated completely after the detachment step and provided that the concentrations of the reactants are kept constant, steady state conditions with regard to the oxide surface species are established (Table 3). If, furthermore, the system is far from dissolution equilibrium, the back reaction can be neglected, and constant dissolution rates occur.

The scheme of Figure 8a corresponds to steady state conditions (Table 3). We can now apply the general rate law (equation (23b), the rate of the ligand-promoted dissolution R_L, is proportional to the concentration of surface sites occupied by L (metal-ligand complex, >ML) or to the surface concentration of ligands, C_L^s (mol m^{-2}); i.e., P_j in (24) is proportional to the surface area concentration of L,

$$R_L = k_L' \{>ML\} = k_L' \, C_L^s \qquad (28)$$

where k_L' is the reaction constant. C_L^s, the surface concentration of the ligand (oxalate) [mol m^{-2}], was determined experimentally by generalizing the quantity of oxalate that was removed from the solution by adsorption and the specific surface area of the Al_2O_3. The relationship between surface concentration and solute concentration is also obtained from the ligand exchange equilibrium constant (Sigg and Stumm, 1981). As shown in Figure 8b, the experimental results are in accord with equation (28)

However, we have to reflect on one of our model assumptions (Table 3). It is certainly not justified to assume a completely uniform oxide surface. The dissolution is favored at a few localized (active) sites where the reactions have lower activation energy. The overall reaction rate is the sum of the rates of the various types of sites. The reactions occurring at differently active sites are parallel reaction steps occurring at different rates (Table 3). In parallel reactions the fast reaction is rate determining (Lasaga, 1981). We can assume that the ration (mole fraction, χ_a) of active sites to total (active plus less active) sites remains constant during the dissolution; that is, the active sites are continuously regenerated after Al(III) detachment, and thus steady state conditioins are maintained. The reaction constant k_L' in equation (28) includes χ_a, which is a function of the particular material used (cf. (25)). In the activated complex theory the surface complex is the precursor of the activated complex (Figure 6) and is in local equilibrium with it. The detachment corresponds to the desorption of the activated surface complex.

TABLE 3: Model Assumptions (from Stumm and Wieland, 1990)

1. Dissolution of slightly soluble hydroxous oxides:
     Surface process is rate-limiting
     Back reactions can be neglected if far away from equilibrium
2. The hydrous oxide surface, as a first approximation, is treated like a cross-linked polyhydroxo-oxo acid:
     All functional groups are identical
3. Steady state of surface phase:
     Constancy of surface area
     Regeneration of active surface sites
4. Surface defects, such as steps, kinks, and pits, establish surface sites of different activation energy, with different rates of reaction:

     Active sites $\xrightarrow{\text{faster}}$ Me(aq)   (a)

     Less active sites $\xrightarrow{\text{slower}}$ Me(aq)   (b)

     Overall rate is given by (a):
          Steady-state condition can be maintained if a constant mole fraction, χ_a, of active sites to total (active and less active) sites is maintained, i.e., if active sites are continuously regenerated
5. Precursor of activated complex:
     Metal centers bound to surface chelate, or surrounded by n protonated functional groups
     $(C_H^S/S) \ll 1$

THE PROTON-PROMOTED DISSOLUTION REACTION

The dissolution reaction under acid conditions requires protons, which get bound to the surface oxide ions and weaken critical bonds; thus detachment of the metal species into the solution results. Another part of the consumed protons replaces the metal ions, leaving the solid surface and thus maintaining the charge balance.

A scheme for the dissolution reaction of a trivalent oxide is given in Figure 9 Although this representation cannot account for individual crystallographic structures, it attempts to illustrate a typical sequence of the reaction steps which occur on the surface. The adsorption of protons at the surface is very fast (Hachiya et al., 1984); thus surface protonation is faster than the detachment of the metal species, so it can be assumed that the concentration of protons at the surface is in equilibrium with the solution. Surface protonation may be assumed to occur at random. But the protons may move fast from one functional group to another and occupy terminal hydroxyl as well as bridging oxo or hydroxo groups. Tautomeric equilibria may be assumed. The detachment process (step 4 in Figure 9) is (far from equilibrium) the slowest of the consecutive steps. If the steady state conditions are maintained, the following equation describes the overall reaction rate:

$$R_H = k_4(D) \qquad\qquad (29)$$

Under steady state conditions, i.e., if the original surface sites are regenerated completely after the detachment step (Table 3) and if it is assumed that surface protonation equilibria are retained and kept constant by controlling the solution pH, one may write

$$d\{A\}/dt = d\{B\}/dt = d\{C\}/dt = d\{D\}/dt = 0$$

If the fraction of surface that is covered with protons is smaller than 1 ($\chi_H \ll 1$), the surface density of singly, doubly, and triply protonated surface sites (B,C, and D, respectively (see Figure 9)) can be described as probability functions of the surface protonation C_H^s.

Figure 9

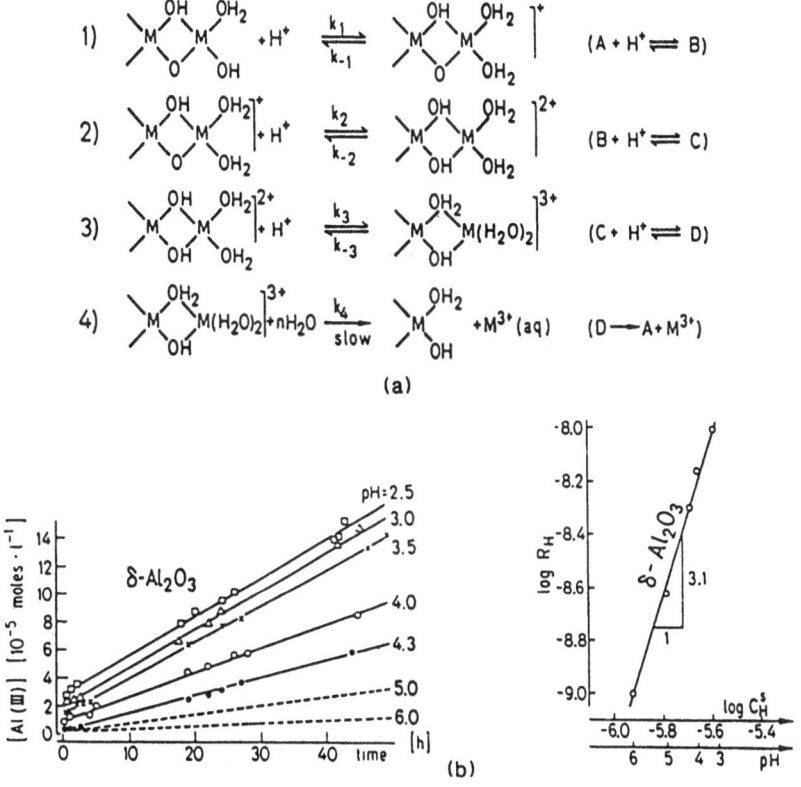

(a)

(b)

Figure 9: (a) Schematic representation of the proton-promoted dissolution process at a M_2O_3 surface site. Three preceding fast protonation steps are followed by a slow detachment of the metal from the lattice surface. (b) The reaction rate derived from indivi-dual experiments is proportional to the surface protonation to the third power (from Furrer and Stumm, 1986).

A weakening of the critical metal-oxygen bonds occurs as a consequence of the protonation of the oxide ions neighboring a surface metal center. The concentration (activity) of D should reflect that three of such oxide or hydroxide ions have to be protonated. If there is a certain number of surface-adsorbed (bound) protons whose concentration C_H^s (mol m^{-2}) is much lower than the density of surface sites, S (mol m^{-2}), the probability of finding a metal center surrounded with three protonated oxide or hydroxide ions is proportional to $(C_H^S/S)^3$. Thus, as has been derived from lattice statistics by Wieland et al. (1988), the activity of D is related to $(C_H^S)^3$, and the rate of proton-promoted dissolution, R_H (mol m^{-2} h^{-1}), is proportional to the third power of the surface protonation:

$$R_H \propto \{D\} \propto (C_H^S)^3$$

that is,

$$R_H = k_H' \{>MeOH_2^+\}^3 \tag{30a}$$

For another oxide, for which the dissolution mechanism requires only two preceding protonation steps, the rate would be proportional to $(C_H^S)^2$. Generally, for all oxides, which dissolve by an acid-promoted surface controlled process, the following rate equation may be postulated

$$R_H \propto \{MeOH_2^+\}^j \tag{30b}$$

where j is an integer if dissolution occurs by one mechanism only. If more than one mechanism occur simultaneously, the exponent j will not be an integer.

ENHANCEMENT OF DISSOLUTION BY DEPROTONATION

The dissolution rate of most oxides increases both with increasing surface protonation and with decreasing surface deprotonation, equivalent to the binding of OH$^-$ ligands; thus in the alkaline range the dissolution rate increases with increasing pH (Chou and Wollast, 1984; Schott, 1990; Brady and Walther, 1990).

EXPERIMENTALLY ACCESSIBLE PARAMETERS

Surface protonation and deprotonation are experimentally directly accessible from alkali-metric or acidimetric surface titrations. The surface concentrations $\{>MOH_2^+\}$ or $\{>MO^-\}$ are nonlinearly related to H$^+$ by surfacce complex formation equilibria or by semi-empirical relations; in other words,

$$\{>MOH_2^+\} \propto [H^+]^n \qquad \{>MO-\} \propto [H^+]^{-m} \tag{31}$$

Many authors have shown the empirical rate law

$$R_H = k_H [H^+]^n \tag{32}$$

where n is typically between 0 and 0.5. These observations are relevant for the assessment of the impact of acid rain on weathering rates. As equation (32) suggests, the rate of weathering does not increase linearly with the acidity of water. A tenfold increase in [H$^+$] leads to an increase in the dissolution rate by a factor of about 2-3. Similarly, in the alkaline range the empirical rate law holds:

$$R_{OH} = k_{OH} [H^+]^{-m} \tag{33}$$

ACTIVATION ENERGY OF THE RATE-DETERMINING STEP

The activated complex theory provides a model to bridge the gap between thermodynamic information (surface coordination and lattice or site energy) and kinetic information. The rate constant k is related to the free energy of conversion ($\Delta G^0 \ddagger$ or $\Delta H^0 \ddagger$) of a suitable surface complex (precursor to an activated surface complex) (Figure 6); this energy can be compared according to the Arrhenius theory with the activation energy E_a of the rate-determining step. The fact that for many minerals a consistent relationship was found between the apparent activation energy E_{app} (from the temperature dependence of the overall dissolution rate) and the dissolution rate (Wieland et al., 1988) lends support to the inference that E_{app} is often a good approximation of E_a. Thus one may postulate that the rate constant k_H is mainly a measure of the metal-oxygen bond broken in the rate-determining step of a detachment reaction.

Redox Processes On Oxide Surfaces

Surface ligands can catalyze redox reactions (Wehrli et al., 1989). Surface OH^- groups, as we have seen, act as σ-donor ligands which increase the electron densitiy at the metal center. Higher oxidation states will generally form more stable surface complexes, i.e., the specific adsorption of Cu(II) should lower its redox potential with regard to $Cu(I)_{(aq)}$. Coordinated donor ligands accelerate the ligand exchange kinetics especially in the transposition ("trans effect"). Bridged surface ligands may mediate electron transfer; otherwise slow electron transfer reactions between metal ions can occur rapidly if the redox couple is linked by a bridging ligand. The most effective bridging ligand in pure aqueous solutions is OH^-. Electron transfer between hexaquoions probably occurs via a slow o.s. reaction. Increasing the pH accelerates many exchange reactions between transition-metal ions. Similar to dissolved OH^- groups at the mineral-water interface may act as bridging ligand between the redox couples. Hydrous oxide surfaces can catalyze the photooxidation of chemically adsorbed organic compounds.

We restrict this discussion to an exemplification of reductive dissolution of Fe(III) (hydr)oxides.

REDUCTIVE AND CATALYTIC DISSOLUTION OF FE(III) (HYDR)OXIDES

Changes in oxidation state affect the solubility of metal (hydr)oxides. The oxides of transition elements become more soluble on reduction, whereas other oxides such as Cr_2O_3 or V_2O_3 become more soluble on oxidation. The reduction of surface metal centers in reducible metal oxides typically leads to easier detachment of the reduced metal ions from the latttice surface. This is readily accounted for by the larger lability of the reduced metal-oxygen bond in comparison to the nonreduced metal-oxygen bond; for example, the Madelung energy of the Fe^{II}-O bond in a crystalline lattice is much smaller than that of the Fe^{III}-O bond.

Stone and Morgan (1987) comprehensively surveyed the reductive dissolution of metal oxides. We give here some examples of recent work in our laboratory on the reductive dissolution of Fe(III) (hydr)oxides (Banwart et al.,1989; Sulzberger, 1988; Suter et al., 1988; Zinder et al., 1986). In each case investigated, the dissolution rate was controlled by surface chemical reactions. The rate law can be accounted for by the following reaction sequence: (1) surface complex formation with reductant, (2) Electron transfer leading to some Fe(II) in the lattice surface, and (3) detachment of Fe(II) from the mineral surface into the solution. The detachment (step 3) is rate-determining; that is, the precursor of the activated complex is a surface >Fe(II) species (with surrounding bonds that are partially protonated), whose concentration is proportional to that of the precursor, the surface complex. Correspondingly, the rate of dissolution was found to be proportional to the concentration of the surface bound reductant:

$$R_{reductive} \propto \{>\text{Fe-O-reductant}\} \tag{34}$$

Figure 10 illustrates the reaction scheme that accounts for the reductive dissolution of Fe(III) (hydr)oxides by ascorbate. Figure 11 gives experimental results illustrating the zero-order dissolution rate with varying ascorbate concentrations and its dependence on the concentration of the ascorbate surface complex.

Figure 10

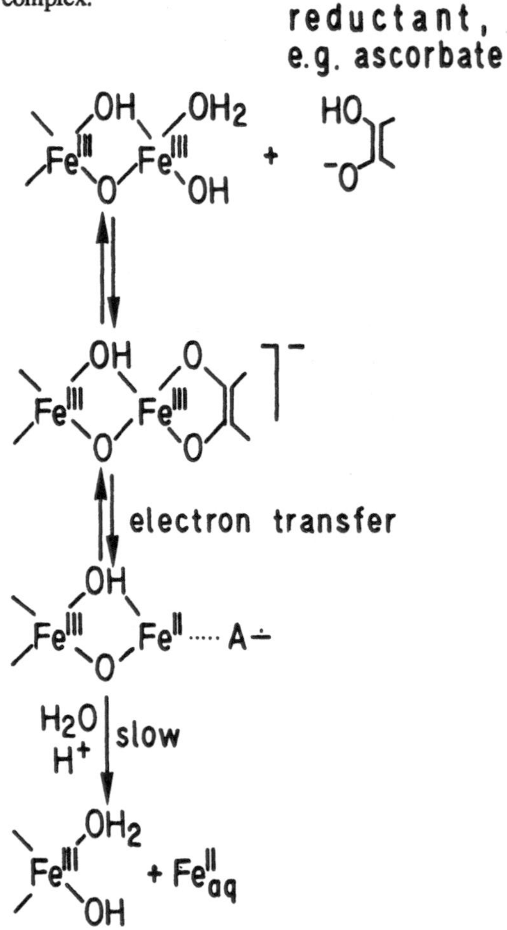

Figure 10: Reaction sequence for the dissolution of Fe(III) (hydr)oxides by a reductant such as ascorbate. The fast adsorption of the reductant is followed by steps that involve the electron transfer; >Fe(III) is reduced to >Fe(II) and ascorbate oxidized to a radical that is desorbed from the surface. The resulting Fe(II)-O bond at the surface is more labile than the Fe(III)-O bond. Then Fe(II) becomes detached from the surface and the original surface structure is reconstituted (from Banwart et al., 1989).

Iron(III) (hydr)oxides can be dissolved [as Fe(III)] most effectively by a catalytic mechanism that involves a bridging ligand and Fe(II). Thus, as exemplified in Figure 12 ligands such as oxalate bind via ligand exchange at the surface >Fe^{III}OH. Subsequently, a ternary complex >Fe^{III}LFe^{II} is formed. The efficiency of a bridging ligand depends on its surface complex formation properties, the stability of the ternary complex as a function of pH, the electronic properties of the bridging ligand (a potentially conjugatable electronic arrangement facilitates the electron transfer), and its ability to transport the successor complex away from the surface before the reverse process occurs (Suter et al., 1988; Wehrli et al., in press). It is interesting to note that Fe(II) binds to hematite and goethite at pH 3-4 only if a ligand such as oxalate is present. The effect of the ligand is caused by an increase in concentration of adsorbed Fe(II); furthermore, oxalato-Fe(II) is, thermo-dynamically speaking, a much stronger reductant than Fe(II). In addition, oxalate binds the reductant Fe(II) at a greater distance from the surface than OH$^-$; this facilitates the detachment of the oxidized surface complex. A detailed discussion of similar inner-sphere electron transfer reactions at the pyrite-water interface is given by Luther (1990).

Figure 11

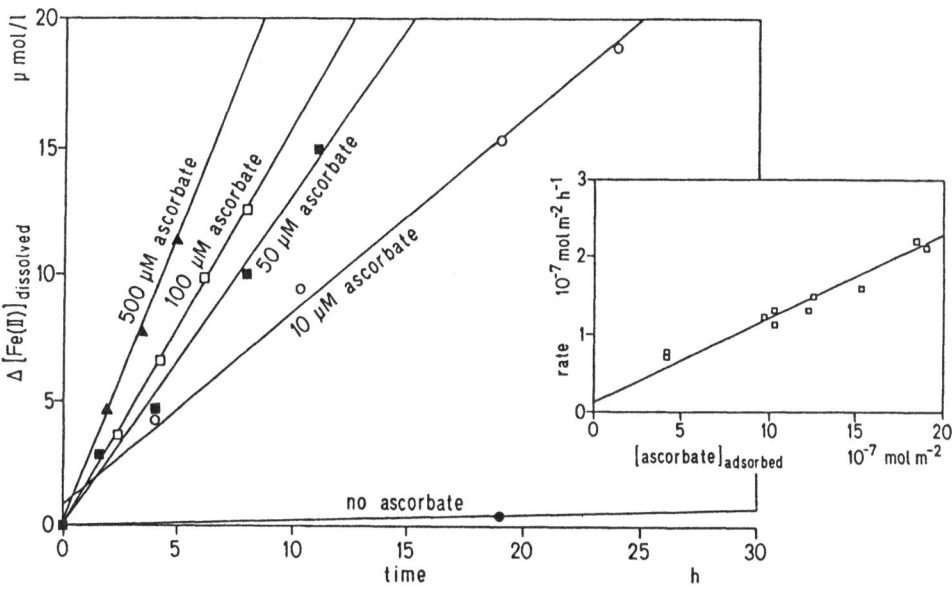

Figure 11: Representative results on the dissolution of hematite by ascorbate at pH3. As shown by the inset, the rate of dissolution is proportional to the surface concentration of ascorbate (from Banwart et al., 1989)

Figure 12 gives representative experimental results (Suter et al., 1988) for the dissolution of hematite in the presence of oxalate nd Fe(II). The total amount of Fe(III) dissolved during an experiment is much higher than the amount of Fe(II) present; the concentration of Fe(II) remains constant. There is no reduction, although the reduction of the surface Fe(II) has a strong effect on the kinetics of the reaction. As shown by Sulzberger (1990) the effect described can become important by establishing an auto-catalytic mechnism in the light-induced dissolution of Fe(III) oxides in the presence of surface ligands such as oxalate; the Fe(II) produced photochemically acts as a catalyst for the further dissolution.

468

Figure 12

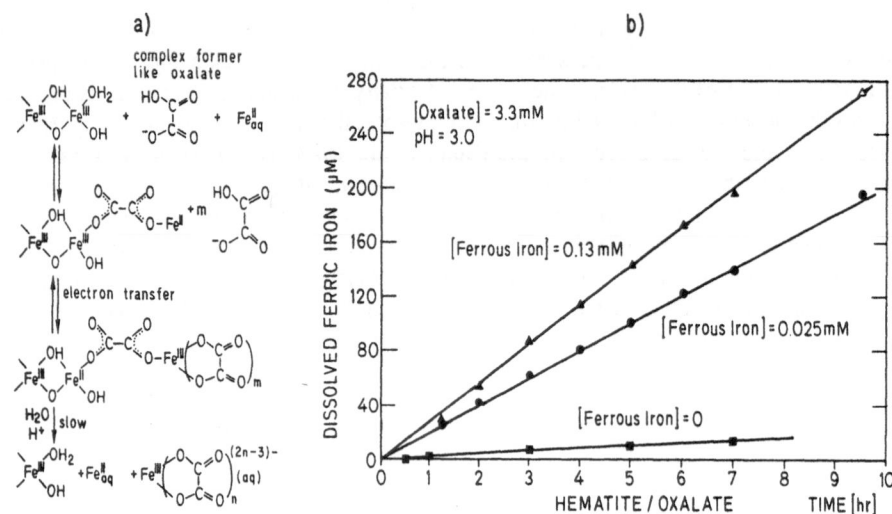

a)

complex former
like oxalate

b)

Figure 12: (a) Catalytic dissolution of Fe(III) (hydr)oxides. In the presence of a suitable bridge-building ligand, Fe(II) catalyzes the dissolution of Fe(III). Ligands such as oxalate adsorb via ligand exchange at the surface. Subsequently a ternary complex with Fe(II) is formed. The electron transfer from the Fe(II) on the solution site to the >Fe(III) at the lattice surface reduces the latter to >Fe(II), which then becomes detached, replacing the Fe(II) in solution. The Fe(III), now bound to the successor cpomplex, goes into solution. (b) Representative experimental results (from Suter et al., 1988).

Light-Induced Dissolution of Hematite in the Presence of Oxalate

The reaction pathways involved were described in much detail by Sulzberger (1990) and by Sulzberger and Siffert (in preparation). We restrict our discussion to a brief overview of the possible mechanism involved, following the explanation given by Sulzberger (1990).

REACTION PATHWAYS INVOLVED IN THE LIGHT-INDUCED DISSOLUTION OF HEMATITE IN THE PRESENCE OF OXALATE

Apart form the photoredox reaction occurring at the surface of hematite and leading to dissolved iron(II), Fe^{II} is also produced through photolysis of dissolved iron(III) trioxalato complexes. Dissolved iron(III) is formed via thermal pathways, where Fe^{II} acts as a catalyst.

The various elementary steps involved in the surface photoredox reaction, leading to dissolution of hematite are outlined in Figure 13. An important step is the formation of a hypothetical bidentate, mononuclear surface complex. Electron transfer occurs via an electronically excited state (indicated by an asterisk) which is either a ligand-to-metal charge-transfer transition of the surface complex and/ or $\leftarrow O^{-II}$ charge-transfer transition of hematite. The oxalate radical undergoes a fast decarboxylation reaction yielding CO_2 and the $CO_2^{\cdot}$ radical, which is a strong reductant that can reduce a second surface iron(III) in a thermal reaction. Thus, two surface iron(II) and two CO_2 may theoretically be formed per absorbed photon. However, the quantum yield of this surface redox reaction is less than two (Siffert, 1989) because of loss reactions such as thermal deactivation from the excited state. For the sake of simplicity this thermal reaction of the $CO_2^{\cdot-}$ radical is omitted in Figure 13. We assume that detachment of Fe^{II} from the ctrystal lattice is the rate-limiting step of the overall reaction. After detachment of the surface group, the surface of hematite is reconverted into its original configuration; hence, the surface concentration of active sites and of adsorbed oxalate does not change throughout the experiment.

Figure 13

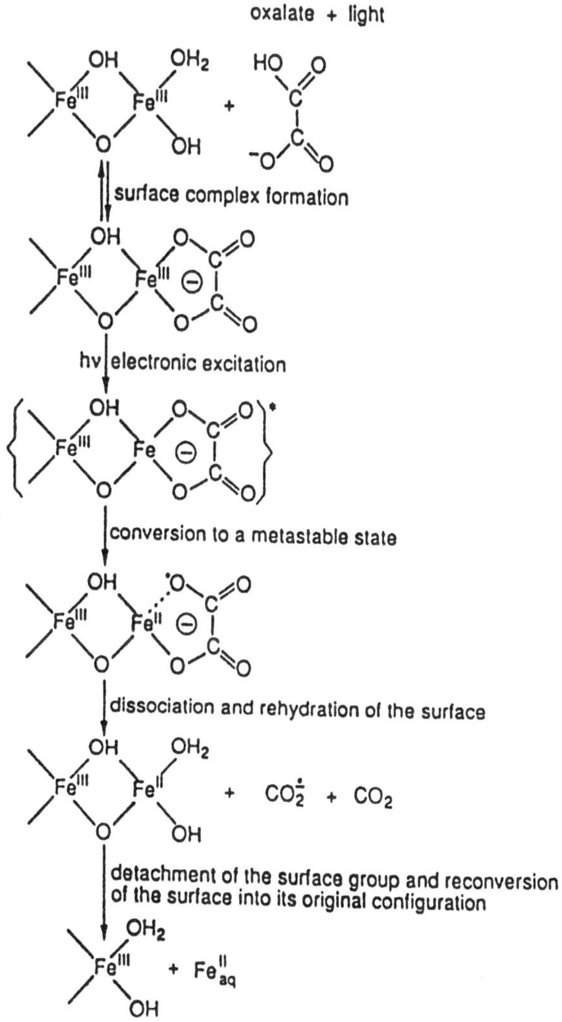

oxalate + light

surface complex formation

hv electronic excitation

conversion to a metastable state

dissociation and rehydration of the surface

$+ \quad CO_2^{\cdot-} \quad + \quad CO_2$

detachment of the surface group and reconversion of the surface into its original configuration

$+ \quad Fe^{II}_{aq}$

Figure 13: Schematic representation of the various steps involved in the light-induced reductive dissolution of hematite in the presence of oxalate (from Sulzberger, 1990).

Part of the photochemically formed iron(II) is readsorbed on the hematite surface, with oxalate acting as a bridging ligand. The adsorbed Fe^{II} is thermodynamically a suitable reductant of surface iron(III). We assume that electron transfer from the adsorbed iron(II) to the surface of iron(III) occurs through the bridging ligand (see Figure 12):

The surface iron(II) thus formed is detached from the surface of hematite as the rate-determining step (Suter et al., 1988). This pathway is a thermal reductive dissolution of hematite, which leads, however, to an increase of the concentration of dissolved iron(III) while the concentration of dissolved iron(II) remains constant; Fe^{II} acts as a catalyst for the dissolution of iron(III) hydroxides (see Figure 8).

Acknowledgements

Our work on surface coordination has been supported by the Swiss National Science Foundation, and has been critically dependent on advice and research contributions by Steven Banwart, Gerhard Furrer, Herbert Motschi, Christoph Siffert, Laura Sigg, Barbara Sulzberger, Daniel Suter, Bernhard Wehrli, and Erich Wieland.

It may not be sufficiently evident from this chapter how much our ideas have been influenced by James J. Morgan, Paul Schindler and Garrison Sposito.

References

Banwart, S., Davies, S., and W. Stumm (1989) 'The role of oxalate in accelerating the reductive dissolution of hematite (α-Fe$_2$O$_3$) by ascorbate'. Colloids Surf. **39**, 303-309

Brady, P.V. and Walther, J.V. (1990) 'Controls on silicate dissolution rates in neutral and basic pH solutions at 25°C', Chem. Geol, in press.

Brown, G.E., Jr., Parks, G.A., and Chisholm-Brause, C.J. (1989) 'In situ X-ray absorption spectroscopic studies of ions at oxide-water interfaces'. Chimia **43**, 248-256.

Chisholm-Brause, C.J., Brown G.E., Jr., and Parks, G.A. (1989) 'EXAFS investigation of aqueous Co(II) adsorbed on oxide surfaces in-situ', Physica B **158,** 646-648.

Chou, L. and R. Wollast (1984) 'Study of the weathering of albite at room temperature and pressure with fluidized bed reactor'. Geochim. Cosmochim. Acta **48**, 2205-2217.

Dzombak, D.A., Morel, F.M.M. (1990) Surface Complexation Modeling; Hydrous Ferric Oxide. Wiley-Interscience, New York.

Furrer, G. and Stumm W. (1986). 'The coordination chemistry of weathering.
I. Dissolution kinetics of α-Al$_2$O$_3$ and BeO.' Geochim. Cosmochim. Acta **50**, 1847-1860.

Hachiya, K., Sasaki, M., Ikeda, T., Mikami, N., Yasunaga, T. (1984) 'Static and kinetic studies of adsorption-desorption of metal ions on a γ-Al$_2$O$_3$ surface. II. Kinetic study by means of pressure jump technique'. J. Phys. Chem. **88**, 27-31.

Hayes, K.F. and Leckie, J.O. (1987) 'Modelling ionic strength effects on cation adsorption at hydrous oxide/ solution interfaces'. J. Coll. Interf. Sci. **115**, 564-572.

Hayes, K.F., Roe, A.L., Brown, G.E., Jr., Hodgson, K.O., Leckie, J.O., Parks, G.A. (1987) 'In situ X-ray absorption study of surface complexes: Selenium oxyanions on α-FeOOH', Science **238**, 783-786.

Hohl, H. and Stumm, W. (1976) 'Interaction of Pb^{2+} with hydrous γ-Al$_2$O$_3$'. J. Colloid Interface Sci. **55**, 281-288.

Hohl, H., Sigg, L., and Stumm, W. (1980) 'Characterization of surface chemical properties of oxides in natural waters; the role of specific adsorption determining the specific charge'. In: Particulates in Water, Advances in Chemistry Series, ACS **189**, 1-31.

Ikeda, T., Sasaki, M., Hachiya, K., Astumian, R.D., Yasunaga, R., and Schelly, Z.A. (1982) 'Adsorption-desorption kinetics of acetic acid on silica-alumina particles in aqueous suspensions, using the pressure-jump relaxation method'. J. Phys. Chem. **86**, 3861-3866.

Lasaga, A.C. (1981) 'Transition state theory'. Rev. Mineral. **8**, 135-169.

Luther, G.W., III, (1990) 'The frontier-molecular-orbital theory approach in geochemical processes'. In W. Stumm, Ed., Aquatic Chemical Kinetics. Wiley-Interscience, New York, Chapter 6, 173-198.

Lyklema, J. (1987) 'Electrical double layers on oxides: disparate observations and unifying principles', Chemistry and Industry, 741.

Motschi, H. (1987) 'Aspects of the molecular structure in surface complexes; spectroscopic investigations'. In W. Stumm, Ed., Aquatic Surface Chemistry. Wiley-Interscience, New York, 111-124.

Motschi, H. and Rudin, M. (1984) '^{27}Al ENDOR study of VO^{2+} adsorbed on δ-alumina. Direct Evidence for inner-sphere coordination with surface functional groups'. Coll. Polymer Sci. **262**, 579-583.

Parks, G. A (1967) in 'Equilibrim Concepts in Natural Water Systems', Advances in Chemistry Series **67**, Am. Chem. Soc., Washington, 121.

Schindler, P.W. and Kamber, H.R. (1968) 'Die Azidität von Silanolgruppen'. Helv. Chim. Acta **51**, 1781-1786.

Schindler, P.W. and Stumm, W. (1987) 'The surface chemistry of oxides, hydroxides and oxide minerals', in: W. Stumm (Ed.): Aquatic Surface Chemistry, Wiley-Interscience, New York.

Schott, J.(1990) 'Modelling of the dissolution of strained and unstrained multiple oxides: The surface speciation approach'.In W. Stumm, Ed., Aquatic Chemical Kinetics, Wiley-Interscience, Chapter 12, 337-366.

Siffert, C. (1989) 'L'effet de la lumière sur la dissolution des oxydes de Fer(III) dans les milieux aqueux'. Ph. D. thesis, ETH Zürich, No. 8852.

Sigg, L. and Stumm, W. (1981) 'The interactions of anions and weak acids with the hydrous goethite (a-FeOOH) surface', Colloids and Surfaces **2**, 101.

Sposito, G. (1984) 'The Surface Chemistry of Soils', Oxford University Press, New York.

Sposito, G (1989) 'Surface reactions in natural aqueous colloidal systems', Chimia **43**, 169-176.

Stone, A.T. and Morgan, J.J. (1987) 'Reductive dissolution of metal oxides'. In W. Stumm, Ed., Aquatic Surface Chemistry, Wiley-Interscience, New York, 222-254.

Stumm, W. and Morgan, J.J. (1981) Aquatic Chemistry, Wiley-Interscience, New York.

Stumm, W. and Wieland, E. (1990) 'Dissolution of oxide and silicate minerals: rates depend on surface speciation'. In W. Stumm, Ed., Aquatic Chemical Kinetics, Wiley-Interscience, New York., Chapter 13, 367-400.

Stumm, W. and Wollast, R. (1990) 'Coordination chemistry of weathering: Kinetics of the surface-controlled dissolution of oxide minerals', Reviews of Geophysics **28**/1, 53-69.

Stumm, W., Huang, C.P., Jenkins, S.R. (1970) 'Specific chemical interaction affecting the stability of dispersed systems', Croat. Chem. Acta **42**, 223.

Stumm, W., Kummert, R., and Sigg, L. (1980) 'A ligand exchange model for the adsorption of inorganic and organic ligands at hydrous oxide interfaces', Croat. Chem. Acta **53**, 291.

472

Sulzberger, B. (1988) 'Oberflächen-Koordinationschemie und Redox-Prozesse: Zur Auflösung von Eisen(III)-Oxiden unter Lichteinfluss'. Chimia **42**, 257-261.

Sulzberger, B. (1990) 'Photoredox reactions at hydrous metal oxide surfaces: A surface coordination chemistry approach'. In W. Stumm, Ed., Aquatic Chemical Kinetics, Wiley-Interscience, New York, Chapter 14, 401-429.

Sulzberger, B. and Siffert, C. 'Light-induced dissolution of hematite in the presence of oxalate', in preparation.

Suter, D., Siffert, C., Sulzberger, B. and Stumm, W. (1988) 'Catalytic dissolution of iron(III) (hydr)oxides by oxalic acid in the presence of Fe(II)'. Naturwissenschaften **75**, 571-573.

Wehrli, B. (1989) 'Monte Carlo simulations of surface morphologies during mineral dissolution', J. Colloid and Interface Sci. **132**, 230-242.

Wehrli, B. and Stumm W. (1988) 'Oxygenation of vanadyl(IV); effect of coordinated surface-hydroxyl groups and OH⁻'. Langmuir **4**, 753-758.

Wehrli, B., Sulzberger, B., and Stumm, W. (1989) 'Redox processes catalyzed by hydrous oxide surfaces'. Chem. Geol. **78**, 167-179.

Wieland, E., Wehrli, B., and Stumm, W. (1988) 'The coordination chemistry of weathering: III. A potential generalization on dissolution rates of minerals' Geochim. Cosmochim. Acta **52**, 1969-1981.

Zeltner, W.A., Yost, E.C., Machesky, M.L., Tejedor-Tejedor, M.I., and Anderson, M.A. (1986) 'Characterization of anion binding on goethite using titration calorimetry and cylindrical internal reflection-Fourier transform infrared spectroscopy', in J. A. Davis and K.F. Hayes, Eds., Geochemical Processes at Mineral Surfaces, Am. Chem. Soc., Washington, 142-161.

Zinder, B., Furrer, G., and Stumm, W. (1986) 'The coordination chemistry of weathering. II. Dissolution of Fe(III) oxides'. Geochim. Cosmochim. Acta **50**, 1861-1870.

DISCRETE CHARGES ON BIOLOGICAL MEMBRANES

RICHARD T. MATHIAS, GEORGE J. BALDO, KANDIAH
MANIVANNAN, AND STUART MCLAUGHLIN
Department of Physiology and Biophysics, Health Sciences Center
State University of New York at Stony Brook
Stony Brook, NY, 11794, USA

ABSTRACT. The main objective of this chapter is to derive an exact expression for the potential due to a single charge at an interface, then examine the validity of the limiting Debye-Hückel expression when the dielectric constant of the membrane phase has a finite value characteristic of a biological membrane.

INTRODUCTION

The principal components of biological membranes—lipids, proteins and sugars—all have fixed charges. The biological importance of these charges has been reviewed several times in the past few years (Honig et al., 1986; McLaughlin, 1989; Sharp and Honig, 1990; Green and Andersen, 1991; Cafiso, 1991). We recommend these reviews to readers interested in a detailed discussion of the biological literature.

For many purposes, it is adequate to assume the charge on a membrane is smeared uniformly over the interface and use the Gouy-Chapman theory to describe the average potential. There are many monographs and reviews concerned with the Gouy-Chapman theory; references may be found in McLaughlin (1977; 1989). There are also excellent reviews of the discrete-ion effect at metal-solution interfaces (Barlow and MacDonald, 1967; Levine et al., 1967). In this chapter we consider the potential produced by a single fixed charge (or an array of discrete charges) at a membrane-solution interface. Cole (1969), Brown (1974), Nelson and McQuarrie (1975), Sauve and Ohki (1979) have considered discrete charges at a membrane-solution interface. In most of these biological treatments, the authors assume the potential in the aqueous phase adjacent to a single fixed charge can be described by an expression that is identical to twice the Debye-Hückel expression for the potential produced by a charge in a bulk aqueous phase. This limiting expression arises when the dielectric constant of the membrane phase approaches zero; it was obtained by Wagner (1924) and used by Onsager and Samaras (1934) to calculate the surface tension of electrolyte solutions.

Our objective in this chapter is to derive an exact expression for the potential due to a single charge at an interface, then examine the validity of the limiting Debye-Hückel expression when the dielectric constant of the membrane phase has a finite value characteristic of a biological membrane, $2 < \epsilon_m < 4$. We attempt to give the biologically oriented reader an intuitive appreciation for the rather complicated mathematical expressions required to describe the potential produced by a single fixed point charge by presenting several graphical representations of the potential in

R. Guidelli (ed.), Electrified Interfaces in Physics, Chemistry and Biology, 473–490.
© 1992 *Kluwer Academic Publishers.*

both the membrane and aqueous phases.

Stillinger (1961) derived formulae to describe the potential produced by a single charge at an interface; we follow his mathematical treatment but present the results in a slightly different form. We first solve the linear form of the Poisson-Boltzmann equation for a single fixed point charge located at the interface between two semi-infinite dielectric phases, the membrane and the aqueous phase. The linear equation is valid if the potential is small with respect to $kT/e \approx 25$ mV. The electrolyte is confined to the aqueous phase. We make all the assumptions inherent in the standard Debye-Hückel theory of electrolyte solutions: specifically, the discrete nature of the ions in the aqueous phase and any interactions (correlation effects) between them in the ion atmosphere adjacent to the fixed charge are ignored in this mean field treatment, and water is assumed to be a structureless medium with a uniform dielectric constant. There is now some theoretical and experimental justification for this simple model, which we consider below. We solve the problem in general, then compute the potential for two values of the dielectric constant of the membrane phase, assuming it is 1/20th or equal to the dielectric constant of the aqueous phase. One important result is that the potential in the aqueous phase adjacent to a single fixed charge on the interface can be described very accurately by the limiting relation (twice Debye-Hückel theory) for values of the dielectric constant applicable to biological membranes. We then consider the finite size of the membrane and show there is no significant interaction between charges located on opposite sides of the membrane. Finally, we show that the average potential produced by an array of discrete charges equals that predicted from Gouy-Chapman theory, which assumes the charges are smeared uniformly over the membrane-solution interface.

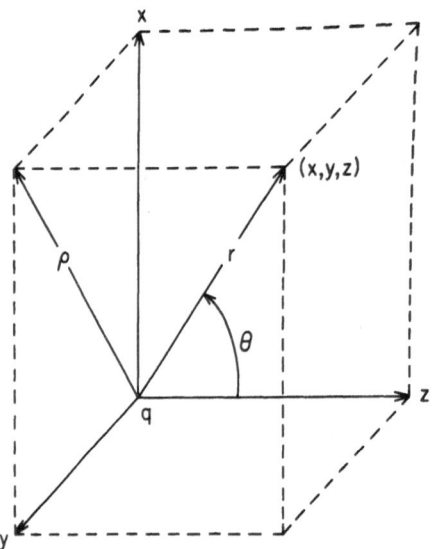

Figure 1. Sketch of the co-ordinate system. The membrane-solution interface is located at $z = 0$ in the x-y plane and a single charge, q, is at the origin. There are no ions in the membrane phase ($z < 0$). The electrostatic problem we consider has cylindrical symmetry, and we express our calculations of the potential in terms of the two variables ρ and z. The angle θ is also indicated because in Figs. 3 and 4 we illustrate the dependence of the potential on r for $\theta = 0, 90$ and 180 degrees.

ANALYSIS

Fig. 1 illustrates the co-ordinate system. The membrane-solution interface is in the x-y plane at z = 0 and a single fixed point charge is located at the origin. The monovalent electrolyte is confined to the aqueous phase (z > 0) and the potential in the semi-infinite membrane phase, ψ_m, is described by the Laplace equation:

$$\nabla^2 \psi_m = 0 \tag{1}$$

The potential in the aqueous phase, ψ_a, is described by the linearized form of the Poisson-Boltzmann equation used by Debye and Hückel:

$$\nabla^2 \psi_a - \kappa^2 \psi_a = 0 \tag{2}$$

where $1/\kappa$ is the Debye length[1], which is about 1 nm for a 0.1 M salt solution.

Consider first the case where $\kappa \to 0$ (i.e., the aqueous phase contains no electrolyte). The potential in both the membrane and aqueous phases is described by the Laplace equation, and this problem is solved in standard texts (e.g., Jackson, 1975; page 147). The potential at a distance r from the fixed charge q in both the membrane and aqueous phase is

$$\psi_m = \psi_a = \frac{2q}{4\pi\epsilon_0(\epsilon_m + \epsilon_a)r} \qquad \kappa = 0 \tag{3}$$

where ϵ_0 is the permittivity of free space, ϵ_m is the relative permittivity or dielectric constant of the membrane phase, and ϵ_a is the dielectric constant of the aqueous phase. We assume ϵ_a is 78 in all our numerical calculations. Eq. 3 predicts that when $\epsilon_m << \epsilon_a$ the potential in both the membrane and aqueous phases is twice the value predicted from Coulomb's law for a single charge q in a bulk aqueous phase. The equipotential profiles predicted by Eq. 3 are illustrated in the lower portion of Fig. 2.

It may not be immediately apparent why the equipotential surfaces illustrated in the lower portion of Fig. 2 should be perfect spheres since the dielectric constant of the membrane phase differs from that of the aqueous phase. A derivation of Eq. 3 using the method of images may be helpful (e.g., Jackson, 1975; page 147). In this method one calculates the potential in the aqueous phase due to a single point charge q located at z = d by considering all space to be filled with a medium of dielectric ϵ_a and adding an image charge of magnitude $q(\epsilon_a - \epsilon_m)/(\epsilon_a + \epsilon_m)$ at z = −d. When d = 0, the image charge is superimposed on the real charge and the sum of the two potentials yields Eq. 3 for the potential in the aqueous phase. A similar analysis yields an identical expression for the potential in the membrane phase.

If the aqueous phase contains electrolyte, the potential does not have spherical

---

[1] The reciprocal length parameter of Debye-Hückel theory is defined as $\kappa = (2e^2n/\epsilon_a\epsilon_0 kT)^{1/2}$, where e is the electronic charge, n is the number of monovalent cations (anions) per unit volume in the bulk aqueous phase, k is the Boltzmann constant, and T is the temperature. $n(number/m^3) = 1000Nc$ where N is Avogadro's number and c is the molar concentration of electrolyte.

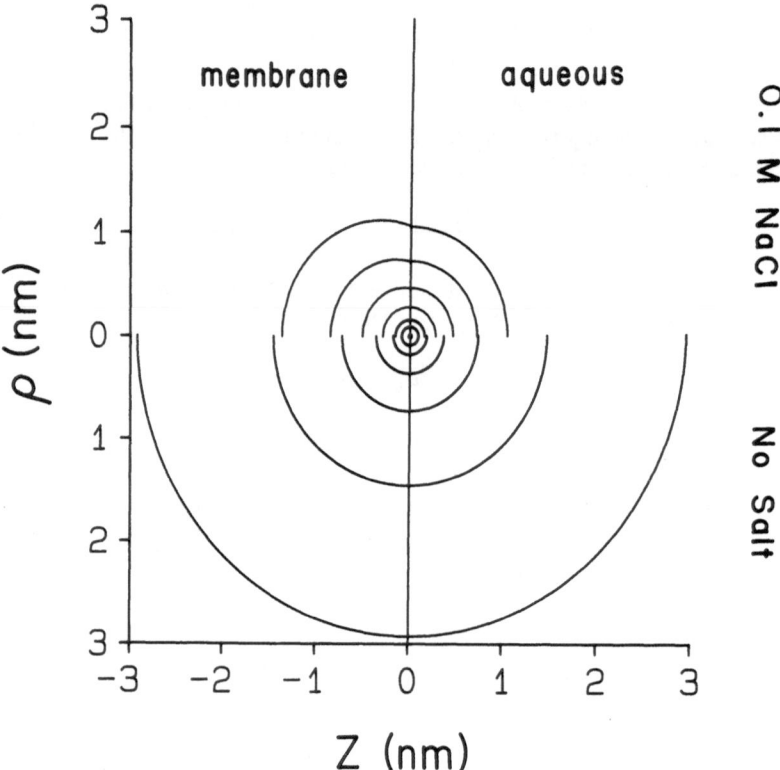

Figure 2. Equipotential profiles when the aqueous phase (z > 0) contains either no electrolyte (lower half of figure) or a 0.1 M monovalent salt solution, for which the Debye length is about 1 nm (upper half of figure). For both cases we assume ϵ_m/ϵ_a = 0.05 (i.e. ϵ_a = 78, ϵ_m = 3.9) and draw profiles for potentials of 400, 200, 100, 50, 25 and 12.5 mV. The equipotential profiles have exact spherical symmetry when no salt is present (Eq. 3), as illustrated in the lower portion of the figure. The spherical symmetry is retained (approximately) in the aqueous phase even when salt is present (Eq. 10) as illustrated in the upper portion of the figure. Note that addition of electrolyte to the aqueous phase "screens" the surface charge and reduces the potential in both the aqueous and membrane phases.

symmetry. It does, however, have cylindrical symmetry, and Eqs. 1 and 2 can be written as a function of the two variables ρ and z (see Fig. 1):

$$\frac{\partial^2 \psi_m}{\partial \rho^2} + \frac{1}{\rho}\frac{\partial \psi_m}{\partial \rho} + \frac{\partial^2 \psi_m}{\partial z^2} = 0 \tag{4}$$

$$\frac{\partial^2 \psi_a}{\partial \rho^2} + \frac{1}{\rho}\frac{\partial \psi_a}{\partial \rho} + \frac{\partial^2 \psi_a}{\partial z^2} - \kappa^2 \psi_a = 0 \tag{5}$$

The standard boundary conditions are (e.g., Jackson, 1975; page 146):

$$\frac{\partial \psi_a(\rho,0)}{\partial \rho} = \frac{\partial \psi_m(\rho,0)}{\partial \rho} \tag{6}$$

$$\frac{\epsilon_a \partial \psi_a(\rho,0)}{\partial z} - \frac{\epsilon_m \partial \psi_m(\rho,0)}{\partial z} = \sigma = \frac{q\delta(\rho)}{2\pi\rho} \tag{7}$$

where σ is the surface charge density and $\delta(\rho)$ is the one dimensional Dirac delta function. If we invoke the condition ψ_a, $\psi_m \rightarrow 0$ as $r \rightarrow \infty$, Eq. 6 reduces to

$$\psi_a(\rho,0) = \psi_m(\rho,0) \tag{8}$$

In other words, the potential is continuous at the membrane-solution interface. For $r << 1/\kappa$ the potentials in both the membrane and aqueous phases should approach[2] the value predicted by Eq. 3:

$$\psi_m \rightarrow \psi_a \rightarrow \frac{2q}{4\pi\epsilon_0(\epsilon_m + \epsilon_a)r} \qquad \kappa r \rightarrow 0 \tag{9}$$

The solutions to Eqs. 4 and 5 that satisfy the boundary conditions are

$$\psi_a(\rho,z) = \frac{2q}{4\pi\epsilon_0(\epsilon_m + \epsilon_a)}\left[\frac{e^{-\kappa r}}{r} + \kappa\int_0^\infty A(\beta)J_0(\beta\kappa\rho)e^{-\sqrt{\beta^2+1}\,\kappa z}d\beta\right] \tag{10}$$

---

[2] Eq. 9 follows from the integral form of Gauss' law if we consider a sphere of radius r centered on the fixed point charge. If $r << 1/\kappa$, there is negligible space charge (mobile counterions) within the sphere: it contains only the fixed charge q. The problem reduces to the case for which Eq. 3 is valid.

$$\psi_m(\rho,z) = \frac{2q}{4\pi\epsilon_0(\epsilon_m + \epsilon_a)}\left[\frac{1}{r} + \kappa\int_0^\infty M(\beta)J_0(\beta\kappa\rho)e^{-\beta\kappa|z|}d\beta\right] \tag{11}$$

where J_0 is a Bessel function of the first kind and

$$A(\beta) = -\frac{\epsilon_m}{\epsilon_a}\frac{\beta M(\beta)}{\sqrt{\beta^2+1}} \tag{12}$$

$$M(\beta) = -\frac{1 - \beta/\sqrt{\beta^2+1}}{1 + (\epsilon_m/\epsilon_a)(\beta/\sqrt{\beta^2+1}\ )} \tag{13}$$

Stillinger (1961) derived mathematically equivalent expressions for the potential, but expressed them in a different form. The equipotential profiles predicted by Eqs. 10 and 11 for $\epsilon_m/\epsilon_a = 0.05$ are shown in the upper portion of Fig. 2 for the case where the aqueous phase contains 0.1 M monovalent salt. When salt is present, there is an ion atmosphere of counter charges (net value –q) in the aqueous phase adjacent to the charge q. A comparison of the upper and lower portions of Fig. 2 illustrates that the counterions in the aqueous phase reduce the potential in both the membrane and aqueous phases. For example, the 0.1 M electrolyte screens the charge and reduces the potential in the aqueous phase to $kT/e \simeq 25$ mV for r = 0.7 nm (upper portion of Fig. 2) whereas the potential does not fall to 25 mV until r = 1.4 nm when no electrolyte is present (lower portion of Fig. 2). Fig. 2 also illustrates that salt does not affect the potential near the origin. When r << $1/\kappa$, the potential produced by the fixed charge q dominates the potential produced by the diffuse cloud of counter charges and the potential is described by Eq. 3. This statement is true for all values of ϵ_m and ϵ_a.

The equipotential profiles for $\epsilon_m/\epsilon_a = 0.05$ are also illustrated in Fig. 3A and 3B. To the resolution of this graph, the potential in the aqueous phase is indistinguishable from the potential predicted by the limiting equation:

$$\psi_a \to \frac{2qe^{-\kappa r}}{4\pi\epsilon_0\epsilon_a r} \qquad \epsilon_m/\epsilon_a \to 0 \tag{14}$$

It is apparent by inspection that Eq. 14 follows from Eq. 10 in the limit that $\epsilon_m = 0$. Recall that Eq. 14 is the Debye-Hückel solution for the potential a distance r from a charge 2q in a bulk aqueous solution. This expression, valid in the limit that $\epsilon_m = 0$, has also been derived by Vrij (1966) and by Stigter and Dill (1986), following Wagner (1924) and Onsager and Samaras (1934), using the method of images. Their simple derivation enables one to visualize why the potential in the aqueous phase has hemispherical symmetry in the limit $\epsilon_m = 0$ even though the ions are confined to the aqueous phase and the dielectric constants of the membrane and aqueous phases have different values. In the method of images one calculates the potential in the aqueous phase by replacing the membrane phase of dielectric constant ϵ_m by a salt-free phase of dielectric ϵ_a and

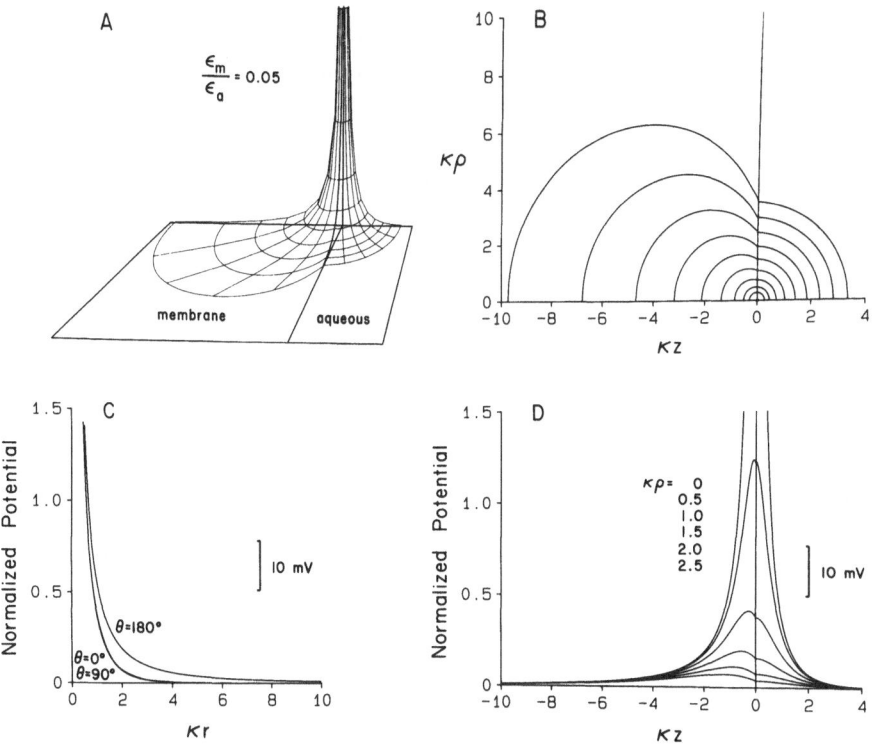

Figure 3. The potential predicted by Eqs. 10 and 11 when $\epsilon_m/\epsilon_a = 0.05$ (i.e. $\epsilon_a = 78$, $\epsilon_m = 3.9$) and the aqueous phase contains 0.1 M salt, T = 25°C. **A.** Sketch of the potential in the membrane and aqueous phases. The equipotential profiles are also illustrated in panel B. **B.** Equipotential profiles of 100, 50, 25, 12, 6, 3, 1.6, 0.8 and 0.4 mV. The potential in the aqueous phase is described to a good approximation by Eq. 14, which is twice the Debye-Hückel expression for the potential produced by a charge q in a bulk aqueous solution. The Debye length is about 1 nm in a 0.1 M monovalent salt solution ($1/\kappa = .96$ nm) and the normalized distances on the abscissa (κz) and ordinate ($\kappa\rho$) are thus approximately equal to the distances in nanometers. **C.** Dependence of the normalized potential on normalized distance from the origin (κr) for either $\rho = 0$ ($\theta = 180°$, r into the membrane; $\theta = 0°$, r into the aqueous phase) or $z = 0$ ($\theta = 90°$, r along the surface of the membrane). The potential is normalized (see Eqs. 10 and 11) by the factor $(2q\kappa)/4\pi\epsilon_0(\epsilon_a + \epsilon_m)$, which is equal to 36.6 mV for $1/\kappa = .96$ nm, a monovalent fixed charge $q = 1.6 \ 10^{-19}$ coulombs, $\epsilon_a = 78$ and $\epsilon_m = 3.9$. **D.** The normalized potential plotted as a function of normalized distance from the membrane-solution interface for different values of $\kappa\rho$.

adding image charges. The image charges have the same magnitude as the real charges because we assume $\epsilon_m = 0$. The image charge corresponding to the fixed point charge q at the membrane-solution interface is superimposed on it. Each hemispherical shell of uniform charge density in the aqueous diffuse double layer produces an equivalent image hemispherical shell of the same density, which completes the symmetry.

Fig. 3B and 3C illustrate that the hemispherical symmetry predicted by the limiting Eq. 14 is maintained to a good approximation even for a membrane with a dielectric of $\epsilon_m \approx 4$. Thus, Eq. 14 should adequately describe the potential produced by a fixed charge at the membrane-solution interface of a phospholipid bilayer, which has a dielectric constant of about 2.

Now consider the potential within the membrane phase. The potential profiles are illustrated in Figs. 2 and 3. Deep within the membrane phase, where the magnitude of $z \gg 1/\kappa$, Eq. 11 reduces to Eq. 15:

$$\psi_m \simeq \frac{2q}{4\pi\epsilon_0(\epsilon_a + \epsilon_m)} \frac{|z|(1 + \epsilon_m/\epsilon_a)/\kappa}{(\rho^2 + z^2)^{3/2}} \qquad |z| \gg 1/\kappa, \quad 0 \leq \frac{\epsilon_m}{\epsilon_a} \leq 1 \quad (15)$$

This equation has a simple interpretation. From the analysis in Jackson (1975, p. 147), one can show that Eq. 15 describes the potential produced within the membrane by an unscreened dipole consisting of one charge q located on the interface at the origin and the other charge −q located in the aqueous phase a distance $(1 + \epsilon_m/\epsilon_a)/\kappa$ from q along the z-axis. In the limit we are considering here, $\epsilon_m/\epsilon_a \ll 1$, the potential deep within the membrane may be calculated by replacing the diffuse, hemispherically symmetrical, ion atmosphere by a single charge, −q, located a distance $1/\kappa$ from the interface. (This is an intuitively appealing result. The potential at the surface of a uniformly charged interface predicted by the linearized form of the Gouy-Chapman theory is equivalent to the potential predicted by replacing the diffuse double layer by a single layer of counterions located a distance $1/\kappa$ from the interface.) The validity of Eq. 15 can be seen in Fig. 3B by comparing the 1.6 and 0.4 mV profiles, which cross the κz axis at about −5 and −10. Thus a 2-fold increase in κz produces a 4-fold decrease in the potential, as predicted for a dipole.

We can summarize the results obtained so far quite simply. If we are considering a charge at the surface of a phospholipid bilayer or the bilayer component of a biological membrane, the dielectric constant of the interior of the membrane is $2 < \epsilon_m < 4$. If $\epsilon_m < 4$, Eq. 14 is an excellent approximation to the exact expression for the potential throughout the aqueous phase, including the membrane-solution interface. It also describes the potential within the membrane close to the fixed charge. Deep within the membrane the dipole expression, Eq. 15, describes the potential. In the next section we show that the potential deep within the membrane does not depend strongly on the value of the membrane dielectric constant, even when it is increased to a value as high as $\epsilon_m = \epsilon_a$.

The above analysis should be applicable to a charge at the surface of a phospholipid bilayer or the bilayer component of a biological membrane. However, if the charge is on a protein at the surface of a membrane, the dielectric constant could approach that of the aqueous solution. Jordan (1987), for example, has considered the potential produced by single charges located near the mouth of a channel in a

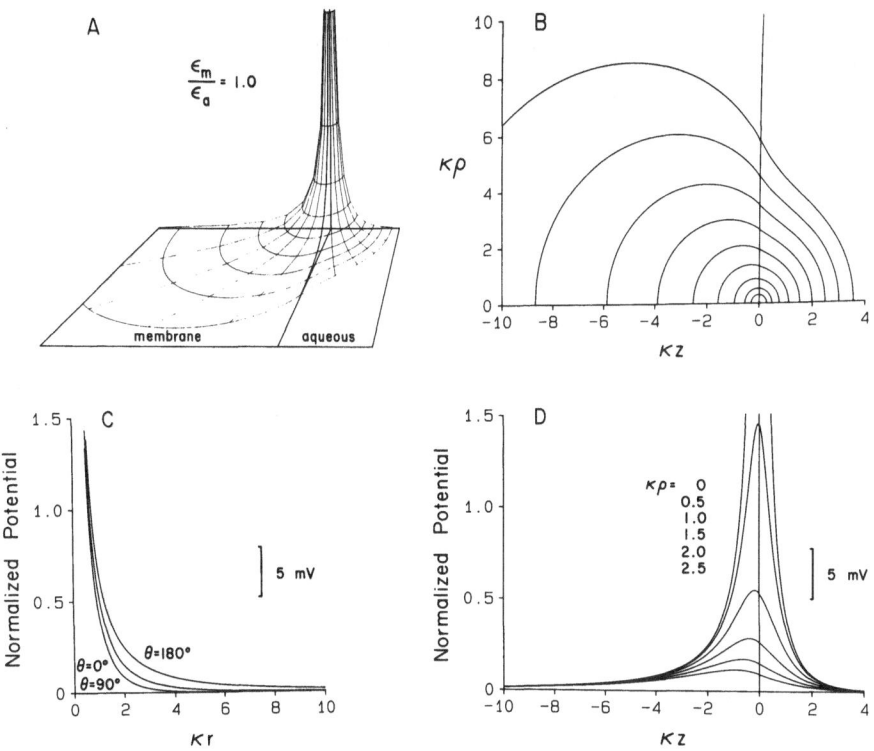

Figure 4. The potential predicted by Eqs. 10 and 11 when $\epsilon_m/\epsilon_a = 1$ (i.e. $\epsilon_m = 78$) and the aqueous phase contains 0.1 M salt. **A.** Sketch of the potential in the membrane and aqueous phases. **B.** Equipotential profiles of 50, 25, 12, 6, 3, 1.6, 0.8, 0.4, and 0.2 mV. For $r \ll 1/\kappa$, the equipotential curves are circles or the equipotential surfaces are spheres, as in Fig. 3B. However, the potential at a given value of r ($r \ll 1/\kappa$) is half as large in Fig. 4B as in Fig. 3B. **C.** Dependence of the normalized potential on the normalized distance from the origin (κr) for either $\rho = 0$ ($\theta = 0°$ and $\theta = 180°$) or $z = 0$ ($\theta = 90°$). **D.** The normalized potential plotted as a function of normalized distance from the membrane-solution interface for different values of $\kappa\rho$.

membrane. Fig. 4 illustrates the equipotential curves when $\epsilon_m/\epsilon_a = 1$. Fig. 4B illustrates that for r << $1/\kappa$ the equipotential curves are circles, as in Fig. 3B. Eq. 9, however, indicates the potential is half as large in Fig. 4B as in Fig. 3B. At the membrane-solution interface both the potential and the electric field are continuous when $\epsilon_m/\epsilon_a = 1$ (Fig. 4B). Thus, the hemispherical symmetry apparent in the aqueous phase when ϵ_m << ϵ_a (Fig. 3B) is lost when $\epsilon_m = \epsilon_a$ (Fig. 4B). The loss of hemispherical symmetry is also apparent in Fig. 4C: the $\theta = 0°$ and $\theta = 90°$ curves are not identical.

The potential deep within the membrane, d >> $1/\kappa$, is still described by Eq. 15 when $\epsilon_m/\epsilon_a = 1$. Note that Eq. 15 is independent of ϵ_m. In other words, even though the dielectric constant of the membrane phase has increased 20-fold and the average value of the dielectric constant, $(\epsilon_m + \epsilon_a)/2$, has increased about 2-fold, the potential deep within the membrane is essentially unchanged. The interpretation of the approximate expression Eq. 15 is not as straightforward as the case when ϵ_m << ϵ_a, but may be visualized as follows. The increase in the average dielectric constant is exactly compensated for by the increase in the effective distance between the charge q and the counterions in the diffuse atmosphere. This increase in average distance is apparent from a comparison of Fig. 3B and 4B. (Recall that the charge density is proportional to the potential in this linear theory.) There is the same net charge of –q in the ion atmosphere, but in Fig. 4B the average counter charge is a distance $2/\kappa$ from q whereas in Fig. 3B the average counter charge is located a distance $1/\kappa$ from q.

For z = 0, the potential has a simple analytical representation when $\epsilon_m/\epsilon_a = 1$. Eqs. 10 and 11 reduce to

$$\psi_m(\rho,0) = \psi_a(\rho,0) = \frac{q}{4\pi\epsilon_0\epsilon_a} \left\{ \frac{(2/\kappa)^2/2}{\rho^3} - \left\{ \frac{2/\kappa^2}{\rho^3} + \frac{2/\kappa}{\rho^2} \right\} e^{-\kappa\rho} \right\} \qquad \frac{\epsilon_m}{\epsilon_a} = 1 \quad (16)$$

For large values of $\kappa\rho$,

$$\psi_m(\rho,0) = \psi_a(\rho,0) \simeq \frac{q}{4\pi\epsilon_0\epsilon_a} \frac{(2/\kappa)^2/2}{\rho^3} \qquad \kappa\rho >> 1 \qquad \frac{\epsilon_m}{\epsilon_a} = 1 \quad (17)$$

Both Eq. 17 and Eq. 15 (with $\epsilon_m/\epsilon_a = 1$) describe the potential within the membrane produced by an unscreened dipole consisting of one charge q located on the interface at the origin and the other charge –q located in the aqueous phase a distance $2/\kappa$ from q along the z axis, provided the distance from the fixed charge is r >> $1/\kappa$. Thus the potential everywhere within the membrane is approximately described by this simple dipole approximation for distances $\kappa r > 3$, as revealed by an examination of Fig. 4B. In summary, increasing the membrane dielectric from 4 to 80 (20-fold) increases the average dielectric constant approximately 2-fold and increases the potential close to the fixed charge q by a factor of 2 in both the membrane and aqueous phases, but does not significantly affect the potential in the membrane phase when the distance from the fixed charge is more than a few Debye lengths.

Fig. 5 illustrates that the finite thickness of the membrane has little effect on the

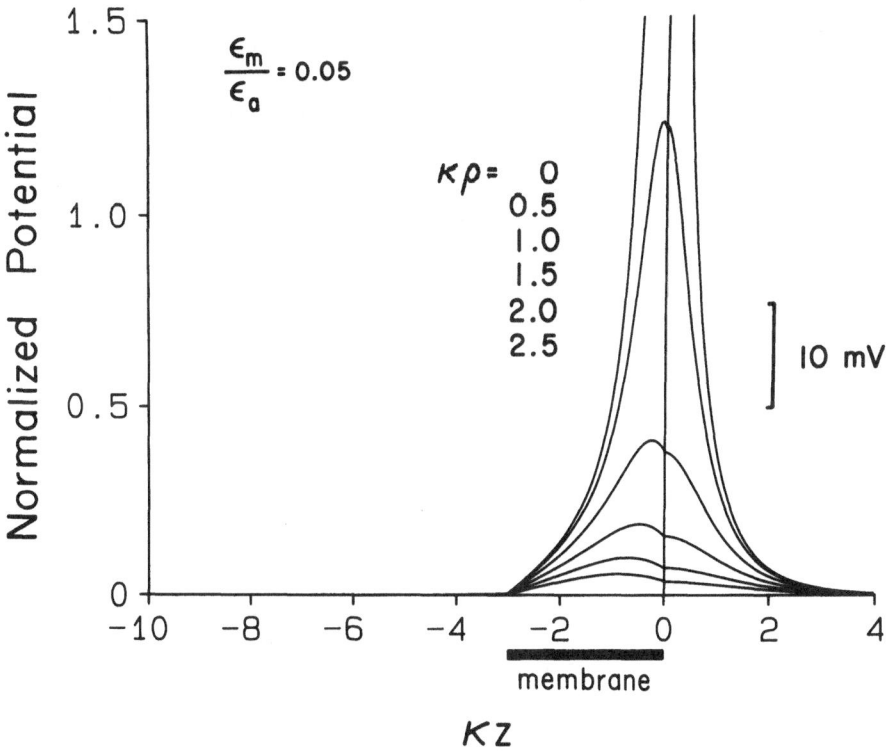

Figure 5. The effect of the finite thickness of the membrane on the potential profiles. We solved Eqs. 4–9 with the additional boundary conditions that

$$\frac{\epsilon_a \partial \psi_a(\rho, -\delta)}{\partial z} = \frac{\epsilon_m \partial \psi_m(\rho, -\delta)}{\partial z}$$

and that $\psi_a(\rho, -\delta) = \psi_m(\rho, -\delta)$. $\epsilon_m / \epsilon_a = 0.05$. Fig. 5 is similar to Fig. 3D, except that we consider the membrane to have a finite thickness of $\delta = 3$ Debye lengths (about 3 nm). Both aqueous phases contain 0.1 M monovalent salt. The potential in the right hand aqueous phase is similar in Figs. 5 and 3D. The potential in the left hand aqueous phase is essentially zero.

potential profiles in either the aqueous phase adjacent to the fixed charge at the right hand interface or within the membrane, except at the left hand interface where the potential is essentially zero. This indicates that charges on the two interfaces should have virtually no electrostatic interaction. This theoretical prediction has been confirmed experimentally by A. Cifu, R. Koeppe, and O. Andersen (personal communication). In their elegant experiments they were unable to detect any electrostatic repulsion between triple negatively charged o-Pyromellityl gramicidin A half channels located on opposite sides of a bilayer membrane formed in 0.05 M salt. Our calculations thus extend the conclusion that there is little interaction between the diffuse double layers on the two sides of a biological membrane if the charges are considered to be uniformly smeared over the interface (Chandler et al., 1965).

We now consider the potential produced in the aqueous phase by a number of fixed charges at a membrane solution interface. The inner or cytoplasmic surface of a typical biological membrane contains negatively charged phospholipids spaced about 2 nm apart. This distance is about twice the Debye length in a physiological salt solution, so the potential produced by these fixed charges will have a very punctate appearance (see, e.g., Fig. 3A). Nevertheless it is easy to show that the average potential produced by an array of discrete fixed point charges is exactly the same as the potential produced by a uniformly charged interface with the same net charge/area.

Consider first a single charge q at the membrane-solution interface. At an arbitrary location in the aqueous phase, the space charge density, as predicted from Eq. 2 and the Poisson equation, is equal to $-\kappa^2 \epsilon_0 \epsilon_a \psi_a(\rho,z)$. The total or net charge in the region between the interface and a plane parallel to the interface located a distance z from the interface is

$$Q_a(z) = -\kappa^2 \epsilon_0 \epsilon_a \int_0^z \int_0^\infty \psi_a(\rho,z) 2\pi\rho \, d\rho \, dz \tag{18}$$

If we insert Eq. 10, the general expression for $\psi_a(\rho, z)$, into Eq. 18 and integrate, we obtain

$$Q_a(z) = -q(1 - e^{-\kappa z}) \tag{19}$$

Comparing Figs. 3B and 4B, it is apparent that the distribution of space charge in the aqueous phase changes markedly as ϵ_m/ϵ_a changes. Eq. 19 indicates that the redistribution of space charge occurs entirely in the ρ direction. For one fixed charge q at the interface, the total charge in the slab of thickness dz located at a distance z from the interface is the derivative of $Q_a(z)$ times dz or $-q\kappa e^{-\kappa z}dz$. Now consider an interface where, on average, there is one charge q for each element of area A. In other words, the average surface charge density is

$$\sigma = q/A \tag{20}$$

For this linear problem the counterions of each fixed charge contribute the same total charge to the slab of thickness dz. As the total area goes to infinity, the total number of counterions goes to infinity proportionally, so the average space charge density in the slab is $-\sigma\kappa e^{-\kappa z}$. Recalling from Eq. 2 and the Poisson equation that the average space charge density is equal to $-\kappa^2\epsilon_0\epsilon_a<\psi_a(z)>$, we obtain

$$<\psi_a(z)> \; = \; (\sigma/\kappa\epsilon_0\epsilon_a)e^{-\kappa z} \tag{21}$$

Thus the average potential at any distance z from the plane is independent of the dielectric constant of the membrane phase, even though ϵ_m does affect markedly the potential adjacent to each individual fixed charge for $r < 1/\kappa$ as well as the distribution of charge around q.

We obtain an expression identical to Eq. 21 if we assume the charges are smeared uniformly over the interface to produce a uniform surface charge density σ (Eq. 20) and apply the linearized form of the Poisson-Boltzmann equation (linearized Gouy-Chapman theory). The Gouy-Chapman potential, ψ_{GC} is

$$\psi_{GC} \; = \; \psi(0)e^{-\kappa z} \tag{22}$$

where the surface potential, $\psi(0)$, is given by

$$\psi(0) \; = \; \sigma/(\kappa\epsilon_0\epsilon_a) \tag{23}$$

Note that Eq. 23 also describes the potential produced by a parallel plate capacitor with the spacing between the plates $= 1/\kappa$.

We can make a heuristic argument that the average of the discrete potentials should be independent of ϵ_m and equal the smeared charge Gouy-Chapman potential. In Figs. 3B and 4B we illustrated the potential profiles by drawing equipotential curves. We did not illustrate the field lines (lines of force, which indicate the direction of the electric field and cut the equipotential curves orthogonally). However, it should be apparent from Fig. 3B that all the field lines starting into the membrane phase from the fixed charge q at the origin must eventually bend back and reach the aqueous phase, which contains all the counterions. This statement is valid for all values of ϵ_m and ϵ_a.

One might suspect that this simple theoretical treatment would not be applicable to a real membrane-solution interface. However, a statistical mechanical description of an ensemble of discrete charges indicates the average potential can be predicted from a smeared charge model, even for an inhomogeneous dielectric profile near the surface (Vorotynsev & Ivanov, 1988; 1989a,b). Furthermore, the work of Marcelja and his colleagues demonstrates the Gouy-Chapman theory provides a surprisingly accurate description of the potential adjacent to a surface with discrete mobile monovalent surface charges, even when one takes into account the discrete nature of the charges in the aqueous phase, their electrostatic interactions, and the finite size of the ions (Langner et al., 1989).

DISCUSSION

The analysis above raises two questions: is this treatment applicable to cell membranes and what is the biological significance of fixed charges on membranes?

In most applications of the Poisson-Boltzmann equation to biological systems, the molecular structure of water has been ignored. Moreover, any variation in dielectric constant within the aqueous phase that might be produced by the surface is also ignored. Eqs. 21 and 22 predicts that the potential decays with distance from the interface according to $e^{-\kappa z}$. (Recall that the Debye length, $1/\kappa$, is proportional to the square root of ϵ_a.) This expression has been tested by measuring the force required to bring two negatively charged phospholipid bilayer membranes from infinity to a distance z. Both measurements with the Israelachvili surface force apparatus (Marra, 1986) and X-ray diffraction measurements (Loosley-Millman et al. 1982; Evans and Parsegian, 1986) demonstrate this equation describes the experimental data obtained for z > 2 nm. These measurements were made in monovalent salt solutions with concentrations ranging from 0.001 to 0.1 M (10 > $1/\kappa$ > 1 nm). Force-distance measurements obtained with these techniques for z < 2 nm are difficult to interpret because a large repulsive "hydration force" of unknown origin overshadows the electrostatic force. However, fluorescence measurements with probes located 0, 1, and 2 nm from the surface of negatively charged phospholipid bilayer membranes also are consistent with the predictions of Eq. 21 (Winiski et al., 1988; Langner et al., 1989). Although these experiments are less definitive than the surface force measurements, all the available data support the assumption that water adjacent to a phospholipid membrane may be treated as a uniform dielectric medium when calculating the electrostatic potential.

The above analysis is a mean field theory. In it we assume the discrete nature of the ions in the aqueous phase, the electrostatic forces between them, and their finite size all may be ignored. An extensive series of Monte Carlo computer "experiments" suggests these assumptions are valid for the charge densities and salt concentrations encountered in biological systems (Carnie and Torrie, 1984); in these calculations the surface charge is assumed to be uniformly smeared over the interface. A statistical mechanical treatment of a membrane with discrete mobile surface charges also supports these assumptions, provided the charges are monovalent (Langner et al., 1989). Experimental results on membranes obtained with a variety of fluorescent, NMR and EPR probes indicate this mean field Gouy-Chapman theory is applicable to many biological problems (McLaughlin, 1989).

There are some examples where both the discreteness-of-charge calculations considered above and the Gouy-Chapman theory are inadequate. First, in surface force measurements with negatively charged phospholipid bilayers the two surfaces leap into contact from a distance of about 2 nm in the presence of divalent cations (Marra, 1986). The mean field theories described above cannot describe this striking observation. More sophisticated theories involving interactions between the ions in the aqueous phase, however, can account for the result (Jonsson and Wennerstrom, 1985). Second, the analysis above assumes the fixed charges are located precisely at the membrane-solution interface. If the fixed charges are buried within the low dielectric interior of the membrane, they produce an anomalously large potential within the membrane. These large potentials are observed with lipid soluble ions such as tetraphenylborate (Andersen et al., 1978) and should also be produced by charges on intrinsic proteins that span the membrane. Andersen et al. (1978) and Tsien and Hladky (1982) have presented theoretical descriptions of these potentials. Third, most of the fixed charges on the outer surface of a biological membrane are located on gangliosides or glycoproteins and are several Debye lengths from the membrane-solution interface. Although the above analysis is not applicable, the Poisson-Boltzmann equation with the appropriate boundary conditions does describe the electrostatic potential produced by charges located about 1 nm from the membrane-solution interface on gangliosides (Langner et al., 1988). Extensive calculations (Heinrich et al. 1982) suggest that the potential in the glycocalyx (thickness about 5 nm) adjacent to an erythrocyte is < 5 mV. Fourth, it is incorrect to assume the ions that accumulate in the aqueous diffuse double layer are point charges if their diameter is larger than the Debye length. Thus the interaction of peptides (Kim et al., 1991; Mosior and McLaughlin, 1991), and proteins such as melittin (Kuchinka and Seelig, 1989; Beschiaschvili and Seelig, 1990; Stankowski and Schwarz, 1990) with membranes cannot be described by the simplest form of the Gouy-Chapman theory. The Gouy-Chapman analysis has been extended theoretically (Carnie and McLaughlin, 1983) and tested experimentally (Alvarez et al., 1983) for a simple large divalent cation but the theory has not been developed well for larger more complex ions.

What is the biological relevance of the analysis we presented above? In our model we assume the fixed charges are located precisely at the membrane-solution interface, so our calculations will be most relevant to the potentials produced by charged phospholipids. The cytoplasmic surface of a plasma membrane typically contains between 10 and 30% negatively charged phospholipids, mainly phosphatidylserine for a mammalian cell membrane (Op den Kamp, 1979). Two examples of the possible biological importance of these negatively charged phospholipids relate to their ability to bind to positively charged amino acid residues on proteins. First, many intrinsic membrane proteins, proteins that span the phospholipid bilayer, have clusters of positively charged residues on their cytoplasmic surface (von Heijne, 1990a). For example in glycophorin, one of the major proteins in erythrocytes, 4 of the first 6 cytoplasmic residues after the single membrane-spanning region are positively charged (Ross et al., 1982). Experiments suggest that clusters of positive charges on several different intrinsic proteins are important in determining their orientation in the membrane (Nilsson and von Heijne, 1990; Hartmann et al., 1989). Hartmann et al. (1989) postulated the positively charged residues might sense the local electric potential produced by the negatively charged phospholipids on the cytoplasmic surface of the membrane and bind to these lipids. Hubbell (1990), however, postulated the lipid asymmetry in the rod outer segment disk membranes arises because of transmembrane charge asymmetry in rhodopsin, the major protein in these membranes. Second, several cytoplasmic proteins interact with negatively charged lipids on membranes (Adams and Pollard, 1989; Benfenati et al., 1989; Geisow and Walker, 1986; Klee, 1988); protein kinase C, an important component of the calcium/phospholipid second messenger system, is one example (Kikkawa et al. 1989; Parker et al., 1989). Like many of these proteins, protein kinase C bears a net negative charge. Thus to understand how these proteins interact electrostatically with negatively charged lipids it will be necessary to know the location of the positively charged residues on the protein. The complicated problem of solving the electrostatic potential profile around a protein has been greatly advanced in recent years by new computer methodologies (Sharp and Honig, 1990): programs now exist that can rapidly solve the Poisson-Boltzmann equation when the structure of the protein, and thus the location of all its charges, is known. As the structures of protein kinase C and the other membrane-binding proteins become available, it should be possible to combine this numerical approach with a discretness-of-charge analysis to understand the detailed electrostatic mechanisms by which these proteins interact with membranes.

ACKNOWLEDGEMENTS

Supported by NIH grants EL 36075 (R.M.), EY 06391 (R.M.), GM 24971 (S.McL.) and NSF grant DMB 9044656.

REFERENCES

Adams, R. J. and Pollard, T. D. (1989) Binding of myosin I to membrane lipids, *Nature* 340, 565–568.

Alvarez, O., Brodwick, M., Latorre, R., McLaughlin, A., McLaughlin, S., and Szabo, G. (1983) Large divalent cations and electrostatic potentials adjacent to membranes: Experimental results with hexamethonium, *Biophys. J.* 44, 333–342.

Andersen, O. S., Feldberg, S., Nakadomari, S., Levy, S., and McLaughlin, S. (1978) Electrostatic interactions among hydrophobic ions in lipid bilayer membranes, *Biophys. J.* 21, 35–70.

Barlow, C. A. and Macdonald, J. R. (1967) Theory of discreteness of charge effects in the electrolyte compact double layer, in P. Delahay (ed.), *Advances in electrochemistry and electrochemical engineering*, Interscience Publishers, New York, pp. 1–199.

Benfenati, F., Greengard, P., Brunner, J., and Bähler, M. (1989) Electrostatic and hydrophobic interaction of synapsin I and synapsin I fragments with phospholipid bilayers, *J. Cell Biol.* 108, 1851–1862.

Beschiaschvili, G. and Seelig, J. (1990) Melittin binding to mixed phosphatidylglycerol/phosphatidylcholine membranes, *Biochemistry* 29, 52–58.

Brown, R. H. (1974) Membrane surface charge: Discrete and uniform modelling, *Prog. Biophys. Mol. Biol.* 28, 343–370.

Cafiso, D. S. (1991) Lipid bilayers: Membrane-protein electrostatic interactions, *Current Opinion in Structural Biology*, in press.

Carnie, S. and McLaughlin, S. (1983) Large divalent cations and electrostatic potentials adjacent to membranes, *Biophys. J.* 44, 325–332.

Carnie, S. L. and Torrie, G. M. (1984) The statistical mechanics of the electrical double layer, *Adv. Chem. Phys.* 56, 141–253.

Chandler, W. K., Hodgkin, A. L., and Meves, H. (1965) The effect of changing the internal solution on sodium inactivation and related phenomena in giant axons, *J. Physiol.* 180, 821-836.

Cole, K. S. (1969) Zeta potential and discrete vs. uniform surface charges, *Biophys. J.* 9, 465–469.

Evans, E. A. and Parsegian, V. A. (1986) Thermal-mechanical fluctuations enhance repulsion between bimolecular layers, *Proc. Natl. Acad. Sci. USA* 83, 7132–7136.

Geisow, M. J. and Walker, J. H. (1986) New proteins involved in cell regulation by Ca^{2+} and phospholipids, *TIBS* 11, 420–423.

Green, W. N. and Andersen, O. S. (1991) Surface charges and ion channel function, *Annu. Rev. Physiol.* 53, 341–351.

Hartmann, E., Rapoport, T. A., and Lodish, H. F. (1989) Predicting the orientation of eukaryotic membrane-spanning proteins, *Biochemistry* 86, 5786–5790.

Heinrich, R., Gaestel, M., and Glaser, R. (1982) The electric potential profile across the erythrocyte membrane, *J. theor. Biol.* 96, 211–231.

Honig, B. H., Hubbell, W. L., and Flewelling, R. F. (1986) Electrostatic interactions in membranes and proteins, *Annu. Rev. Biophys. Biophys. Chem.* 15, 163–193.

Hubbell, W. L. (1990) Transbilayer coupling mechanism for the formation of lipid asymmetry in biological membranes, *Biophys. J.* 57, 99–108

Jackson, J. D. (1975) *Classical electrodynamics*, John Wiley & Sons, New York.

Jönsson, B. and Wennerström, H. (1989) Computer simulation studies of the electrical double layer, in S. H. Chen and R. Rajagapalon (eds.), *Statistical thermodynamics of micellar and microemulsion systems*, Springer-Verlag, Berlin/New York.

Jordan, P. C. (1987) How pore mouth charge distributions alter the permeability of transmembrane ionic channels, *Biophys. J.* 51, 297–311.

Kikkawa, U., Kishimoto, A., and Nishizuka, Y. (1989) The protein kinase C family: Heterogeneity and its implications, *Annu. Rev. Biochem.* 58, 31–44.

Kim, J., Mosior, M., Chung, L. A., Wu, H., and McLaughlin, S. (1991) Binding of peptides with basic residues to membranes containing acidic lipids, *Biophys. J.*, in the press.

Klee, C. B. (1988) Ca^{2+}-dependent phospholipid- (and membrane-) binding proteins, *Biochemistry* 27, 6645–6653.

Kuchinka, E. and Seelig, J. (1989) Interaction of melittin with phosphatidylcholine membranes. Binding isotherm and lipid head-group conformation, *Biochemistry* 28, 4216–4221.

Langner, M., Winiski, A., Eisenberg, M., McLaughlin, A., and McLaughlin, S. (1988) The electrostatic potential adjacent to bilayer membranes containing either charged phospholipids or gangliosides, in R. W. Ledeen, E. L. Hogan, G. Tettamanti, A. J. Yates, and R. K. Yu (eds.), *New trends in ganglioside research: Neurochemical and neuroregenerative aspects*, Liviana Press, Padova, pp. 121–131.

Langner, M., Cafiso, D., Marcelja, S., and McLaughlin, S. (1990) Electrostatics of phosphoinositide bilayer membranes: Theoretical and experimental results, *Biophys. J.* 57, 335–349.

Levine, S., Mingins, J., and Bell, G. M. (1967) The discrete-ion effect in ionic double-layer theory, *J. Electroanal. Chem.* 13, 280–329.

Loosley-Millman, M. E., Rand, R. P., and Parsegian, V. A. (1982) Effects of monovalent ion binding and screening on measured electrostatic forces between charged phospholipid bilayers, *Biophys. J.* 40, 221–232.

Marra, J. (1986) Direct measurement of the interaction between phosphatidylglycerol bilayers in aqueous electrolyte solutions, *Biophys. J.* 50, 815–825.

McLaughlin, S. (1977) Electrostatic potentials at membrane-solution interfaces, *Current Topics in Membranes and Transport* 9, 71–144.

490

McLaughlin, S. (1989) The electrostatic properties of membranes, *Annu. Rev. Biophys. Biophys. Chem.* 18, 113–136.

Mosior, M. and McLaughlin, S. (1991) Peptides that mimic the pseudosubstrate region of protein kinase C bind to acidic lipids in membranes, *Biophys. J.*, in the press.

Nelson, A. P. and McQuarrie, D. A. (1975) The effect of discrete charges on the electrical properties of a membrane. I., *J. theor. Biol.* 55, 13–27.

Nilsson, I. and von Heijne, G. (1990) Fine-tuning the topology of a polytopic membrane protein: Role of positively and negatively charged amino acids, *Cell* 62, 1135–1141.

Onsager, L. and Samaras, N. N. T. The surface tension of Debye-Hückel electrolytes, *J. of Chem. Phys.* 2, 528–536.

Op den Kamp, J. A. F. (1979) Lipid asymmetry in membranes, *Annu. Rev. Biochem.* 48, 47–71.

Parker, P. J., Kour, G., Marais, R. M., Mitchell, F., Pears, C., Schaap, D., Stabel, S., and Webster, C. (1989) Protein kinase C — a family affair, *Molecular and Cellular Endocrinology* 65, 1–11.

Ross, A. H., Radhakrishnan, R., Robson, R. J., and Khorana, H. G. (1982) The transmembrane domain of glycophorin A as studied by cross-linking using photoactivatable phospholipids, *J. Biol. Chem.* 257, 4152–4161.

Sauve, R. and Ohki, S. (1979) Interactions of divalent cations with negatively charged membrane surfaces. I. Discrete charge potential, *J. theor. Biol.* 81, 157–179.

Sharp, K. A. and Honig. B. (1990) Electrostatic interaction in macromolecules: Theory and applications, *Annu. Rev. Biophys. Biophys. Chem.* 19, 301–332.

Stankowski, S. and Schwarz, G. (1990) Electrostatics of a peptide at a membrane/water interface. The pH dependence of melittin association with lipid vesicles, *Biochim. Biophys. Acta* 1025, 164–172.

Stigter, D. and Dill, K. A. (1986) Interactions in dilute monolayers of long-chain ions at the interface between *n*-heptane and aqueous salt solution, *Langmuir* 2, 791–796.

Stillinger, F. H. (1961) Interfacial solutions of the Poisson-Boltzmann equation, *J. Chem. Phys.* 35, 1584–1589.

Tsien, R. Y. and Hladky S. B. (1982) Ion repulsion within membranes, *Biophys. J.* 39, 49–56.

von Heijne, G. (1990) The signal peptide, *J. Membrane Biol.* 115, 195–201.

Vorotyntsev, M. A. and Ivanov, S. N. (1988) Potential drop at the insulator/electrolyte solution interface, *Elektrokhimiya* 24, 805–807.

Vorotyntsev, M. A. and Ivanov, S. N. (1989a) Energy of the image forces and the interaction of an ion with a charged group at the insulator/electrolyte solution interface, *Elektrokhimiya* 25, 550–554.

Vorotyntsev. M. A. and Ivanov, S. M. (1989b) Ionic adsorption isotherms at homogeneous insulator/electrolyte solution interfaces, *Elektrokhimiya* 25, 554–557.

Vrij, A. (1966) The equation of state of ionized insoluble monolayers, *A. Vlaam. Acad. Wet., Lett. Schone Kunsten Belg.* A26, 13–26.

Wagner, C. (1924) Die oberfiächenspannung verdünnter elektrolytlösungen, *Physik. Zeitschr.* XXV, 474–477.

Winiski, A. P., Eisenberg, M., Langner, M., and McLaughlin, S. (1988) Fluorescent probes of electrostatic potential 1 nm from the membrane surface, *Biochemistry* 27, 386–392.

STRUCTURAL REARRANGEMENTS IN LIPID BILAYER MEMBRANES

YURY CHIZMADZHEV
Frumkin Institute of Electrochemistry
Academy of Sciences of the USSR
31 Leninsky Prospect, Moscow V-71, 117071
U S S R

ABSTRACT. We will examine local structural rearrangements in lipid bilayers, which are of great functional significance for cell biology and biotechnology. The choice of this topic will permit interesting analogies to be drawn with classical electrochemistry and chemical kinetics. Formation of two types of intermediate structures, inverted pores and stalks, will be dwelt upon. These structures are responsible for the sharp increase in permeability of membranes and for their fusion. While permeability is induced by the electric field, fusion is induced by close contact between membranes. The phenomenology of these events and their theoretical interpretation will be examined. As a result, a molecular mechanism will be formulated and experimental evidence in its support will be provided.

1. Introduction

During the last decade considerable research work has been devoted to the sharp increase in permeability and breakdown of lipid bilayers and cell membranes, as induced by an electric field. Nowdays, the mechanism of this phenomenon has been substantially elucidated, due to the efforts of many researchers of different countries - Kinozita, Tsong, Zimmermann, Benz, Dimitrov, Deuticke, Neumann, Petrov, Weaver, Sowers, Teissie, etc., just to name a few. This list is far from complete, since an exhaustive review of this subject is outside the scope of the present contribution. Rather, our presentation will be mainly based on our own published results, which are hopefully recognized by the scientific community. There are good reasons to believe that the phenomenon under discussion originates from the formation in the bilayer of new structures, namely inverted pores having a hydrophylic inner surface. Planar lipid membranes bordered by their meniscus are metastable systems with respect to these pores, whose radii change under the action of heat fluctuations. Until the pore radii are less than a critical value, the system is stable. As this critical value is exceeded, the resulting supercritical pore undergoes an irreversible growth which ultimately leads to membrane breakdown. The whole phenomenon is governed by laws known from physical chemistry of surfaces. However, some specific features are associated with

R. Guidelli (ed.), Electrified Interfaces in Physics, Chemistry and Biology, 491–507.
© 1992 *Kluwer Academic Publishers.*

electric-field induced effects, which amply justifies the inclusion of this phenomenon among those concerning electrified interfaces.

Let us now turn to the problem of membrane fusion. This process also proves to be accompanied by certain structural rearrangements. Using a model of two planar bilayers brought into contact by an excess of hydrostatic pressure, it was shown that, after a time lag, spontaneous monolayer fusion takes place, leading to a trilaminar structure. The process starts with the formation of a stalk between two adjacent monolayers. At the beginning the area of the stalk increases linearly in time, and ultimately tends to a saturation value. The trilaminar structure is rather stable. However, an electric field induces in the bilayer supercritical pores which lead to its breakdown, and therefore to total membrane fusion. Formation of both the stalk and hydrophylic pores involves monolayer bends, whose directions, however, are different. Therefore, in terms of membrane mechanics, both processes are described by the same bending elasticity modulus. This provides an excellent opportunity to reduce the number of unknown parameters by comparing the data for the rates of the first and second stage. Thus, electroporation and membrane fusion, although completely different at first sight, turn out to be related and lend themselves to a unified description. It is also significant that this description is akin to that accepted in electrochemical kinetics. This problem deserves to be dwelt upon in more detail because it will help us to establish a correlation between classical electrochemical problems and an exotic system such as a lipid bilayer. The literature on membrane electroporation and fusion uses two approaches. One is based on linear analysis of stability [1, 2] while the other is strictly nonlinear in nature (for a review with a complete bibliography see Ref. 3; in this contribution we shall confine ourselves to a minimum of references). The latter approach considers certain intermediate structures which cannot be treated as small perturbations. Such a description is similar to that accepted in chemical kinetics and based on the concept of the activated complex (transition state). A good example is provided by any reaction of electron transfer in a polar medium. Here the structural rearrangement consists in the repolarization of the solvent. While at large distances the dynamics of the process is conveniently described by a phonon approximation [4, 5], at short distances one has to consider nonlinear effects. However, even in a linear approximation the transition to an activated state is a multiphonon process. If we wish to continue with analogies, we may note that the formalism used in the quantitative description of structural rearrangements in membranes coincides in practice with that used in the kinetic theory of phase transitions.

All membrane processes to be described in the following take place when the membrane is subject to a strong electric field. Since, as a rule, membranes themselves bear a surface charge and are in contact with electrolytic solutions, diffuse layers are formed at the resulting interfaces. All this makes it possible to investigate the occurring events by electrical measurements. Therefore, in what follows we will rely on those concepts of membrane electrostatics described elsewhere in this volume. We will be primarily interested in the molecular mechanisms of electroporation and fusion on lipid bilayers, which are models of cell membranes. A natural question which can be raised is whether it is possible to extrapolate the results obtained with these model systems to cell membranes. This question is rather topical because electroporation and fusion have already been applied in biotechnology and gene engineering. Among other things, electroporation has permitted the introduction of foreign DNA into bacterial and animal cells by a method which exceeds all traditional techniques by its efficiency and universal use; it is also a very efficient method to obtain hybrid cells. Far from dwelling on all possible applications, we will confine ourselves to providing some arguments in favour of the view that the mechanisms of electroporation and electrofusion in cell systems basically coincide with those found in lipid bilayers. After these introductory remarks we will now turn our interests to experimental facts and their theoretical analysis.

2. Membrane Electroporation

We will consider the results obtained on planar bilayer lipid membranes (BLMs). The experimental setup is a Teflon cell divided into two compartments by a diaphram with a hole about 1 mm in diameter. The cell is filled with an electrolyte whose composition and concentration can be changed (Fig.1).

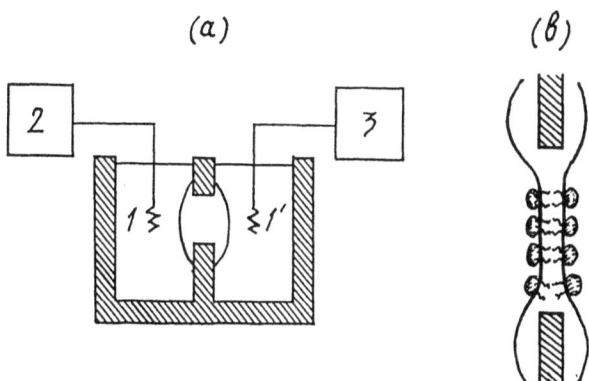

Figure 1. Experimental set-up: 1 and 1', reversible electrodes; 2, generator; 3, measuring device. (b) Bilayer lipid membrane formed spontaneously after spreading a drop of a solution of phospholipids in an organic solvent.

In 1962 Mueller *et al.* [6] showed that, upon applying a drop of a solution of phospholipids in an organic solvent to the hole, the drop spreads out spontaneously giving rise to a black membrane about 50 Å thick, surrounded by a meniscus. BLMs are good dielectrics, with a dielectric constant ε_m of about 2-3, specific resistance ρ_m of about 10^8 Ω/cm^2, and specific capacitance C_m of about 1 $\mu F/cm^2$. They are usually formed with phospholipids whose polar heads are negatively charged. As a result, the electric potential across a symmetrical membrane varies as shown in Fig. 2.

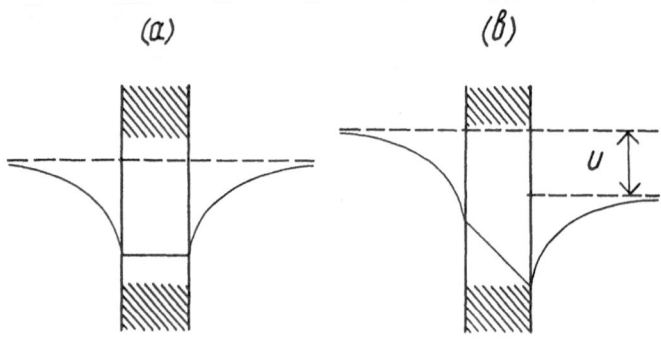

Figure 2. Potential-distance profile in a symmetrical system (a) and upon application of an external electric field U (b).

The potential drops across the two diffuse layers are determined, as usual, by the surface charge density and the ionic strength of the solution. For the sake of simplicity the figure does not show the surface dipole potentials produced by the polar heads of the phospholipid, because these potentials remain constant with changing the ionic strength of the solution. Upon applying a potential difference U between the two reversible electrodes 1 and 1' of Fig.1, the whole potential difference U is practically located inside the membrane, because its capacitance is about two orders of magnitude less than that of the two diffuse layers over the usually adopted range of ionic strengths.

A typical oscillogram of the current following a potential step U is show in Fig. 3a. The capacitive current due to the membrane charging is followed by a background current which, at a random instant of time, is replaced by high-amplitude fluctuations; these are followed by an irreversible increase in current, which ultimately reaches a saturation value. The latter value is determined by the ohmic resistance of the electrolytic cell, since the membrane is now broken and does not contribute to the total resistance. This process can be conveniently characterized by the membrane lifetime under the applied potential U. This lifetime is random and varies notably from experiment to experiment. Therefore, one may

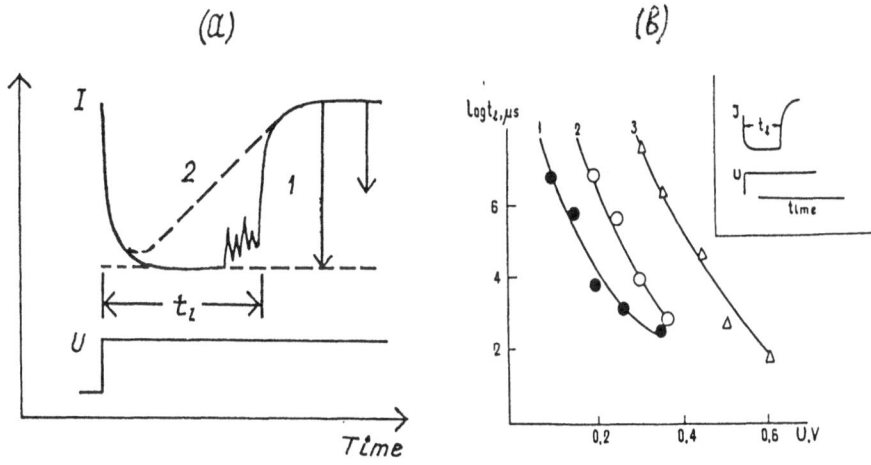

Figure 3. (a) Current vs. time curves following the potential step U: 1, irreversible breakdown; 2, reversible electroporation. (b) BLM mean lifetime vs. U: (●) lysophosphatidylcholine (LPC); (Δ) phosphatidylethanolamine (PE); (o) phosphatidylcholine (PC).

only refer to a membrane mean lifetime, which depends upon U as shown in Fig. 3b. The dependence upon E is extremely sharp: five orders of magnitude along the abscissas as U is changed from 0.1 to 0.6 V.

An altogether different and peculiar behaviour [8] is exhibited by membranes modified by UO_2^{++} ions or containing oxidized cholesterol (Fig. 3a, curve 2). Immediately after the initial flow of capacitive current, the current begins growing gradually and reversibly, reaching the same saturation value. However, if immediately after reaching the plateau the potential difference U is dropped, the system returns to its initial state; this implies that at the current plateau the membrane is mechanically stable, even though it has such a high conductance that the current is limited by the ohmic resistance of the electrolytic cell rather than by that of the membrane. If the potential difference U is dropped after a longer time interval, the membrane breaks down. Hence, the process consists of a reversible stage of increase in conductance, which is followed by an irreversible stage of breakdown.

2.1. IRREVERSIBLE ELECTRIC BREAKDOWN

Let us first discuss the mechanism of irreversible membrane breakdown. The first attempt to explain this phenomenon was made by Crowley [9], who regarded the breakdown as an electromechanical collapse of the membrane. At a glance this approach seems quite attractive. The electric field compresses the membrane, which at first resists mechanically. However, as the membrane thickness decreases, the force of compression increases faster than that of resistance, until a critical thinning is reached beyond which the collapse takes place. This thinning is estimated to be 30% of the initial thickness. This interpretation implies that during the breakdown (for instance, during the fluctuation period of curve 1 in Fig. 3) the capacitance should increase significantly. On the contrary, experiments have shown that the capacitance remains constant [7]. This experimental observation, as well as other findings which we shall not dwell on here, indicate that the breakdown is a local phenomenon, due to the occurrence and progress of some conducting structural defect. It is natural to postulate that this defect is an aqueous pore. Before discussing its structure, let us consider the change in membrane free energy during the formation of such a pore:

$$\Delta F = -\pi r^2 \sigma + 2\pi r \gamma - \pi r^2 C_m \left(\frac{\varepsilon_w}{\varepsilon_m} - 1 \right) \frac{U^2}{2} \tag{1}$$

Here r is the pore radius, σ is the membrane tension, γ is the pore linear tension and ε_w is the dielectric constant of water. The physical significance of the terms in eqn.(1) is quite transparent. The first term expresses the decrease in surface energy during pore formation; the second term, which is proportional to the linear tension, measures the energy increase due to pore edge formation; finally, the third term accounts for the decrease in energy due to the replacement of the hydrophobic medium with $\varepsilon_m \approx 2$ by a polar medium with $\varepsilon_w = 80$ inside the pore. Thus, the stress provided by the meniscus tends to increase the pore size; conversely, the linear tension promotes pore closing while the potential difference U across the membrane, with the resulting electric field, leads to an increase in pore radius. The field effect can be regarded as a tendency of the polar solvent to displace the hydrophobic medium from the region of high electric field. The free energy depends parabolically upon the pore radius, with the maximum

$$\Delta F^* = \frac{\pi \gamma^2}{\sigma + CU^2/2}$$

lying at:

$$r^* = \frac{\gamma}{\sigma + CU^2/2} \tag{2}$$

where $C = C_0(\varepsilon_w/\varepsilon_m - 1)$.

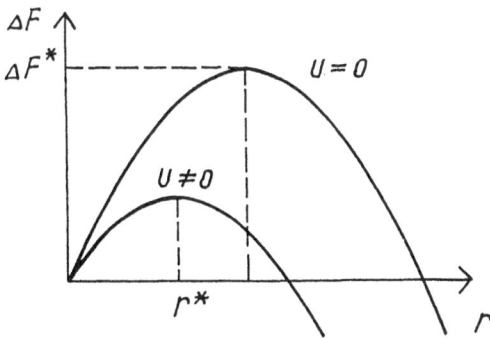

Figure 4. Work of pore formation vs. pore radius at different U values.

From the very form of the curve in Fig. 4 it is apparent that the planar BLM is a metastable system. In other words, the pores whose radius is less than the critical radius r* tend to close. If, however, as a result of heat fluctuations, the size of one of the pores exceeds the critical value, its radius begins to increase indefinitely, ultimately leading to membrane breakdown. The initial barrier ΔF^* at U=0 is rather high, and hence the membrane possesses a long lifetime. However, when an electric field is applied, the height of the barrier decreases, and so does the value of the critical radius. This qualitative picture accounts for the observed dependence of the mean lifetime on E (Fig. 3b). The quantitative theory (see Ref. 3 for a review) uses approaches developed in the description of nucleation phenomena. The resulting formalism considers pore diffusion in the radius space in the presence of an external restoring force (r<r*). If one does not strive to calculate the preexponential factor, which is not of paramount significance, one can obtain an expression for the mean lifetime from simple considerations based on absolute reaction rate theory. Thus, upon assuming that the pores are distributed according to the Boltzmann equilibrium law up to the critical size, we obtain:

$$P^* \approx \exp(-\Delta F^*/kT) \tag{3}$$

where P* is the probability of finding a pore of critical radius. The mean lifetime is then given by:

$$\bar{t}_1 \approx \exp(\Delta F^*/kT) \tag{4}$$

where ΔF^* is expressed by eqn.(2). The exact relationship for the case of a high barrier is:

$$\bar{t}_1 = \frac{(kT)^{3/2}}{4\pi cSD\gamma\left(\sigma+\frac{CU^2}{2}\right)^{1/2}} \exp\left[\frac{\pi\gamma^2}{kT\left(\sigma+\frac{CU^2}{2}\right)}\right] \tag{5}$$

where c is the pore concentration, S is the membrane area and D is the pore diffusion coefficient in the radius space. In practice, γ is an adjustable parameter to be determined by fitting experimental curves, whereas σ is found by independent means. Equation (5) compares satisfactorily with experimental results (see Fig. 3b).

2.2. REVERSIBLE ELECTROPORATION

While the above formalism describes membrane breakdown in a natural way, reversible electroporation requires special consideration. In the preceding section we did not touch on the problem of pore molecular organization, but this problem must now be considered, as will become clear in the following. There are several pieces of experimental evidence pointing to the pores of relatively large size (r>0.5 nm) being hydrophilic, namely having their inner surface covered with polar heads of phospholipids (Fig. 5).

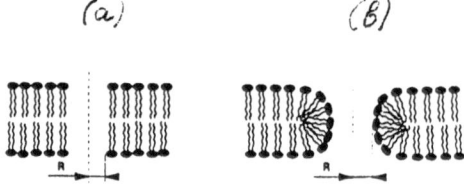

Figure 5. Structures of hydrophobic (a) and hydrophilic (b) pores.

Among these pieces of evidence we may cite a cation-anion pore selectivity which is about 10, closeness of γ to $\sigma\delta$ (δ is the membrane thickness) and, finally, a quite definite dependence of lifetime on the molecular geometry of the phospholipid molecules. The latter piece of evidence will be dwelt upon in more detail, since its consideration will subsequently be used to elucidate the mechanism of membrane fusion. In recent years, use has been frequently made of a classification of phospholipid molecules on the basis of their efficient molecular geometry (Fig. 6): inverse cones (e.g.,LPC, I in Fig. 6), cylinders (PC, II in Fig. 6), and cones (PE, III in Fig. 6). It is physically evident that a hydrophilic pore will be most conveniently formed by molecules shaped as inverse cones, such as lysolecithins.

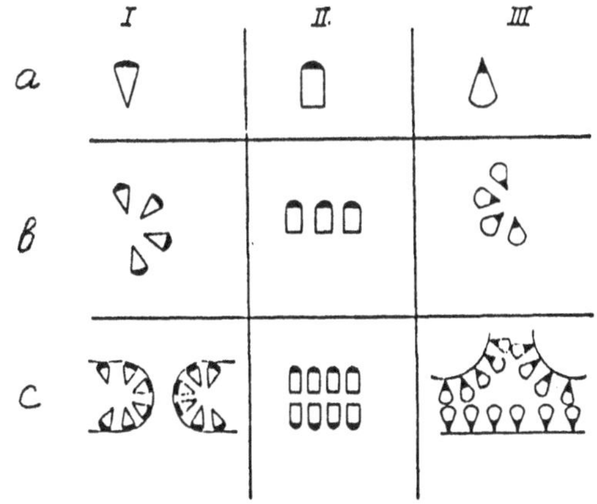

Figure 6. Efficient geometry of phospholipid molecules: I, LPC; II, PC; III, PE.

Correspondingly, the shortest lifetime should be observed in the case of lysoform-containing membranes if, certainly, the pores do have the hydrophilic edge which is formed as a result of an intramembrane-directed monolayer bend. Correspondingly, the longest lifetime should be characteristic of PE membranes. This prediction is supported by the experimental data reported in Fig. 3b. This result casts some doubt on the correctness of the energy curve in Fig. 4, at least in the region of small radii. Indeed, this curve was obtained upon assuming that linear tension does not depend on pore radius. However, if pore formation involves a monolayer bend (i.e., the turn of phospholipid molecules), then it will require excessive energy losses over the range of small radii, where the curvature is large. Another effect to be accounted for is hydrational repulsion of pore edges, which is rather significant at distances less than 1 nm. Therefore, there are good reasons to believe that the energy curve for a hydrophilic pore should have the form shown by the solid curve 1 in Fig. 7a.

It is practically impossible to calculate the rising branch of curve 1 quantitatively over the range of small radii, though this is not actually required since hydrophilic pores of small radius are so energetically unfavourable that they are not realized at all. Calculations show [10] that this region has another energetic branch which conforms to hydrophobic pores (Fig. 5a) and extends to far lower radii, up to the point of intersection (see dashed

(a)   (b)

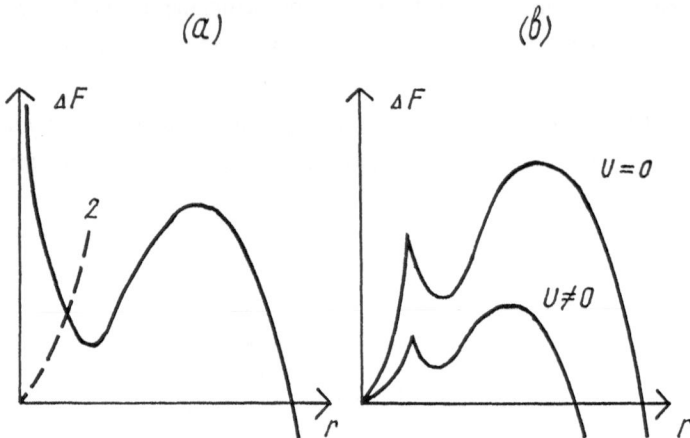

Figure 7. (a) Energy of hydrophylic (1) and hydrophobic (2) pores vs. their radius, r. (b) Ditto, at different applied potentials.

curve 2 in Fig. 7a). The estimate of curve 2 takes into account the hydrophobic attraction of pore edges which results in the equation:

$$F_2(r) = 2\pi\delta r\sigma(r), \tag{6}$$

where $\sigma(r) = \sigma(\infty)I_1(r/\rho)/I_0(r/\rho)$.

Here $I_n(x)$ is a modified Bessel function of the n-th order and ρ is the order parameter. When a potential is applied, the energy curves are modified as shown in Fig. 7b. The above considerations make it possible to explain reversible electroporation qualitatively and also, after simple calculations, to obtain a quantitative description of the phenomenon as a whole with an estimate of all kinetic parameters of the process - height of the intermediate barrier, critial radius, number of pores, etc. Indeed, the smooth current rise as shown in curve 2 of Fig. 3 is attributable to the increase in the number of pores after the first barrier and the gradual increase in their radii when the applied potential changes from U=0 to U≠0 (see Fig. 7b).

All stages of the process where studied quantitatively using the experimental signal depicted in Fig. 8. In all cases, both the current dynamics and the current-voltage characteristics were measured. Comparison between experimental behaviour and theoretical

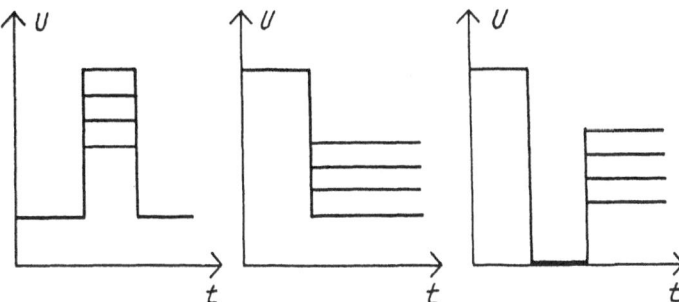

Figure 8. Experimental signal.

predictions points out that the system has three characteristic times. The first and shortest time is less than 1 µs and conforms to the conductance of a pore with a fixed radius; the second, about 10 ms, corresponds to a redistribution over the radii; the third, more than 10 s, describes pore closure. The typical number of pores in these experiments proved to be 10^4-10^5; the first barrier is located at $r^* \approx 0.5$ nm whereas the energetic minimum is located at $r_{min} \approx 1$ nm. Additional evidence in favour of this picture of electroporation will be provided when considering the mechanism of membrane fusion and a comparative analysis of data on electroporation and fusion.

3. Mechanism of Lipid Bilayer Fusion

Experimental studies of fusion of planar lipid bilayers make convenient use of a cell having two partitions with coaxial holes on which membranes are formed. The distance between the partitions is variable. The cell has three compartments with electrodes in all of them. Hydrostatic pressure can be varied and, thus, the two membranes can be blown towards each other up to a close approach (Fig. 9). The apparatus makes it possible to record transient and capacitive currents and to measure surface charges of all membrane monolayers before and after fusion. The area of black membranes can be recorded optically. This setup was suggested in [11], it was used in [12], and in our version it was modified so as to measure some additional quantities [13]. The phenomenology of the process is illustrated in Fig. 10. Upon applying a pressure excess ΔP the membranes pass to the state of plane-parallel contact (b). The distance between the two bilayers is determined by a balance of all forces, namely molecular and coulombic forces as well as

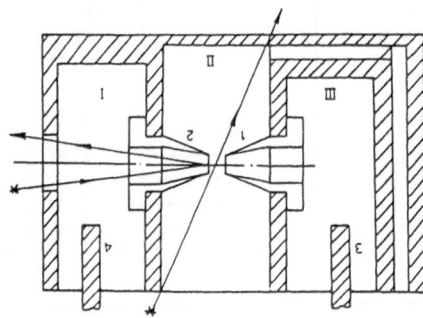

Figure 9- Design of an experimental cell. 1 and 2 are openings in the walls of Teflon discs, 3 and 4 are calibrated Teflon rods connected to micrometer screws, I, II and III are compartments of the cell. By shifting compartment III within compartment II it is possible to change the distance between openings 1 and 2. Visual monitoring is carried out through the glass windows in the front and side walls of the cell in both transmitted and reflected light [13].

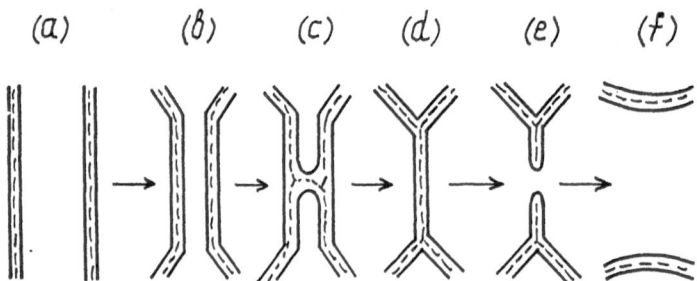

Figure 10. Stages of fusion of membrane bilayers.

hydrostatic and hydrational pressures. After a time lag, a strap is formed spontaneously between the adjacent monolayers (c), whose expansion leads to the trilaminar structure (d). State (d) is very stable, but after application of a potential step U electroporation takes place (e), followed by an irreversible breakdown of the contact bilayer and the resulting formation of a membrane tube (f). If pressure ΔP is applied to this tube, it breaks down and the system returns to its initial state (a). Thus, the cycle is complete and can be repeated many times.

The stage of monolayer fusion leading to the formation of the trilaminar structure is of special interest. *A priori* one might think that its formation is due to an interpenetration of the two bilayers. However, direct experiments have shown that this is not the case. In these experiments one of the membranes was made from neutral lipids and the other from charged ones. The surface charge on the two sides of the contact membrane was then measured, and it was found that one of monolayers is neutral whereas the other has the same charge as the original charged bilayer. Therefore, the interpenetration mechanism is not operative; rather, monolayer fusion takes place via the formation of a bridge, a strap between the adjacent bilayers, called "stalk". The probability of stalk formation and evolution is determined by its energy which, in turn, depends on the efficient molecular geometry of phospholipid molecules. In terms of mechanics of *continua*, this implies a dependence upon the spontaneous curvature of monolayers. Energy curves are shown in

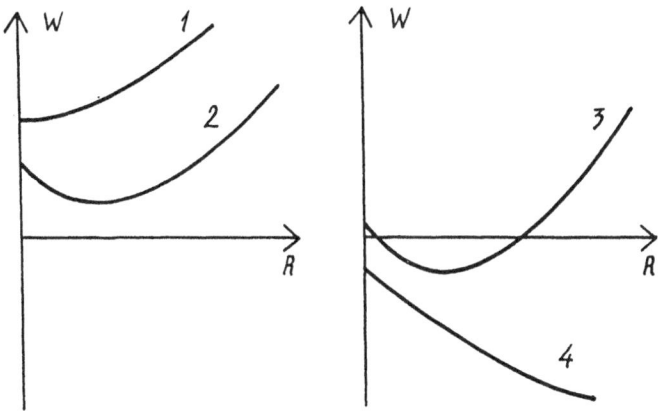

Figure 11. Energy of a stalk as a function of its radius R for different values of the spontaneous curvature, which increases progressively from curve 1 to 4.

Fig. 11 [14]. Curve 4 refers to the case in which phospholipid molecules are efficiently shaped as a cone, which is optimal for stalk formation (see Fig. 6). This theoretical approach predicts that the expectation time of monolayer fusion, t_{mf}, should be a minimum for phosphatidylethanolamine and a maximum for lysolipids. The t_{mf} value is determined experimentally from the time dependence of the capacitive current I_c during stages (b) -> (c)->(d). A typical curve is shown in Fig. 12. Over the time range of the first plateau the capacitance is low, since it corresponds to the system of parallel bilayers separated by the aqueous slab. However, as is apparent from the figure, the capacitance starts increasing at time t_{mf}, first linearly, and ultimately attaining saturation. Over the time range of the second plateau the capacitance corresponds to a single bilayer. Thus, an experiment so conducted

504

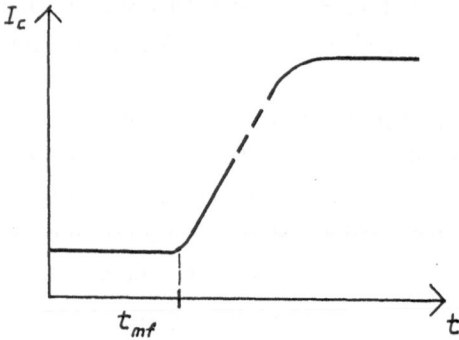

Figure 12. Capacitive current vs. time during stalk formation and growth.

permits us to measure t_{mf} and, at the same time, to obtain information on stalk evolution. Even without reporting any concrete data, it should be stressed that the predictions of the theory are in qualitative agreement with experimental data. Thus, t_{mf} is indeed a minimum for lipid cones and a maximum for inverted cones.

Let us now briefly dwell on the quantitative evidence of the whole theory, which relies on the assumption that local structural rearrangements in membranes play a role as intermediate states leading to electroporation and fusion. Attention should be focused on the circumstance that fusion proceeds through a sequence of two intermediate structures, i.e. a stalk which is a trilaminar structure still in embryo, and a pore which promotes breakdown of the contact bilayer and completes fusion. It should also be noted that stalk formation is possible as membrane segments approach each other at distances of $\approx$ 5-10 Å. This approach is possible during heat fluctuations which promote semistalk formation, with hydrophobic tips which attract each other at short distances (for a review see Ref. 13).

Let us now return to the description of crucial experiments carried out having in mind that, from the viewpoint of membrane mechanics, stalks and pores are related deformations which differ only in the direction of the monolayer bends. Therefore, they are described by the same elastic bending modulus B and depend on the spontaneous curvature ξ. The experimental strategy consists in measuring the mean lifetime of a bilayer in an electric field, t_1, as well as the expectation time of monolayer fusion, t_{mf}, and in comparing them with theory. Measurements were carried out on membranes of mixed composition, with different concentrations, c_1, of lysolipids . Thus, the first stage consists in measuring $t_1(U,c_1)$. Fitting of experimental curves by the use of eqn. (5) permits one to determine the linear tension $\gamma(c_1)$. According to the theory, γ is given by:

$$\gamma = (\pi B/\delta) - 2\pi B\xi_2 - (2\pi B/\delta)(\xi_1 - \xi_2)\delta A c_0 \qquad (7)$$

Here ξ_1 is the spontaneous curvature of a monolayer of given composition when the concentration of lysoforms in the aqueous solution is c_0. A is the membrane-solution partition coefficient (i.e., $Ac_0 = c_1$), and ξ_2 is the spontaneous curvature of the initial monolayer in the absence of lysoforms. On the other hand, t_{mf} is given by:

$$\log t_{mf} = const + 6.15(B\delta/kT)(\xi_1 - \xi_2)Ac_0 \qquad (8)$$

Thus, eqns. (7) and (8) express the dependence of γ and t_{mf} on c_0. If we denote the quantity $B(\xi_1 - \xi_2)\delta A$ by the symbol ξ for the sake of brevity, the slopes of the $\gamma(c_0)$ and $t_{mf}(c_0)$ functions are given by:

$$d\gamma/dc_0 = -2\pi\xi/\delta \qquad (9)$$

and

$$d\log t_{mf}/dc_0 = 6.15\ \xi/(kT) \qquad (10)$$

Thus, both slopes are expressed as a function of the same unknown mechanical parameter ξ, whereas the other parameters are known. Therefore, by measuring $d\gamma/dc_0$, we may estimate ξ and substitute its value into eqn. (10). By this procedure we obtain the theoretical slope $d\log t_{mf}/dc$, which can be compared with the corresponding experimental value, without having to use adjustable parameters. The result of this procedure is reported in Fig. 13. From Fig.13b it is apparent that agreement is excellent, thus supporting the starting assumptions.

In conclusion, the possibility of extrapolating these strategies to cell membranes should be envisaged. As concerns electroporation, at a qualitative level such a possibility is quite realistic. Certainly, some differences with respect to planar BLMs exist, due to the fact that the cell is a vesicular system and its membrane is not in contact with a meniscus. This specific problem was analyzed theoretically, so that the formalism to describe electroporation of vesicular lipid bilayers is already available. Another distinctive feature of cell membranes is the presence of proteins, glycocalyx and cell (or membrane) skeleton. Such a complex structure of cell membranes leads to a change in characteristic times of all stages, and in particular to a deceleration of membrane recovery. However, pore formation, as such, must necessarily occur in lipid domains.

The situation is more complicated in the case of cell electrofusion. Attempts to

506

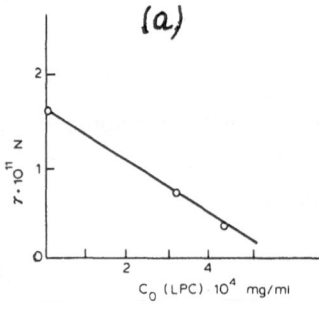
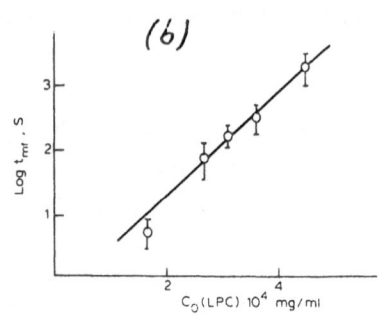

Figure 13. Linear tension γ as a function of LPC concentration in aqueous solution; bilayers from PE in squalene [13]. (b) Expectation time of monolayer fusion as a function of LPC concentration in aqueous solution [13].

reveal trilaminar structures in cell suspensions during electrical treatments have not yet been successful. Another fusion mechanism was considered [15], via interactions of coaxial pores induced at adjacent membranes by an electric pulse. The final choice of the mechanism requires further investigations.

References

(1) Steinchen, A., Gallez, D. and Zanfeld, A. (1982) "A viscoelastic approach to the hydrodynamic stability of membranes", J. Coll. Interface Sci. **85**, 5-15.

(2) Dimitrov, D. S. and Jain, R. K. (1984) "Membrane stability", Biochim. Biophys. Acta **779**, 437-468.

(3) Chizmadzhev, Yu.A. and Pastushenko, V. F. (1988) "Electric breakdown of bilayer lipid membranes", in J. B. Ivanov (ed.), *Thin Lipid Films*, Marcel Dekker Inc., New York and Basel, pp. 1059-1120.

(4) Marcus, R. A. (1968) "Electron transfer at electrodes and in solution: comparison of theory and experiment", Electrochim. Acta **13**, 995-1004.

(5) Dogonadze, R.R. and Kuznetsov, A.M. (1983) "Quantum electrochemical kinetics: continuum theory", in B. E. Conway, J. O' M. Bockris, E. Yeager, S. U. M. Khan and R. E. White (eds.), *Comprehensive Treatise of Electrochemistry*, Plenum Press, New York and London, pp. 1-40.

(6) Mueller, P., Rudin, D. C., Tien, H. T. and Wescott, W. C. (1962) "Reconstitution of cell membrane structure *in vitro* and its transformation into an excitable system", Nature **194**, 979-980.

(7) Abidor, I. G., Arakelyan, V. B., Chernomordik, L. V. Chizmadzev, Yu. A., Pastushenko, V. F. and Tarasevich, M. R. (1979) "Electric breakdown of bilayer lipid membranes" Bioelectrochem. Bioenerg. **6**, 37-52.

(8) Abidor, I. G., Chernomordik, L. V., Sukharev, S. I. and Chizmadzev, Yu. A. (1982) "The reversible electrical breakdown of bilayer lipid membranes modified by uranyl ions", Bioelectrochem. Bioenerg. **9**, 141-148.

(9) Growley, J. M. (1973) "Electrical breakdown of bimolecular lipid membranes as an electrochemical instability" Biophys. J. **13**, 711-724.

(10) Glaser, R. W., Leikin, S. L., Chernomordik, L. V., Pastushenko, V. F. and Sokirko, A. I. (1988) "Reversible electrical breakdown of lipid bilayers: formation and evolution of pores", Biochim. Biophys. Acta **940**, 275-287.

(11) Liberman, E. A. and Nenashev, V. A. (1968) "Interaction of artificial lipid membranes", Biofizika **13**, 193-196 (in Russian).

(12) Neher, E. (1974) "Asymmetric membrane resulting from the fusion of two black lipid membranes" Biochim. Biophys. Acta **373**, 327-336.

(13) Chernomordik, L. V. Melikyan, G. B. and Chizmadzhev, Yu. A. (1987) "Biomembrane fusion: a new concept derived from model studies using two interacting planar lipid bilayers", Biochim. Biophys. Acta **906**, 309-352.

(14) Markin, V. S., Kozlov, M. M. and Borovjagin, V. L. (1984) " On the theory of membrane fusion. A stalk mechanism" Gen. Physiol. Biophys. **5**, 361-377.

(15) Kuzmin, P. I., Pastushenko, V. F., Abidor, I. G., Sukharev, S. I., Barbul, A. I. and Chizmadzhev, Yu. A. (1988) "Theoretical analysis of cell electrofusion", Biol. Membrany **5**, 600-612 (in Russian).

EVALUATION OF THE SURFACE POTENTIAL AT THE MEMBRANE-SOLUTION INTERFACE OF PHOTOSYNTHETIC BACTERIAL SYSTEMS.

A.CORAZZA, B.A.MELANDRI, G.VENTUROLI and R.CASADIO
Laboratory of Biochemistry and Biophysics,
Dept. of Biology, University of Bologna, Via Irnerio, 42
I-40126 Bologna, Italy

ABSTRACT. In chromatophores from photosynthetic bacteria, the dependence of the Nernst redox potential of electroactive centers of membrane proteins upon the ionic strength of the electrolyte solution is detected by using redox potentiometry in combination with fast kinetic spectroscopy. The thermodynamic parameter, as measured in the bulk, differs from that at the redox site embedded in the membrane by a value equal to the electrostatic potential at the solution-membrane interface. In order to reconcile experimental results with this notion at different ion concentrations, surface potentials are calculated by using the Gouy-Chapman theory for the diffuse double layer. Therefrom, the average charge density at the membrane surface is estimated at $-(2.0\pm0.4)$ $\mu C/cm^2$, in agreement with previously reported values for biomembranes of similar composition. On the basis of mathematical computations, it is shown that more refined models considering the sphericity of the chromatophore system or the presence of a layer of fixed charges on the surface (Stern model) are not required to account for our results. It is therefore concluded that, in spite of its theoretical limitations, the Gouy-Chapman model of the diffuse double layer is sufficiently accurate to describe electrostatic interactions with the electrolyte solution even in the case of rather rough surfaces such as those of biological membranes.

1. Introduction

1.1. AN OVERVIEW

Electrical phenomena occurring at the boundary of biological membranes with the adjacent electrolyte solution play a key role in promoting and/or modulating many processes of physiological relevance, including adhesion, fusion and division of cells, antigen-antibody recognition, gating of ion channels and redox reactions at the water-membrane interfaces (for reviews, see Refs.1-4). Among these processes, electron and proton transfer reactions in photosynthesis and respiration are extensively characterized, being at the basis of energy transduction and conservation in biomembranes of different origins as chloroplasts, mitochondria and bacteria (for a review, see Ref.5).

An understanding of the electrostatic properties of the membrane surface seems

509

R. Guidelli (ed.), Electrified Interfaces in Physics, Chemistry and Biology, 509–532.
© 1992 *Kluwer Academic Publishers.*

therefore essential to describe the physiological behaviour of biomembranes. The description of the interfacial region between membranes and solution is generally carried out within the framework of membrane electrochemistry (6), a science which deals with the application of the basic principles and models of surface electrochemistry (7-8) to biological membranes. It should be pointed out, however, that a biological membrane is a rather complex and structured system, quite different from a planar interface as considered in surface electrochemistry. It comprises lipids, mainly phospholipids, and proteins in a proportion which depends on the type of membrane. Lipid molecules, due to their amphipacity, are organized in a bilayer structure with the polar head groups located at the interfaces and the nonpolar tails constituting a hydrophobic region of low dielectric constant in the membrane core. Proteins, in turn, are embedded in the lipid bilayer to different extents, depending on their properties and functions, so as to expose even large hydrophilic domains to the aqueous environment. As a consequence, the two membrane surfaces can be rather asymmetrical with respect to their composition and charge density, when we consider that different lipid molecules and proteins are differently distributed between the outer and inner layers, and that most of the lipid polar head groups and exposed aminoacidic residues are ionized at physiological pH values. In addition, membrane proteins may protrude with their hydrophilic portions beyond the lipid surface at distances as large as, or even larger than, the lipid bilayer thickness (≈ 5 nm). The resulting topology is rather fractured and difficult to be modeled by plane geometry and to be described by the terminology of metal-electrolyte interfaces in terms of inner and outer Helmholtz planes (for a general overview on these subjects, see Refs.9-12).

One strategy to characterize the water-membrane interface has consisted in making use of synthetic membranes of known lipid composition, in the form of homogeneous vesicle populations. In these systems, the surface charge density is then a known parameter, and its relation to the electrostatic potentials at the membrane surface (surface potential) or at any distance from the membrane surface have been measured as a function of the ionic strength of the bulk solution (for a review see Refs.1,2). Electron paramagnetic and fluorescent probes, adsorbed and covalently bound, respectively, at the membrane interface have been employed (13-16). By these techniques it has been clearly shown that, in spite of its limitations (smeared charge distribution, point ions and ideal surface of the membrane), the Stern-Gouy-Chapman model of the double layer is adequate to describe the interfaces in model membrane systems. This has also been recently confirmed in a qualitative way by analyzing Zn^{2+} distribution in a solution in contact with a charged phospholipid monolayer with X-ray standing waves (17). Apparently, the reason why the Gouy-Chapman model works so well even for the structered interfaces of lipid bilayers can be justified on the grounds of statistical mechanics (18).

The question that is left open and that we will address here is then whether and to what extent membrane proteins modify the modeling of these electrified interfaces.

1.2. DETERMINATION OF SURFACE POTENTIALS IN BIOLOGICAL MEMBRANES

When the Gouy-Chapman model of the double layer is tested in biomembranes, neither the surface charge density nor the surface potential can be directly evaluated, since the two parameters are both unknown at a fixed ionic strength of the bathing solution. The relation between the surface charge density σ and the resulting electrostatic potential ψ_o arising at the electrolyte-membrane interface is obtained by combining the Poisson, Boltzmann and Gauss equations (19):

$$\sigma = \{2\varepsilon_r\varepsilon_o RT\, \Sigma_i C_i [\exp(-z_i F\psi_o/RT)-1]\}^{1/2} \tag{1}$$

Here C_i is the concentration of the i-th ion in the bulk suspending medium, ε_r and ε_o are the relative and vacuum dielectric constants respectively, z_i is the charge carried by the given ion, F is the Faraday constant, R is the gas constant, and T is the absolute temperature. When the bathing solution contains only one z,z-valent symmetrical electrolyte, Eqn.(1) reduces to:

$$\sigma = (8\varepsilon_r\varepsilon_o RTC_i)^{1/2}\, \sinh(|z|F\psi_o/2RT) \tag{2}$$

The existence of a diffuse double layer at the external surface of any closed membrane system, such as cells or biomembranes therefrom derived, was inferred first from measurements of electrophoretic mobility, which provide the value of the electrokinetic potential (for reviews, see Refs.1,3,20). From these data, it was possible to establish that most biomembranes are negatively charged, both on the inner and outer surfaces, due to the predominance of ionizable carboxyl groups of proteins at neutral pH values (pK about 4.6 (21)). The electrokinetic potential only provides a qualitative estimate of the electrostatic potential at the membrane interface, because of the potential drop across the region enclosed between the membrane surface and the hydrodynamic shear plane, and since it is derived upon assuming that the whole double layer is mobile. The σ values so derived, albeit underestimating the true value by a factor of 2-3, are nonetheless sufficient to locate the order of magnitude of this parameter between -1 and -20 $\mu C/cm^2$, depending on the type of membrane.

Different approaches were then adopted to solve Eqn.(1). The general concept underlying these strategies, reviewed in the literature, is that any phenomenon occurring at interfaces should be dependent on the membrane surface potential, which, in turn, depends on the bulk electrolyte concentration. Indeed, upon considering that negatively charged surfaces attract positive ions (counterions) while repelling negative ones (coions), the local concentration of cations near the surface is higher than that in the bulk, according to the Boltzmann distribution law. Hence the charges at membranes are screened and the ψ_o

values modulated. Several phenomena were shown to be affected when changing the total concentration of the ions in the bathing solution. Some of these phenomena have been employed to evaluate the surface charge density, e.g., the binding of fluorescent cations as detected from their fluorescence quenching, and the electrochromism of the electric field-sensing endogenous pigments of photosynthetic biomembranes (for reviews see Refs. 20,22). In these cases, Eqn. (2) was solved simultaneously for two conditions (routinely when the same effect was detected by imposing the same ψ_0 value with monovalent and divalent cations, as predicted by the Gouy-Chapman model for symmetrical electrolytes), and the σ value was found to range from -0.07 to -2 $\mu C/cm^2$. These procedures, however, provide only a qualitative evidence for the validity of the double layer model, since the observed phenomena were not quantified in terms of the surface potential.

A more stringent test to verify which model best describes the electrostatic properties at interfaces, can be indirectly obtained by considering any physical law in which a measurable parameter is explicitly related to ψ_0. In this case, by detecting how the concentration of electrolytes affects the given phenomenon, it is possible to discriminate between different models, since the expected dependence will be verified only for certain values of the surface potential. This approach has been so far used to determine the effect of the surface potential on reactions involving charged electron donors and acceptors at interfaces (23-27) and on the thermodynamic equilibrium of redox couples involved in electron transfer reactions within the membrane phase (28-29).

The surface charge density can also be estimated for biomembranes containing proteins of known sequence and transmembrane topological organization. Models relating σ and ψ_0 can then be tested by measuring the surface potential. This was done for photoreceptor disk membranes, which contain only the well characterized protein rhodopsin; it was shown that the potentials detected with spin labelling techiques could be accounted for by the accepted model of rhodopsin, the known lipid composition of this membrane, and the Stern-Gouy-Chapman model (30). Unfortunately, the situation is by far more complicated with most biomembranes, for which indirect approaches are to be adopted.

1.3. THE CHROMATOPHORE MEMBRANE OF PHOTOSYNTHETIC BACTERIA

When grown photoheterotrophically, photosynthetic bacteria develop in their cytoplasmic membrane a rather complex system of invaginations, which, upon mechanical rupture of the cell, originates closed vesicles, referred to as chromatophores, whose polarity is inside out with respect to the original cell. Chromatophore membranes contain both the photosynthetic apparatus and the ATPase complex capable of coupling the light-induced transmembrane electrochemical potential difference for protons ($\Delta\tilde{\mu}_{H^+}$) to ATP synthesis (for a review see Ref.5). For this reason, and since they constitute a simpler system than the whole photosynthetic cell, chromatophores have been extensively employed in studies of bioenergetics aimed to characterize the early steps of bacterial photosynthesis and the mechanisms of energy conservation (31). The main physico-chemical properties of these membranes are listed in Tables 1 and 2. A typical preparation of chromatophores

TABLE 1. PROPERTIES OF CHROMATOPHORE MEMBRANES

Buoyant density (at 20° C)	1.140 g cm^{-3}

Particle density on freeze-fracture faces:

Concave face	2800 ± 200 (n° μmol μg^{-2})
Convex face	1500 ± 100 (n° μm^{-2})
Electrical capacitance	$0.5\text{-}0.6 \; \mu F \; cm^{-2}$

has an average diameter of 50 ± 25 nm, with an inner osmotic volume which is consequently rather small (32). The lipid to protein ratio of the chromatophore membrane is similar to that of other energy conserving membranes of different origins, and its lipid content includes mainly phospholipid molecules (33). Both phosphatidylglicerol and phosphatidylethanolamine, although to a minor extent, contribute to the negative charge of the inner and outer interfaces at neutral pH values (34,35). A clear asymmetrical distribution of the membrane proteins is detected from the electron micrographs of a freeze-cleaved population, which show that membrane particles are more densely distributed in the inner surface than in the outer surface by a factor of two (33).

TABLE 2. COMPOSITION OF CHROMATOFORE MEMBRANES

Bacteriochlorophyll	100 ± 10 nmol mg_p^{-1}
Carotenoids	60 ± 5 nmol mg_p^{-1}
Phospholipids	0.8 ± 0.2 μmol mg_p^{-1}
Phosphatidylglycerol	30% (of the total)
Phosphatidylethanolamine	40%
Phosphatidylcholine	30%
Intrinsic membrane proteins	0.7 ± 0.1 mg mg^{-1}

Data are summarized from Refs.(32) and (33). mg_p: mg of proteins.

The topological and functional organization of the photosynthetic apparatus of chromatophores is shown in Fig.1. Two main supermolecular membrane protein complexes contain the electroactive redox centers, namely the so called reaction center (RC) and the ubiquinol - cytochrome c_2 oxidoreductase complex (bc_1 complex). Upon photoactivation, a bacteriochlorophyll dimer is photooxidized: one electron is delivered to the primary acceptor quinone molecule Q_A and subsequently to a second quinone molecule Q_B. The secondary acceptor quinone molecule Q_B is fully reduced to quinol and concomitantly protonated at the outer interface. This quinone molecule is in equilibrium with a thermodynamically homogenous quinone pool in the bilayer, which delivers electrons to a redox site (Q_Z) on the opposite side of the membrane, while protons are released at the inner interface of the vesicle. Q_Z is located in the bc_1 complex, which also comprises one c-type and two b-type cytochromes. According to the accepted mechanism, whenever a quinol molecule is oxidized at the Q_Z site, the first electron cycles back to the bacterichlorophyll dimer via the iron-sulfur cluster and the bound c_1 and soluble c_2 cytochromes; conversely, the second one is delivered to the Q_C site, which faces the outer surface, in an electrogenic fashion through b-type cytochromes. The generation of a transmembrane electrical potential difference, as also monitored by the electrochromic shift of the endogenous carotenoids (36), is then due to a charge separation from cytochrome c_2 to Q_A and from cytochrome b_{566} to Q_C (indicated by black arrows in Fig.1). Since two electrons are necessary in order to fully reduce the secondary acceptor, the full cycle is completed upon a second reduction of the Q_Z site, with a stoichiometry of two protons translocated through the membrane per electron cycling (37,38).

The functional sequence of electron donors and acceptors was elucidated over a period of about twenty years by studying redox equilibria and electron transfer reactions of the electroactive centers by means of redox potentiometry combined with EPR, as well as conventional and fast kinetic spectroscopy. Most of our knowledge of the topological arrangment of the RC protein complex and its cofactors has become recently available after its crystallization (<3 Å resolution) (39,40). Predictions of the possible transmembrane configuration of the bc_1 complex (as shown in Fig.1) have been confirmed upon sequencing of its genes (41), and by EPR studies of a similar supermolecolar complex, derived from mitochondria and reconstituted in phospholipid vesicles (42).

1.4. REDOX POTENTIOMETRY AS A TOOL TO EVALUATE SURFACE POTENTIALS IN BIOMEMBRANES

The application of redox potentiometry to biomembranes dates back to the seventies, when it was developed mainly by the work of Wilson and Dutton (43). As depicted in Fig.1, the measuring electrode in the bulk monitors the redox potential of any redox center embedded

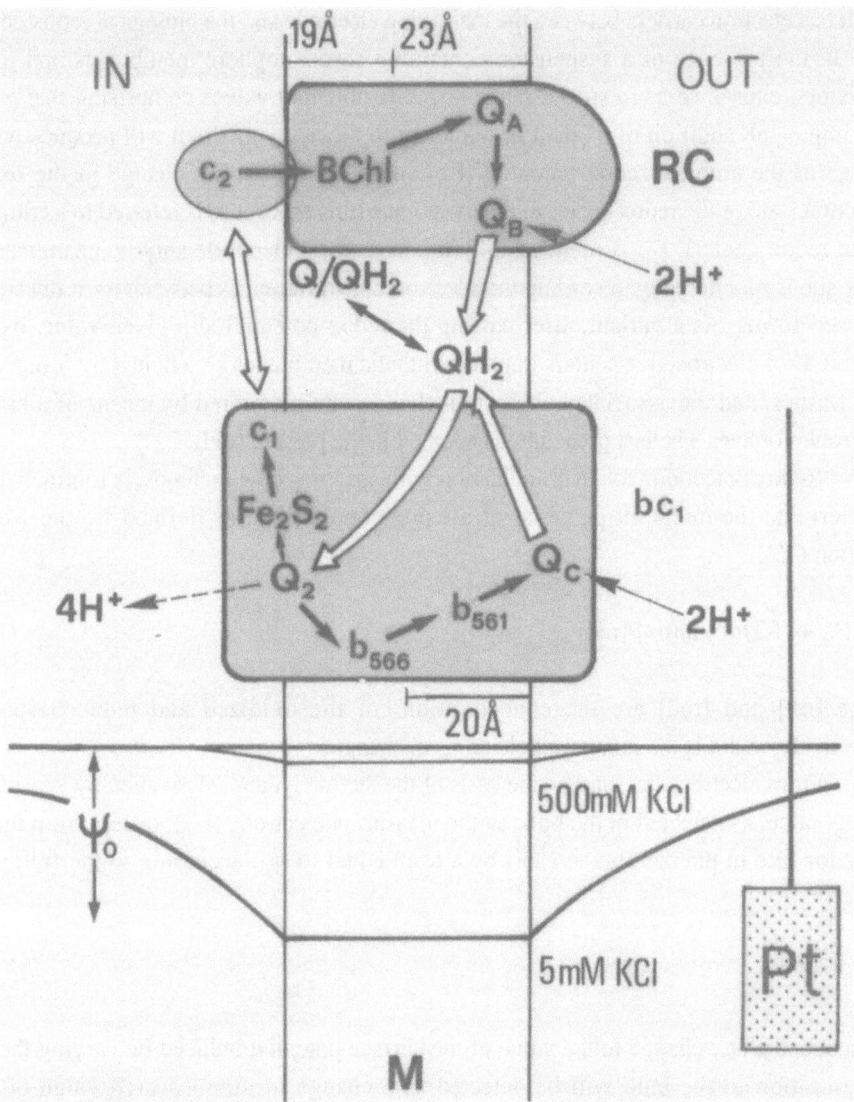

*Figure 1. Topological organization of the electron transfer components in the photosynthetic apparatus of the bacterium **Rhodobacter sphaeroides**. RC: Reaction center complex; bc_1: ubiquinol-cytochrome c_2 oxidoreductase complex; Q/QH_2: thermodynamically homogenous pool of Q_{10} molecules. B: bulk phase; M: membrane phase; Pt: platinum electrode. The figure shows how the potential in the membrane phase differs from that in the bulk, both in the presence of a 5 and of a 500 mM monovalent salt.*

in the membrane phase, as long as the redox potential in the bathing solution is in equilibrium with that at the redox site. This involves an electrochemical equilibrium between the membrane-solution-electrode interfaces, which is established by using redox mediators, namely molecules which partition between the membrane and the aqueous phase and act as electron carriers between the measuring electrode and the biological redox couple (44). In the presence of a suspension containing chromatophore membranes and redox mediators, chosen so as to span a range of redox potential values comprising that of the biocomponent, addition of oxidant or reductant to the assay medium will produce both a change of the ambient redox potential (E_h) and a corresponding change in the oxido-reduction state of the redox center in the membrane (this procedure is referred to as titration of the redox center). E_h is monitored by the measuring electrode and the change at the redox site is monitored by recording the spectroscopic differences between its reduced and oxidized forms. As a variant, after poising the redox potential at a given value, oxido-reductions of electroactive centers in photosynthetic membranes are elicited by firing short light flashes, and the associated absorption changes are monitored by means of a kinetic spectrophotometer. The last procedure is adopted in the present study.

Redox potentiometry, in combination with spectroscopic methods, is routinely used to determine the midpoint potential of a redox couple (E_m), as defined by the Nernst equation (45):

$$E_h = E_m + (RT/nF) \ln[ox]/[red] \tag{3}$$

where [ox] and [red] are the concentrations of the oxidized and reduced species respectively, and n is the number of electrons exchanged.

From electrostatics, and on the basis of the thermodynamic derivation by Walz (46), the E_m value, as detected in the bulk by the measuring electrode (E_m^B), differs from that at the redox site in the membrane (E_m^M) by a term equal to ψ_o, according to the following relation:

$$E_m^B = E_m^M + \psi_0 \tag{4}$$

Consequently, any change in the value of the surface potential induced by varying the ion concentration in the bulk will be detected as a change in the measured value of E_m^B. Equation (4) is therefore a constraint which will permit us to estimate the values of the surface potential and of the surface charge density on the basis of a chosen electrostatic model.

When redox reactions occurring in biological membranes involve the simultaneous transfer of electrons and protons, the process is electroneutral, and the effect of the surface potential on the apparent midpoint potential value vanishes. Many redox couples involved in photosynthetic electron-transfer chains are characterized by protonation equilibria of the reduced form over the physiological pH range, while the oxidized form is fully deprotonated. In these cases, the redox behaviour of the center is described by the following equation (47):

$$E_{m,pH} = E_m(dep) + (RT/F) \ln[1+10^{(pK_{red}-pH)}]$$ (5)

where $E_m(dep)$ is the redox midpoint potential of the deprotonated form. When the effect of the surface potential is considered, the pH in the bulk (pH^B) will also differ from the pH at the redox center (pH^M):

$$pH^B = pH^M - 0.43(F/RT)\psi_o$$ (6)

By considering the pK at the redox center(pK^M_{red}) in a similar way, the dependence of $E_{m,pH}$ upon the surface potential as expressed by Eqn.(4) is modified as follows:

$$E^B_{m,pH} = E^M_m(dep) +\psi_o+(RT/F) \ln(1+10^{[pK^M_{red}-0.43(F/RT)\psi_o-pH^B]})$$ (7)

As a consequence, E^B_m depends upon pH but not upon the surface potential over a range of pH values less than pK_{red}; beyond this value, however, the effect of the surface potential on the resulting pH-independent bulk midpoint potential of the redox couple will be detected.

2. Materials and Methods

Chromatophores from photosynthetic bacteria were prepared by mechanical rupture of the cells at 100 Kg/cm^2, with an Aminco French press. The membranes were then collected after repeated washing, in order to remove the membrane wall and the cytoplasmic contents, and were stored at 258 K in a glycerol-containing buffer (routinely 50 mM 3-[N-morpholino]propanesulfonic acid (MOPS) at pH 7, in a 60% (v/v) mixture with glycerol) (48). For each preparation bacteriochlorophyll (BChl) and endogenous Mg^{2+} ion contents were determined with procedures described in detail elsewhere (49,50).

The assay medium for redox potentiometry of the reaction center contained, in a final volume of 3 ml, chromatophores corresponding to 40μM BChl, MOPS buffer at pH 7, 17 μM carbonyl cyanide-p-trifluoromethoxyphenylhydrazone (CCCP), 10 μM antimycin and 0.7 μM myxothiazol. CCCP was added to abolish proton concentration differences across the membrane; antimycin and myxothiazol prevented electrogenic electron transfer through the bc$_1$ complex. 100 μM ferro- and ferricyanide ($E_{m,7}$= 430 mV), 10 μM diaminodurene (DAD) ($E_{m,7}$ = 275 mV) and p-benzoquinone (p-BQ) ($E_{m,7}$=280 mV) were added as redox mediators. Different ionic strengths were obtained by modulating the buffer and the salt concentrations, which included the concentrations of the endogenous Mg^{2+} ions and of the oxidizing or reducing agents added to perform redox titrations. The experimental conditions of the chromatophore preparations did not permit us to impose ionic strengths less than 5 mM. Salts added to the bulk solution were NaCl, $MgCl_2$ and $CaCl_2$, depending on the experimental conditions. When other redox centers were investigated, the assay compositions were slightly different with respect to the presence of redox mediators, type of buffer and of the electron-transfer inhibitor. In each circumstance, before starting the redox titration, thermodynamic equilibrium of the membrane-solution-electrode interfaces was attained by long periods of incubation in the dark (routinely 5 hrs) under nitrogen and continuous stirring. Moreover, it was verified that the titration of the redox species was fully reversible when the ambient redox potential was lowered and/or increased by stepwise additions of titrants.

The apparatus employed to perform redox potentiometry in combination with fast kinetic spectroscopy is extensively described in Ref. 51. The sample was contained in an anaerobic stirred cuvette, under a continuous nitrogen flow, and redox potentials were measured using a Radiometer 26 pH-meter and Radiometer K401 calomel and P101 Pt electrodes . The flash-induced absorption changes were measured using a single-beam spectrophotometer with a minumum resolution time of 0.5 µs and a bandwidth of 1.5 nm. Light-induced redox changes of the primary donor of the reaction center were detected at 542 nm, and flash excitation was provided by a xenon lamp (3.25 J discharge energy) screened through two layers of Wratten 88A gelatin filter, giving a saturating (>90%) flash of 15 µs duration at half-maximal intensity. Digitalization of the signals was performed by a Datalab DL905 transient recorder, interfaced to an Olivetti M24 computer. Non-linear least-squares fitting procedures were used in order to determine the E_m^B values from the redox titrations at the different ionic strengths, by means of the Marquardt algorithm (52). Equation (1) was numerically solved in combination with Eqn.(4), for each experimental condition, by using the Newton-Raphson routine (53), included in a computer programme run in an Olivetti M24 PC.

3. Results

3.1. REDOX POTENTIOMETRY OF THE PRIMARY DONOR OF THE REACTION CENTER AT DIFFERENT IONIC STRENGTHS

Absorption changes of the primary donor of RC,as elicited by a train of single turnover saturating flashes at controlled bulk redox potential, are shown in Fig.2A. At a fixed value of the bulk redox potential,the extent of the absorption change detected at 542 nm, which is proportional to the amount of reduced primary donor before flash excitation, varies with the

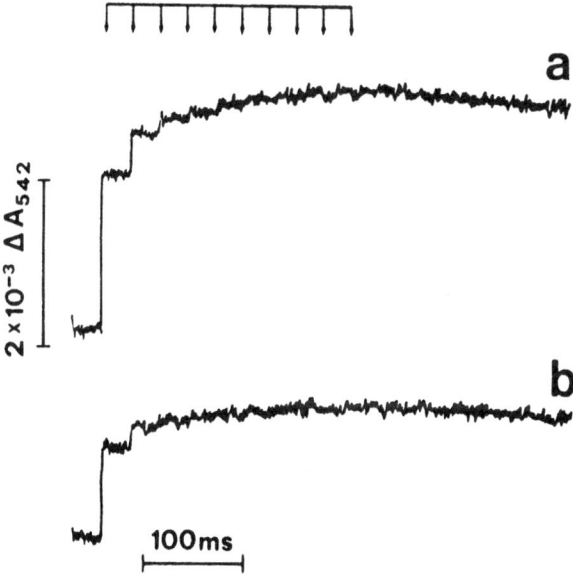

Figure 2A. Absorbance change of the primary donor of the reaction center complex at different salt concentrations. E_h=420 mV.

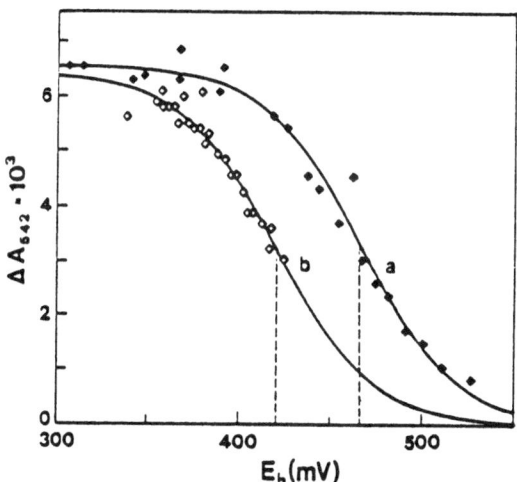

Figure 2B. Evaluation of the bulk redox midpoint potential of the primary donor of the reaction center at different ionic strengths. Curve fitting was performed by using a non-linear least-squares fit algorithm to the Nernst equation (Eqn.(3)). Dotted lines indicate $E_m{}^B$ values at 200 (a) and 2 mM (b) NaCl, respectively.

electrolyte concentration in the assay medium: it diminishes when the ionic strength is decreased. Redox titrations of the RC donor were performed at different ionic strengths. The results were employed to determine the apparent redox midpoint potential of the redox species as a function of the electrolyte concentration, as shown in Fig.2B. The extent of the absorption changes obtained from experiments similar to those shown in Fig.2A were plotted as a function of the ambient redox potential, and the data were fitted to the Nernst equation in order to determine E_m^B at a fixed ionic concentration. A large decrease of the E_m^B value was detected upon decreasing the ionic strength of the solution: specifically, from 467 to 421 mV when the concentration of the salt added to the assay medium was decreased from 200 to 2 mM. This effect was brought progressively about by a stepwise decrease of the bulk ionic strength (data not shown). This result is in agreement with what predicted by Eqn.(4), when considering a negative charge distribution on the chromatophore membrane.

3.2. EVALUATION OF THE SURFACE CHARGE DENSITY OF PHOTOSYNTHETIC CHROMATOPHORES

As discussed in Section 1.4, the dependence of the bulk midpoint potential of the RC primary donor upon the ion concentration of the bathing solution can be used to evaluate the surface charge density of the membrane, provided that a suitable electrostatic model of the water-membrane interface is adopted. According to Eqn.(4), the bulk redox midpoint potential of the electroactive center, when plotted against the surface potential, yields a straight line of unit slope. This dependence is satisfied by the E_m^B values determined above, when the surface potentials at the different ionic concentrations are estimated from Eqn.(1) by assuming the Gouy-Chapman model of the diffuse double layer. The fitting of the experimental data to Eqn.(4) was carried out by solving Eqn.(1) numerically with respect to ψ_0 for an assumed σ value. The best fit (shown in Fig.3) was achieved for a surface charge density of $-(2.0\pm0.4)$ $\mu C/cm^2$. By this procedure, Eqn.(4) is used as a constraint which allows both an estimate of σ and a check of its validity over a large range of ionic concentrations. Noticeably, in Fig.3 the linear dependence of E_m^B upon ψ_0 is verified when both monovalent and divalent cations are used. In this respect, Mg^{2+} and Ca^{2+} behave similarly. From these experiments the redox midpoint potential of the primary donor of the reaction center can be estimated at 495 mV for $\psi_0=0$.

As shown in Fig.1, other redox centers are present in the membrane of photosynthetic chromatophores, whose properties can be investigated with respect to the ionic composition of the bulk solution. This will be of help in characterizing the effects observed as a function of the location of the electroactive center in the membrane region. We therefore monitored the redox behaviour of other two redox centers of the

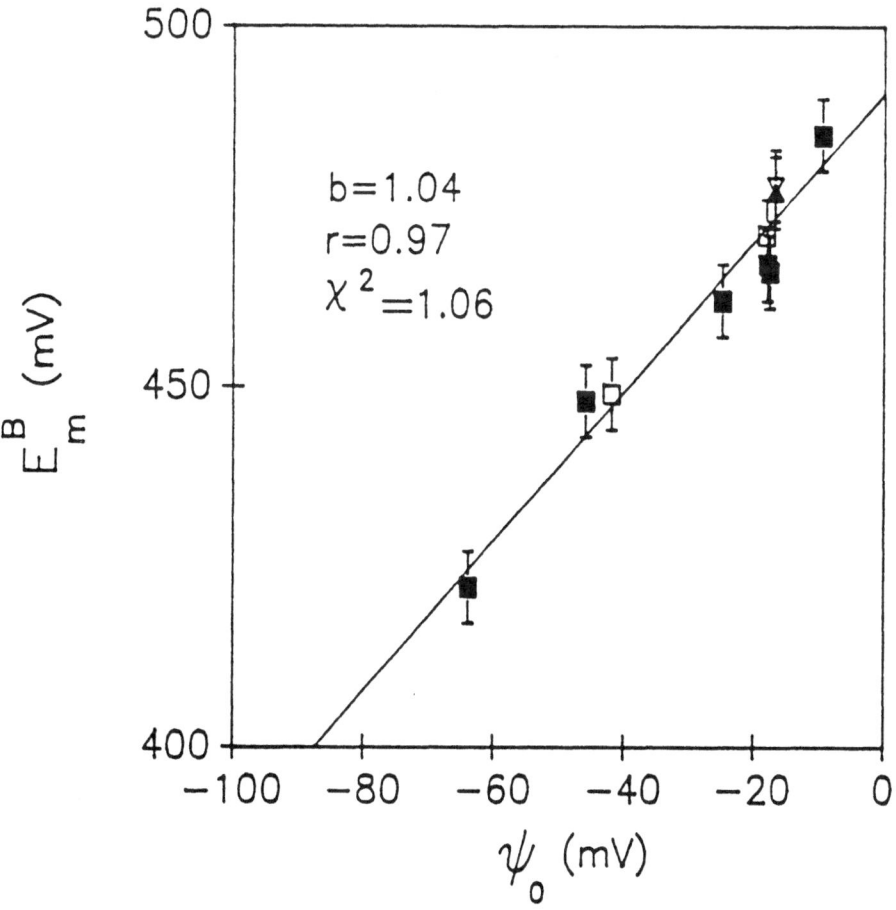

Figure 3. Bulk redox midpoint potential of the primary donor of the reaction center as a function of the electrostatic potential at the solution-membrane interface. Data were least-squares fitted to Eqn.(4) upon calculating ψ_o values from Eqn.(1). The best fit shown in the figure was obtained for $\sigma = -(2.0 \pm 0.4)$ $\mu C/cm^2$. b: angular coefficient of the straight line; r: linear regression coefficient; χ^2: reduced chi-square. Each experiment was performed at different ionic strengths, as obtained by adding salts at different concentrations to the medium: ($\blacksquare$) NaCl; ($\square$) KCl; ($\blacktriangle$) CaCl$_2$, ($\triangledown$) MgCl$_2$.

chromatophore membrane with the same experimental approach, namely the heme group of cytochrome b_{561} in the bc_1 complex and the primary acceptor of electrons of RC, the Q_A/Q_A^- couple. In these cases, however, redox reactions are also coupled to protonation-deprotonation equilibria of the reduced species, as mentioned in Section 1.4 . As long as the proton activity is much higher than K_{red}, the dependence of the bulk redox midpoint potential of both redox centers upon bulk pH is unaffected by a change of ionic strength; at pH values higher than the pK values, ($pK^M_{red}= 9.2$ and 9.4 for the cytochrome b_{561} and Q_A/Q_A^- couples, respectively), the observed redox midpoint potential becomes pH-independent and affected by the ionic concentration of the solution. The redox behaviour of both species, over the whole pH range explored, can be quantitatively accounted for by Eqn.(7), provided that ψ_o is calculated from Eqn.(1) and the same σ value as determined above is assumed. These results add to the validity of the estimate of the surface charge density in the chromatophore membranes and indicate that this feature is not affected by the location of the redox center investigated.

4. Discussion

Our estimate of the surface charge density has been obtained by using a quite indirect approach, based on the validity of redox potentiometry, as applied to membrane systems, and on the effects of electrostatic potentials at the water-membrane interface upon the measured redox midpoint potential of electroactive centers. A reasonable question is then to which extent this value is the only possible value that can accomodate our experimental results in the framework of an assumed electrostatic model. This will be discussed in the following, by considering separately different factors which should in principle be included in a realistic description of the water-membrane interface of the chromatophore system.

4.1. SENSITIVITY OF THE GOUY-CHAPMAN MODEL TO THE ASSUMED VALUE OF THE SURFACE CHARGE DENSITY

When the σ value assumed to calculate ψ_o from Eqn.(1) is halved or doubled, the dependence of the measured E^B_m on the surface potential is still linear. However, the least-squares fitting to a straight line as used to accomodate the data to Eqn.(4) is now characterized by a much higher reduced χ^2 value (as shown in Fig.4). This indicates that the procedure adopted to estimate the value of the surface potential is rather sensitive to the assumed σ value and that Eqn.(4) is best verified for certain ψ_o values, namely for a certain σ value.

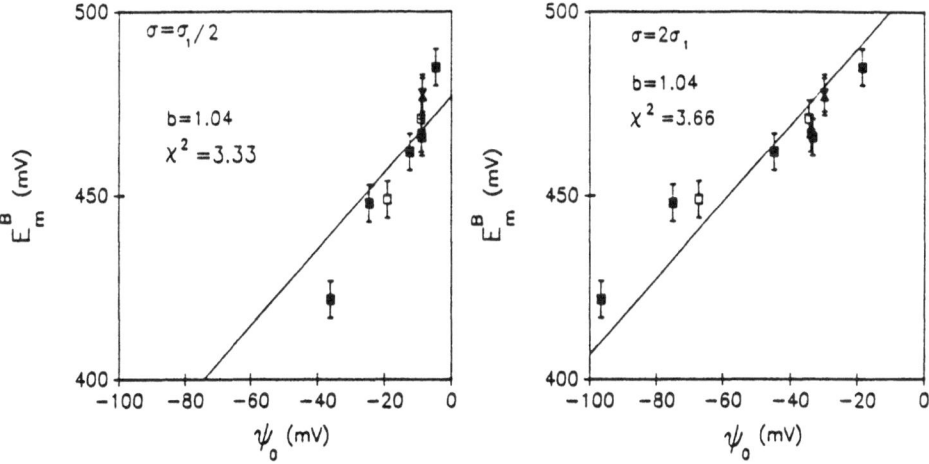

Figure 4. Sensitivity of the linear relationship between $E_m{}^B$ and ψ_o to the values of the surface charge density. Data were fitted to a straight line with a slope equal to that shown in Fig.3 using a non-linear least-squares fitting procedure.

4.2 COMPARISON BETWEEN THE GOUY-CHAPMAN THEORY AND THE STERN-GOUY-CHAPMAN MODELS

As outlined in Section 1.1, electrostatic phenomena at the interfaces of model systems are best described in terms of the Stern-Gouy-Chapman model, by considering ion interactions with the charges at the membrane surface. In this case, values of surface potential can still be calculated through Eqn.(1) or (2), upon including the Langmuir adsorption isotherm to describe ion-membrane interactions (for a review, see Ref.11):

$$\sigma = \sigma_{max}/ (1 + kC_o) \tag{8}$$

where C_o is the ion concentration at the membrane surface (related to that in the bulk by the Boltzmann law) and k is an intrinsic association constant. k values ranging from 1 to 12 M^{-1} and from 0.6 to 6 M^{-1} have been measured, respectively, for the binding of Ca^{2+} and Na^+ ions to the polar head groups of negatively charged phospholipids (54,55). Figure 5 shows the dependence of the surface potential upon the concentration of monovalent and divalent ions as predicted by the two models, upon setting the surface charge density equal to -2 and -8 $\mu C/cm^2$. It is evident that the difference between the Gouy-Chapman and Stern models is enhanced when the ion association constant k increases at costant σ, or else when

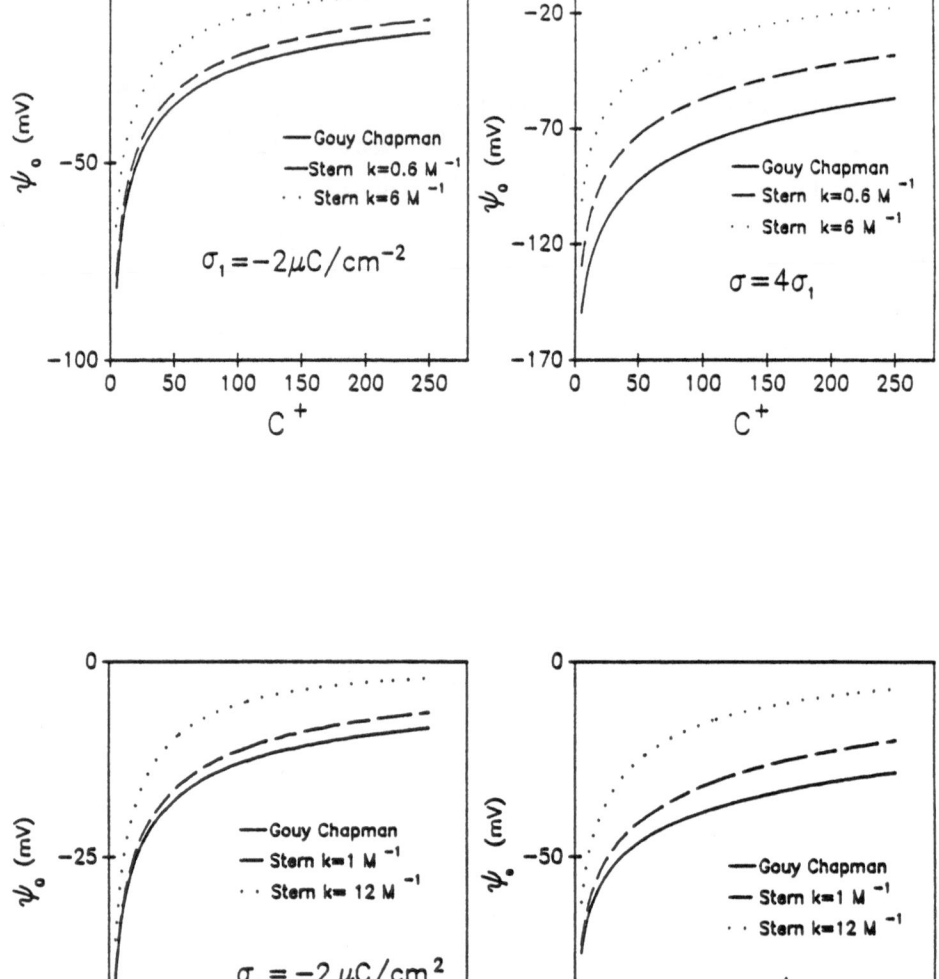

Figure 5. Comparison between the Gouy-Chapman and the Stern-Gouy-Chapman models. C values are in millimoles/litre.

σ increases at constant k. For the lowest σ and k values, the surface potentials values are similar in either models. When the Stern model is adopted, the dependence of the experimental bulk redox midpoint potential of the reaction center upon ψ_0 can still be fitted to a straight line with a correlation coefficient of 0.97 (coincident with the fit shown in Fig.3), provided that σ^{max} concides with the value obtained from the simple Gouy-Chapman model, and that the association constants are 0.6 and 12 M^{-1} for monovalent and divalent ions, respectively. The mathematical analysis performed for different choices of σ and k values shows that, in principle, the Stern model is also adequate to fit the experimental data to the expected dependence upon the surface potential values, as long as the surface charge density and the ion interaction with the membrane are low. We therefore conclude that this effect may as well be neglected, especially when we consider that the association constants yielding the best fit are rather fictitious, since they only refer to the lipid component of the membrane.

4.3. THE EFFECT OF THE SPHERICAL VERSUS PLANAR GEOMETRY UPON THE DETERMINATION OF SURFACE POTENTIAL

A realistic electrostatic model for the interface of the chromatophore membrane should include the spherical geometry of the system. This feature can be accounted for by using the approximate analytical relationship between the surface charge density and the surface potential as obtained by solving the spherical Poisson-Boltzmann equation (56). For spherical particles of radius a, suspended in an electrolytic solution, this expression reads:

$$\sigma = (\varepsilon_0 \varepsilon_r \kappa \ RT/F) \ I \qquad (9)$$

where κ is the the Debye-Hückel reciprocal length. I is a first-order solution in $(\kappa a)^{-1}$ for the spherical equation, which for symmetrical univalent electrolytes reads:

$$I = 2 \ \sinh(F\psi_0/2RT) + (4/\kappa a) \ \tanh(F\psi_0/4RT) \qquad (10)$$

In Fig.6, the ψ_0 values calculated from Eqns.(9) and (10) upon using the σ values previously estimated, are compared with those (solid curve) calculated from the planar model. Although the average radius of a chromatophore vesicle is 25 nm (see Table 1), two limiting values of 20 and 15 nm are considered. The ψ_0 values differ by a small percentage (<10%) only at low ionic strenghts (<20 mM). The difference is more pronounced when the particle radius is smaller than that of chromatophores. We may therefore conclude that the σ values as obtained upon assuming a planar model for the interface in these membrane systems are not appreciably affected when the sphericity of the particle is taken into account.

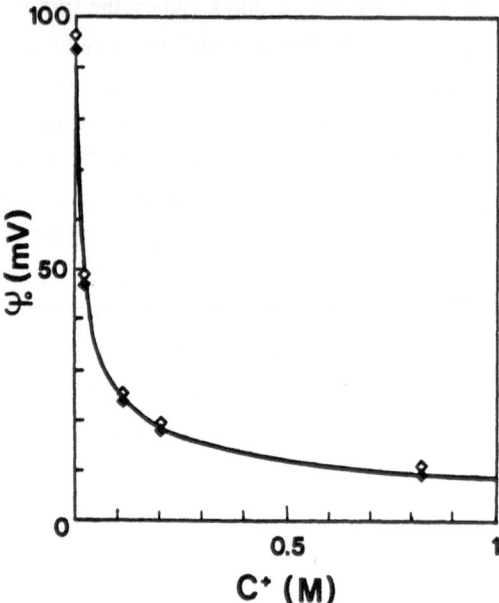

Figure 6. Comparison between the linear and spherical models of interfaces in evaluating the surface potential. The dependence of ψ_o upon the concentration was estimated by using Eqn.(2) (solid line) as well as Eqns.(9) and (10) with a particle radius equal to 15 ($\diamond$) and 20 nm ($\blacklozenge$).

4.4. ARE THE INNER AND OUTER SURFACE CHARGE DENSITIES EQUAL IN CHROMATOPHORES?

In a membrane characterized by different inner and outer surface charge densities (σ^i and σ^o, respectively), a redox center located at a distance x from the outer interface experiences an electrostatic potential which is due to the relative contributions of the inner and outer surface potentials. Assuming a linear potential drop across the membrane dielectric, the electrostatic potential at the redox center is:

$$\Delta\psi_o = \psi_o{}^i + (\psi_o{}^i - \psi_o{}^o)x/d \tag{11}$$

where $\psi_o{}^i$ and $\psi_o{}^o$ are the inner and outer surface potentials, respectively, and d is the membrane thickness. For a redox site located in the center of the bilayer ($x = d/2$), as we may reasonably assume for the primary donor of the reaction center, $\Delta\psi_0$ values were calculated from the Gouy-Chapman model for different ratios of the inner to the outer surface charge density. The results are shown in Fig.7. Different $\Delta\psi_0$ values are obtained when the σ^i/σ^o ratio is changed from 0.5 to 2. This difference, when it is referred to the ψ_0 values used to fit the experimental data in Fig.2, is equal to or greater than 10 mV, which is

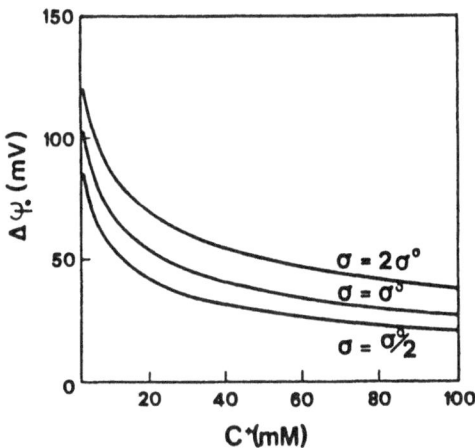

Figure 7. Electrostatic potential at the center of the bilayer for different ratios of the inner (σ^i) to the outer (σ^o) surface charge density, as a function of the bulk concentration of monovalent cations.

above the lower resolution limit of our experimental procedure. Hence, in the presence of a large asymmetry in charge density between the inner and outer interfaces, the dependence of the bulk redox midpoint potentials of the reaction-center donor upon ψ_0 would not be fitted by a straight line, whereas Fig.2 shows that this is not the case. The idea of two opposite interfaces with similar surface charge densities is indeed consistent with the finding that the redox behaviour of redox sites located at different depths within the membrane dielectric, such as the cytochrome b_{561} and Q_A/Q_A^- couples, can be accounted for by postulating the same σ value. This result, which is unexpected in view of the topology of chromatophore membranes, can be reconciled with the asymmetrical organization of the membrane components by assuming that an asymmetrical distribution of the phospholipid polar head groups -well documented in biomembranes of similar origin (57)- is compensated for by an asymmetrical distribution of membrane proteins between the two surfaces of the bilayer, with a resulting rather similar surface charge density.

528

5. Conclusions

The above analysis lends support to the description of the electrostatic properties at the solution-membrane interface of chromatophores in terms of the Gouy-Chapman model of a charged interface. This occurs in spite of the physical complexity of the system. One obvious limit of the Gouy-Chapman model stems from the assumption of a uniform, smeared out charge distribution, whereas a real membrane surface contains discrete charges. As previously discussed (58), the smeared out charged model is justifiable when the intercharge spacing is short with respect to the Debye length of the solution. With our approach we estimate a surface charge density of about one electronic charge per 9 nm^2. If this value is regarded as realistic, then the resulting surface intercharge spacing should be comparable with Debye lengths in the range from 4 to 0.3 nm, which corresponds to the range of the ionic strengths adopted. This range would seem to include discreteness-of-charge effects. However, no evidence for the discrete nature of the charges on the membrane surface has been found whenever the predictions of this model have been tested in artificial phospholipid vesicles by measuring the partitioning of charged amphiphiles (13,16), or in photoreceptor disc membranes (30).

It should also be considered that in chromatophores the topological asymmetry between the inner and outer membrane surfaces is enhanced by the presence of the membrane-bound ATPase complex, which interacts with its membrane counterpart *via* a stalk; its large hydrophilic portion (the average diameter of the overall size of the ATPase from *Escherichia coli*, as detected by cryoelectron microscopy (59), is about 10 nm) protrudes some ten nm beyond the outer surface. From 10 to 50 copies of ATPase complexes can be found in a single chromatophore vesicle (60). Evidently, our data permit us to conclude that the presence of this large protein complex does not invalidate the description of the outer interface, as compared with the inner one, *via* the uniform, smeared out charge model.

References

(**1**) McLaughlin, S. (1977) "Electrostatic potentials at membrane-solution interfaces", Curr. Top. Membr. Transp. **9**, 71-144.
(**2**) McLaughlin, S. (1989) "The electrostatic properties of membranes", Ann. Rev. Biophys. Biophys. Chem. **18**, 113-136.
(**3**) Pethig, R. (1986) "Ion, electron and proton transport in membranes: a review of the physical processes involved", in Gutmann, F. and Keyzer, H., *Modern Bioelectrochemistry*, Plenum Press, New York & London.

(4) Thorne, S. W. and Duniec, J. J. (1983) "The physical principles of energy transduction in chloroplast thylakoid membranes", Quat. Rev. Biophys. **16**, 197-278.

(5) Skulachev, V. P. (1988) *Membrane Bioenergetics*, Springer-Verlag, Berlin.

(6) Krysinski, P. and Ti Tien, H. (1986) "Membrane electrochemistry", Progr. Surf. Science, 2 3(4), 317-412.

(7) Davies, J. T. and Rideal, E: K. (1963) *Interfacial Phenomena*, Academic Press, New York.

(8) Aveyard, R. and Haydon, D. A. (1973) *An Introduction to the Principles of Surface Chemistry*, Cambridge University Press, London.

(9) Gennis, R. (1989) *Biomembranes. Molecular Structure and Function*, (Cantor, C. R. ed.), Springer-Verlag, New York.

(10) Kotyk, A., Jana'cek, K. and Koryta, J. (1988) *"Biophysical Chemistry of Membrane Functions"*, J. Wiley and Sons, Chichester.

(11) Starzak, M. E. (1984) *The Physical Chemistry of Membranes*, Academic Press, Inc, Orlando.

(12) Jones, M.N. (1975) *Biological Interfaces*, Elsevier, Amsterdam.

(13) Winiski, A. P., Eisenberg, M., Langner, M. and McLaughlin, S. (1988) "Fluorescent probes of electrostatic potential 1 nm from the membrane surface", Biochemistry **27**, 386-392.

(14) Langner, M., Cafiso, D., Marcelja, S. and McLaughlin, S. (1990) "Electrostatics of phosphoinositide bilayer membranes. Theoretical and experimental results", Biophys. J. **57**, 335-349.

(15) Sundberg, S., A. and Hubbel, W. L. (1986) "Investigation of surface potential asymmetry in phospholipid vesicles by a spin label relaxation method", Biophys. J. **49**, 553-562.

(16) Hartsel, S. C., Cafiso, D. S. (1986) "A test of discreteness-of-charge effects in phospholipid vesicles: measurements using paramagnetic amphiphiles", Biochemistry **25**, 8214-8219.

(17) Bedzyk, M. J., Bommarito, M. G., Caffrey, M. and Penner, T. L. (1990) "Diffuse-double layer at a membrane aqueous interface measured with X-ray standing waves", Science **248**, 52-56.

(18) Carnie, S. L. and Torrie, G. M. (1984) "The statistical mechanics of the electrical double layer", Adv. Chem. Phys. **56**, 141-253.

(19) Grahame, D. C. (1947) "The electrical double layer and the theory of electrocapillarity", Chem. Rev. **41**, 441-501.

(20) Barber, J. (1980) "Membrane surface charges and potentials in relation to photosynthesis", Biochim. Biophys. Acta **594**, 253-308.

(21) Cantor, R. C. and Schimmel, P. R. (1980) *Biophysical Chemistry. Part I: The Conformation of Biological Macromolecules*, W. H. Freeman and Company, San

530

Francisco, p. 49.

(2 2) Barber, J. (1982) "Influence of surface charges on thylakoid structure and function", Ann. Rev. Plant Physiol. **33**, 261-295.

(2 3) Itoh, S. (1978) "Membrane surface potential and the reactivity of the system II primary electron acceptor to charged electron carriers in the medium", Biochim. Biophys. Acta **504**, 324-340.

(2 4) Itoh, S. (1979) "Surface potential and reaction of membrane-bound electron transfer components. I. Reaction of P-700 in sonicated chloroplasts with redox reagent", Biochim. Biophys. Acta **548**, 579-595.

(2 5) Itoh, S. (1979) "Surface potential and reaction of membrane-bound electron transfer components. II. Integrity of the chloroplast membrane and reaction of P-700", Biochim. Biophys. Acta **548**, 596-607.

(2 6) Yerkes, C. T. and Babcock, G.T. (1981) "Surface charge asymmetry and a specific calcium ion effect in chloroplast photosystem II",Biochim.Biophys.Acta**634**,19-29.

(2 7) Pethig, R., Gascoyne, P. R. C., McLaughlin, J. A. and Szent-Györgyi, A. (1984) "Interaction of the 2,6-dimethoxy semiquinone and ascorbyl free radicals with Ehrlich ascites cells: a probe of cell-surface charge", Proc. Natl. Acad. Sci. USA **81**, 2088-2091.

(2 8) Itoh, S. (1979) "Effects of surface potential and membrane potential on the midpoint potential of cytochrome c-555 bound to the chromatophore membrane of *Chromatium vinosum*", Biochim. Biophys. Acta **591**, 346-355.

(2 9) Matsuura, K., Takamiya, K., Itoh, S., Nishimura, M. (1979) "Effects of surface potential on the equilibrium and kinetics of redox reactions of membrane components with external reagents in chromatophores from *Rhodopseudomonas sphaeroides* ", J. Biochem. **87**, 1431-1437.

(3 0) Tsui, F. C., Sundberg, S. A. and Hubbel, W. L. (1990) "Distribution of charge on photoreceptor disc membranes and implications for charged lipid asymmetry", Biophys. J. **57**, 85-97.

(3 1) Ort, D. R. and Melandri, B. A. (1982) "Mechanism of ATP synthesis", in Govidje (ed.), *Photosynthesis: Energy Conversion by Plants and Bacteria*, Academic Press, New York, vol. I, pp. 537-588.

(3 2) Casadio, R., Venturoli, G. and Melandri, B. A. (1988) "Evaluation of the electrical capacitance in biological membranes at different phospholipid to protein ratios. A study in photosynthetic bacterial chromatophores based on electrochromic effects", Eur. Biophys. J. **16**, 243-253.

(3 3) Casadio, R., Venturoli, G., Di Gioia, A., Castellani, P., Leonardi, L., Melandri, B. A. (1984) "Phospholipid-enriched bacterial chromatophores. A system suited to investigate the ubiquinol-mediated interactions of protein complexes in photosynthetic oxidoreduction processes", J. Biol. Chem. **259**, 9149-9157.

(34) Tocanne, J. F. and Tiessié, J. (1990) "Ionization of phospholipids and phospholipid-supported interfacial lateral diffusion of protons in membrane model systems", Biochim. Biophys. Acta **1031**, 111-142.

(35) Tsui, F. C., Ojcins, D. M. and Hubbel, W. L. (1986) "The intrinsic pK_a values for phosphatidylethanolamine in phosphatidylcoline host bilayers", Biophys. J. **49**, 459-468.

(36) Junge, W. and Jackson, B. J. (1982) "The development of electrochemical potential gradient across photosyntetic membranes", in Govindjee (ed.), *Photosyntesis: Energy Conversion by Plants and Bacteria*, Academic Press, New York, vol. I, pp. 589-646.

(37) Crofts, A. R. and Wraight, C. A. (1982) "The electrochemical domain of photosynthesis", Biochim. Biophys. Acta **726**, 149-185.

(38) Robertson, D.E. and Dutton, P. L. (1988) "The nature and magnitude of the charge separation reactions of ubiquinol-cytocrome c_2 oxidoreductase", Biochim. Biophys. Acta **935**, 273-291.

(39) Deisenhofer J., Huber, R. and Michel, H. (1989) "The structure of the photochemical reaction center of *Rhodopseudomonas viridis* and its implications for function", in Fasman, G. D. (ed.), *Prediction of Protein Structure and the Principle of Protein Conformation*, Plenum Press, New York.

(40) Feher, G., Allen, J. P., Okamura, M. Y. and Rees, D. C. (1989) "Structure and function of bacterial photosyntetic reaction centers", Nature **339**, 111-116.

(41) Gabellini, N. (1988) "Organization and structure of the genes for the cytochrome bc_1 complex in purple photosyntetic bacteria. A phylogenetic study describing the homology of the bc_1 subunits between prokaryotes, mitochondria and chloroplast", J. Bioenerg. Biomembr. **20**, 59-83.

(42) Ohnishi, T., Schägger, H., Meinhardt, S. W., Lo Brutto, R., Link, T. A. and von Jagow, G. (1989) "Spatial organization of the redox active centers in the bovine heart ubiquinol-cytochrome c oxidoreductase", J. Biol. Chem. **264**, 735-749.

(43) Dutton, P. L. and Wilson, D. F. (1974) "Redox potentiometry in mitochondrial and photosynthetic bioenergetics", Biochim. Biophys. Acta **346**, 165-212.

(44) Wilson, G. W. (1978) "Determination of oxidation-reduction potentials", Meth. Enzym. **65**, 396-410.

(45) Clark, W. M. (1960) *Oxidation-Reduction Potentials of Organic Systems*, The Williams and Wilkins Company, Baltimore.

(46) Walz, D. (1979) "Thermodynamics of oxido-reduction reactions", Biochim. Biophys. Acta **505**, 279-353.

(47) Dutton, P. L. (1978) "Redox Potentiometry: determination of midpoint potentials of oxidation reduction components of biological electron transfer systems", Meth. Enzym. **65**, 411-435.

532

(48) Baccarini Melandri, A., and Melandri, B. A. (1971) "Partial resolution of the photophosphorylation system of *Rhodopseudomonas capsulata* ", Meth. Enzymol. 23, 556-561.

(49) Clayton, R. K. (1963) "Towards the isolation of a photochemical reaction center in *Rhodopseudomonas sphaeroides* ", Biochim. Biophys. Acta 75, 312-323.

(50) Orange, M. and Rhein, H. C. (1951) "Microestimation of magnesium in body fluids", J. Biol. Chem. 189, 379-386.

(51) Venturoli, G., Fernández-Velasco, J. G., Crofts, A. R. and Melandri, B. A. (1986) "Demonstration of a collisional interaction of ubiquinol with the ubiquinol-cytocrhome c_2 oxidoreductase complex in chromatophores from *Rhodobacter sphaeroides* ", Biochim. Biophys. Acta 851, 340-352.

(52) Bevington, P. R. (1969) *Data Reduction and Error Analysis for the Physical Sciences*, McGraw-Hill Inc, New York.

(53) Press, W. H., Flannery, P. P., Teukolsky, S. A., Vetterlin, W. T. (1986) *Numerical Recipes*, Cambridge University Press, London.

(54) McLaughlin, S., Mulrine, N., Gresalfi, T., Vaio, G. and McLaughlin, A. (1981) "Adsorption of divalent cations to bilayer membranes containing phosphatidylserine", J. Gen. Physiol. 77, 445-473.

(55) Eisenberg, M., Gresalfi, T., Riccio, T. and McLaughlin, S. (1979) "Adsorpion of monovalent cations to bilayer membranes containing negative phospholipids", Biochemistry 23, 5213-5223.

(56) Ohshima, H., Healy, T. W. and White, L. R. (1982) "Accurate analytic expression for the surface charge density/surface potential relationship and double-layer potential distribution for a spherical colloidal particle", J. Colloid Interface Science 90, 17-26.

(57) Al-Bayatti, K. K. and Takemoto, J. Y. (1981) "Phospholipid topography of the photosynthetic membrane of *Rhodopseudomonas sphaeroides*" , Biochemistry 20, 5489-5495.

(58) Nelson, A. P. and McQuarrie, D. A. (1974) "The effect of discrete charges on the electrical properties of a membrane", J. Theor. Biol. 55, 13-27.

(59) Lücken, U., Gogol, E. P., Capaldi, R. C. (1990) "Structure of the ATP synthase complex (ECF_1F_o) of *Escherichia coli* from cryoelectron microscopy", Biochemistry 29, 5339-5343.

(60) Norling, B., Srid, A., Tourikas, C. and Nyrén, P. (1989) "Amount and turnover rate of the F_oF_1-ATPase and stoichiometry of its inhibition by oligomicin in *Rhodospirillum rubrum* chromatophores", Eur. J. Biochem. 186, 333-337.

ELECTRICAL CURRENTS OF THE LIGHT DRIVEN PUMP BACTERIORHODOPSIN. THE ROLE OF ASP 85 AND ASP 96 ON PROTON TRANSLOCATION

E. BAMBERG[¥], H-J. BUTT[¥], J. TITTOR* , D. OESTERHELT*

[¥]*Max-Planck-Institut für Biophysik*
Kennedyallee 70, D-6000 Frankfurt/Main 70
**Max-Planck-Institut für Biochemie*
D-8033 Martinsried bei München

ABSTRACT. The functional properties of the light-driven proton pump of bacteriorhodopsin and two point-mutated analogues have been studied on planar lipid membranes and on gels where membrane fragments containing the ion pumps have been oriented in an electric field. It is shown that Asp 85 and Asp 96 play a central role in proton pumping of bacteriorhodopsin.

1. Introduction

The functional properties of bacteriorhodopsin (BR), the light-driven proton pump of *Halobacterium halobium* (Oesterhelt and Stoeckenius, 1973), have been studied in the past by a variety of biophysical methods. Optical spectroscopy demonstrated the photoisomerization of the retinylidene chromophore in the primary reaction and a sequence of dark reactions re-establishing the initial state. The intermediate products of this so-called photocycle have been denoted by J, K, L, M, N and O and have lifetimes in the range from femto- to milliseconds. The formation of the M-intermediate is accompanied by the deprotonation of the C=N double bond between the retinal moiety and the lysine residue 216 and by the proton release to the outer aqueous phase. Decay of the M-intermediate occurs under re-protonation of the chromophore and re-uptake of a proton from the aqueous phase. These are the key reactions by which vectorial proton transfer is linked to the photocycle of the molecule. Due to the electrogenic nature of proton translocation in BR, the concomitant electrical phenomena were investigated in detail by electrical methods under both stationary and time-resolved conditions and could be correlated with the photocycle (Drachev et al., 1974, 1976; Dancshazy and Karvaly, 1976; Herrmann and Rayfield, 1978; Keszthelyi and Ormos, 1980; Fahr et al., 1981. The electrical properties were recently reviewed by Trissl, H.W., 1990).

Charge translocation in BR and two mutated analogues has been studied on planar lipid membranes. The method consists in the adsorption of membrane fragments containing

R. Guidelli (ed.), Electrified Interfaces in Physics, Chemistry and Biology, 533–549.
© 1992 *Kluwer Academic Publishers.*

534

the ion pump on a planar lipid film. In a short circuit experiment the displacement and pump currents were detected *via* the capacitive coupling between the membrane fragment and the underlying planar film.

Alternatively, the transport properties of BR in purple membranes can be studied on a gel, where the membrane fragments are oriented in an electric field (Eisenbach et al., 1977 and Dèr *et al.*,1985). Control experiments were carried out with the same purple membrane suspension immobilized in agar gel, and the photoresponse was found to be the same as in the polyacrylamide gel. In kinetic experiments the sample was excited with the appropriate wavelength by a dye laser pulse of 10 ns duration and an energy of up to 500 μJ. Current amplifiers of different time resolutions (0.15, 1 and 100 μs) were connected successively to the electrodes in order to obtain a high signal-to-noise ratio for each time range. Absorption changes in the samples were monitored with monochromatized light from a 250 W halogen tungsten lamp; the light passed an interference filter before its intensity was measured by a photomultiplier. The kinetics of the photocycle and of the photocurrent was recorded simultaneously in two channels of a transient recorder.

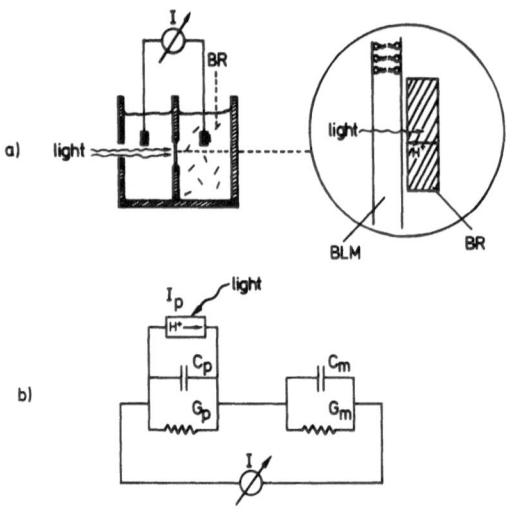

Figure 1-(a) Experimental set-up for the measurement of photocurrents and schematic representation of the arrangement of bacteriorhodopsin (BR) patches adsorbed on the black lipid membrane (BLM). (b) Equivalent circuit for the BR-BLM system. C_m, C_p and G_m, G_p denote the capacitance and the conductance of the purple membrane and of the lipid bilayer respectively.

The composed membrane system consisting of a purple membrane associated with a planar lipid membrane is described in terms of an equivalent circuit diagram. A schematic representation is given in Fig. 1.

Pump currents were measured under stationary conditions and the kinetics of these currents was determined and related to the photocycle of BR obtained by flash photolysis. The spectroscopic behaviour of BR is reviewed by Stoeckenius and Bogomolni, 1982.

2. Description of the Compound Membrane and Stationary Currents

When a purple membrane suspension is added to one aqueous side of the black film, photosensitivity develops in course of minutes. A schematic representation of the experimental situation is given in Fig. 1. The recording of the short-circuit photocurrent as obtained about 1 h after BR addition is shown in Fig. 2. The first measurement (Fig. 2A) was carried out in the absence of protonophores. Under these conditions the black lipid film has an extremely low conductance (of the order of 10 nS cm^{-2}) and is virtually impermeable to protons. After switching on the light the current rises within less than 20 ms and thereafter declines to a low level with a time constant of the order of 50 ms. The sign of the photocurrent was always the same, corresponding to a proton-transfer towards the BR-free side. In the second part of the experiment (Fig. 2B) a protonophore was added to the membrane system. It is apparent from figure 2B that, in the presence of the protonophore, a large stationary photocurrent is observed after the decline of the early transient. The highest quasi-stationary photocurrents at a light intensity of about 2 W/cm^2 were of the order of 0.1 μA/cm^2. If a continuous monolayer of purple membrane were present on the planar film, the total charge corresponding to 1 H$^+$ per BR would be about 1.4 μC cm^{-2}. If a current of 100 nA cm^{-2} flows for several minutes, the transferred charge is far in excess of 1.4 μC cm^{-2}. This means that a continuous function of the proton pump is observed in these experiments.

In Fig. 3 the action spectrum of the stationary photocurrent j_∞, as obtained from measurements with a series of narrow-band interference filters, is compared with the optical absorption spectrum of the purple membrane. The action spectrum is slightly blue-shifted (by about 15 nm), but otherwise it closely agrees with the absorption spectrum.

Both the initial current j_0 (which is observed after switching on the light) and the stationary current j_∞ saturate with increasing light intensity. This is shown in Fig. 4, where the reciprocal values of j_0 and j_∞ are plotted as a function of the reciprocal light intensity I. Both j_0 and j_∞ may be represented by the equation:

$$j = j_{sat} \frac{I_{light}}{I_{light} + I_{light,1/2}} \tag{1}$$

536

where j_{sat} is the saturation current and $I_{light,1/2}$ is the half-saturation intensity, which is approximately the same for j_0 and j_∞ ($I_{light,1/2} = 0.5$ W cm^{-2}).

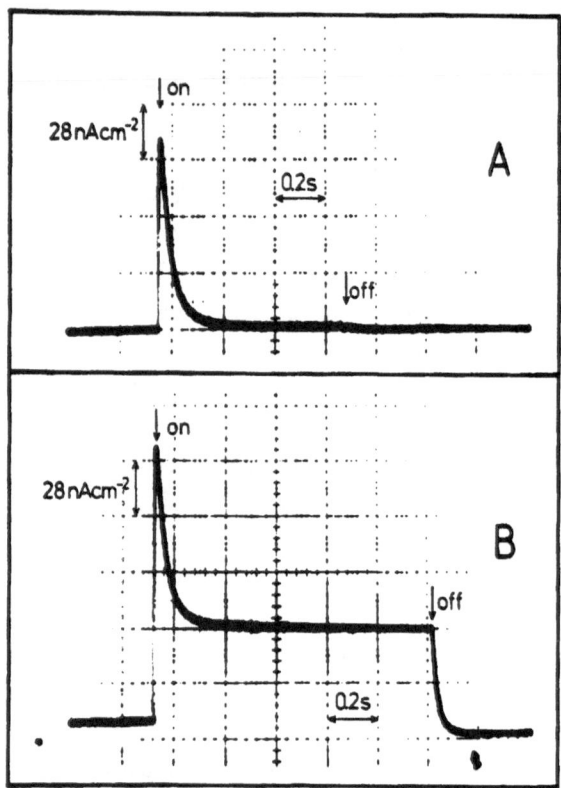

Figure 2. Short-circuit photocurrent after addition of purple membrane fragments to one aqueous compartment. The aqueous phase contained 0.1 M MgCl$_2$ and 0.5 mM Tris, pH 7.0. The light was filtered through a K4 filter (Balzers, λ_{max} 550 nm). The area of the black film was 1.8 mm^2. A) Photocurrent in the absence of protonophore. The sign of the photocurrent corresponds to a proton transfer towards the bacteriorhodopsin-free side. B) After addition of the protonophore to the bacteriorhodopsin-free compartment. After switching off the light the current falls to a slightly negative value and approaches the zero level (horizontal trace on the left) within a few seconds.

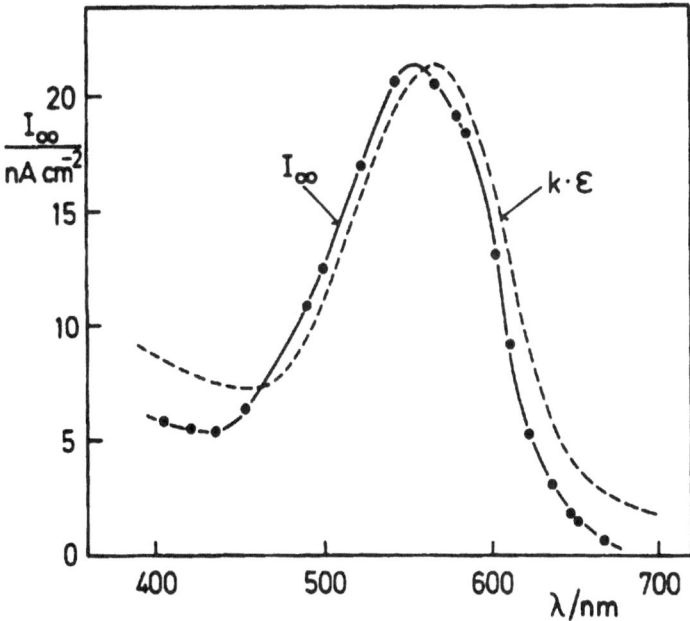

Fig.3 Action spectrum of the stationary photocurrent j_∞, as measured with a series of narrow-band interference filters. The experimental conditions were similar to those of Fig. 2. Correction factors accounting for the emission spectrum of the lamp and the transmission of the filters were obtained by calibration with a bolometer. The action spectrum was normalized to equal quantum flux density. The dashed line represents the extinction coefficient of the purple membrane, normalized to equal peak height with a factor K.

The observation that the stationary photocurrent increases strongly upon addition of protonophores (Fig. 2) argues against the possibility that the purple membrane is incorporated into the black lipid film. The most likely interpretation of the findings represented in Fig. 2 consists in the assumption that purple-membrane sheets are attached to the black lipid film in a preferential orientation (Fig. 1a). If the proton permeability of the underlying black lipid film is low, only a transient capacitive photocurrent occurs which creates a voltage drop between the intermediate layer and the aqueous phase. On the other hand, after incorporation of proton-permeable channels or protonophores into the black lipid film, a permanent photocurrent can flow between the aqueous phases. This method may be analyzed on the basis of an equivalent circuit (Fig. 1b) similar to that used by Herrmann and Rayfield for the interpretation of their experiments with purple membrane vesicles. Light absorbed by bacteriorhodopsin drives a pump current j_p which tends to

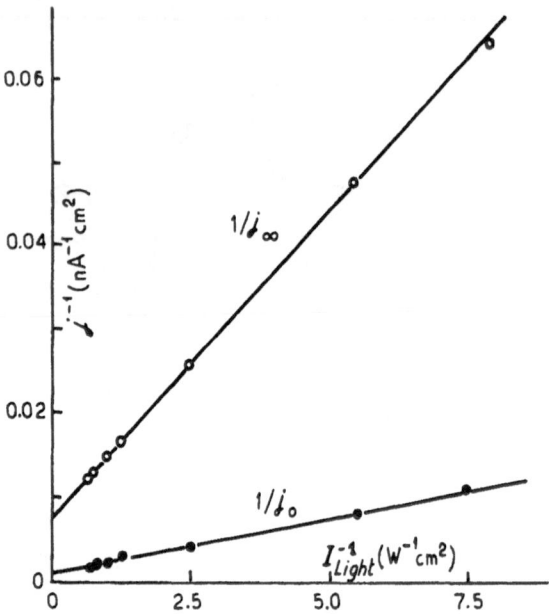

Figure 4. Reciprocal values of the initial current j_o and of the stationary current j_∞ as a function of the reciprocal of the light intensity I_{light}. All measurements were performed with the same membrane in the presence of protonophores; the aqueous phases contained 0.1M $MgCl_2$ and 0.5M Tris, pH 7.0.

build up a voltage U_p across the purple membrane. In general, the pump current j_p depends on the voltage U_p; as a first approximation, j_p may be represented as a linear function of U_p:

$$j_p = j_{po} \left(1 - \frac{U_p}{U^*} \right) \tag{2}$$

j_{po} is the pump current for $U_p = 0$ and U^* is a constant. (The term U_p/U^* also accounts for gradients of proton activity built-up by the pump, which should be roughly proportional to U_p.) Under short-circuit conditions the relation $U_p = U_m$ holds, where U_m is the voltage across that part of the black lipid film which is covered by the purple membrane. G_m and C_m are the conductance and the capacitance of the covered part of the black lipid film, and G_p and C_p are the corresponding values of the purple membrane. (Under short-circuit conditions in which the total voltage across the membrane vanishes, the covered parts of the membrane may be omitted in the circuit analysis.) G_p not only contains the conductance across the purple membrane sheet, but also accounts for any leakage pathway within the contact layer between the black lipid film and the purple membrane. Simple circuit analysis may be used to calculate the time course of the externally measured current j. The result reads (Bamberg et al.,1979):

$$j(t) = j_\infty + (j_0 - j_\infty) \exp\left(-\frac{t}{\tau}\right) \tag{3}$$

$$j_0 = j_{po} \frac{C_m}{C_m + C_p} \tag{4}$$

$$j_\infty = j_{po} \frac{G_m}{G_m + G_p + j_{po}/U^*} \tag{5}$$

$$\tau = \frac{C_m + C_p}{G_m + G_p + j_{po}/U^*} \tag{6}$$

j_0, j_∞, G_m, G_p, C_m and C_p are referred to the unit area of the black lipid film.

After switching off the light, the time course of the current is given by τ_{off}:

$$\tau_{off} = \frac{C_m + C_p}{G_m + G_p} \tag{7}$$

The derivation of equations (3)-(7) implies the assumption that the intrinsic time constants of the pump are small compared with the time constants and τ_{off} of the equivalent circuit. As the turnover rate of the pump is of the order of 100 s^{-1} (Lozier et al.,1975), this assumption seems reasonable within the time scale of the experiments represented in Fig. 2.

Equations (3)-(6) account approximately for the observed photoresponse of the system. The undoped planar film has a low conductance G_m and therefore, according to Eqn. (5), the stationary membrane current j_∞ should be small. The conductivity G_m can be increased by addition of the uncoupler carbonylcyanide-p-trifluoromethosyphenyl-hadrazone (FCCP), which acts as a proton carrier. From previous studies (Cohen et al.,1978) it is known that the FCCP-induced proton conductance of the bilayer is proportional to the aqueous FCCP concentration. According to Eqn.(5), $1/j_\infty$ should be a linear function of $1/G_m$ (and therefore of $1/c$). A linear relationship between $1/j_\infty$ and $1/c$ was indeed obtained (data not shown).

According to Eqn.(6) the time constant τ for the decay of the photocurrent should depend on the pump current j_{po} and therefore on the light intensity I_{light}. The relationship between j_{po} and I_{light} may be expressed by (cf. Eqn. 1):

$$j_{po} = j_{sat,po} \frac{I_{light}}{I_{light} + I_{light,1/2}} \tag{8}$$

where $j_{sat,po}$ is the value of j_{po} in the limit $I_{light} -> \infty$. Equation (6) may be written as:

$$\frac{1}{\tau} = \frac{G_m + G_p}{C_m + C_p} + \frac{j_{sat,po}/U^*}{C_m + C_p} \frac{I_{light}}{I_{light} + I_{light,1/2}} \tag{9}$$

Figure 5 shows that the reciprocal decay time $1/\tau$ tends to saturate with increasing the light intensity I_{light}, as predicted by Eqn. (9).

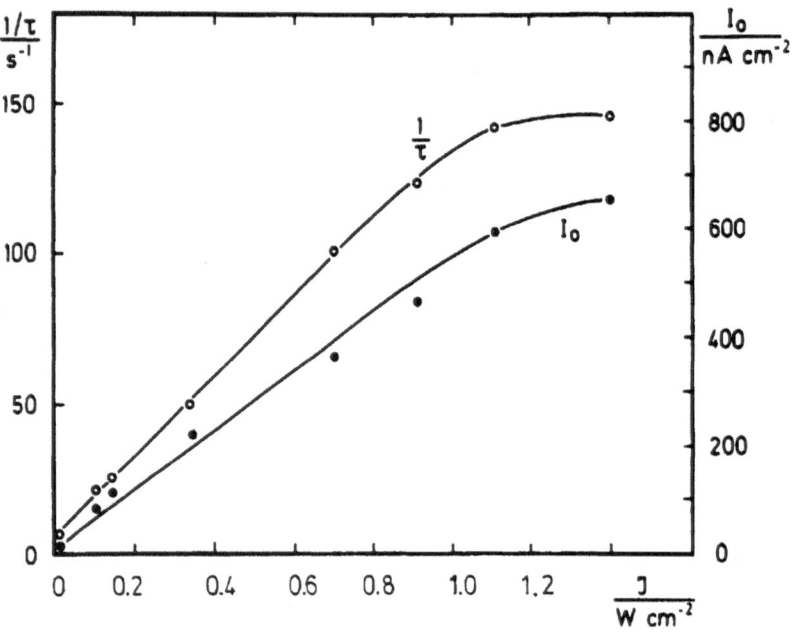

Figure 5. Reciprocal decay time $1/\tau$ of the photocurrent as a function of the light intensity I_{light}. The aqueous phase contained 0.1 M $MgCl_2$ and 0.5 mM Tris, pH 7. The figure also shows the corresponding values of the initial current j_o, as measured in the same experiment.

3. Kinetics of the Photocurrents

As shown in the previous section, the RC time τ_{off} of the compound membrane system is about 0.25 s. Optimum adjustment of the measuring system yields a time resolution of about 1 μs (Fahr *et al.*,1981). Therefore, relaxations of the photocurrent should be observable in a time window of 1 μs - 0.2 s. The time dependent photocurrent after excitation with a short laser flash is given by:

$$j(t) = \frac{C_m}{C_m + C_p} \sum_i j_{po,i} \exp(-\frac{t}{\tau_i}) \tag{10}$$

The time course of the photocurrent is shown in Fig. 6a-c. The polarity of the current is defined as indicated in Fig.1, i.e. j>0 corresponds to a movement of positive charges towards the supporting membrane. During the first 10 μs after the flash, a negative current is observed which rises steeply toward positive values and thereafter declines to zero. (The finite slope of the falling phase of j(t) within the first 5 μs is an artifact caused by the time constant of the circuit, which was about 3 μs in this case.) The shape of j(t) for t>5 μs can be represented with high accuracy by an expression consisting of a sum of four exponential terms:

$$j(t) = \sum_{i=1}^{4} a_i \exp\left(-\frac{t}{\tau_i}\right) \tag{11}$$

The analysis was carried out with four exponential functions throughout. In almost all experiments the calculated fitted curve was identical with the experimental curve for times t > 10 μs, within the width of the drawing line in a plot such as that in Fig. 6. Comparing these results with the photocycle of BR a good agreement with the optically observed transitions K-> L-> M-> BR was obtained. It should be noted that two time constants are assigned to the L->M transition (Fahr et al.,1981, Keszthelyi and Ormos, 1980).

The lipid bilayer system is advantageously compared with the oriented membrane in gel with respect to sensitivity and versatility. Other pumps like the Na+K+ATPase, Ca++ATPase and the H+K+ATPase have been studied by this method (Fendler et al. 1985, V.d.Hijden et al. 1990, Hartung et al. 1987). On the other hand, the oriented purple membranes in gel can be studied spectroscopically and electrically at the same time, with a high time resolution. In the following paragraph application of both methods to two point mutated bacteriorhodopsins will be described.

4. Spectroscopic and Transport Properties of BR Asp[8 5]->Glu and Asp[9 6] ->Asn

The proton transfer reactions in BR during the photocycle involve not only changes of the retinal moiety, but also protonation changes of amino-acid side chains in the protein. Fourier transform infrared spectroscopy (FTIR) demonstrated that four aspartates undergo

542

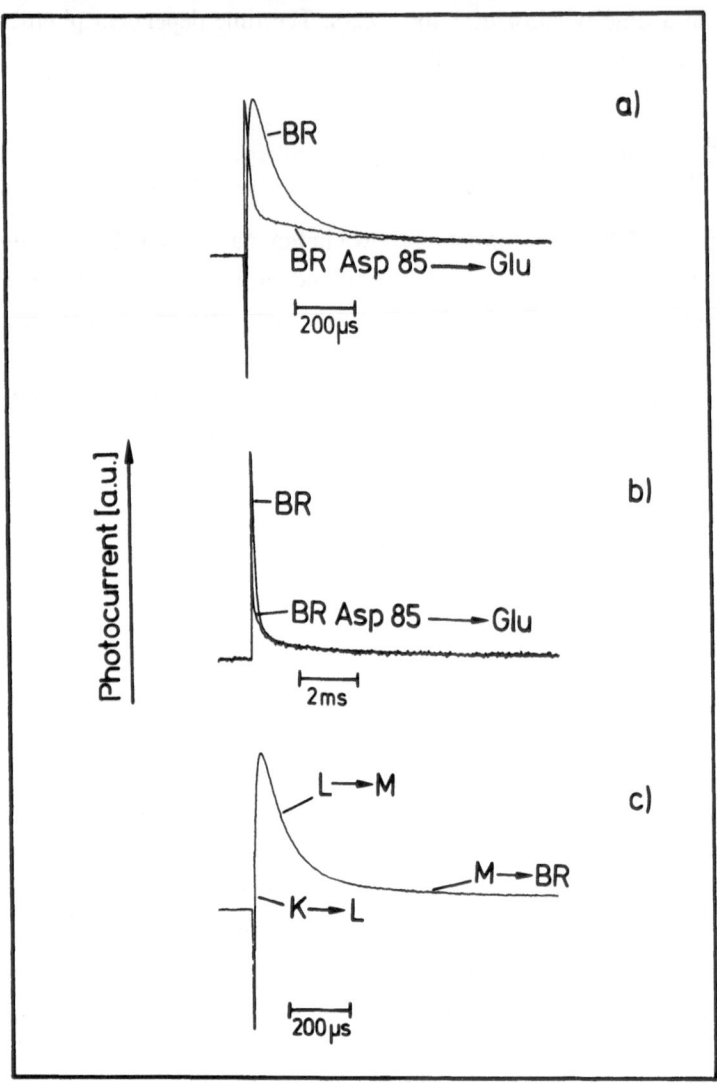

Figure 6. Time resolved currents of wild-type BR and of BR Asp[85]-> Glu obtained with purple membranes oriented in polyacrylamide gel in 20 mM KCl of pK 8.0. The wavelength of the laser light was adjusted to 550 nm for the mutant and to 570 nm for the wild-type BR, respectively. The signals are averages of 15 measurements. (a) Photocurrent in the microsecond time range. (b) Photocurrent in the millisecond time range. (c) Indications of phases in the photocurrent of BR that could correlate to the spectroscopic transitions shown in the photocycle represented in Fig. 7.

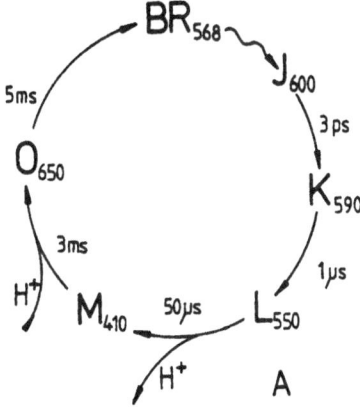

Figure 7. Photocycle of bacteriorhodopsin with the intermediates BR, K, L, M and O.

protonation changes, two of them being protonated in the initial state and two being deprotonated (Engelhard *et al.*,1984; Eisenstein *et al.*, 1987). Indeed, site-specific mutagenesis of aspartic acids indicated the importance of aspartic acid residues 85, 96, 115 and 212 for proper function (Mogi *et al.*, 1988, Holz *et al.*, 1989, Butt *et al.*, 1989). In an alternative approach to site-specific mutagenesis, mutation of halobacterial cells and selection of phototrophic-negative mutants allowed the isolation of bacteriorodopsin point-mutations (Soppa and Oesterhelt, 1989). One advantage of this method is that BR becomes available in its natural lipid environment, the purple membrane. Two of these mutants are especially interesting: one carries an asparagine residue instead of aspartic acid at position 96, whereas in the other one aspartic acid 85 is replaced by a glutamic acid (Soppa *et al.*, 1989). The aspartic acid 96 -> asparagine exchange was investigated in detail by FTIR spectroscopy and allowed the identification of this residue as a protonated aspartic acid in the initial state, which becomes deprotonated after formation of the M state (Gerwert *et al.*, 1989). Here we will show, on the basis of a detailed investigation of both mutated BRs, that aspartic acid 96 is indeed involved in the reprotonation process of the Schiff base in BR, and its lack causes retardation of the photocycle. Aspartic acid 85, on the other hand, seems to be involved in the deprotonation of the Schiff base and its replacement by glutamic acid accelerates the deprotonation reaction.

All experiments were carried out with BR isolated as purple membranes; the photocycle was measured with a time resolution of 0.5 µs in gels or in aqueous solution.

Mutant strain 384 produces a BR that contains a glutamic acid in position 85 instead of the aspartic acid of the wild-type. The absorption spectrum of BR Asp[85]->Glu at pH 6.5

is red-shifted, exhibiting a maximal absorbance at λ=610 nm. Increasing the pH causes a decrease of absorption at 610 nm and an increasing dominance of a shoulder at 550 nm (see Fig.8). The results of a kinetic study of the photocycle of BR Asp[85] ->Glu, when excited by 540 nm laser pulses, are shown in Table I. The M-formation is characterized by two relaxation times in wild-type BR of 50 and 150 µs, respectively. This has been interpreted as reflecting the formation of two different M-intermediates in a 1:1 ratio (F. Siebert, personal communication). In the case of BR Asp[85] ->Glu, two relaxation times of 5 and 150 µs in an amplitude ratio of 9:1 were observed. The main component of M-formation in the mutant is accelerated by a factor of 10 compared to the corresponding time constant in the wild-type.

M-formation and decay of BR Asp[85] ->Glu show pH-independent kinetics between pH 5 and 9 within a factor of 2. It is especially important to note that the mutant protein behaves like the wild-type in the M decay.

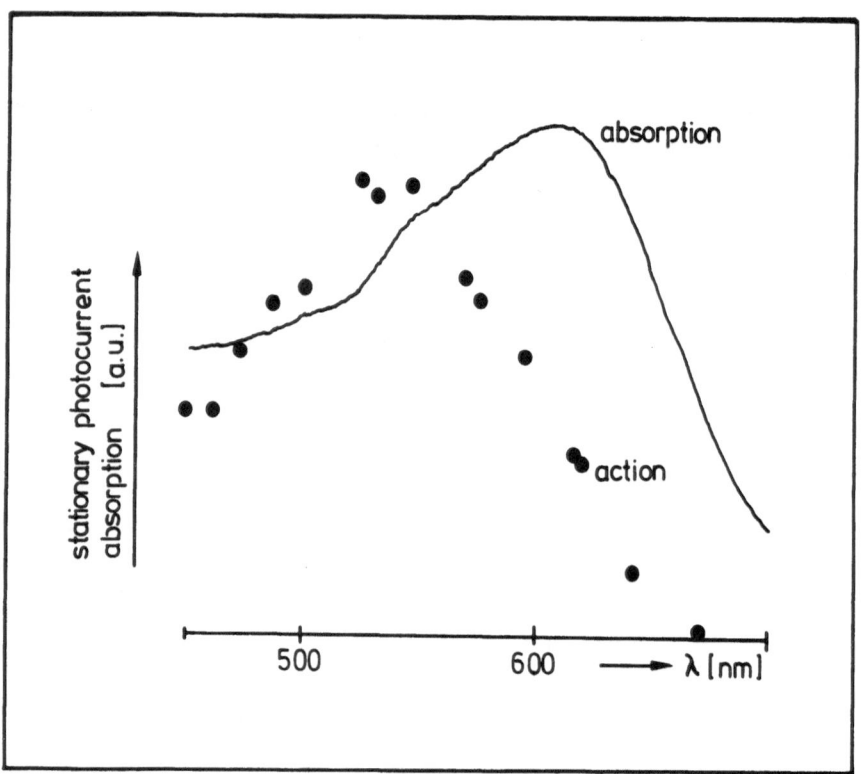

Figure 8. Absorption spectrum of BR Asp[85]->Glu (pH 7.5) and action spectrum of the stationary photocurrents on planar lipid membranes in the mutant.

Table I. Measurements on wild-type BR and BR Asp⁹⁶->Asn were carried out in 10 mM Mops (pH 6.2, 20 mM KCl) and those on BR Asp⁸⁵->Glu in 10 mM Tris (pH 8.4, 20 mM KCl), unless otherwise specified. The time constant corresponding to the M decay for the photocurrent in BR Asp⁹⁶-> Asn was obtained from the plot of dark interval vs. peak current.

Sample	Observables	relaxation time (τ)/displaced charge (ΔQ)				
BR wild type	absorption change at 419 nm (τ)			50 μs, 150 μs increase		3 ms, 8 ms decrease
	photocurrent (τ)	$\leq$ 150 ns	1.2 μs	50 μs	200 μs	4 ms
	displaced charges ΔQ ($nA \cdot ms$)	−1.8	− 0.8	6	6	41
BR Asp96 → Asn	absorption change at 419 nm (τ)			100 μs, 100 μs increase		450 ms decrease
	photocurrent (τ)	$\leq$ 150 ns	1.1 μs	50 μs	300 μs	500 ms
	displaced charges ΔQ ($nA \cdot ms$)	−1.3	− 0.6	6	15	ca. 20*
BR Asp85 → Glu	absorption change at 419 nm (τ)			5 μs, 150 μs increase		4 ms 15 ms decrease
	photocurrent (τ)	$\leq$ 150 ns	400 ns	6 μs	150 μs	2 ms 16 ms
	displaced charges ΔQ ($nA \cdot ms$)	−0.2	− 0.1	1.0	1.7	7 5

Mutant 326 contains a BR that carries an asparagine residue at position 96 of the aminoacid sequence, instead of an aspartic acid. The absorption spectrum of this mutated BR is identical with that of wild-type BR, and M-formation has the same kinetic characteristics as wild-type BR, within a factor of 2-3. The M-decay, however, is drastically slowed down to 450 ms at pH 6.2 (see Table I).

5. Photoelectical Experiments with the Mutated Bacteriorhodopsin Asp$^{8\,5}$->Glu

The ion transport properties of BR and its mutated analogues were studied by measuring the displacement currents and the pump currents on purple membranes attached to planar lipid membranes and on purple membranes oriented and immobilized in polyacrylamide gel. Both methods gave identical results.

In Fig.8 the absorption spectrum of BR Asp^{85}->Glu at pH 7.5 is compared with its action spectrum. The action spectrum was determined by measuring the stationary photocurrents at different wavelengths, using the planar lipid bilayer system. Stationary currents were obtained upon addition of the protonophore 1799 to permeabilize the planar lipid film underlying the purple membrane sheets. The action spectrum was also recorded by measuring stationary photocurrents from the gel-embedded oriented purple membrane sheets. Independent of pH, the resulting photoresponse was a maximum at 550 nm, and thus confirmed the result shown in Fig.8 for the lipid membrane system. Obviously, the shoulder at 550 nm in the absorption spectrum of BR Asp^{85}->Glu is due to a species that is active in proton pumping, but the main component whose maximum adsorption lies at 610 nm is inactive in proton translocation.

The measured photocurrent kinetics of BR Asp^{85}->Glu is shown in Fig. 6. The overall shape of the time-resolved photocurrent signal is similar to that of the wild-type BR. A fast and negative current is followed by a positive charge displacement which has been assigned to the proton translocation during the steps L->M and M->BR ground-state of the catalytic cycle. In BR Asp^{85}->Glu the positive signal of the photocurrent rises much faster than in the wild-type. As listed in Table I, the relaxation times of the mutant protein are decreased similarly as those measured for the photocycle kinetics. A time constant of 50 µs is found for wild-type BR, but only of 6 µs for the mutant. A second kinetic component of 150 µs remains unaffected by the mutation. It is interesting to observe that, in addition to the relaxation time of 6 µs, the mutant has another relaxation time of only 400 ns in its positive part, replacing the negative 1.2 µs component of the wild-type. Control experiments showed that this relaxation time is indeed a property of the mutant protein, and not an artefact of the measuring system. Further results of these experiments are summarized in Table I, which reports the relaxation times with their concomitant displaced charges. The main conclusion to be drawn is that replacement of Asp[85] by Glu accelerates the electrogenic step linked to deprotonation of the Schiff base in BR. The photocurrent of the wild-type BR is shown in Fig. 6c for comparison. In addition, a simplified scheme of the photocycle with the intermediates K, L, M and O is shown in Fig.7. The subscript denotes the absorption wavelength of the particular intermediate.

6. Photoelectrical Experiments with the Mutated Bacteriorhodopsin Asp^{96}-> Asn

BR Asp[96]->Asn has an action spectrum of ion translocation identical with its absorption maximum at 570 nm (not shown). Figure 9 compares the photocurrent under stationary illumination of purple membranes of the mutant Asp[96] ->Asn, oriented in a polyacrylamide gel, with the photocurrent for the wild-type. Since the overall photocycle time of the mutant BR under the adopted conditions is 450 ms, a photocurrent smaller than that for the wild-type is expected. Therefore, the concentration of wild-type BR in the gel was lowered so as to adjust its response to that of the mutant. Under these conditions the stationary currents of the two BRs are comparable, but a conspicuous difference exists in the shape of the electrical signals. After the onset of illumination, mutant Asp[96]->Asn shows a peak current which decays to a stationary value (Fig. 9a), while the signal from the wild-type sample lacks this peak current (Fig. 9b). The height of the peak in the mutant preparation depends upon the length of the dark period between two subsequent illuminations. Figure 9a shows two different peak values, after a long dark interval (first illumination) and after intermediate dark times (all other illuminations). Shortening or extending the dark interval modulates the peak current, from an almost negligible amplitude at short intervals to a maximum peak height at long dark intervals. By plotting the interval time against the corresponding peak height the exact value of its recovery time could be obtained. A value of 500 ms was determined, reflecting the cycle time of the mutant.

The kinetics of the photocurrent of mutant BR Asp[96]->Asn was also measured after a laser flash, and was correlated with its photocycle kinetics. Comparing the mutant with the wild-type, the fast part of the kinetics up to M formation is changed by at most a factor of 3. A drastic effect, however, is that no current relaxation occurs in the millisecond range. Since the decay of M in the mutant increases to 450 ms, no direct observation of the electric relaxation is possible, because the amplitude of the current during this process is too small. This does not mean, however, that no proton transport occurs in the mutated BR. As already demonstrated by the stationary current, such a transport must occur.

Asp[96] is a suitable candidate for reprotonation of the Schiff base, either directly or *via* participation of other residues. The protonated Asp[96] residue in the interior of the membrane explains the substantial pH independence of the M decay in wild-type BR as

548

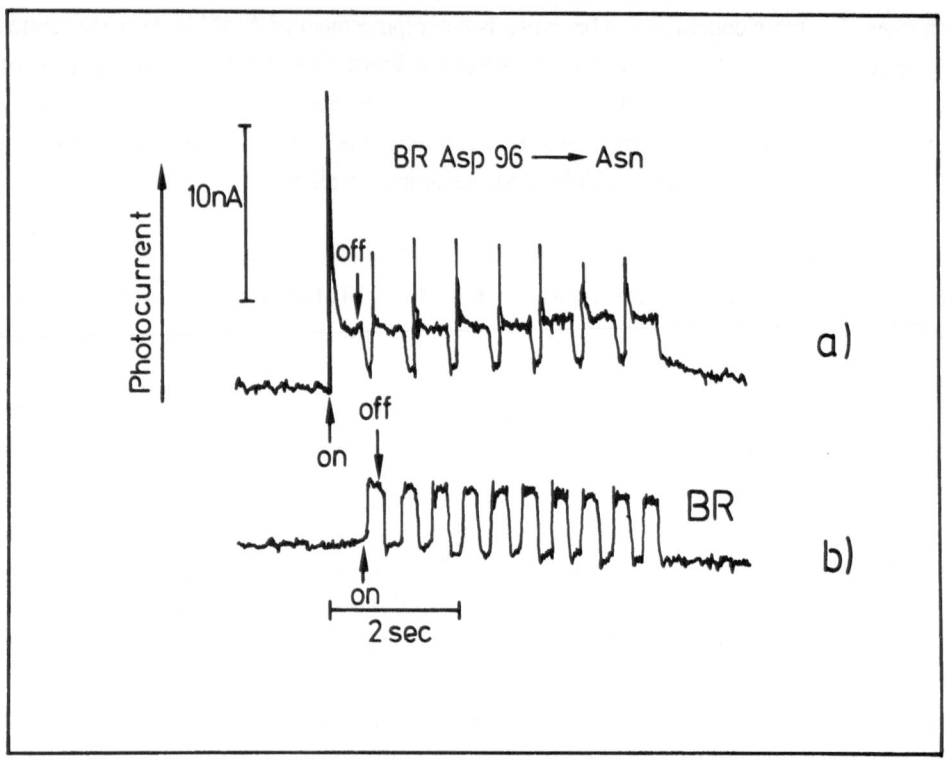

Figure 9. Photocurrent of BR Asp^{96}->Asn (a) and wild-type BR (b) resulting from repetitive exposure to continuous light. Polyacrylamide gel, 10 mM buffer of pH 6.0 and 20 mM KCl; optical density was 0.5 for mutant BR and 0.05 for wild-type BR.

well as the dramatic increase in rate at low pH values upon removal of the internal proton donor. It will certainly be interesting to investigate in detail isotope effects and spectroscopic properties of intermediates following M in these mutants.

References

Bamberg, E., Apell, H-J., Dencher, N.A., Sperling, W., Stieve, H., Lauger, P.(1979) Biophys.Struct.Mechanism **5**,277-291.
Cohen, F.S., Eisenberg, M., McLaughlin, S.,(1978) J.Membr. Biol. **37**, 361
Dancshazy, Z. and Karvaly, B. (1976) FEBS Lett. **72**, 136-138.

Dèr, A., Hargittai, P., Simon, J. (1985) J. Biochem. Biophys. Methods **10**, 295-300.

Drachev, L.A., Jasaitis, A.A., Kaulen, A.D., Kondrashin,A.A., Libermann, E.A., Nemecek, I.B., Ostroumov, S.A., Semenov, Y.Y., Skulachev, V.P.(1974) Nature **249**,321-324.

Eisenbach, M., Weissmann, Ch., Tanlly, G., Caplan, S.R. (1977) FEBS Lett. **81**, 7-80.

Eisenstein, L., Lin.S.-L., Dollinger, G., Odashima, K., Termini, J., Kuuno, K., Ding, W.-D., Nakanishi,K. (1987) J.Am.Chem.Soc. **109**, 6860-6862.

Engelhard, M., Gerwert, K., Hess, B., Kreutz, W., Siebert, F. (1984) Biochemistry **24**, 400-407.

Fahr, A., Lauger, P., Bamberg E. (1981) J.Membrane Biol. **60**, 51-62.

Fendler, K., Grell, E., Haubs, M.,Bamberg, E., EMBO J., **4**, 3079 (1985).

Fendler, K., Grell, E., Bamberg, E., FEBS Lett. **224**, 83 (1987).

Fendler, K., Gartner, W., Oesterhelt, D., Bamberg, E. (1987) Biochim. Biophys. Acta **893**, 60-68.

Gerwert, K., Hess, B., Soppa, J., Oesterhelt, D. (1989) Proc. Natl. Acad. Sci. USA **86**,4943-4947.

Herrmann, T.R., Rayfield, G.W. (1978) Biophys. J. **21**,111-125.

Holz, M., Drachev, L.A., Mogi, T., Otto, H., Kaulen, A.D., Heyn, M.P., Skulachev, V.P. Khorana, H.G. (1989) Proc. Natl.Acad.Sci. USA **86**, 2167-2171.

Karvaly, L., Dancshazy, Z., (1977) FEBS Lett. **76**, 36-40.

Keszthelyi, L., Ormos, P. (1980) FEBS Lett. **109**,189-193.

Lozier, R.H., Bogomolni, R.A., Stoeckenius, W.(1975) J.Biophys. **15**, 955.

Mogi,T., Stern,L.J., Marti,T., Chao,B.H., Khorana,H.G. (1988) Proc. Natl. Acad.Sci. USA **85**, 4148-4152.

Oesterhelt,D., Stoeckenius,W. (1973) Proc.Natl.Acad.Sci.USA,**70**,2853-2857.

Oesterhelt,D., Hess,B. (1973) Eur.J.Biochem.,**37**,316-326.

Ormos,P., Dancshazy,Z., Karvaly,B. (1978) Biochim.Biophys.Acta **5503**,304-315.

Soppa,J., Otomo,J., Straub,J., Tittor,J., Meessen,S., Oesterhelt,D. (1989) J. Biol. Chem., in press.

Soppa,J., Oesterhelt,D. (1989) J.Biol.Chem., in press.

Stoeckenius,W., Bogomolni,R.A., Ann.Rev. Biochem., **52**, 587 (1982).

Trissl,H.-W (1990) Photochem.Photobiol. **5**.

THE ELECTROCHEMICAL RELAXATION AT THYLAKOID MEMBRANES

W. JUNGE, A. POLLE, P. JAHNS,
G. ALTHOFF AND G. SCHÖNKNECHT
Universität Osnabrück
Postfach 4469
D-4500 Osnabrück
Deutschland

SUMMARY. The primary processes of photosynthesis by green plants take place in thylakoids of chloroplasts. About one fourth of the useful work derived from light quanta is used for the generation of a transmembrane electrochemical potential difference of the proton, which, in turn, is used by a proton translocating ATP synthase to yield ATP. This article describes the electrochemical relaxation involving proton pumps, the ATP synthase, lateral and transversal conductances for protons and other ions. It relies mainly on flash spectrophotometry for the pulsed generation of the electrochemical potential difference and time-resolved detection of pH- and voltage-transients by appropriate dyes. The pore-forming antibiotic gramicidin serves as standard. The electrochemical analysis is complicated by the folding pattern of thylakoid membranes with tightly apposed and extended membrane domains which differ in their protein composition. In this article we discuss the following aspects: 1.) The single channel conductance of gramicidin in this protein-rich membrane compares well with figures reported for pure lipid bilayers. The area related dimerization constant is about ten-fold larger because of upconcentration in lipid domains. 2.) The whole contents of thylakoid membranes, a total area of several 100 μm^2 with more than 10^8 molecules of chlorophyll and 10^5 of ATP synthase, forms one contiguous entity. This capacitor can be discharged by a single gramicidin dimer. 3.) There is no evidence for surface-enhanced proton diffusion along this membrane. In a membrane patch with only 10^5 molecules of chlorophyll and in stacked thylakoids the resistance to lateral proton flow between pumps and ATP synthases along tightly appressed membranes dissipates a few percent of the transmembrane protonmotive force. Thus the effective membrane unit for protonic coupling between pumps and ATP synthases is smaller than the electrically coupled one. 4.) The area specific electric capacitance of the membrane is by more than one order of magnitude smaller than the specific chemical (buffering) capacitance for protons. The greater damping power for pH-transients as compared to the one for electric transients may be why thylakoids (subject to fluctuating light intensity) use ΔpH rather than $\Delta\psi$ (like mitochondria) to drive ATP synthesis.

1. Introduction

Chloroplasts of higher plants carry an inner lamellar system, thylakoids, that is folded into densely stacked membranes (grana) and interconnecting membranes (stroma lamellae). The photosynthetic apparatus consists out of four large protein complexes: photosystem II oxidizes water, photosystem I reduces NADP, the cytochrome b_6,f-complex mediates electron transfer between the former, and CF_0CF_1 synthesizes ATP at the expense of the transmembrane electrochemical potential difference of the proton (Mitchell, 1961; Mitchell, 1966). Fig. 1 illustrates this structure. The protein complexes are seggregated according to the folding pattern

R. Guidelli (ed.), Electrified Interfaces in Physics, Chemistry and Biology, 551–564.
© 1992 *Kluwer Academic Publishers.*

of the membrane (Andersson & Anderson, 1980). Apposed membrane portions contain photosystem II, exposed membrane portions contain photosystem I and the ATP synthase. The upper inserts show an artist's view of the complicated folding of thylakoids (left) and an electron micrograph of one granum. Each of the stacked disk-like structures (diameter about 0.5 μm) contains about 100 photosystem II molecules, but as the disks are interconnected by stroma lamellae the functional unit of the electrochemical events is much larger (see below). The inner lumen and the outer partitions between stacked thylakoid membranes are very narrow, only about 5 nm wide slabs.

The middle of Fig. 1 shows the electron transport chain. In reaction centres the primary event is a very rapid transmembrane transfer of electrons from the inner to the outer surface of the membrane (some 100 ps). This is followed by proton uptake from the outer phase (the chloroplast stroma) and proton release into the inner phase (thylakoid lumen), as indicated. The lower portion of Fig. 2 shows an equivalent circuit for cyclic proton flow between pumps and ATP synthases.

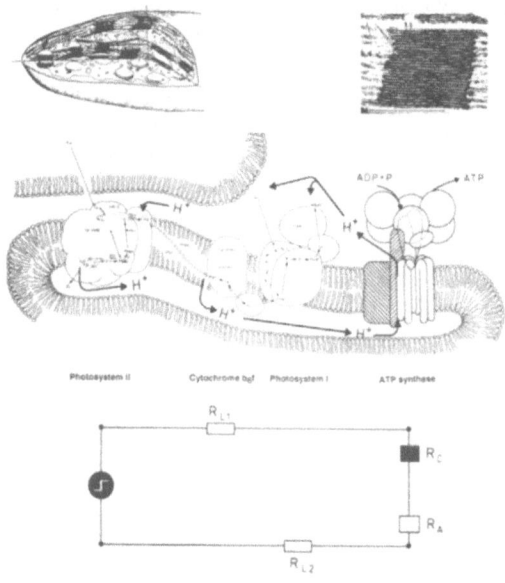

FIGURE 1. Schematic drawing of the thylakoid membrane with the three large electron transfer complexes and the ATP synthase. Apposed membrane proteins with photosystem II and exposed portions with photosystem I and the synthase are apparent. Sites of light driven proton binding from the outer medium (the stroma) and of proton release into the lumen, lateral proton flow, and transmembrane proton backflow over the ATP synthase are indicated. The insert in the upper right shows an electron micrograph of stacked thylakoid membranes. Each of the disk-like structures contains at least 100 pieces of any protein complex, but the true function unit is larger as disks are interconnected with each other. The lower portion shows a equivalent circuit for cyclic proton flow between pumps and the ATP synthase (adapted from (Junge, 1989)).

Photoinduced protolytic reactions have been studied at high time resolution by laser-flash spectrophotometry. Thylakoids are a favourable object for such studies. Photosystems are to be stimulated for a single turnover by short laser flash (<10 ns). Transients of the transmembrane voltage can be measured at ns time resolution by electrochromism of intrinsic pigments (Junge &

Witt, 1968; Witt, 1979; Junge & Jackson, 1982). Hydrophilic pH-indicating dyes are practically selective for pH-transients in the large external bulk phase (Polle & Junge, 1986) while membrane adsorbed indicators (mainly neutral red) react to pH transients at both surfaces of the membrane. In the presence of appropriate non-permeating buffers the latter can be selective for pH transients in the very narrow (only 5 nm wide) lumen (Auslaender & Junge, 1975; Junge et al.1979; Hong & Junge, 1983; Junge et al.1986). In stacked thylakoids, however, this selectivity is expressed only after the time interval necessary for the lateral relaxation of pH-differences along the narrow slabs between appressed membranes (Lavergne & Rappaport, 1990; Jahns et al.1991). While the response time of electrochromism so far has been instrument-limited and in the ns time range, pH indicating dyes respond in the time range of 100 μs (around neutral pH). With these three observables, namely pH(lumen), pH(bulk medium) and transmembrane voltage, the spectrophotometric techniques have allowed "complete tracking of proton flow" (Schoenknecht et al.1986; Junge, 1987).

2. The lateral relaxation of pH-transients between appressed thylakoid membranes

In stacked thylakoids certain sites of proton pumping, namely photosystem II, are laterally seggregated from ATP synthases. The center of a thylakoid disk may be 300 nm away from the nearest synthase. Cyclic proton flow between pumps and ATP synthases is illustrated in Fig.2 (top). A flash of light generates an electric potential difference (in some 100 ps) and a pH-difference (0.1-1ms) around each photosystem II. Initially the electric and the chemical portion of the protonmotive force are localized. The spreading of the electric potential difference is very fast. With the cable equation, the dimension of partitions and of lumen slabs and with the concentrations of ions present one arrives at an upper limit for the electric relaxation time of some 100 ns. The lateral relaxation of the chemical portion of the protonmotive force is much slower, indeed. Activation of photosystem II molecules by a flash of light generates an alkalinization jump in the *partition* region between opposing membranes in a stack. When the ATP synthase is inactive (e.g. too low concentrations of ADP present) the *transmembrane* relaxation of a pH-difference is much slower (20 s) than the *lateral* one (100 ms).

The lateral relaxation by diffusion of protons, of hydroxyl anions and of mobile buffers can be measured by hydrophilic dyes as pH-transients in the bulk phase. The expected pH-profile at different times is illustrated in the middle of Fig. 2, the observed transient alkalinization in the bulk (Polle & Junge, 1986) and a theoretical fit (Junge & Polle, 1986) are shown at the bottom. The rise of alkalinization in the external phase occurs in about 100 ms as contrasted with the expectation of about 1 μs by solving Fick's equation under the assumption of the same diffusion coefficients as in bulk water.

The dramatic increase of the relaxation time is understood by theory of diffusion in domains with fixed and mobile buffers (Junge & Polle, 1986; Junge & McLaughlin, 1987). The relaxation is determined by an "effective diffusion coefficient", D_{eff}, made up of the "true" diffusion coefficients of H^+, OH^- and of mobile buffers (index i):

(1) $\qquad D_{eff} = (2.3/\text{ß}_{tot}) \cdot (D_H \cdot [H^+] + D_{OH} \cdot [OH^-] + \sum(D_i \cdot \text{ß}_i/2.3))$

wherein the ß denote the respective buffer capacities of mobile ($D \neq 0$) and immobile ($D = 0$) buffers. The buffering capacity is here defined for very small perturbations as $\text{ß} = -d[H^+]_{total}/dpH$, wherein the nominator denotes the total concentration change of protons, both, bound and free. It is important to note that the diffusion of protons between two captures by buffering groups is still governed by D_H.

In experiments with thylakoids the behaviour of the effective diffusion coefficient has followed the expectation (Equ.1). Both the predicted minimum around neutral pH and the acceleration by added mobile buffer have been observed (Polle & Junge, 1989). The delay over the expected relaxation in the absence of buffers by a factor of 10^5 is compatible with the estimated amount of buffer intrinsically present in partitions (Junge & Polle, 1986). To account for this very large factor it has to be assumed that one half of the total buffering power of the thylakoid suspension (no extra buffers added) resides in partitions. This certainly overestimates the buffering capacity in this domain. Accordingly, the diffusion coefficients of hydrogen- and hydroxyl-ion are rather less than larger than in bulk water (Polle & Junge, 1989).

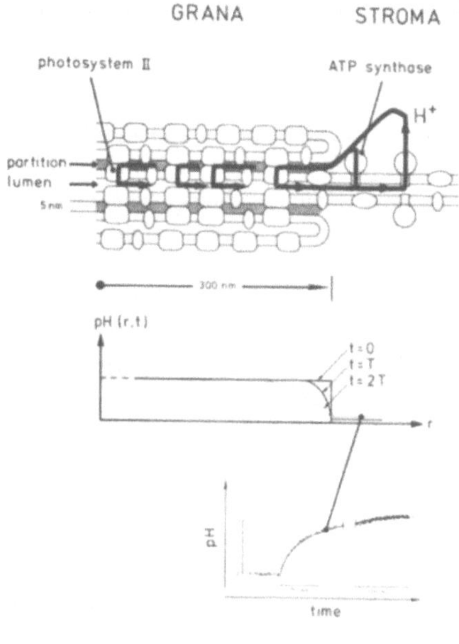

FIGURE 2. Schematic side view of stacked thylakoid membranes with appressed grana lamellae and interconnecting stroma lamellae. The narrow space between the outer surface of thylakoid membranes in a granum, called partition, is hatched. The *arrows* illustrate (a) light-driven proton pumping by photosystem II which is directed from the partition into the lumen, (b) opposite directed proton flow which is coupled to ATP synthesis, (c) lateral proton flow through the lumen, and (d) lateral backflow through partitions. The drawing in the middle illustrates the box shaped pH profile induced by excitation with a single flash of light at time zero and its evolution in time (t = T, t = 2T). The bottom trace is a reproduction of original data on the flash induced, and photosystem II-related pH rise in the stroma compartment and a theoretical curve (Junge & Polle, 1986; Polle & Junge, 1986).

It is worth noting that Equ.1 predicts that all buffers slow down the relaxation, since their diffusion coefficient is always smaller than the one of the proton (Eigen, 1963). This slowing down by buffering groups of a diffusive relaxation does not imply the diminuation of the steady proton flux. In the steady state the rates of proton binding and release by fixed buffers are equal. The steady flux is given by Fick`s first law and depends on the true diffusion coefficients (see 2.2).

2.1 THE CONTROVERSY OVER SURFACE ENHANCED DIFFUSION OF PROTONS

A controversy over the validity of Mitchell's theory of a chemiosmotic mechanism of ATP synthesis (Mitchell, 1966; Mitchell, 1961) is focussed on whether or not the ATP synthase operates between two aqueous bulk phases separated by the coupling membrane or between more restricted domains, which are not in equilibrium with the respective bulk phase (see (Williams, 1988; Williams, 1961)). In this respect, surface-enhanced diffusion of protons at lipid monolayers has been hypothesized (Haines, 1983). Experiments with lipid monolayers bridging a barrier between two aqueous compartments led Prats and his collegues to the conclusion of 20-fold enhanced proton diffusion at the lipid/water interface (Prats et al.1986; Prats et al.1987). Moreover the authors claimed the absence of equilibration of the surface pH with the bulk. Their experimental layout has been criticized for stirring artifacts and also for neglect of equilibration with the bulk (Kasianowicz et al.1987). In an alternative approach the conductance between two electrodes immersed in ultrapure water (20 mm apart) has been measured, without a barrier and without stirring. The difference-conductance (plus/minus lipid monolayer) has revealed a sharp increase above a certain compression of the monolayer (Morgan et al.1991). This is not necessarily due to enhanced diffusion but it may be attributable to the upconcentration of protons in the diffuse double layer, which does matter, of course, in an electrophoretic experiment as this one. Experiments by laser-flash spectrophotometry on proton diffusion in the ultrathin layer between bilayers in osmotically compressed multi-lamellar lipid vesicles give no indication of enhanced diffusion (Gutman et al.1989). Our above experiments with thylakoids give no indication either (Polle & Junge, 1989). Thus it is very likely that there is no enhanced diffusion at the surface of *biological membranes* and there is certainly no inhibition of rapid equilibration with the surface and the bulk.

2.2 LOSSES OF PROTONMOTIVE FORCE BY LATERAL PROTON FLOW

In a comparison of the efficiencies of photosystem II and photosystem I to drive ATP synthesis slightly lower figures have been observed for the former (Haraux & de Kouchkowsky, 1982). This might be understood in terms of the losses of the protonmotive force during lateral flow of protons and hydroxyl anions between photosystem II in the appressed membrane portions and the ATP synthases in the exposed ones (see FIG.2). A total flux of protons, I (moles/s), over the boundary of a thylakoid disk requires a drop in proton concentration between the center (suffix c) and the fringe (suffix f) of the disk. For a disk that is homogeneously filled with pumps the concentration drop has been calculated (Junge & Polle, 1986):

(2) $[H^+]_c - [H^+]_f = I/4\Pi hD$

It depends on the thickness of the disk-shaped slap between membranes, h, and on the diffusion coefficient, D. A proton flux, which is equivalent to the highest rate of ATP synthesis in a model thylakoid (radius 300 nm, area per clorophyll molecule 2.2 nm), namely, 1.3×10^{-19} mol s^{-1}, and assuming a thickness of 5 nm and the diffusion coefficient as in bulk water, $D_{H+} = 9.3 \times 10^{-9}$ m^2 s^{-1}, implies a drop proton concentration of 0.23 mM.[27] The magnitude of the corresponding pH drop, $DpH = -D[H^+]/2.3 \times [H^+]$, depends on the pH in the medium. It decreases toward more acid pH. A similar relation holds in the alkaline pH domain, where die diffusion of OH⁻ dominates. Taking these results together and assuming the same diffusion coefficients as in bulk water the pH drop has been calculated. At the outer surface of stacked thylkoids it amounts to 0.14 pH units, if the outer side is kept at pH 8. It is below 0.01 pH units at the internal side and at pH 4 (Junge & Polle, 1986). If the diffusion coefficients in these narrow spaces are lower than in bulk water, greater losses are expected. It is probable that ohmic losses of protonmotive force are small, but they are not negligible in tightly stacked thylakoids.

2.3 CONSEQUENCES FOR THE SPECIFICITY OF NEUTRAL RED FOR pH-TRANSIENTS IN THE THYLAKOID LUMEN

Neutral red is a hydrophobic pH-indicating dye which has a long history in microscopy as vital stain. It adsorbs to biomembranes. Because of this property it has been used as specific indicator of pH-transients in the lumen of thylakoids (Auslaender & Junge, 1975). The rationale of the experiments is illustrated in Fig. 3 (Jahns et al.1991). Since neutral red is adsorbed to both sides of the membrane it responds to pH-transients in both compartments. Its selectivity is elicited only by addition of a non-permeant buffer, typically bovine serum albumin, to eliminate pH-transients in the outer bulk phase. The well-behavedness of this indicator includes the following features: Although it is a redox-active compound (Prince et al.1981), its transient absorption changes in thylakoids are fully quenched by permeating buffers (Junge et al.1979). Its response to pH-transients and its binding to the membrane both vary as function of pH and salt composition in exactly the way expected for an indicator sensing the surface-pH, with $pH_s = pH_{bulk} - \psi_s/59mV$ (Hong & Junge, 1983). Although amphiphilic, the dye is only mildly protonophoric (Junge et al.1986). Hence the response to rapid pH-transients is first owed to dye molecules that are already in-place. A time resolution for pH-transients in the thylakoid lumen down to 100 μs is obtained (Foerster & Junge, 1985). A slow redistribution of the dye across the membrane and in response to a pH-difference occurs only in the range of several seconds (Junge et al.1986).

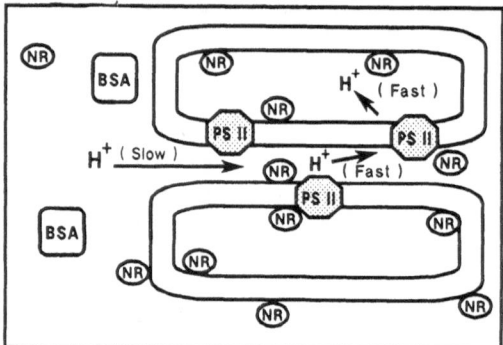

FIGURE 3. Scheme illustrating the interpretation of absorption transients of neutralred as indicator of pH-transients in the lumen of thylakoids (Lavergne & Rappaport, 1990; Jahns et al.1991). It is assumed that neutralred is adsorbed to the membrane surface at both sides of thylakoids in a grana stack. However, the non-permeating buffer, bovine serum albumin (BSA) is practically excluded from the narrow external partitions. Initially the dye responds rapidly to pH-transients at both sides of the membrane. The transient at the outer side is more slowly quenched by BSA (see above). Only thereafter is neutralred fully selective for the transient in the lumen phase.

In stacked thylakoids the well-behavedness of the indicator is blurred by the above described slow lateral relaxation of the alkalinization jump in partitions. This has been first pointed out by Lavergne and Rappaport (1990) and worked out in greater detail by Jahns et al. (Jahns et al.1991). Because of the tight appression of stacked thylakoid membranes the non-permeating buffer BSA is not present in the partition slab. Accordingly, the transient of neutral red also contains a smaller contribution from neutral red molecules at the outer side of the membrane.

Only after lateral relaxation in partitions, i.e. after some 100 ms, is it selective for pH-transients in the lumen. This is mended when thylakoids are unstacked. Then the added buffer has more direct access to the sites of proton uptake at the outer side of the membrane. These results have stimulated a reevalutation of the stoichiometry and kinetics of proton release during the four-stepped reaction of the water oxidase (photosystem II) with the result of a non-integer pattern of proton/electron ratio during the four successive transitions (Jahns et al.1991).

3 The transmembrane electric relaxation in the presence of the cation channel gramicidin

Gramicidin is perhaps the best characterized pore-forming antibiotic. It conducts alkali-cations across lipid bilayers as a head-to-head dimer (Veatch et al.1975; Durkin et al.1990). When the respective ATP synthase is inactive photosynthetic membranes are rather proton tight and only moderately pervious to other cations and to anions. The nature of the intrinsic *leak conductance* is a matter for itself and not further pursued here. In thylakoids the relaxation time of the electric potential difference (generated by one flash of light) ranges around 100 ms at 10 mM KCl and 3 mM $MgCl_2$. The capacitor equation links the relaxation time, τ, or the decay rate, k, to the specific conductance, G, and the specific electric capacitance, C, of the membrane:

(3) $\qquad \tau = C/G = 1/k$

where C comes in F/m^2 and G in S/m^2.

The area specific capacitance of thylakoids is about 10^{-2} F/m^2 (Arnold et al.1985). Thus a relaxation time of 100 ms gives rise to a specific leak conductance of 10^{-1} S/m^2. The total area of one thylakoid in a stack (It is not isolated from the rest, but see below.) is that of closed disk with radius 300 nm, namely $566 \cdot 10^{-15}$ m^2. The presence of one ion channel with time averaged conductance of only 57 fS accounts for the observed leak conductance and a powerful ion channel like gramicidin (several pS) causes a considerable acceleration of the electric relaxation. There are two limiting cases, in small vesicles (thylakoid fragments, chromatophores of photosynthetic bacteria) a single gramicidin dimer causes a drastic acceleration of the relaxation of the transmembrane voltage. In a typical spectrophotometric experiment the observed decay of the electrochromic absorption changes reflects the voltagedecay in more than 10^{10} vesicles. It is determined by the Poisson statistical ensemble properties (vesicles with 0 channels reveal a slow decay, others with 1, 2, ... channels a very fast one). On the other hand, with the large membrane area of interconnected, folded and stacked thylakoids one single dimer per large vesicle causes only little effect, the average number of channels is large and the ensemble behaves homogeneously. This has been used for a spectrophotometric determination of the unit conductance of gramicidin and its dimerization constant in photosynthetic membranes and furthermore to evaluate the electric unit size of chromatophores and of thylakoids. It has emerged that almost all thylakoid membranes within a chloroplast (see Fig. 1, top) are part of one electrically connected entity.

3.1 GRAMICIDIN IN SMALL VESICLES (BACTERIAL CHROMATOPHORES, FRAGMENTED THYLAKOIDS): POISSON's DISTRIBUTION

It was claimed that the addition of gramicidin to chromatophores "inhibited" the generation by flash light of carotenoid absorption changes (Fleischmann & Clayton, 1968). These absorption transients are of electrochromic origin (Jackson & Crofts, 1969; Junge & Jackson, 1982). Seemingly, it did not accelerate their decay which was expected (Saphon et al.1975). Fig. 4 explains why (Althoff et al.1991). The transient traces in the upper half show the generation of transmembrane electric potential by flash light at time zero and a rather slow decay of the voltage by low leak conductance of the membrane. In the presence of gramicidin at increasing amounts the extent of the voltage is seemingly diminished while the decay rate is less affected. The same experiment but now at thousand-fold higher time resolution (lower half of Fig. 4) reveals a very rapid decay, which has excaped detection at lower time resolution. The slow decay is attributed to vesicles without any gramicidin dimer and the fast one to those with 1, 2, ... dimers.

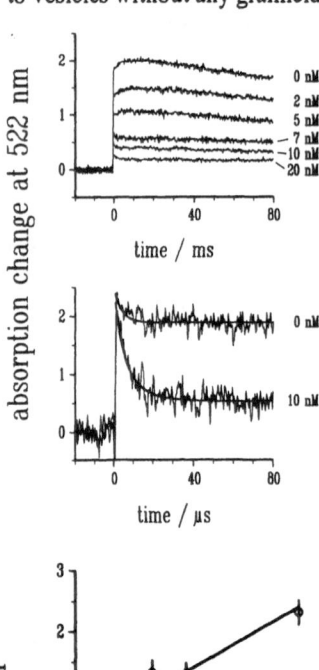

FIGURE 4. Electrochromic absorption changes after single flash-excitation. Chromatophores were incubated for 30 min in the presence of different gramicidin (monomer) concentrations. The ordinate indicates $-\Delta I/I$ and was scaled up by a factor of 1000. Top: 40 transients with 200 µs time resolution were averaged. Gramicidin concentration as indicated. Bottom: 120 transients with 200 ns time resolution were averaged. Gramicidin concentration as indicated. The transients were fitted according to Eq. (6) with the following parameters: n=0.27, k= 0.24 µs^{-1} for 0 nM and n = 1.35, k = 0.11 µs^{-1} for 10 nM (Althoff et al.1991).

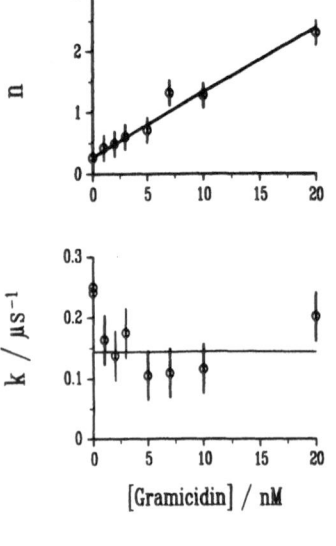

FIGURE 5. Results obtained by fitting traces as shown in Fig. 4, lower part. Top: Average number of conducting gramicidin dimers per chromatophore, n, as function of gramicidin (monomer) concentration in the chromatophore suspension. Solid line by linear regression of datapoints (see text). The error bars indicate the standard deviation from the mean of the fitted n values. Bottom: Decay rate caused by a single gramicidin dimer, k, as function of the gramicidin (monomer) concentration. The solid line indicates the mean k value of the data points in the range from 1 nM to 20 nM gramicidin.

The theory for this situation was developed previously ((Schmid & Junge, 1975), see also (Apell & Läuger, 1986)). In a vesicle with m ohmic channels the voltage decays exponentially.

(4) $U(t) = U_0 \cdot \exp(-m \cdot k \cdot t)$

wherein k denotes the decay rate caused by one channel and m the number of channels in this vesicle. In a population of many identical vesicles doped with n channels, in the average, the probablity to find m channels is given by Poisson's distribution:

(5) $P(m) = (n^m \cdot \exp(-n))/m!$

Merging these equations yields the double-exponential decay law for the average behaviour of a very large ensemble:

(6) $U`(t) = U`_0 \cdot \exp(-n) \cdot \exp(n \cdot \exp(-kt))$

It has been shown elsewhere (Lill et al.1987) that the incorporation of the vesicle size-distribution into this equation does not lead to much different behaviour in a wide range of sizes. Equation 5 has only two essential fit parameters, n, the average number of open channels per vesicle, and k, the decay rate caused by one channel per vesicle, which is related to the specific conductance by Equation 2.

The decay of the electrochromic absorption changes as documented in Fig. 4 has been analyzed in terms of Equation 5 and the fit parameters have been plotted as function the the gramicidin concentration (see Fig. 5). The detailed analysis is found elsewhere (Althoff et al.1991). Broadly speaking it has yielded the following results: 1.) The average number of channels (i.e. gramicidin dimers) per vesicle depends linearly on the concentration of monomers. This implies complete solvation and complete dimerization of the antibiotic in the chromatophore membrane. 2.) The average number of bacteriochlorophyll-molecules per vesicle is 770. They cover an area of $10^4 \, nm^2$. 3.) With $1.1 \, \mu F/cm^2$ for the specific capacitance of the chromatophore-membrane (Casadio et al.1988) the single-channel conductance of one gramicidin-dimer is 15 pS. At the given K^+-concentration (115 mM) this compares well with its conductance in lipid bilayer membranes as determined by electrophysiological techniques (Hladky & Haydon, 1972; Neher et al.1978). A similar analysis has been carried out with small vesicles from thylakoid membranes, resealed fragments which were obtained by EDTA treatment. The membrane area of these vesicles is $1 \, \mu m^2$ and they contain about $5 \cdot 10^5$ molecules of chlorophyll (Lill et al.1987)(Schoenknecht et al.1990).

3.2 GRAMICIDIN IN THYLAKOID MEMBRANES: THE VERY LARGE SIZE OF THE ELECTRIC UNIT IN CHLOROPLASTS

When the envelope of intact chloroplasts is broken by hypoosmolar shock and when broken thylakids are resuspended in salt containing medium (say at 3 mM MgCl) the complicated folding pattern with grana of stacked membranes, interconnected by stroma lamellae is conserved. In the absence of salts, however, they swell into spherical blebs with more than 10 μm wide diameter (DeGrooth et al.1980; Stolz & Walz, 1988). These blebs contain up to $2 \cdot 10^8$ molecules of chlorophyll covering a membrane area of 400 μm^2 (Stolz & Walz, 1988). Electron microscopic evidence suggests that these blebs represent the total contents of thylakoid membranes within one chloroplast. We have asked whether or not this large membrane area encloses one continuous internal space even when folded into the complicated topology with stacked and interconnecting membranes which is shown in the upper left of Fig.1. This has been approached by asking for the maximum size of the membrane area which is to be electrically discharged by a single dimer of gramicidin. Due to the larger membrane area and thereby the larger electric capacitance of integral thylakoids the effects of small concentrations of gramicidin are less dramatic as for instance in chromatophores (Fig.4). Still the addition of 3 pM gramicidin on 20 µM chlorophyll has caused an appreciable acceleration of the electrochromic transient. Accordingly, the electric unit contains at least 10^7 chlorophyll molecules. The detectability of the minimum action of a single dimer per large vesicle depends on the magnitude of the intrinsic leak conductances of the membrane. An analysis that takes these leak conductances into account has pushed the chlorophyll contents of the electric unit up to at least $7 \cdot 10^7$ molecules (Schoenknecht et al.1990). Thus it is probable that the whole contents of $2 \cdot 10^8$ molecules of chlorophyll of one chloroplast is organized in one coherent membrane enclosing one electrically coherent lumen space.

3.3 GRAMICIDIN IN THYLAKOID MEMBRANES: DIMERIZATION CONSTANT AND SINGLE-CHANNEL CONDUCTANCE IN A MEMBRANE RICH IN PROTEIN

In the above estimates on the electric unit size of photosynthetic membranes the (time averaged) single channel conductance of gramicidin dimers is read out straightforwardly from experiments with small vesicles where Poisson's distribution dominates the decay of electrochromic transients. There is good coincidence with the unit conductance determined by patch clamp. What about the distribution between membrane and water and the dimerization constant in a membrane with high protein contents as in thylakoids? The equilibration of gramicidin between thylakoids, its distribution between membranes and water, its dimerization constant in the membrane and the single-channel conductance has been studied with thylakoids (Schönknecht et al., submitted). When two suspensions of thylakoids are mixed, one with gramicidin and one without, it takes more than 20 min until the initially bipartite distribution of the antibiotic over vesicles is homogeneous. When gramicidin is added to a suspension under vigorous stirring, its distribution is homogeneous, right from the beginning. In a suspension with 20 µM chlorophyll (or a total protein contents of 200 µg/ml) less than 1% of the antibiotic is free in the suspending medium. There is evidence for higher concentration in appressed membranes than in stroma lamellae. As the dimer is the conducting species one expected a quadratic dependence of the decay of the electric potential difference on the concentration of gramicidin. This has only been observed at very low concentrations. Above a few hundred pM further increase is linear. This is indicative of complete dimerization. The concentration range with quadratic-to-linear transition gives information on the dimerization constant. For thylakoids from spinach it is calculated to be $5 \cdot 10^{10}$ m^2/mol, ten times larger than reported for phosphatidyl choline bilayers (Veatch et al.1975). About 60 % of the total surface of thylakoids is covered by protein and one-half of the lipid is supposedly in protein annuli (Murphy, 1986). This leaves about 20 % of the area to free lipid. If there is further upconcentration of gramicidin in appressed membranes, the factor of 10 in favour of dimerization is plausible. Since the dimerization of gramicidin is complete above a certain concentration, and as the dimer represents the "open-state" of this channel, the single-channel conductance can be derived from the above experiments. When the negative surface potential is shielded the figures range around 0.5 pS at 10 mM NaCl, in very good agreement with published data obtained on bilayers made from neutral lipids (Finkelstein & Andersen, 1981). In the absence of screening cations, however, the conductance in thylakoid membranes is strongly modulated by the surface potential at the electro-positive side of the membrane (the lumen side), with differences between plant species (see Schönknecht, 1990).

4   The transmembrane relaxation of the protonmotive force

When thylakoids are excited with a short flash of light an electrochemical potential difference of the proton is generated by vectorial electron transport plus protolytic reactions. The main pathways of discharge are the proton-specific pathways through the ATP synthase plus leak conductances for other ions. The intact enzyme conducts protons in a reaction that is coupled to ATP synthesis (see (Junesch & Graeber, 1987) for a review, and (Junge, 1987) for an analysis of proton flow under flashing light). If the channel portion of this enzyme, CFO, is exposed by removal of the catalytic F1-portion, the channel acts as an extremely proton-selective pathway with a conductance of at least 10 fS (Junge et al.1986) and probably much higher (Althoff et al.1989; Lill et al.1987). It is extremely selective for protons even at pH 8 and against a background of 300 mM of other cations like Na^+ (Althoff et al.1989). Fig. 6 illustrates the relaxation behaviour of the membrane in the presence of the large and specific proton conductance through CFO, plus the comparatively small leak conductance to other ions (left column) or with added gramicidin to drastically increase the conductance for K^+ (right column). The traces in the upper row show pH-transients in the medium, the ones in the middle row show electrochromic signals and the ones at the bottom pH-transients in the lumen of destacked thylakoids (see also (Schoenknecht et al.1986)). The scheme at the top helps to read these traces. Pairs of traces are superimposed, with one obtained before and the other one after addition of a blocking agent to proton transfer through CFO. The pair in the upper left, for instance, shows the transient proton uptake from the medium (upward rising) when the proton channel is blocked, on one hand, and the rapid overcompensation of this uptake, by proton outflux through the open channel, on the other. Since the electron acceptor has been chosen as to eliminate proton uptake at one of the two sites at the outer surface of the membrane (see Fig. 6, top), there are two protons released in the lumen per proton taken up from the medium and per two electrons charging the membrane electrically. Therefore, the initial alkalinization was overcompensated through the open channel.

We consider now the initial events as viewed by all three observables: A short flash of light acidifies the thylakoid lumen (bottom, left, with DCCD present), it charges the membrane electrically (middle) and alkalinizes the medium (top). With the blocking agent to the proton channel present, the pH-difference across the membrane relaxes more slowly than the electric potential difference, in more than 10 s (Schoenknecht et al.1986). This reflects the dominance of the conductance for ions other than the proton. With the channel conducting, however, all three observables decay rapidly. Broadly speaking, the differences between the traces with and without the blocking agent reflect: proton intake by CFO from the lumen (bottom), charge passage across the membrane (middle) and proton arrival in the medium (top) (see (Lill et al.1987; Schoenknecht et al.1986) for details). The first two observables have the same relaxation time and they are coincident in their extent (number of charges versus protons, not documented here, but see (Althoff et al.1989)). This relaxation reflects the discharge of the specific capacitance of the membrane by the dominating proton conductance. From the relaxation time the specific conductance of CFO is obtained by Equ. 3.

An analysis of the density of CFO-molecules which were exposed by removal of their CF1-counterpart has revealed the time-averaged conductance per channel, 10 fS (Schoenknecht et al.1986). This is equivalent to about 6000 protons at a driving force of 100 mV. We just mention that furthergoing studies demonstrate that not all exposed channels are conducting, so that the conductance of the active channels is much greater (Lill et al.1987; Althoff et al.1989).

562

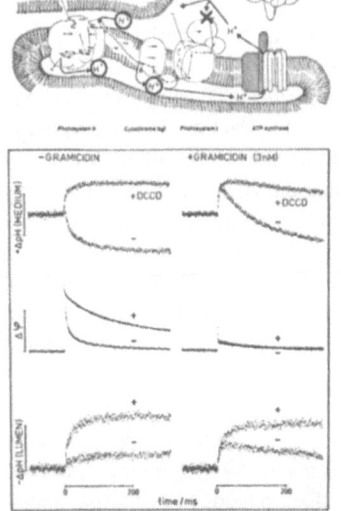

FIGURE 6. Flash-light induced transients of the pH in the outer medium (top traces), the transmembrane voltage (middle traces) and the pH in the lumen (bottom traces). Thylakoids were treated with EDTA to remove the catalytic portion of the ATP synthase in order to expose the channel portion CFO. The traces in the left column were obtained in the absence , those in the right column in the presence of the K⁺-pore gramicidin (3 nM). Pairs of traces without and with DCCD (10 µM) to block proton conduction through CFO for details, see Schoenknecht, 1990).

The traces in the right column result from a similar experiment except for the addition of gramicidin to create a specific conductance for K⁺ in great excess of the proton conductance. While the electric potential relaxes more rapidly by orders of magnitude the relaxation of the pH-transients is slowed down. In the framework of the linearized NernstPlanck Equation this can be understood as follows: With two dominating conduction pathways, for protons through CFO and for K⁺ through gramicidin, the relaxation is governed by only five variables: the specific electric capacitance, C in $F \cdot m^{-2}$; the specific buffering capacitance (mainly of the lumen) for protons, C_h in $F \cdot m^{-2}$; the specific storage capacitance for K⁺, C_k in $F \cdot m^{-2}$; the specific conductance for protons, G_h in $S \cdot m^{-2}$; and the specific conductance for K⁺, G_k in $S \cdot m^{-2}$ (Junge and Schönknecht, in preparation). The definition of C, G_h and G_k is straightforward and the *chemical or storage capacitances* are:

(7) $C_h = \beta \cdot F^2 \cdot V / (A \cdot 2.3 RT)$ and

(8) $C_k = [K^+] \cdot F^2 \cdot V / (A \cdot RT)$

where ß denotes the differential buffering capacity as defined by:

(9) $\beta = - dpH / d[H^+_{total}]$

[K⁺] is the average potassium ion concentration, V and A are volume and surface area of vesicles, F is the Faraday and RT are as usual. If the specific conductance for protons is larger than the one for any other ion (Fig. 6, left) one observes the discharge of the electric capacitance via the proton conductance and the relaxation time is given by Equ. 3. If the conductance for K⁺ is much larger than the proton conductance (Fig. 6, right) the relaxation time of the pH-transient is governed by the chemical or storage capacitance for protons (Equ.s 7 and 9). If C_k greatly exceeds C_h the relaxation time is given simply by Equ. 10:

(10) $\tau = C_h / G_h$

If C_k is of same order of magnitude as C_h the relaxation is faster than given by Equ. 10.

Equ. 10 has been used to evaluate the specific buffering capacitance of thylakoid membranes around neutral pH (Schoenknecht, 1990). With about 7 $\mu F/cm^{-2}$, it is much larger than the specific electric capacitance, 1 $\mu F/cm^{-2}$ (Arnold et al.1985). The resulting figure for the buffering capacity ß compares well with published values (Walz, 1974; Haraux & de Kouchkowsky, 1979; Junge et al.1979) as determined by other techniques. The buffering capacitance increases from neutral to acid pH. It is obvious that the chemical storage capacitance of thylakoids for protons is larger than the electric capacitance of the membrane. Accordingly, transients of the chemical potential difference of the proton are more strongly damped than those of the electric potential difference. In this respect it is noteworthy that the major driving force for ATP synthesis in mitochondria is an electric potential difference while it is pH-difference in thylakoids. While mitochondria generate the electrochemical potential difference of the proton by substrate driven respiratory chain with only slow fluctuations of the substrate level, photosynthetic organisms (under water or under a cannopy of leaves) are subject to rapidly fluctuating supply of light. It is attractive to consider the greater damping of pH-transients as compared with electric transients as advantageous for the steady function of the ATP synthase.

ACKNOWLEDGEMENTS

This work has been supported by the Deutsche Forschungsgemeinschaft (SFB 171, TP A2 and B3) and by the Fonds der Chemischen Industrie.

References

Althoff, G., Lill, H., & Junge, W. (1989) *J. Membr. Biol. 108*, 263.

Althoff, G., Schönknecht, G., & Junge, W. (1991) *Eur. Biophys. J. 19*, 213.

Andersson, B. & Anderson, J.M. (1980) *Biochim. Biophys. Acta 593*, 427.

Apell, H.J. & Läuger, P. (1986) *Biochim. Biophys. Acta 861*, 302.

Arnold, W.M., Wendt, B., Zimmermann, U., & Korenstein, R. (1985) *Biochim. Biophys. Acta 813*, 117.

Auslaender, W. & Junge, W. (1975) *FEBS Lett. 59*, 310.

Casadio, R., Venturoli, G., & Melandri, B.A. (1988) *Eur. Biophys. J. 16*, 243.

DeGrooth, B.G., Van Gorkom, H.J., & Meiburg, R.F. (1980) *Biochim. Biophys. Acta 589*, 299.

Durkin, J.T., Koeppe, R.E., & Andersen, O.S. (1990) *J. Mol. Biol. 211*, 221.

Eigen, M. (1963) *Angew. Chem. 75*, 489.

Finkelstein, A. & Andersen, O.S. (1981) *J. Membrane Biol. 59*, 155.

Fleischmann, D.E. & Clayton, R.K. (1968) *Photochem. Photobiol. 8*, 287.

Foerster, V. & Junge, W. (1985) *Photochem. Photobiol. 41*, 183.

Gutman, M., Nachliel, E., & Moschiach, S. (1989) *Biochemistry 28*, 2936.

Haines, T. (1983) *Proc. Natl. Acad. Sci. U. S. A. 80*, 160.

Haraux, F. & de Kouchkowsky, Y. (1979) *Biochim. Biophys. Acta 546*, 455.

Haraux, F. & de Kouchkowsky, Y. (1982) *Biochim. Biophys. Acta 679*, 235.

Hladky, S.B. & Haydon, D.A. (1972) *Biochim. Biophys. Acta 274*, 294.

Hong, Y.Q. & Junge, W. (1983) *Biochim. Biophys. Acta 722*, 294.

Jackson, J.B. & Crofts, A.R. (1969) *FEBS Lett. 4*, 185.

Jahns, P., Lavergne, J., Rappaport, F., & Junge, W. (1991) *Biochim. Biophys. Acta 1057(3)*, 313.

Junesch, U. & Graeber, P. (1987) *Biochim. Biophys. Acta 893*, 275.

Junge, W. (1987) *Proc. Natl. Acad. Sci. U. S. A. 84*, 7084.

Junge, W. (1989) *Ann. N. Y. Acad. Sci. 574*, 268.

Junge, W., Auslaender, W., McGeer, A.J., & Runge, T. (1979) *Biochim. Biophys. Acta 546*, 121.

Junge, W. & Jackson, J.B. (1982) in *Photosynthesis, Volume 1, 589-646. Edited by: Govindjee. Academic: New York, N. Y (*

564

Junge, W. & McLaughlin, S. (1987) *Biochim. Biophys. Acta 890*, 1.

Junge, W. & Polle, A. (1986) *Biochim. Biophys. Acta 848*, 265.

Junge, W., Schoenknecht, G., & Foerster, V. (1986) *Biochim. Biophys. Acta 852*, 93.

Junge, W. & Witt, H.T. (1968) *Z. Naturforsch. ,B: Anorg. Chem. ,Org. Chem. ,Biochem. ,Biophys. ,Biol. 23*, 244.

Kasianowicz, J., Benz, R., Gutman, M., & McLaughlin, S. (1987) *J. Membr. Biol. 99*, 227.

Lavergne, J. & Rappaport, F. (1990) in *Current Research in Photosynthesis* (Baltscheffsky, M., Ed.) pp I,873-I,876, Kluwer Academic, Dordrecht.

Lill, H., Althoff, G., & Junge, W. (1987) *J. Membr. Biol. 98*, 69.

Mitchell, P. (1961) *Nature (London) 191*, 144.

Mitchell, P. (1966) *Physiol. Rev. 41*, 445.

Morgan, H., Taylor, D.M., & Oliveira, O.N. (1991) *Biochim. Biophys. Acta 1062(2)*, 149.

Murphy, D.J. (1986) *Biochim. Biophys. Acta 864*, 33.

Neher, E., Sandblom, J., & Eisenmann, G. (1978) *J. Membrane Biol. 40*, 97.

Polle, A. & Junge, W. (1986) *Biochim. Biophys. Acta 848*, 257.

Polle, A. & Junge, W. (1989) *Biophys. J. 56*, 27.

Prats, M., Teissie, J., & Tocanne, J.F. (1986) *Nature (London) 322*, 756.

Prats, M., Tocanne, J.F., & Teissie, J. (1987) *Eur. J. Biochem. 162*, 379.

Prince, R.C., Linkletter, S.J.G., & Dutton, P.L. (1981) *Biochim. Biophys. Acta 635*, 132.

Saphon, S., Jackson, J.B., Lerbs, V., & Witt, H.T. (1975) *Biochim. Biophys. Acta 408*, 58.

Schmid, R. & Junge, W. (1975) *Biochim. Biophys. Acta 394*, 76.

Schoenknecht, G. (1990) Dissertation, Universität Osnabrück.

Schoenknecht, G., Althoff, G., & Junge, W. (1990) *FEBS Lett. 277*, 65.

Schoenknecht, G., Junge, W., Lill, H., & Engelbrecht, S. (1986) *FEBS Lett. 203*, 289.

Stolz, B. & Walz, D. (1988) *Mol. Cell. Biol. (Life Sci. Adv.) 7*, 83.

Veatch, W.R., Mathies, R., Eisenberg, M., & Stryer, L. (1975) *J. Mol. Biol. 99*, 75.

Walz, D,Goldstein,L.and Avron,M. (1974) *Eur. J. Biochem. 47*, 403.

Williams, R.J.P. (1988) *Annu. Rev. Biophys. Biophys. Chem. 17*, 71.

Williams, R.P.J. (1961) *Theor. Biol. J. 1*, 1.

Witt, H.T. (1979) *Biochim. Biophys. Acta 505*, 355.

EVALUATION OF THE ELECTRIC FIELD IN A PROTEIN BY DYNAMIC MEASUREMENTS OF PROTON TRANSFER

R. YAM, S. KIRYATI, E. NACHLIEL and M. GUTMAN
Laser Laboratory for Fast Reactions in Biology
Department of Biochemistry
The George S. Wise Faculty of Life Sciences
Tel Aviv University, Israel

ABSTRACT. The binding of a proton to a defined site on a protein is a function of the electrostatic interaction between the proton and all charged moieties of the protein. A long observation time, as in equilibrium measurements, reflects the total charge of the complex and is compatible with a centrosymmetric location of the net charge in the protein. Short observation times (microsecond to subnanosecond dynamics) reveal the local heterogeneity of the electric field in the immediate vicinity of the interaction site. This coupling between time resolution of kinetics and spatial refinement of the electric force field is demonstrated in this study using the complex between lysozyme and 8-hydroxy pyrene 1,3,6-trisulfonate as a model.

1. Introduction

On a molecular scale, the surface of a protein (or a membrane) is a non-homogeneous structure. Positive and negative charges are unevenly distributed; dipole fields of buried carbonyls or polypeptide helixes point in various directions, and even the hydrophobicity of the surface varies with the location. Besides this local arrangement of components, we observe properties stemming from the topography of the dielectric boundary. The dielectric discontinuity at the surface creates image charges, and the interaction between them and ions in solution disturbs the electric force field, generating local regions where the effective dielectric constant is lower than in bulk water (1). The low dielectric constant of small aqueous pockets in proteins also modulates the self energy of charged particles entrapped in them (2,3). Thus, the potential field, resulting from both charge-charge interactions and self-energy, becomes very heterogeneous in structure. The structured potential field leads to the appearance of electrosteering effects: guidance of a charged ligand along a favored trajectory towards its binding site. This mechanism, first described for the interaction of the superoxide anion with superoxide dismutase (4), has been investigated either by steady-state kinetics, which has a low time resolution, or by molecular dynamics calculations.

R. Guidelli (ed.), Electrified Interfaces in Physics, Chemistry and Biology, 565–575.
© 1992 *Kluwer Academic Publishers.*

In the present communication we shall describe a general experimental system using the common ion H$^+$ as a probe. We shall study its interaction with a specific site by subnanosecond kinetics and analyze the results by numerical integration of the chemical rate equations, producing microscopic molecular parameters. The quantitative interpretation of these parameters permits us to evaluate the electric field at the site under study.

2. Materials and Methods

The probing process in our measurement is the reversible proton dissociation from pyranine (8-hydroxy pyrene 1,3,6-trisulfonate), a photoexcitable dye which by pulse excitation becomes temporarily a very strong acid (pK$_o$=7.7, pK*=1.4) (5). We can measure the protonation dynamics of the ground state anion (ΦO$^-$) by microsecond transient absorption spectroscopy (5) or the geminate recombination of the proton with the excited anion (ΦO*$^-$) by picosecond fluorometry (6).

In the present study we carried out these measurements on the dye when it is adsorbed on lysozyme (7), a small positively charged protein (19 charged nitrogenous bases and 10 carboxylates) (12).

Upon applying an amidation procedure we neutralized the negative charges of seven carboxylates, in order to evaluate, on the same structure, how the net charge increment affects the dynamics of protonation of the same chromophoric moiety. For details of experimental and computational procedures the reader is referred to Refs.5-8.

3. Results and Discussion

3.1. THE EFFECT OF TOTAL CHARGE ON THE pK OF PYRANINE ADSORBED ON LYSOZYME

The pK of pyranine as measured in a dilute aqueous solution is 7.7. Adsorption of the dye on native lysozyme lowers its pK to 7.45 $\pm$ 0.05 (see Table I). The shift in pK is explained by the Kirkwood-Tanford formalism (9) which states that the pK of a moiety on a polyelectrolyte body is the sum of its intrinsic energy of dissociation plus the interaction energy of the moiety with all charges and dipoles of the supporting body, and is further modulated by the accessibility of each site to the solvent. The calculation of the interaction energies is laborious and necessitates a precise knowledge of the three-dimensional structure of the protein (10). For this reason we cannot interpret quantitatively the pK difference between the free and the bound dye. On the other hand we can investigate the pK shift caused by elimination of the negative charges, assuming that no major conformational change is caused by amidation.

As is apparent from Table I, the incremental charge, calculated from the pK shift, is in accord with the common knowledge of the number of neutralized carboxylates (7). Thus, for the purpose of analysing a pK shift of an unknown structure, it is better to employ the centrosymmetric model regarding the protein as a low dielectric sphere with all charges located at its center (11).

TABLE I

The effect of carboxylic acid neutralization on the lysozyme-pyranine complex

time (min)	pK	ΔpK	ΔZ^a	Z^b
0	7.45±0.05	-	-	9.0
15	6.5 ±0.05	0.95	5.3	14.3
45	6.3 ±0.15	1.15	6.5	15.5
180	6.1 ±0.05	1.35	7.6	16.6

[a] *The charge increment was calculated according to Tanford and Kirkwood (1957).*
[b] *The total charge was taken as that of the neutral protein at pH=5.5 plus the charge increment, Δz, calculated in column 4.*

3.2. TIME RESOLVED REACTION OF PROTON WITH PYRANINE ANION

Figure 1 depicts the transient deprotonation of pyranine adsorbed on lysozyme by a brief exciting laser pulse (7ns FWHM, λ= 337 nm, 0.5 Mwatt/cm^2). This perturbation triggers a synchronized dissociation of proton from the hydroxyl moiety of the dye, and the ground state anion so formed is detected by its strong absorption at 457 nm (5). The rise of the transient is unresolved and the relaxation corresponds to the reprotonation of ΦO^-.

A qualitative comparison between Fig.1A (native protein) and 1B (fully amidated protein) reveals that the incremental positive charge delayed the reprotonation of the chromophore. This qualitative observation is subject to a precise quantitative analysis.

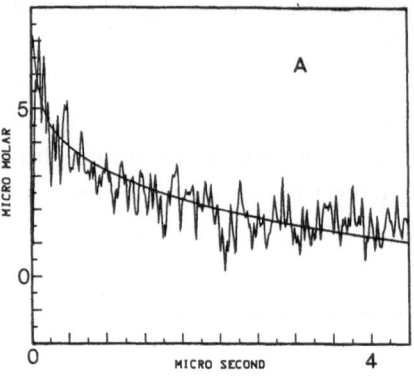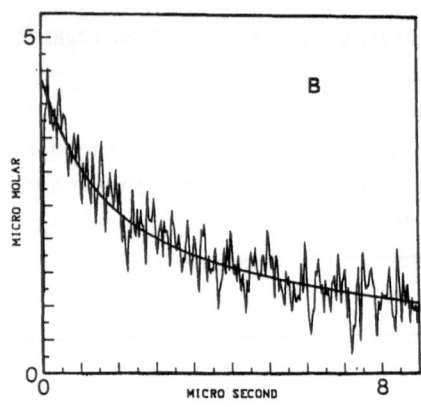

Figure 1. Time resolved measurement of the pyranine anion reprotonation in the dye-lysozyme complex. The complex (20 μM dye in 100 μM lysozyme) was excited by a short laser pulse (λ=337 nm, 7 ns full width at half maximum) which photodissociated the proton leaving the anion (ΦO⁻) bound to the protein. The reprotonation was monitored at λ= 457 nm (ε= 24,000 M⁻¹cm⁻¹). The experimental signal is a time resolved average of 512 successive excitation pulses delivered at a rate of 10 pps. The continuous curve is a simulated dynamic using the differential rate equation given in Yam et al. (1988).
A. Dynamics of reprotonation of pyranine native-lysozyme complex.
B. Dynamics of deprotonation of the dye adsorbed on the enzyme subjected to amidation for 180 min.
Note the changes in the scale units of the two axes.

The reaction of the proton with the protein-dye complex is described by a set of parametric, first-order coupled non-linear differential rate equations (8). Their numerical integration reconstructs the observed dynamics, as demonstrated by the smooth curves depicted in Fig. 1, which are superimposed on the experimental curves. The parameters used for reconstructing the dynamics are the first and second order rate constants of all proton transfer reactions taking place in the experimental system (for details see Ref. 9), e.g., protonation of the dye, protonation of carboxylates on the protein and proton exchange between surface groups. While all these rate constants have been determined (12), in the present case we shall focus our attention on the rate of the direct reaction between H^+ and the bound dye, and we shall analyze the effect of the total charge (protein-dye complex) on the rate constant of this reaction (see second column in Table II).

TABLE II

Rate constants of protonation of pyranine anion on native and modified lysozyme

Modification time (min)	$k_{dc}{}^a$ $(10^{10}M^{-1}s^{-1})$	Z^b	$\dfrac{k_{dc}{}^c}{k_{en}}$	δ^d	$Z_{eff}{}^e$
0	10.0 ±0.3	+5	1.6	-1.05	-1
15	6.25±0.25	+10.3	1.01	0	0
40	4.25±0.25	+11.5	0.68	0.68	+0.65
180	3.0 ±0.2	+12.6	0.48	1.3	+1.23

a As obtained from simulated dynamics exemplified in Fig.1. The rates were measured at a low ionic strength $I<10^{-3}$.
b The charge of the protein-pyranine anion. Taken as that calculated according to Tanford and tabulated in Table I, adjusted by the -4 charge of the dye.
c k_{en} encounter was calculated according to Shoup (1981), correcting for the rotational diffusion of the protein.
d calculated as described in the text.
e The effective charge corresponding to the measured dimension of the Coulomb cage assuming $\varepsilon = 78$.

The rate constant for a second-order diffusion controlled reaction (at a vanishingly small ionic strength) is given by the Debye-Smoluchowski equation:

$$k_{dc} = k_{en}\delta/[\exp(\delta)-1] \tag{1}$$

The encounter rate (k_{en}) is the collision rate between the two diffusing species, assuming no long-range interaction between them. This value is corrected for the rotational diffusion of the protein and the size of the pyranine target on its surface (13). This rate is assumed to be unaffected by the amidation of the protein.

The electrostatic contribution to the chemical dynamics is a function of the δ term. This is the ratio of two lengths:

$$\delta = R_c/R_o$$

R_c is the radius of the Coulomb cage, a sphere where the electrostatic potential of a charged particle is greater than kT:

$$R_c = Z_1 Z_2 e_o^2/\varepsilon \, kT$$

R_o is the radius of the reaction sphere. It is the distance at which the covalent bond between H^+ and O^- is formed and broken. For pyranine in dilute aqueous solution we have R_c =28.3 Å and R_o=6 Å (6).

The reconstructed dynamics gave us the value of k_{dc}, which was analyzed through Eqn.(1) to provide δ and the corresponding charge affecting the dynamics of interaction between the dye and the proton (Z_{eff}). For a native protein δ is negative, indicating that the proton is electrostatically attracted by the dye, in spite of the centrosymmetric charge Z=+5 of the dye-protein complex. The effective charge Z_{eff} = -1 is compatible with the net charge of the oxyanion left on the dye after proton discharge.

Gradual elimination of negative charges from the protein increases the effective charge in the immediate vicinity of the dye, and electrostatic repulsion takes place (see Table II). Yet, in contrast with equilibrium measurements, the effective charge is smaller than the total charge, reflecting the inhomogeneity of the electric field (compare columns III and VI).

3.3. PROTON-ANION GEMINATE RECOMBINATION OF THE PROTEIN SURFACE

Figure 2 depicts the earliest events observed after excitation of pyranine to its first electronic singlet state (ΦOH^*). The excited species can relax by two competing pathways. There is a regular fluorescence decay with a time constant τ = 6 ns (λ = 430 nm) and, in parallel, a very fast proton dissociation (τ = 110 ps) yielding an excited anion (ΦO^{*-}). The anion decays with a 6 ns time constant, emitting at a longer wavelength (λ = 515 nm).

The trace reported in Fig. 2 (Curve A) was obtained from a dilute pyranine solution. The initial decay of ΦOH^* emission is due to its rapid dissociation to ΦO^{*-} + H^+. The shallow long tail, dominating after about 0.5 ns, is a non-exponential decay attributed to recombination between the dissociation products with ΦOH^* formation . As the reaction proceeds at an almost neutral pH ([H^+] = 3.10^{-6}M), the observed reprotonation is too fast to be due to a proton coming from the bulk. It is the proton ejected from ΦOH^* which recombines with its sibling in a process of geminate reaction (naturally,this proton has been exchanged with the water).

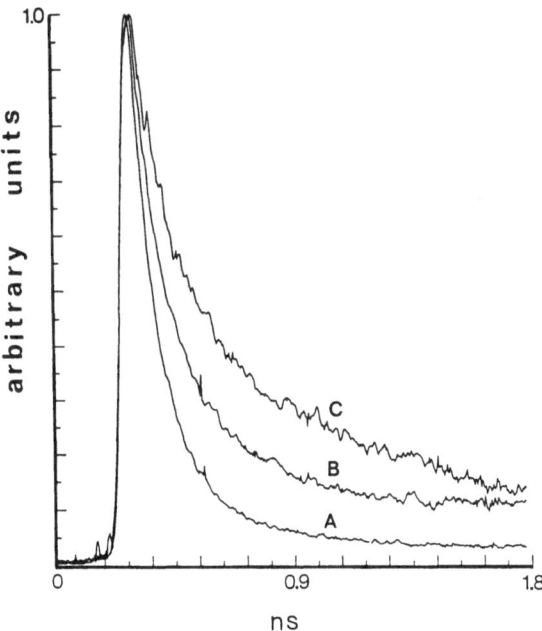

Figure 2. Time resolved fluorescence emission of ΦOH* adsorbed on lysozyme. The dye was excited by the third harmonics of a Yag laser (λ= 355 nm, 25 ps full width at half maximum), and the emission of ΦOH* was monitored by a streak camera and multichannel analyser as described by Pines et al., 1988. Curve A was obtained from 45 μM dye, dissolved in water (pH=5.4). Curve B was obtained from 45 μM dye in the presence of 270 μM native enzyme. Curve C was obtained from 45 μM dye and 140 μM enzyme subjected to amidation conditions for 180 min.

Curves B and C depict the same experiment, but the dye is now adsorbed either on the native protein (curve B) or on a totally amidated one (curve C). The effect of dye binding to the protein upon the dynamics is well distinguished: the initial decay is somewhat slower and the size of the tail is notably increased.

The observed dynamics is subject to quantitative analysis using Agmon's formalism (6). This procedure divides the reaction space into concentric shells and applies the Debye-Smoluchowski operator to calculate the transition probability between the shells. A reiterated computation reconstructs the time-space integral of the proton location, which appears as a reconstructed dynamics of the observed fluorescence signal. Such a reconstructed dynamics is shown in Fig. 3, which refers to a 50 ns long observation of the ΦOH* emission of the dye-native protein complex. For clarity, the amplitude is reported on a logarithmic scale, demonstrating the accuracy of the reconstructed dynamics; it actually follows the experimental recording up to the level at which the amplitude merges with the experimental background.

The adjustable parameters which yield the reconstructed dynamics in Fig. 3 are listed in Table III. Their magnitude is subject to quantitative interpretation.

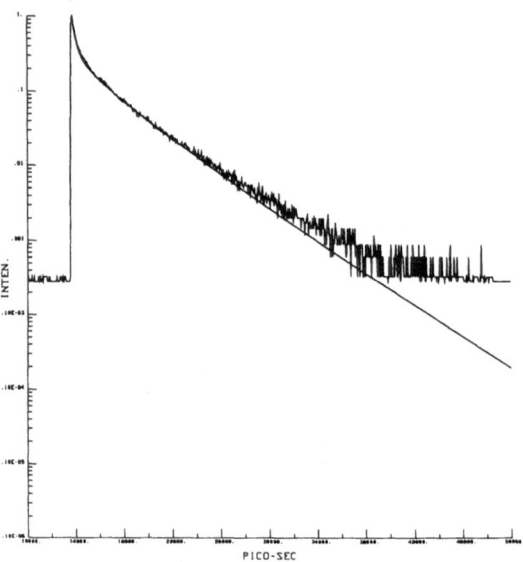

Figure 3. Theoretical reconstruction of the fluorescence decay curve of the ΦOH* form of pyranine adsorbed on native lysozyme. The reaction rate was measured by a single proton counting system at the wave length of ΦOH* emission over a 50 ns time window. The amplitude is reported on a logarithmic scale. The solid curve is a reconstructed dynamics using Agmon's formalism (6) and the molecular parameters listed in Table III.

The two rate constants, k_f and k_r, are those for proton transfer from the hydroxyl group to the water molecules at the surface of the reaction sphere, and for the reverse reaction. Both constants are reduced by the same factor, indicating a partial masking of the reaction sphere surface by the contact region with the protein. The proton can only be discharged to the aqueous phase, and not to the proteinaceous matrix. On the other hand, the constancy of the reaction sphere radius implies that the solvation shell around the dye, in the direction towards the water, does not differ from that in bulk water.

TABLE III

Molecular parameters characterizing geminate recombination between proton and pyranine adsorbed on lysozyme

Parameter	Protein dye complex	Dilute pyranine solution
k_f (rate of proton transfer to water) (s^{-1})	$2.8 \ 10^{10}$	$7.0 \ 10^{10}$
k_r (rate of proton transfer from water)(Å/ns)	2.8	6.0 ± 0.5
R_o - radius of reaction sphere (Å)	6 ± 0.3	6.0 ± 0.1
D_{H+} - diffusion coefficient of proton (cm^2/sec)	$4 \ 10^{-5}$	$9.3 \ 10^{-5}$
ε_{eff} calculated as $\varepsilon_{eff} = Z_1 Z_2 e_o^2 /(R_c kT)$	13.6	78

The diffusion coefficient of proton at the dissociation site is significantly smaller than in bulk water. This points to an enhanced ordering of water at the protein surface. Proton diffusion is controlled by the frequency of the randomization of the percolating network of hydrogen bonds. A slow rate of randomization forces a proton to shuttle to and fro in a given structure of water molecules, thus decreasing its ability to execute a random walk process (14).

Finally, we find that the effective dielectric constant is lower than that of bulk water. A decrease of the dielectric constant in the immediate vicinity of a dielectric boundary is actually to be expected (2). On a flat surface the local dielectric constant is $\varepsilon_{eff} = 38$. The value we measured implies that dissociation takes place at a site where water is surrounded by low dielectric walls.

The presence of a surrounding structure is also indicated by the effect of the neutralization of negative charges upon the geminate recombination (see curves B and C in

Fig. 2). One would expect that an increase in the positive charge of the protein, or in the effective charge near the site (Table II), should repel the proton from the immediate vicinity of the dye. Yet, the experimental observation points to a higher probability of geminate recombination, as revealed by a longer tail of the fluorescence decay dynamics (7). The reconciliation between this observation and the effective positive charge reported in Table II is brought about by placing the extra positive charges on the edge of the cleft in which the dye is enclosed. This positioning will repel protons coming from the bulk, in accord with the data in Table II, and will simultaneously create a potential barrier at the opening of the cavity, thus delaying the coming out of the proton.

Acknowledgement

This research was supported by the United States-Israel Binational Science Foundation, grant no. 87 0035, and the US Navy, Office of Naval Research, grant no. N00014-89-J-1622.

References

(1) Honig, B.H., Hubbell, W.L and Flewelling, R.F. (1985) Electrostatic interactions in membranes and proteins. Ann. Rev. Biophys. Chem. **1 5**, 163-193.

(2) Gilson, M.K., Rashin, R.A., Fine, R. and Honig, B.H.(1985) On the calculation of electrostatic interactions in proteins. J. Mol. Biol. **183**, 505-516.

(3) Jordan, P.C. (1982) Electrostatic modeling of ion pores energy barriers and electric field profiles.Biophys. J. **3 9**, 157-164.

(4) Head-Gordon, T. and Brooks, C.L. (1987) The role of electrostatics in the binding of small ligands to enzymes. J. Phys. Chem. **9 1**, 3342-3349.

(5) Gutman. M. (1984) The pH jump: Probing of macromolecules and solutions by a laser-induced, ultrashort proton pulse. Theory and applications in Biochemistry. Method. Biochem. Anal. **3 0**, 1-105.

(6) Pines,E., Huppert, D. and Agmon, N. (1988) Geminate recombination in excited-state proton transfer reactions: Numerical solution of the Debye-Smoluchowski equation with backreaction and comparison with experimental results. J. Chem. Phys. **8 8**, 5620-5630.

(7) Yam, R., Nachliel, E., Kiryati, S., Gutman, M. and Huppert, D. (1991) Proton transfer dynamics in the non homogeneous electric field of a protein. Biophys. J. **5 9**, 4-11.

(8) Yam, R., Nachliel, E. and Gutman, M. (1988) Time resolved proton-protein interaction. Methodology and kinetic analysis. J. Am. Chem. Soc. **110**, 2636-2640.

(9) Tanford, C. and Kirkwood, J.G. (1957) Theory of protein titration curves. I. General equations for impenetrable spheres. J. Am. Chem. Soc. **79**, 5333-5339.

(10) Matthew, J.B., Hannania, G.I.H. and Gurd, F.R.N. (1974) Electrostatic effects in hemoglobin: hydrogen ion equilibria in human deoxygenated oxyhemoglobin A. Biochemistry **18**, 1919-1928.

(11) Tanford, C. and Roxby, R. (1972) Interpretation of protein titration curves: Application to lysozyme. Biochemistry **11**, 2192-2198.

(12) Yam, R. Ph.D. thesis (1991) Tel Aviv University, Israel.

(13) Shoup, D., Lipari, G. and Szabo, A. (1981) Diffusion controlled bimolecular reaction rates. The effect of rotational diffusion and orientation constraint. Biophys. J. **36**, 697-714.

(14) Gutman, M. and Nachliel, E. (1990) The dynamic aspect of proton transfer processes. Biochim. Biophys. Acta **1015**, 391-414.

LIST OF PARTICIPANTS

Aloisi, G.	Dipartimento di Chimica, Universita' di Firenze, Via G. Capponi, 9, 50121, Firenze, Italy
Arnold, W. M.	Lehrstuhl für Biotechnologie, Universität Würzburg, Röntgenring, 11, 8700, Würzburg, Germany
Avranas, A.	Department of Chemistry, Laboratory of Physical Chemistry, Aristotelian University, 54006, Thessaloniki, Greece
Baars, A.	Van't Hoff Laboratory for Physical and Colloidal Chemistry, Padualaan 8, 3508 TB, Utrecht, The Netherlands
Bamberg, E.	Max Planck Institute of Biophysics, Kennedy-Allee 70, 6000, Frankfurt am Main, Germany
Bewick, A.	Department of Chemistry, University of Southampton, Highfield, S09 5NH, Southampton, England
Borsari, M.	Dipartimento di Chimica, Universita' di Modena, Via Campi, 183, 41100, Modena, Italy
Boselli, V.	Montefluos C.R.S., Gruppo Ausimont, Via S. Pietro, 50, 20021, Bollate (Milano), Italy
Carla', M.	Dipartimento di Fisica, Universita' di Firenze, Largo E. Fermi, 2, 50125, Firenze, Italy
Casadio, R.	Dipartimento di Biologia Ev. Sp., Universita' di Bologna, Via Irnerio, 42, 40126, Bologna, Italy
Chiaradia, P.	Dipartimento di Fisica, Universita' Tor Vergata, Via O. Raimondo, 00173, Roma, Italy
Chizmadzhev, Y.	The Frumkin Institute of Electrochemistry.: Academy of Sciences of the USSR, Leninsky Prospect 31, 117071, Moscow, U.S.S.R.
Corazza, A.	Dipartimento di Biologia Ev.Sp.: Istituto di Botanica, Via Irnerio, 42, 40100, Bologna, Italy
Correia, J.P.	Departamento Energias Renovaveis, Azinhaga dos Lameiros a, Estrada do Paco do Lumiar, 1699, Lisboa Cedex, Portugal
Cosma, P.	Dipartimento di Chimica, Campus Universitario 4, trav. 200 Re David, 70126, BARI, Italy
Del Sole, R.	Dipartimento di Fisica, Universita' Tor Vergata, Via O. Raimondo, 00173, Roma, Italy
de Oliveira Damasceno, O.	ENSEEG, Domaine Universitaire, BP 75, 38402, Saint-Martin-d'Hères, France
de Wit, A.R.	Debye Research Institute, Univ. Utrecht, Gecondenseerde Materie, P.O. Box 80,000, 3508 TA, Utrecht, The Netherlands
Fernandes,J.S.	Departamento de Engenharia Quimica, Universidade Tecnica de Lisboa, Av. Rovisco Pais, P-1096, Lisboa Cedex, Portugal
Fonseca Ricardo, A.P.	Departamento Energias Renovaveis, Azinhaga dos Lameiros, Estrada do Paco do Lumiar, 1699, Lisboa Cedex, Portugal
Fontanesi, C.	Dipartimento di Chimica, Universita' di Modena, via Campi, 183, 41100, Modena, Italy
Frenay, J.	Faculté des Sciences Appliquées, Metallurgie des Metaux non Ferreux, Armand Stévart, 2, B-4000, Liege, Belgium
Galli, G.	IBM Research, Zürich Research Laboratory, Säumerstrasse 4, CH-8803 Rüschlikon, Switzerland

578

Galun, E. Department of Materials Research, The Weizmann Institute of Science, 76100, Rehovot, Israel

Giorgi, L. ENEA, FARE/ICHI/ELC, CRE Casaccia, Via Anguillarese, 301, 00060, S.Maria Galeria (Roma), Italy

Goossens, H. Rijksuniversiteit Gent, Lab. voor Fysische Scheikunde, Krijgslaan 281, B-9000, Gent, Belgium

Guidelli, R. Dipartimento di Chimica, Universita' di Firenze, Via G. Capponi, 9, 50121, Firenze, Italy

Gutman, M. Tel Aviv University George S. Wise, Department of Biochemistry, Ramat Aviv, 69 978, Tel Aviv, Israel

Hale, J.M. Orbisphere Laboratories, 114 rte de Thonon, CH-1222 Vesenaz, Genève, Switzerland

Hamelin, A. LEI, CNRS, 1 Place A. Briand, 92195, Meudon Cedex, France

Hauser, A.K. Department of Chemical Engineering, The University of Texas at Austin, TX 78712, Austin, U.S.A.

Heberle, J. Freie Universität Berlin, FB Physik/AG Biophysik, Arnimallee 14, D-1000, Berlin 33, Germany

Herron, M. Southampton University, Department of Chemistry, Highfield, S09 5NH, Southampton, England

Hiesgen, R. Institut für Solarenergie, forschung SMBH, Sokelantstrasse 5, D-3000, Hannover 1, Germany

Imer, F. Yildiz Üniversitesi, Fen-Edebiyat, Fak Kimya Bölümü, 80270, Sisli-Istanbul, Turkey

Jarzabek, G. Institute of Physical Chemistry, Polish Academy of Sciences, ul. Kasprzaka 44/52, 01-224, Warszawa, Poland

Jäger, A.H. Lehrstuhl fuer Biotechnologie, Röntgering, Universität Würzburg, D-8700, Würzburg, Germany

Junge, W. Department of Biophysics, Osnabruck University, Postfach 4469, D-4500, Osnabruck, Germany

Kadirgan, F. Istanbul Teknik Universitesi, Fen-Edebiyat Fakültesi, Kimya Bölümü, 80626, Maslak - Istanbul, Turkey

Klein, A. Hahn-Meiter-Institut Berlin, Postfach 390128, Glienicker Strasse 100, D-1000, Berlin 39, Germany

Koczorowski, Z. The University of Warsaw, Department of Chemistry, ul. Pasteura 1, 02-093, Warszawa, Poland

Kolb, D.M. Abteilung für Elektrochimie, Universität Ulm, Albert-Einstein-Allee, 11, Postfach 4066, D-7900 ULM, Germany

Kruijt, W.S. Van't Hoff Laboratory for Physical and Colloidal Chemistry, Padualaan 8, 3508 TB, Utrecht, The Netherlands

Kucernak, A.R. UK Atomic Energy Authority, Harwell Laboratory, Didcot, OX11 ORA, Oxfordshire, England

Lama, F. CNR Frascati, Istituto Struttura della Materia, Via Enrico Fermi, 38, 0044, Frascati (Roma), Italy

Lehnert, W. Institut für Physik. Chemie, Universität Bonn, Wegelerstrasse, 12, 5300, Bonn 1, Germany

Maack, J. MPI für Biophysikalische Chemie, Abt. 110, Am Faßberg, D-3400, Göttingen, Germany

Mayer, T. Hahn-Meitner-Institut Berlin, Postfach 390128, Glienicker Strasse 100, D-1000, Berlin 39, Germany

McLaughlin, S. Department of Physiology & Biophysics, Health Sciences Center, SUNY, Stony Brook, 11794-8661, New York, U.S.A.

Meissner, D. Institut für Solarenergie, forshung, Sokelantstrasse 5, D 3000, Hannover 1, Germany

Meulekamp, E.A. Debye Research Institute, University of Utrecht, Gecondenseerde Materie, Postbus 80.000, 3508 TA Utrecht, The Netherlands

Molero Casado,M. Departamento de Quimica Fisica, Universidad de Sevilla, Facultad de Farmacia, 41012, Sevilla, Spain

Moncelli, M.R. Dipartimento di Chimica, Universita' di Firenze, Via G. Capponi, 9, 50121, Firenze, Italy

Mulkidjanian, A.Ya. Department of Bioenergetics, Moscow State University, Building A, 119899, Moscow, U.S.S.R.

Murakoshi, K. Department of Chemistry, Faculty of Science, Okkaido University, 060, Sapporo, Japan

Nachliel, E. Department of Biochemistry, Tel-Aviv University, Ramat Aviv, 69 978, Tel Aviv, Israel

Naumann, R. E. Merck, AZL-ASE, Postfach 4119, D-6100, Darmstadt, Germany

Neglia, A. Dipartimento di Fisica, Universita' La Sapienza, Piazzale Aldo Moro, 2, 00185, Roma, Italy

Nelson, A. Marine Biological Association, The Laboratory, Citadel Hill, PLI 2PB, Plymouth, England

Nichols, R.J. Fritz-Haber-Institut der Max-Planck-Gesellschaft, Faradayweg 4-6, D-1000, Berlin 33, Germany

Nore, B.F. Arrhenius Laboratory, Department of Biochemistry, University of Stockholm, S-10691, Stockholm, Sweden

Oliveira Fialho, M.C. Universidade de Lisboa, Departamento de Quimica, Rua Ernesto de Vasconcelos, Ed. C1, Campo Grande - 1700, Lisboa, Portugal

Oskam, G. Debye Research Institute, University of Utrecht, Gecondenseerde Materie, Postbus 80.000, 5508 TA Utrecht, The Netherlands

Ozanam, F. Laboratoire de Physique de la Matière Condensée, Ecole Polytechnique, Route de Saclay, 91128, Palaiseau Cedex, France

Papadopoulos, N. Laboratory of Physical Chemistry, Department of Chemistry, University of Thessaloniki, 54006, Thessaloniki, Greece

Parsons, Roger Department of Chemistry, University of Southampton, Highfield, S09 5NH, Southampton, England

Parsons, Ruby Department of Chemistry, University of Southampton, Highfield, S09 5NH, Southampton, England

Passamonti, P. Universita' di Camerino, Dipartimento di Scienze Chimiche, Via S. Agostino, 1, Camerino, (Macerata), Italy

Perfetti, P. Istituto di Struttura della Materia, CNR, Via Enrico Fermi, Frascati, (Roma), Italy

Peter, L.M. Department of Chemistry, University of Southampton, Highfield, S09 5NH, Southampton, England

Piazza, G. Dipartimento di Chimica G. Ciamician, Universita' di Bologna, Via Selmi, 2, 40126, Bologna, Italy

Posadas, D. INIFTA, Facultad de Ciencias Exactas, Universidad Nacional de la Plata, Sucursal 4, Casilla de Correo 16, 1900, La Plata, Argentina

Poxleitner, M. Max-Planck-Institut für Chemie, Saarstrasse, 23, Postfach 3060, 6500, Mainz, Germany

Pytel, Z. Institute of Physics, Polish Academy of Sciences, Al. Lotnikow 32/46, 02-668, Warsaw, Poland

Ritzoulis, G.   Laboratory of Physical Chemistry, Aristotelian University,
                       Kalavriton 19, 55 1 33, Thessaloniki, Greece
Sakai Nore,Y.   Arrhenius Laboratory, Department of Biochemistry, Stokholm
                       University, 10691, Stokholm, Sweden
Saraç, A. S.   Chemistry Department, Istanbul Technical University, 80626
                       Maslak- Istanbul, Turkey
Sastre De Vicente, M.E. Departamento de Quimica Fund. y Ind., Universidad de La
                       Coruña, Avda Zapateira S/N, La Coruña, Spain
Scalas, E.   Dipartimento di Fisica, Universita' di Genova, Via Dodecaneso,
                       33, 16146, Genova, Italy
Schefold, J.   Institut fur Physikalische Elektronik, Universität Stüttgart,
                       Pfaffenwakl. 47, D-7000, Stüttgart, Germany
Schiffrin, D.   Department of Chemistry, University of Southampton, Highfield,
                       S09 5NH, Southampton, England
Schmickler, W.   Physics Department, Utah State University, Logan, 84332-4415,
                       Utah, U.S.A.
Schwaller, M.A.   I.T.O.D.Y.S.: Université Paris VII - C.N.R.S.: 1, Rue Guy de la
                       Brosse, 75005, PARIS, France
Seitz-Beywl, J.   Max-Planck-Institut für Chemie, Saarstrasse, 23, Postfach 3060,
                       6500, Mainz, Germany
Sesta, B.   Dipartimento di Chimica, Universita' La Sapienza, P.le Aldo Moro,
                       5, 00100, Roma, Italy
Sibille, T.   Montefluos S.p.A.: Gruppo Montedison, Piazza G. Donegani, 5/6,
                       15047, Spinetta Marengo (AL), Italy
Siebert, E.   ENSEEG, Domaine Universitaire, BP 75, 38402, Saint-Martin-
                       d'Hères, France
Silva, C.J.R.   Centro de Quimica Pura e Aplicada, (I.N.I.C.) Universidade do
                       Minho, Avenida Joao XXI, 4700, Braga, Portugal
Solomun, T.   Institute of Phys. & Theor. Chem., Freie Universität Berlin,
                       Takustrasse 3, Berlin, 33, Germany
Sottomayor, M.J.   Departamento de Quimica, Faculdade de Ciencias, Praça Gomes
                       Teixeira, 4000, Porto, Portugal
Souteyrand, E.   Laboratoire de Physique des Liquides et Electrochimie, L.P.15 du
                       C.N.R.S., Tour 22 - 4, Place Jussieu, 75230, Paris Cedex
                       05, France
Souto, R.M.   Departamento Quimica Fisica y Electroquimica, Universidad de La
                       Laguna, 38204, Tenerife, Spain
Stella, A.   Dipartimento di Fisica, Universita' di Pavia, Via A. Bassi, 6,
                       27100, Pavia, Italy
Stradella, L.   Dipartimento di Chimica Inorganica, Chimica Fisica e Chimica dei
                       Materiali, via P. Giuria, 9, 10125, Torino, Italy
Stumm, W.   Swiss Federal Institute of Technology (ETH), EAWAG,
                       Uberlandstrasse 133, CH, 8, Zürich, Switzerland
Szalontai, B.   Institute of Biophysics, Hungarian Academy of Sciences, Odesszai
                       Krt. 62 P.O.B. 521, H-6701, Szeged, Hungary
Trasatti, S.   Dipartimento di Chimica Fisica ed Elettrochimica, Universita' di
                       Milano, Via Venezian, 21, 20133, Milano, Italy
Urbanek, W.   University of Pennsylvania, Department of Electrical Engineering,
                       200 S. 33rd Street, PA 19104-6390, Philadelphia, U.S.A.
Venkateshwaran, S.   Department of Pharmaceutics, The University of Utah, Salt Lake
                       City, 84112, Utah, U.S.A.

Wandelt, K.	Institut für Physikalische Chemie, Universität Bonn, Wegelerstrasse, 12, 5300, Bonn, Germany
Weaver, M.J.	Department of Chemistry, Purdue University, 47907 West Lafayette, Indiana, U.S.A.
Yam, R.	Biochemistry Department, Tel-Aviv University, Faculty of Life Sciences, 69 978, Ramat Aviv, Tel Aviv, Israel
Yoneyama, H.	Osaka University, Department of Applied Chemistry, Suita, 565, Osaka, Japan
Zemel, J.	University of Pennsylvania, Department of Electrical Engineering, 200 S. 33rd Street, PA 19104-6390, Philadelphia, U.S.A.
Zemel, J.N.	University of Pennsylvania, Department of Electrical Engineering, 200 S. 33rd Street, PA 19104-6390, Philadelphia, U.S.A.
Zouboulis, A.I.	Laboratory of Gen. & Inorg. Chem. Techn., Department of Chemistry (114), Aristotelian University, GR-54006, Thessaloniki, Greece

LIST OF CONTRIBUTORS

Ernst Bamberg
Max Planck Institute of Biophyics, 6000 Frankfurt am Main 70, FRG.

Rita Casadio
Department of Biology Ev.Sp., University of Bologna, Via Irnerio 42, 40100, Bologna, Italy.

Yuri Chizmadzhev
Frumkin Institute of Electrochemistry, Academy of Sciences, Leninskii Prospect 31, 117071 Moscow V-71, USSR.

Rodolfo Del Sole
Department of Physics, Rome, University II "Tor Vergata", Via O. Raimondo,00173 Roma, Italy.

Giulia Galli
IBM Research, Zürich Research Laboratory, Säumerstrasse 4, CH-8803 Rüschlikon, Switzerland.

Rolando Guidelli
Department of Chemistry, University of Florence, Via G. Capponi 9, 50121, Florence, Italy.

Menahem Gutman
Department of Biochemistry, Tel Aviv University "George S. Wise", 69 978 Tel Aviv, Israel.

Wolfgang Junge
Department of Biophysics, Osnabruck University, D-4500 Osnabruck, FRG.

Dieter M. Kolb
Abteilung für Elektrochimie, Universität Ulm, Albert-Einstein-Allee, 11, Postfach 4066, D-7900 ULM, Germany.

Stuart McLaughlin
Department of Physiology and Biophysics, School of Medicine, Health Sciences Center, State University of New York at Stony Brook, Stony Brook, NY 11794-8661, USA.

Roger Parsons
Department of Chemistry, The University, Highfield, Southampton S09 5NH, UK.

Paolo Perfetti
Istituto di Struttura della Materia, CNR, Via Enrico Fermi, Frascati, Italy.

Laurence M. Peter
Department of Chemistry, The University, Highfield, Southampton S09 5NH, UK.

584

Wolfgang Schmickler
Physics Department, Utah State University, Logan, 84332-4415, Utah, U.S.A.

Angiolino Stella
Department of Physics, Pavia University, Via A. Bassi 6, 27100 Pavia, Italy.

Werner Stumm
Swiss Federal Institute of Technology (ETH), Federal Institute for Water Resources and Water Pollution Control (EAWAG), Uberlandstrasse 133, CH, 8 Zurich (Switzerland).

Sergio Trasatti
Department of Physical Chemistry and Electrochemistry, Milan University, Via Venezian 21, 20133 Milano, Italy.

Klaus Wandelt
Institute of Physical and Theoretical Chemistry, Bonn University, Wegelerstrasse 12, 5300 Bonn 1, FRG.

Michael J. Weaver
Department of Chemistry, Purdue University, West Lafayette, IN 47907-3699,USA.

SUBJECT INDEX

A b initio simulations, 135
Absolute electrode potential, 19,232-238,247
– standard hydrogen electrode potential, 239
Acceptor impurities in semiconductors, 6
Accumulation layer, 25
Action spectrum of photocurrents, 535,537,544
Activated complex theory, 443
Activation barriers for surface reconstruction, 86
– energy barrier for electron transfer, 429
– – for oxide dissolution, 465
Adatoms, 66
Adiabatic passage through the transition state region, 430
Adsorbate, binding energy of, 417
– dipole moment, 399,403
– – –, linear response formalism for, 403
–, energy of chemical interaction of, 339
– structures, hexagonal and honeycomb, 422
Adsorbed hydrogen, amount of, 296
– –, formation and removal of, 296
– ions, specifically, 29,37
Adsorption coefficient, 225
– energy of Xe on metals, 102
– Gibbs energy, 266,267,338,355
– isotherms, 213,224,348,352
–, non-coulombic, of charged species, 37
– of aliphatic compounds on *sp* metals from water, 338
– of alkali ions, 106
– of anions, 284
– of an oxygen species, 296
– of HSO_4^- on Pt, 305,306
– of ions, 399
– of metals, 101
– of neutral compounds, 264
– of OH on Pt, 305
– of Tl on Pt, 306
– of water from the gas phase, 257

Adsorption of xenon on iridium, 89
– on the steps, 302
– potential, 407,410
– site, 97
Affinity of metals for water, 251,252
Ag, 247,252,255,256,261-265
Al(111), 73
Aliphatic compounds, adsorption behaviour on *sp* metals of, 338
Alkali ions, adsorption of, 106
AlO_2 dissolution, 459
Anionic specific adsorption, 38,284
– – – in disperse systems, 47
Anion induced mobility of surface atoms, 284
Ashcroft pseudopotential, 379
Atomic resolution, 283
ATPase, 512,541,551
ATP synthesis, 561
Au, 247,252,253,255,258,260,261,263
– single crystal faces, 259,262,282-284
– – – –, flame treatment of, 282
– (100) topography, 282-284
Auger Electron Spectroscopy (AES), 69

B acteriorhodopsin (BR), 533
–, Asp^{96}->Asn mutated, 541,547
–, Asp^{85}->Glu mutated, 541,546
–, photocycle of, 533,541-544
–, proton translocation in, 533
Band bending, 22,37,41
– diagram, 180
– edge pinning, 33,59
– theory, 5
Bardeen model, 158
Barrier height for electron tunneling, 394
Ben Naim-Stillinger potential, 322
Bilayer lipid membranes (BLM), 48,493,533
– – –, measurement of ionic surface potential at, 53
– – – fusion, 501
– – – electroporation, 493

Bimetallic catalyst, 102
Binding energy of an adsorbate, 417
Biological membranes, 473,491,511
Bismuth anodization to form amorphous Bi_2O_3, 190
Blocking contact at a semiconductor/electrolyte interface, 183
Body-centered cubic lattice in molecular models, 325
Boltzmann statistics, 9,192
Boltzons, 9,12
–, corrected, 9
Bragg peaks, 75
Buckling model, 124,126
Budd-Vennimenus pressure balance relation, 377
Buffering capacity, 553
Bulk metal deposition, 289
Bulk modulus, 121

Capacitor, parallel plate, 485
Capacity *vs* charge characteristics, 372,388
–, differential, 24,28,221,246,262-264,268,369
– hump, 372
Carbon, phase diagram of, 137
Carotenoid absorption changes, 558
Car-Parrinello (CP) method, 133
Cell membranes, 486
CF_0CF_1, 551
CFO, 561
Charge distribution due to coulombic forces, 21
– transfer at liquid/liquid interfaces, thermodynamics of, 213
– transfer coefficient, 59,403
Chemical potential, 2,377
– –, standard, 15
Chemiosmotic mechanism, 555
Chemisorbed water orientation, 319
Chemisorption process in a gas/solid system, 295
Chloroplasts, 551
Chromatophore membranes of photosynthetic bacteria, 512
Chromatophores, 560
Chronocoulometry, 223
Civilized models, 312

Cleaning cycles in cyclic voltammetry, 298
Coadsorbed anions in Cu underpotential deposition, 289
Collective dipolar motion, 435
Colloids, 43
Compact layer, 28,45,369,454,523
Complex plane in impedence spectroscopy, 193
Conductance due to electron tunneling, 394
Conduction band, 5,181
Contact potential between metals, 17
Coordination chemistry, 443
– – of the oxide/water interface, 443
Correlations between adsorption potential and work function, 407
– between electrochemical and gas phase data, 257,258
– between potential of zero charge and work function, 246-250,254-257
– – – – – – in non-aqueous solvents, 252-253
– – – – – – at single crystal faces, 254-257
Corrugation of the surface, 73,92
Coulomb cage, 570
Coulombic correction factor, 453
 – forces, charge distribution due to, 21
Crystallographic orientation, 254,267,268
Cu bulk deposition on gold single crystals, 290
– /Ru powder samples, 105
Current density, 56,428
– – across metal/solution interfaces, 57
– – across semiconductor/solution interfaces, 59
Cyclic voltammetry (CV), 295
– – for a clean polycrystalline Pt electrode, 295
– voltammograms, first sweep of, 298
Cytochrome b_6, 551

$\mathbf{D}$angling bonds, 68,124
Debye-Hückel theory, 370,473
– relaxation time, 436
– screening length, 31,370,402,475,486,528
– Smoluchowski equation, 569
– – operator, 571
Degenerate semiconductors, 10
Density functional method, 33,120,375,416

Depletion layer, 26
Diamont melting simulation, 137
Dielectric constant for distortional polarization, 319
– –, local, 573
– saturation, 31
Differential capacity, 221,246,262-264,268,309,326,328,369
– – of the diffuse layer, 28
– – of the inner layer, 310
– – of the space charge, 24
– reflectance (DR), 122,124,128,130
Diffuse layer, 2,27
– – capacity, 28
Diffusion coefficient of the proton, 573
– length of minority carriers in semiconductors, 185
– of protons, 553
Diffusive surface reconstructions, 76
Dimeric solute molecules in molecular models, 357
Dimethylsulphoxide (DMSO), 252, 253, 258-260
Dipole formation by partial charge transfer, 107
– moments of simple adsorbates, 403
– potential, 2,49,315,377,422
Discreteness-of-charge effects in membranes, 55,473,474
Disordered systems, 133
Disperse systems, 43
– –, anionic specific adsorption in, 47
– –, specific surface area in, 47
Displacive reconstructions, 76
Dissolution of Al_2O, 459
–, proton promoted, 462
Distortional polarization, 326
Donor impurities in semiconductors, 6
– number of a solvent, 254
Double layer, 245
– – effects in electrochemical reactions, 438
– scattering of electrons, 79
Dynamical LEED, 118,128
– solvent effects in electrochemical kinetics, 427,435

Effective dielectric constant, 573

Electric field modulated absorption, 201
– potential, surface excess of, 326
Electrified interphases, 1,225
– –, control of, 43
– –, in nonequilibrium, 55
– –, thermodynamics of, 213
Electroabsorption, 202
Electrocapillary curves, 216-218
– equation, 218
Electrochemical energy scale, 236
– kinetics, 427
– potential, 1,234
– – difference of the proton, 551
– – of electrons, 5
– – of electrons in redox systems, 12
Electrochromism of pigments, 552
Electrode potential, 229,231, 234, 237, 247
– – of a redox couple, 18
Electrokinetic potential, 46,52
Electrolyte electroreflectance (EER), 196-198
Electromodulation, 201
Electron affinity in semiconductors, 17,41
– current density, 59
– density profile, 375
– – –, trial functions for, 377
– distribution at the surface, 242
– double scattering, 79
– energy loss spectroscopy (EELS), 119
– exchange current density in semiconductors, 61
– gas, inhomogeneous, 33,313,375
– hole pair, 62
– – recombination, 62,187
– spillover, 314,329,363,369
– tail at the surface, 243, 318
– transfers, 427
– transport across a membrane, 552
– tunneling, barrier height for, 394
– – in STM, potential energy surface for, 391
Electronic back donation, 112

Electronic energy, 232-237,243,247,248
– equilibrium, 232, 238
– frequency factor, 431
– transmission coefficient, 430
Electroporation of membranes, 493
– –, irreversible, 496
– –, reversible, 498
Electroreflectance (ER) in the UV-visible, 299
Electrosorption valency, 401
Electrostatic capacity, 224
– self-energy of a three-dimensional lattice, 417,421
Electrostriction, 54
Emersed electrodes, 239,240, 243
Energy bands, 5,180
– gaps, 6,180
– scale, 229,235
Entropy of formation of the inner layer, 311
– surface excess, 314, 326, 328
Equilibrium constants of surface complexes, 450
Equipotential profiles, 476
Euler equation, 376
Excess quantities, 215
Exchange-correlation energy, 34,376,421
– – potential, 34,121
– current density, 59
Excitation energies, 121
Exclusion principle, 7
Ex situ experiments, 257
Extrinsic semiconductors, 181

F$_{e(III)}$ (hydr)oxides, reductive and catalytic dissolution of, 465
Fermi-Dirac distribution law, 7,9,180
– energy, 180
– –, kinetic, 8
– level, 8,97,235,247
– – in solution, 13,182,238
– – pinning, 40,159-162,170
Fermions, 9
Film growth, 111

Flame treatment of gold single crystals, 282
Flash spectrophotometry, 551
Flat-band potential, 26,183
Floating model of Xe adsorption on metal surfaces, 95
Flory-Huggins isotherm, 225
Fluid/fluid interfaces, 213
Fourier transform infrared spectroscopy (FTIR), 198
Franck-Condon principle, 56
Franz-Keldysh effect, 130
Free adsorption energies, 407
– electron approximation, 8
Fresnel formula, 124
Friedel oscillations, 377
Frumkin isotherm, 226,338
– –, generalized, 341
– – plots, 338

GaAs(110), 123,128
Galvani potential, 3
Gärtner equation, 186
Gas/solid system, chemisorption process in, 295
Gauss theorem of electrostatics, 3,23
Geminate recombination of the proton, 566
Generalized Frumkin isotherm, 341
Gibbs adsorption equation, 44,218,222
– Duhem equations, 214
– energy of adsorption, 266,267,338,355
– energy of the adsorption reaction, 296
Gouy-Chapman capacity, 401
– – corrected quantities, 312,320
– – theory, 28,36,45,54,309,320,333,369,402,429,450,453,473,485,512,523
Gramicidin, 551,557
– dimerization constant, 557,570
Green plants, photosynthesis by, 551
Green's functions, 122
GW method, 122

Halobacterium halobium, 533
Hard sphere ions and dipoles, 381,402

Helmholtz layer, 28,45,369,454,523
– – capacity, 32,371,385,405
– –, potential difference across, 40
– plane, inner, 29
– –, outer, 29,52
– Smoluchowski theory, 52
Hematite, light-induced dissolution of, 468
Heteroatoms, 66,102
Heterogeneous catalysis, 111
– nucleation, 101
Hexagonal adsorbate structures, 422
Hex-rot Pt(100) surface modification, 78
Hg/water interface, 243
High resolution electron energy loss spectroscopy (HREELS), 69
Hohenberg-Khon-Sham theorem, 375
Hole current density, 59
– exchange current density in semiconductors, 62
Holes in semiconductors, 6
Honeycomb adsorbate structures, 422
HSO_4^- adsorption on Pt, 305,306
Hydrogen adsorption on platinum, 295
– – region, as an indicator of surface structure, 296
– bond formation, 322,346
– monolayer, 296
–, strongly bound on Pt single crystals, 297,305
–, weakly bound on Pt single crystals, 297
Hydrophilicity, 250,252,263,265
Hydrophobic interactions, 352,454
Hydrous oxides, 443
8-Hydroxy pyrene 1,3,6-trisulfonate, 565

Ideal polarized interface, 4,221
Image charge, 406
– plane in the jellium model, 331,379
– potential in the jellium model, 379
Imperfect surfaces, 91
Independent electron approximation, 7
Inhomogeneous electron gas, 33
Injection of a majority carrier, 188

594

Inner Helmholtz plane, 29
– layer, 28,369
– – differential capacity, 263,264,268,310
– –, entropy of formation of, 311
– –, integral capacity of, 330
– – interfacial tension, 310
– –, potential difference across, 350
– potential, 3
– sphere electrochemical processes, 429
In situ approaches, 246
– infrared spectroscopy, 179,197
Insoluble oxides, 47
Insulator, 6,179
Integral capacity of the inner layer, 330
Intensity modulated photocurrent spectroscopy (IMPS), 193
Interface, 213
–, entropy of formation of, 310
– state (IS) adsorption, 202
– states, 43
– – models, 157
Interfacial capacity, 379
– equilibrium, 237
– structure, 240
– tension, temperature coefficient of, 314
– thickness, 226
Internal reflection spectroscopy, 205
– –, total, 124
Intrinsic semiconductors, 6,180
Inverse photoemission, 119
Inversion layer, 25
Ionophores, 51
Ion scattering spectroscopy (ISS), 69
Ir(100), 77
Ir, polycrystalline electrode of, 293
Ir single crystals, 294
Isoelectric point, 46

Jellium model, 35,313,329,363,375,403,407
– /hard sphere electrolyte model, 369,383

Jellium with pseudopotentials, 416
Jet electrode, 238

Kinematical LEED, 118
Kinetic Fermi energy, 8
Kinks, 66
Kirkwood-Tanford formalism, 566
– variation method, pair correlation approximation of, 324
Kohn-Sham equations, 121-122

Langmuir isotherm, 340,523
Laplace equation, 475
Laser-flash spectrophotometry, 552
Lateral surface Stark effect, 112
Lattice constant, 121
– model, three-dimensional, of water molecules, 324
– –, three dimensional, of water and solute molecules, 351
– statistics, 443
– surface model, 446
Law of mass action for semiconductors, 10
Lennard-Jones 12-6 spherical potential, 321
Ligand-catalyzed dissolution, 460
– exchange, 449
Light driven proton pumps, 533
– induced dissolution of hematite, 468
– ion or atom back-scattering, 118
Linear free energy relations in surface complex formation, 452
– response formalism for the adsorbate dipole moment, 405
– sweep voltammetry, 221,295
Lipid bilayers, 491
– –, electroporation of, 493
– –, fusion of, 501
– monolayers, 555
– vesicles, 49,555
– –, multilamellar, 555
Lipids, 473
Lipophilic ions, 50
Liposomes, 49
Lippmann equation, 216,218

Liquid carbon simulations, 137
– /liquid interphase, 213
Local density approximation (LDA), 34,121
– dielectric constant, 573
– order in statistical mechanical treatments, 318,345
– surface potential, 92
– work function, 92,102
Long-range order, 299
Lorentzian line, 93
Low energy electron diffraction (LEED), 69,74,117
– – – –, dynamical, 118
– – – – I/V measurements, 76
– – – –, kinematical, 118
– – – – pattern, 74
Lysozyme, 565
Lysozyme-pyranine complex, 567

Madelung constant, 418
Majority carriers, 25,42
Mass transport for surface reconstruction, 83
Matrix notation, 72
Mean spherical approximation (MSA), 381,402,432
Membrane breakdown, irreversible, 496
– –, reversible, 498
Membranes, biological, 48,473,491,511
–, cell, 486
–, chromatophore, 512
–, electroporation of, 493
– fusion of, 49,501
–, reconstituted, 49
– /solution interface, 475
Meniscus contact of hemispherical electrodes with the solution, 298
Mercury jet electrode, 239,241
Metal adsorbates, 101
– /electrolyte interphase, 217,222
– – –, a phenomenological approach to, 245
– induced gap states model (MIGS), 43,161
– /semiconductor interface, 42
– single crystal faces, 247,255-258,266,372,415

Metal/solution interface, structure of, 241,246
– /solvent interaction, 254,258,260
– surface electrons, 260
– tip in STM measurements, 278
– /water interactions, 250,251,258, 262-268
– – interaction term, 250
– – interface, 242
Metallocenium/metallocene reactions, 439
Midpoint potential in redox potentiometry, 516
Minority carriers, 25
– – diffusion length, 185
Mo(100), 84
Mobility, induced by anions, 284
– of Pt surfaces, 299
Molecular electromodulation (ME) spectra, 207
– Stark effect, 201
Monolayer of H, 296
– models, 312,317
Monomeric solute molecules in molecular models, 352
Monte Carlo simulation, 320
Mott-Schottky equation, 184
– – plot, 26
Multilamellar lipid vesicles, 555
Multiple internal reflection (MIR), 201

Nernst-Planck equation, 562
Neutral red, as a pH indicator, 553,556
Ni[5(111)x(110)], 112
N_2O as a specific reagent for adsorbed H, 307
Nonadiabatic pathways in electrochemical processes, 430
Noncoulombic adsorption of charged species, 37
Nondegenerate semiconductors, 9
Nonlocal electrostatic models, 432
Nonprimitive models, 312
Nonsteady-state photocurrent techniques, 191
n-type semiconductors, 10
Nuclear frequency factor, 430
– tunneling factor, 430
Nucleation, 101

4-Octanol, adsorption isotherm of, 354
OH adsorption on Pt, 305
Ohmic losses of protonmotive force, 555
Optical dielectric constant, 394,432
– properties, 128
– spectroscopy, 119,122
Ordered adsorption of Cu on Au(111), 288
Organelles, 48
Orientation of solvent molecules, 247,249,258
– of water, 243
Outer Helmholtz plane (OHP), 29,52,429
– potential, 1
Overpotential, 56
Oxidation-reduction cycle (ORC), 286
Oxide, 179
– /electrolyte interface, 443,445
– films, 190
– formation on Au(100), 285
–, redox processes on, 465
– /water interface, coordination chemistry of, 443

Pair correlation functions, 324
Pandey's chain model, 124,127
Parsons-Zobel plot, 36,371
Partial charge transfer, 403
Partition function, 12
Pauli exclusion principle, 7
Pd(100), 98
Phase diagram of carbon, 137
pH dependent charge of a hydrous oxide, 446
– indicating dyes, 553
– of zero net proton charge, 455
Phosphoglycerides, 48
Phospholipid bilayers, 486,491
Phospholipids, 48,495,499
Photoanodic corrosion, 185
Photocurrent conversion efficiency, 186
– doubling, 187,188

Photocurrent due to bacteriorhodopsin, 540
– multiplication processes, 187,195
– quadrupling, 195
– spectroscopy, 179,189
– transient, 192
Photocycle of bacteriorhodopsin, 533,541-544
Photoelectric effect, 16
Photoelectrons, 69
Photoelectron spectroscopy, 69,118
Photoemission measurements of Schottky barrier height, 164,191
– of adsorbed Xenon (PAX), 92,100,110
Photoexcitable dye, 566
Photoexcitation of carriers in semiconductors, 185
Photogeneration of electron-hole pairs, 185
Photoionized Xe atoms, 93
Photosynthesis by green plants, 551
Photosynthetic bacteria, chromatophore membrane of, 512
Photosystem I, 551
– II, 551
Photovoltaic cells, 63
Physical energy scale, 236
Picosecond fluorometry, 566
p-n junction, 36,63
Point of zero charge, 19,36,46,241-248,253-260,267,295,455
Poisson-Boltzmann equation, 23,474,485
– – –, spherical, 525
– distribution, 559
– equation of electrostatics, 3
Polarizability of solvent molecules, 250,263
– of surface electrons, 250,252
Polarized electrode, ideal, 4
Polycrystalline iridium, 293
– substrates, deposition onto, 407
Polyoriented Pt, 301
– surface, 301
Pores in lipid bilayers, 491,498-506
– – –, critical radius of, 497
Potential, chemical, 377
– difference across the Helmholtz layer, 40

Potential energy surface, 391
– of zero charge, 19,36,241-248,253-261,267,295
– – –, absolute value of, 257
– – – of Au single crystal faces, 259
– – –, temperature coefficient of, 259-262,268
Preexponential factor in electrochemical processes, 430
Proteins, 48,473,565
Proton diffusion, 553
– –, surface enhanced, 551,555
–, diffusion coefficient of, 573
–, electrochemical potential difference of, 551
–, geminate recombination of, 566
– promoted dissolution of hydrous oxides, 462
– pumps, 551
– –, light driven, 533
– translocation in bacteriorhodopsin, 533
Protonmotive force, across a membrane, 551
– –, ohmic losses of, 555
– –, transmembrane relaxation of, 561
Pseudopotentials, 35,120,128,316,375,380,415,418
Pt(100), 78
–(111), 98,299,301
– surfaces, mobility of, 299
p-type semiconductors, 10
Purple membrane, 534
Pyranine, 566

Quasi-chemical approximation, 319,345

Random-mixing approximation, 342-345
Rate law for surface-controlled dissolution, 456
Reaction adiabaticity, 430
– center in biomembranes, 518
– zone thickness, 429
Real solvation energy, 238
Reconstruction of Au(100), 283,284
– of Si(111) 2x1, 124,142
Redox couple, permeable, partition in a vesicle of, 51
– Fermi level, 13,182

Redox potentiometry in biomembranes, 514,518
– processes on oxide surfaces, 465
Reductive and catalytic dissolution of Fe(III) (hydr)oxides, 465
Reference electrode, 230,247
Reflectance anisotropy (RA) spectroscopy, 122,128,130
– difference spectroscopy, 122
Reorganization energy of the solvent, 13,58,430
Repeated slab geometry, 119
Rhodobacter sphaeroides, 515
Richardson constant, 163
Ru(001), 92

Scanning tunneling microscope, poket size, 277
Scanning tunneling microscopy (STM), 71,74, 92,110,118,299,370,375
– – –, a jellium model for, 390
– – – images, 300
Scanning tunneling spectroscopy, 172
Schottky barrier, 41,153
– model, 155
Schrödinger equations, one-electron, 34,393
sd-metals, 245,248,250,251,257
Secondary ion mass spectrometry (SIMS), 69
Second harmonic generation (SHG), 299,388
Semiconductors, degenerate, 10
–, nondegenerate, 9
–, *n*-type, 10
–, *p*-type, 10
– /semiconductor interface, 36
– /solution interface, 29,179
– surfaces, 117
– /vacuum interfaces, 41
Shockley levels, 39
Si(111)2x1, 123,124,127,130,144
– (110):H, 127
– /electrolyte interfaces, 201
– formed by photodissolution, 198
– /SiO$_2$ interfaces, 201
Silver iodide sol, 43
Single channel conductance, 551,559

Single crystal faces, 245,247,255-258,266,372,415
Single crystal faces in non-aqueous solvents, 258
– – – of Ag, 247,257
– – – of Au, 247,259,262
– crystals, polyoriented surface of, 301
– –, spherical, 298
– –, underpotential deposition on, 415
– electrode potentials, 229,232-236,243
Slater's exchange, 122,128
Slipping plane, 46
Smoluchowski electron smoothing effect, 76
Solid/fluid interphases, 213,218
– /solution interfaces, structure and reactivity of, 455
Solvent dielectric friction, 436
– donor number, 254
– effects, dynamical, 435
– reorganization, dynamics of, 435
– /solute interactions, 340
– /solvent interactions, 340
Sommerfeld theory of metals, 8
Sonication, 49
Space charge, 2, 21,184
– – differential capacity, 24
– – in intrinsic semiconductors, 27
– – region, 22
Specific adsorption of ions, 29,37
– – of metal ions at membranes, 55
– buffering capacitance of thylakoid membranes, 562
– surface work, 220
Spherical single crystal, 298
sp-metals, 245,248,251,253
Stalks in lipid bilayers, 491,503,504
Standard chemical potential, 15
– hydrogen electrode potential, absolute value of, 239
– overpotential, 431
– rate constant, 428,435
– redox Fermi level, 13
Stark effect, 201
Step charge, 303

Stepped surfaces, 101,302
Steps, 66
–, adsorption on, 302
Step sites, 99
Stern layer, 28,45,369,454,523
Storage capacitance for protons, 562
Structured water, 305
Structure of metal/solution interfaces, 240,241,246
– of solid/solution interfaces, 455
– of surface complexes, 455
– transformation, 87
Sub-bandgap photocurrent, 189
Submonolayer deposition, 222
Subnanosecond kinetics, 566
Sugars, 473
Surface analysis, 68
– band-structure, 120
– charge density, 23,454,520
– – – , variation with pH, 447
– – – waves, 84
– complexes, structure of, 455
– complex formation models, 445,454
– – – equilibrium constant, 450,451
– – – , linear free energy relations in, 452
– concentration, 224
–configuration, 88
– controlled dissolution, rate law for, 456
– corrugation, 73,92
– crystallography, 71
– defects, 66
– dipole layer, 107
– – potential, 2,49,315,377
– – – from electron spillover, 2,315
– electronic structure, 101
– electrons, 251,264
– energy of a metal/adsorbate system, 412
– enhanced proton diffusion, 551,555
– excess quantities, 215,222,224
– – of the electric potential, 326

Surface excess of the entropy, 314
– free energy, 67, 219
– potential, 19,92, 233,241, 511,560
– –, ionic, 2
– –, ionic, control of, 52
– reactions, 111
– reactivity, 443,458
– reconstruction, 2,68,72,133
– relaxation, 2,67,72
– science, 66
– segregation, 68
– states, 22,39,124,128,299
– steps, 97
– stress, 220
– structure, atomic and electronic, 122
– –, hydrogen adsorption region as an indicator of, 296
– tension, 67
– topography, 277,286

Tafel plots, 59
Tamm levels, 39
Temperature coefficient of the interfacial tension, 314
– – of the potential of zero charge, 259,261
– – of the work function, 315
Terrace charge, 303
– sites, 99
Tert-pentanol adsorption on Ga and Hg, 365
Thermal desorption (TDS) measurements, 258,296
Thermal energy atom scattering (TEAS), 80
Thermoemission current, 17,163
Thickness ratio in the inner layer, 402
Thiourea adsorption on Ga and Hg, 365
Thylakoids, 551,553,556
–, capacitance of, 557
–, leak conductance of, 557
–, specific buffering capacitance of, 562
Tight-binding method, 119
Time-dependent fluorescence Stokes shifts (TDFS), 436
TIP4P water-water intermolecular potential, 321

Tip shielding, 290
Tl adsorption on Pt(111), 306
Topography of gold surfaces, 282-284
Total internal reflections, 124
Transition state theory, 436,443,456
Transmembrane potential, 49
– –, control of, 50
– protonmotive force, 551
Trial functions for the electron density profile, 377,408
Tungsten tips for STM measurements, 281,282
Tunneling barrier, 71
– current, 71,278,280,393
– matrix element, 393

Ultra-high vacuum (UHV), 297
Ultramicroelectrodes, 440
Ultraviolet light electron spectroscopy (UPS), 69
Underpotential deposition, 286,406
– –, a jellium model for, 408
– –, on single crystals, 415
– shift, 407,419
Unified defect model, 159
UV-visible electroreflectance (ER), 299,307

Vacancies, 66
Vacuum level, 94
Valence bands, 5,181
Valinomycin, 51
Van der Waals pressure, 332
Vesicles, lipid, 49,555
Vibrating condenser, 238
Vibrational bands, 201
Vicinal surfaces, 97
Vögl's tight-binding Hamiltonian, 120,127
Volta potential difference, 1,234, 235, 241-243,253

Xe adsorption, 89
X-ray diffraction, 299
X_α method, 122,128

X-rays electron spectroscopy (XPS), 69

W(100), (1x1) and c(2x2), surface modifications of, 84,85
Wannier functions, 120
Water adsorption, 257
– orientation, 242,342
Wigner-Seitz cells, 76
– – radius, 377,394,414,418
Wood nomenclature, 72
Work function, 16,20,41,88, 233, 236,240,243,245,247-249,253,255-257,
262,268,315,377,399
– – of Hg, 239
– – of the solution, 18
– –, temperature coefficient of, 315

Zero charge potential, 19,36,241-248,253-260,267,295
– energy level for electrons, 233
Zeta potential, 46,52